Birkhäuser

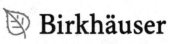

Janusz Czelakowski

The Equationally-Defined Commutator

A Study in Equational Logic and Algebra

Birkhäuser

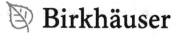

Janusz Czelakowski
Institute of Mathematics and Informatics
Opole University
Opole, Poland

ISBN 978-3-319-36578-7 ISBN 978-3-319-21200-5 (eBook)
DOI 10.1007/978-3-319-21200-5

Mathematics Subject Classification (2010): 08-02, 03C05, 03G27, 06C05, 08A30, 08A35, 08B05, 08B10, 08C15

Springer Cham Heidelberg New York Dordrecht London

Softcover reprint of the hardcover 1st edition 2015

Printed on acid-free paper

Springer International Publishing AG Switzerland is part of Springer Science+Business Media (www.springer.com)

Synopsis

The purpose of this book is to present in a uniform way commutator theory for universal algebra. We are interested in the logical perspective of the research—emphasis is put on an analysis of the interconnections holding between the commutator and equational logic. This book thus qualifies as belonging to abstract algebraic logic (AAL), the area of research that explores to a large extent the methods provided by the general theory of deductive systems.[1] The notion of a commutator equation[2] is introduced, and it plays a central role in the theory to be expounded. This book is therefore concerned with the meanings the term "commutator equation" receives in the models provided by theory and clarifies the contexts in which these meanings occur. Purely syntactic aspects of the theory of the commutator are underlined.

This book is mainly addressed to algebraists and logicians.

---

[1]From the viewpoint of AAL, universal algebra is the study of (reduced) models of equational logics. The latter are defined as structural and finitary strengthenings of the basic equational Birkhoff's logic.

[2]The term "commutator equation" denotes in this book a certain equation from the first-order language of a given algebraic signature. In the literature, this term is also used with the following different meaning: it refers to any equation of the language involving lattice-operation symbols as well as the symbol of the commutator operation; each such equation is interpreted in congruence-lattices, which are additionally augmented with the binary commutator operation (see Section 9.2).

Contents

Chapter 1
Introduction

Commutator theory is a part of universal algebra. It is rooted in the theories of groups and rings. From the general algebraic perspective the commutator was first investigated in the seventies by J. Smith for Mal'cev varieties. (Mal'cev varieties are characterized by the condition that all congruences on their algebras permute.) Further was done by the German algebraists H.P. Gumm, J. Hagemann and C. Herrmann in the eighties. They discovered that congruence-modular varieties (CM varieties, for short) form a natural environment for the commutator. Hagemann and Herrmann's approach is lattice-theoretical. Gumm's approach is based on an analogy between commutator theory and affine geometry which allowed him to discover many of the basic facts about the commutator from the geometric perspective. Freese and McKenzie (1987) summarize earlier results and establish a complementary paradigm for commutator theory in universal algebra. Kearnes and McKenzie (1992) subsequently extended the theory from congruence-modular varieties onto relatively congruence-modular quasivarieties.

In this book the theory of the commutator from the perspective of abstract algebraic logic (AAL) is investigated. AAL offers a very general conception of the commutator defined for any n-dimensional deductive system in the sense of Blok and Pigozzi (1989) (see Czelakowski 2006). The AAL perspective encompasses the above approaches as particular cases. An account of the commutator operation associated with equational deductive systems is presented. In this context an emphasis is put on logical aspects of the commutator for equational systems determined by quasivarieties of algebras. The book is therefore mainly addressed to algebraists and logicians interested in recent advancements in the area of equational logic and in the methods AAL provides.

The focus of this book is on the following problems:

(1) the definition of a commutator equation for an arbitrary class of algebras;
(2) the definition of the equationally defined commutator;
(3) a discussion of general properties of the equationally defined commutator;
(4) a discussion of various centralization relations for relative congruences;

J. Czelakowski, *The Equationally-Defined Commutator*,
DOI 10.1007/978-3-319-21200-5_1

(5) a discussion of the additivity and correspondence properties of the equationally-defined commutator;
(6) the behaviour of the equationally defined commutator in finitely generated quasivarieties;
(7) a discussion of other properties of the equationally defined commutator.

Starting with (1) and (2), the definition of a commutator equation for a class of algebras is introduced. The notion of a commutator equation seems to have not been considered in the literature. Some properties of commutator equations are discussed at length. The crucial notion of the *equationally defined commutator*, based on commutator equations, is defined for any class of algebras.

The importance of quaternary commutator equations (with parameters) in the context of additivity and other properties of the equationally defined commutator is underlined. It is proved that for any relatively congruence-modular quasivariety of algebras (abbreviated RCM), the equationally defined commutator coincides with the one promoted by Kearnes and McKenzie (1992) . An emphasis is put on the role of the equational logics associated with classes of algebras in the study of the equationally defined commutator. More specifically, for each class $\mathbf{K}$ of τ-algebras, the equational consequence operation $\mathbf{K}^{\models}$ associated with $\mathbf{K}$ is defined and its properties are thoroughly investigated, especially in the context of the equationally defined commutator for theories of $\mathbf{K}^{\models}$. (The consequence $\mathbf{K}^{\models}$ operates on the set of equations $Eq(\tau)$ of a given signature τ.)

The commutator for any congruence-modular *variety* is equationally defined by a set of equations derived either from quaternary Day terms or ternary Gumm terms (see, e.g., Proposition 4.2 in Freese and McKenzie (1987) or Theorem 6.4.2 in this book). As to (3), the focus of this book is on providing a uniform treatment of the equationally defined commutator for *any* quasivariety in terms of commutator equations. This definition encompasses congruence-modular varieties as a particular case. It also gives a new insight into the Kearnes-McKenzie's theory because it shows that the commutator defined by them for RCM quasivarieties is equationally defined.

As to (4), the book deals with the issue of equivalence of different concepts of centralization relations provided by the theory in the context of quasivarieties with the relative shifting property. Since the equationally defined commutator of two relative congruences is also determined by appropriate centralization relations, much space is devoted to the discussion of various forms of such relations.

In the context of centralization relations, the focus is on the idea of applying a general notion of an implication for equational logics viewed as a set of quaternary equations having jointly the detachment property relative to a given equational system. In particular, the significance of Day implication systems in commutator theory is highlighted. This idea was outlined in Czelakowski and Dziobiak (1996) and applied in the author's monograph (2001) to various problems in the theory of quasivarieties of algebras. We begin with a certain implication system $\Rightarrow_D$ (viewed as a finite set of quaternary equations in four variables x, y, z, w and possibly some other variables $\underline{u}$ called *parameters*) for the equational logic $(Eq(\tau), \mathbf{Q}^{\models})$

associated with a given quasivariety $\mathbf{Q}$. The system $\Rightarrow_D$ is assumed to possess certain natural properties with respect to $\mathbf{Q}^{\models}$. The set of equations $\Rightarrow_D$ is then called a *Day implication system* for $\mathbf{Q}^{\models}$. More specifically, a finite set of equations in four variables

$$D = \{p_i(x,y,z,w) \approx q_i(x,y,z,w) : i \in I\},$$

more suggestively denoted by $x \approx y \Rightarrow_D z \approx w$, is called a *Day implication system* for $\mathbf{Q}^{\models}$ if the following conditions are met[1]

$$z \approx w \in \mathbf{Q}^{\models}(x \approx y, x \approx y \Rightarrow_D z \approx w), \qquad \text{(iD1)}$$

i.e., $\Rightarrow_D$ has the detachment property relative to $\mathbf{Q}^{\models}$,

$$x \approx y \Rightarrow_D x \approx y \subseteq \mathbf{Q}^{\models}(\emptyset), \qquad \text{(iD2)}$$

i.e., $\Rightarrow_D$ has the identity property relative to $\mathbf{Q}^{\models}$, and

$$x \approx x \Rightarrow_D y \approx y \subseteq \mathbf{Q}^{\models}(\emptyset). \qquad \text{(iD3)}$$

The above conditions are expressed in the standard notation as follows:

the implication $(x \approx y) \wedge \bigwedge_{i \in I} p_i(x,y,z,w) \approx q_i(x,y,z,w) \rightarrow (z \approx w)$ (iD1)*

is $\mathbf{Q}$-valid,

for every $i \in I$, the equation $p_i(x,y,x,y) \approx q_i(x,y,x,y)$ is $\mathbf{Q}$-valid, (iD2)*

for every $i \in I$, the equation $p_i(x,x,y,y) \approx q_i(x,x,y,y)$ is $\mathbf{Q}$-valid. (iD3)*

Condition (iD3) is equivalent to the condition

$$x \approx y \Rightarrow_D z \approx w \subseteq \mathbf{Q}^{\models}(x \approx y, z \approx w). \qquad \text{(iD3)**}$$

The significance of a Day implication for equational logics stems from the fact that $\Rightarrow_D$ characterizes congruence-modular *varieties* of algebras and, more widely, in the context of quasivarieties of algebras, it characterizes the relative shifting property. The notion of a Day implication, although not defined by Day himself, can be shown to be equivalent to a Mal'cev type characterization of congruence-modular *varieties* provided by Day (1969). These and other issues are discussed in Chapter 3.

[1]This understanding of implication departs from the traditional concept according to which implication is viewed as a sentential connective conjoining two sentential formulas and thereby yielding a new formula. AAL provides a more general meaning of the term "implication". In particular, in equational logic, an implication (without parameters) is any set $\Rightarrow$ of quaternary equations having the detachment property with respect to a given consequence operation $\mathbf{Q}^{\models}$.

The presence of a Day implication $\Rightarrow_D$ in the equational logic $\mathbf{Q}^{\models}$ enables one to simplify considerably the description of the centralization relations. (The latter relations are certain congruences defined on the algebras from the quasivariety $\mathbf{Q}$.)

Commutator theory isolates several centralization relations. These relations are strictly conjoined with the properties of the standard or the equationally defined commutator and play an important role in the theory. For example, in many varieties $\mathbf{V}$ and for any algebra $\boldsymbol{A}$ from the class $\mathbf{V}$, the commutator (in whatever meaning of the word) of two congruences Φ and Ψ of $\boldsymbol{A}$ is the least congruence Ξ such that Φ centralizes Ψ with respect to Ξ for a suitably selected centralization relation. In this work seven definitions of centralization relations for relative congruences in quasivarieties of algebras are discussed. The main results (Theorems 4.1.2 and 4.2.2) show that in the presence of a Day implication system, the first four of them are equivalent for quasivarieties and all seven are equivalent for varieties. Some further refinements of relevant notions in this context are also discussed in Chapter 4.

As to (5), the presence of a Day implication system for a quasivariety of algebras is generally too weak to yield a workable commutator theory. Another property, viz. the additivity of the equationally defined commutator is needed. A characterization of the additivity of the equationally defined commutator via quaternary commutator equations is presented. The additivity property already implies the correspondence property for the equationally defined commutator (Theorem 5.1.1). The notion of a generating set of quaternary commutator equations for the additive equationally defined commutator is isolated.

The theory of the commutator for congruence-modular varieties of algebras is developed in Freese and McKenzie (1987). This theory was subsequently extended to relative congruence-modular quasivarieties by Kearnes and McKenzie (1992). Every RCM quasivariety possesses a Day implication system. Moreover, the Kearnes-McKenzie commutator is additive for any RCM quasivariety. It is proved in this book that for RCM quasivarieties, the equationally defined commutator and the one defined by Kearnes and McKenzie (1992) coincide. We may therefore speak of *the* commutator for any RCM quasivariety.

The central part of the book tackles the problem of syntactical characterization of the additivity of the equationally defined commutator. Various conditions equivalent to additivity are presented in Chapter 5, the main part of the book. The property of the *restricted distributivity* of the lattice of theories of the equational logic $\mathbf{Q}^{\models}$ associated with a quasivariety $\mathbf{Q}$ is isolated. This property plays a crucial role in commutator theory. The central theorem of the book—Theorem 5.3.7—states that for any quasivariety $\mathbf{Q}$ for which the lattice of equational theories of $\mathbf{Q}^{\models}$ is distributive in the restricted sense, the equationally defined commutator is additive. Though the restricted distributivity is not a lattice-theoretic property (this property merely reflects a logical aspect of equational systems), it turns out that the lattice of equational theories of the consequence operation $\mathbf{Q}^{\models}$ associated with any RCM quasivariety $\mathbf{Q}$ is distributive in the restricted sense. (But restricted distributivity has a wider scope and there are *non-modular* quasivarieties whose lattices of equational theories obey this law.) It then follows that the equationally defined commutator for any RCM quasivariety is additive. Since the said commutator is the same as the one investigated by Kearnes and McKenzie, the additivity of the Kearnes-McKenzie commutator follows. The above reasoning therefore provides a new proof of additivity property of the commutator for any RCM quasivariety.

Our treatment of the commutator for equational systems was greatly inspired by the above two seminal works. They both contain many deep, highly non-trivial results concerning the commutator. Our contribution to the area of equational logics consists mainly in introducing the notion of the equationally defined commutator and an attempt to disentangle various intricate (often syntactic) characterizations of this commutator and related notions and to put them in a more transparent logical form provided by the conceptual framework of contemporary abstract algebraic logic.

As to (6), it is proved that every finitely generated quasivariety possesses a finite generating set of commutator equations. It follows that the equationally defined commutator of any finitely generated RCM quasivariety has a finite generating set.

As to (7), specialized classes of quasivarieties whose definitions are derived from the properties of the commutator, e.g., abelian or nilpotent quasivarieties, are studied. Each such defining property can be expressed in terms of appropriate sets of quaternary commutator equations.

This book has the following structure. The (hyper)class of quasivarieties of algebras forms the conceptual environment for the equationally defined commutator. There is a natural one-to-one correspondence between equational logics and their model classes, viz. quasivarieties of algebras. We therefore frequently switch from equational logics to their model classes and vice versa. Chapter 2 contains a brief introduction to the theory of quasivarieties and the associated equational logics. In Chapter 3 the theory of the equationally defined commutator for quasivarieties is developed. Chapter 4 provides several definitions of centralization relations and proves their equivalence for some classes of quasivarieties. Chapter 5 contains the basic results that characterize the additivity property of the equationally defined commutator. Chapter 6 is devoted to modularity and RCM quasivarieties. Chapter 7 is concerned with relatively congruence-distributive *subquasivarieties* of quasivarieties whose equationally defined commutator is additive. In Chapter 8 some basic properties of the equationally defined commutator for finitely generated quasivarieties are outlined. Chapter 9 deals with special properties of the lattices of relative congruences and of the equationally defined commutator. The three appendices provide the reader with a more specialized knowledge of some aspects of the expounded theory.

The book is rather self-contained in the matters concerning the equationally defined commutator. As far as universal algebra is concerned, the books by Burris and Sankappanavar (1981), McKenzie et al. (1987), and the yet unpublished book by Ježek (2008) (available on the internet) give a good introduction to the subject. The book by Grätzer (2008) gives an account of classical results in universal algebra. The books by Freese and McKenzie (1987), Hobby and McKenzie (1988), and Kearnes and Kiss (2013) are suggested to the reader wanting to learn more about universal algebra and the commutator.

The classical monograph by Blok and Pigozzi (1989) is recommended for the reader interested in abstract algebraic logic and basic logical issues. The monograph by Czelakowski (2001) provides a comprehensive and detailed exposition of AAL and outlines the theory of quasivarieties from the perspective of AAL (Chapter Q). A survey paper by Font et al. (2003) together with its update (2009) as well as the introductory chapter Font (2014) familiarizes the reader with a broad spectrum of issues AAL deals with. The textbook by Font (2016) is a good introduction to the subject. The book by Font and Jansana (2009) and Jansana shows algebraic semantics for sentential logics from the general perspective of closure systems.

In this book the standard algebraic notation is adopted. In the context of quasivarieties, terminology is modelled after Pigozzi (1988). This terminology is also used in Czelakowski (2001).

The logical notation and terminology does not depart from that used in AAL—see, e.g., Wójcicki (1988) or Czelakowski (2001).

Some of the results of this book were presented in abstracted form in Czelakowski (2006).

The author is greatly indebted to Keith Kearnes and Piotr Wojtylak for valuable and inspiring comments on some aspects of the presented theory.

As a non-native speaker of English, the author is indebted to dr. Matthew Carmody for proof-reading the whole text of the book and for making appropriate linguistic recommendations and amendments.

Chapter 2
Basic Properties of Quasivarieties

This chapter supplies basic facts concerning quasivarieties and the equational systems associated with quasivarieties. Many of these facts are of syntactical character. An equational logic is an extension of the familiar Birkhoff's logic. The narrative structure of the book is strictly linked with the properties of lattices of theories of equational logics. Examining these lattice requires formal tools. They are introduced in this part; some of them are new.

2.1 Quasi-Identities

ω is the set of natural numbers (with zero). An *algebraic signature* is a pair $\tau := \langle F, \boldsymbol{a}\rangle$, where F is a set (of operation symbols), and $\boldsymbol{a} : F \to \omega$ is a function (assigning arity). An *algebra* of signature τ is a pair $\boldsymbol{A} := \langle A, F^A\rangle$, where A is a non-empty set, called the *universe* of $\boldsymbol{A}$, and for each $f \in F$ with $\boldsymbol{a}(f) = m$, there is m-ary operation $f^A : A^m \to A$. The operations f^A are called the *basic* (or *fundamental*) *operations* of $\boldsymbol{A}$. If $\boldsymbol{a}(f) = 0, f$ is also called a *constant* symbol. f^A is then an element of A.

An algebra $\boldsymbol{A}$ of signature τ will often be referred to as a τ-*algebra*.

Hom$(\boldsymbol{A}, \boldsymbol{B})$ is the set of all homomorphisms from a τ-algebra $\boldsymbol{A}$ to τ-algebra $\boldsymbol{B}$.

Let τ be a fixed algebraic signature and let $\boldsymbol{L}_\tau$ be the corresponding first-order language with equality $\approx$. *Var* $= \{v_n : n \in \omega\}$ is the set of individual variables of $\boldsymbol{L}_\tau$. $\boldsymbol{Te}_\tau$ is the algebra of *terms* of $\boldsymbol{L}_\tau$ and $Eq(\tau)$ is the set of *equations* of $\boldsymbol{L}_\tau$.

If $t = t(x_1, \ldots, x_n)$ is a term in at most n individual variables $\underline{x} = x_1, \ldots, x_n$, and $\underline{a} = a_1, \ldots, a_n$ is a sequence of elements of a τ-algebra $\boldsymbol{A}$, then $t^A(a_1, \ldots, a_n)$ is the *value* of the term t for $\underline{a} = a_1, \ldots, a_n$ in $\boldsymbol{A}$. $t^A(a_1, \ldots, a_n)$ is defined in the standard way by induction on complexity of terms. We shall also use the abbreviation $t^A(\underline{a})$ for $t^A(a_1, \ldots, a_n)$ often omitting the superscript 'A' when the algebra is clear from context.

J. Czelakowski, *The Equationally-Defined Commutator*,
DOI 10.1007/978-3-319-21200-5_2

A *quasi-equation* is a formula of the form

$$\alpha_1 \approx \beta_1 \wedge \ldots \wedge \alpha_n \approx \beta_n \rightarrow \alpha \approx \beta,$$

where $\alpha_1, \beta_1, \ldots, \alpha_n, \beta_n, \alpha, \beta$ are terms. $n = 0$ is possible so every equation qualifies as a quasi-equation.

A universally quantified quasi-equation is called a *quasi-identity*. As is customary, the universal quantifiers in quasi-identities are usually not explicitly written. It is left to the context to distinguish quasi-identities from quasi-equations.

Any class of algebras defined by a set of quasi-identities is called a *quasivariety*. If **Q** is a quasivariety, then any set Γ of quasi-identities defining **Q** is called a *base* for **Q**; we then write $\mathbf{Q} = \mathbf{Mod}(\Gamma)$.

The symbols $\boldsymbol{I}$, $\boldsymbol{H}$, $\boldsymbol{S}$, $\boldsymbol{P}$, and $\boldsymbol{P_u}$, respectively, denote the operations of forming isomorphic images, homomorphic images, subalgebras, direct products and ultraproducts. (The class operations $\boldsymbol{S}$, $\boldsymbol{P}$ and $\boldsymbol{P_u}$ are interpreted in the inclusive sense which means that they also comprise isomorphic copies of algebras; for example, $\boldsymbol{P_u}(\mathbf{K})$ is the class of all algebras *isomorphic* to an ultraproduct of a system of algebras from **K**.)

Let **Q** be a class closed under isomorphisms. By a well-known result due to Mal'cev, **Q** is a quasivariety if and only if **Q** is closed under the operations $\boldsymbol{S}$, $\boldsymbol{P}$ and $\boldsymbol{P_u}$.

If **K** is a class of algebras, then $\boldsymbol{Qv}(\mathbf{K})$ is the smallest quasivariety containing **K**; the class **K** is then said to *generate* the quasivariety $\boldsymbol{Qv}(\mathbf{K})$. $\boldsymbol{Va}(\mathbf{K})$ is the variety generated by **K**. The equalities $\boldsymbol{Qv}(\mathbf{K}) = \boldsymbol{SPP_u}(\mathbf{K})$ and $\boldsymbol{Va}(\mathbf{K}) = \boldsymbol{HSP}(\mathbf{K})$ holding for any class **K** are classical results of universal algebra.

A quasivariety **Q** is *finitely generated* if $\mathbf{Q} = \boldsymbol{Qv}(\mathbf{K})$ for a finite set **K** of finite algebras. In this case $\boldsymbol{Qv}(\mathbf{K}) = \boldsymbol{SP}(\mathbf{K})$.

2.2 Rules of Inference

Let τ be a fixed signature. A *rule of inference* r in $Eq(\tau)$ is a set of pairs $\langle \Sigma, \sigma \rangle$, where Σ is a (possibly infinite) set of equations and σ is a single equation. (The pair $\langle \Sigma, \sigma \rangle$ is read: *From the set of equations* Σ *infer the equation* σ.) Any pair $\langle \Sigma, \sigma \rangle$ belonging to r is called an *instance* of r.

Let e be a *substitution* in $\boldsymbol{Te}_\tau$ (i.e., e is an endomorphism of the term algebra). e is fully determined by its values on the set of individual variables of $\boldsymbol{Te}_\tau$. Given a set of equations Σ and an equation $\alpha \approx \beta$, we put: $e\Sigma = \{e\rho : \rho \in \Sigma\}$ and $e(\alpha \approx \beta) := e\alpha \approx e\beta$.

A rule r is *schematic* if there exists a *single* pair $\langle \Sigma_0, \sigma_0 \rangle$ such that r is equal to the set of all instances of $\langle \Sigma_0, \sigma_0 \rangle$, i.e.,

$$r = \{\langle e\Sigma_0, e\sigma_0 \rangle : e \in Hom(\boldsymbol{Te}_\tau, \boldsymbol{Te}_\tau)\}. \qquad (1)$$

The pair $\langle \Sigma_0, \sigma_0 \rangle$ is then called a *scheme* of the rule r. A schematic rule r is *finitary* (or *standard*) if for one its schemes $\langle \Sigma_0, \sigma_0 \rangle$ (equivalently, for all schemes), the set Σ_0 is finite. A schematic rule r is *proper* (or *non-axiomatic*) if Σ_0 is non-empty; otherwise, r is called *axiomatic*.

The equations in $e\Sigma_0$ are called *premises* of the rule r and the equation $e\sigma_0$ is the *conclusion* of r, for any e.

Any schematic rule r is usually identified with its scheme. Therefore any schematic rule r (with a scheme $\langle \Sigma_0, \sigma_0 \rangle$ is often presented in the form

$$\Sigma_0/\sigma_0$$

or, more explicitly, as

$$\{\alpha_i \approx \beta_i : i \in I\}/\alpha \approx \beta \tag{2}$$

where $\Sigma_0 = \{\alpha_i \approx \beta_i : i \in I\}$ and σ_0 is $\alpha \approx \beta$.

In particular, following common practice adopted in metalogic, if r is schematic with a scheme $\langle \Sigma_0, \sigma_0 \rangle$, where Σ_0 is finite, $\Sigma_0 = \{\alpha_1 \approx \beta_1, \dots, \alpha_k \approx \beta_k\}$, and σ_0 is $\alpha \approx \beta$, then the standard rule r is usually presented in the form

$$\alpha_1 \approx \beta_1, \dots, \alpha_k \approx \beta_k/\alpha \approx \beta.\text{[1]} \tag{3}$$

Thus, formally, (3) is the set consisting of ordered pairs

$$\langle \{e\alpha_1 \approx e\beta_1, \dots, e\alpha_k \approx e\beta_k\}, e\alpha \approx e\beta \rangle$$

called *instances* of (3), with e ranging over the set of endomorphisms of $\boldsymbol{Te}_\tau$.

If (2) is axiomatic (and hence the set I is empty), then (2) is written as

$$/\alpha \approx \beta. \tag{4}$$

Birkhoff's rules are the following schematic rules

$$/x \approx x, \qquad \text{(identity axiom)}$$
$$x \approx y/y \approx x, \qquad \text{(symmetry)}$$
$$x \approx y, y \approx z/x \approx z, \qquad \text{(transitivity)}$$

and

$$x_1 \approx y_1, \dots, x_m \approx y_m/f(x_1, \dots, x_m) \approx f(y_1, \dots, y_m), \qquad \text{(functionality)}$$

for each operation symbol f of arbitrary arity m. The identity axiom is also referred to as the *reflexivity* rule. (This rule is axiomatic, with the empty set of premises.)

[1]A similar 'forward slash' notation is sometimes applied to substitutions in the algebra of terms but this should not lead to confusion.

If the signature τ is finite, the above set of standard rules is finite as well. The set of Birkhoff's rules is denoted by *Birkhoff*(τ).

A set of equations Σ is closed with respect to a schematic rule $r : \{\alpha_i \approx \beta_i : i \in I\}/\ \alpha \approx \beta$ if and only if for every endomorphism $e : \boldsymbol{Te}_\tau \to \boldsymbol{Te}_\tau$, $\{e\alpha_i \approx e\beta_i : i \in I\} \subseteq \Sigma$ implies that $e\alpha \approx e\beta \in \Sigma$.

If Σ and Γ are sets of equations, then Σ/Γ denotes the set of rules Σ/σ, where σ ranges over the set Γ.

Let R be a set of standard rules in $Eq(\tau)$. We assume that R includes the set of Birkhoff's rules *Birkhoff*(τ). (This assumption may be overridden, but then one assumes instead that *Birkhoff*(τ) is among secondary rules of the consequence operation C_R^{eq} determined by R—see the definition below.) Let Σ be a set of equations. An *R-proof from* Σ is any finite sequence of equations

$$p_1 \approx q_1, \ldots, p_n \approx q_n \tag{5}$$

satisfying the following condition:

(p1) $p_1 \approx q_1 \in \Sigma$ or $p_1 \approx q_1$ is of the form $x \approx x$ for some variable x,
(p2) for every i $(1 < i \leqslant n)$, either $p_i \approx q_i \in \Sigma$ or $p_i \approx q_i$ is of the form $x \approx x$, or there are indices $i_1, \ldots, i_k$ smaller than i and a rule $r : \alpha_1 \approx \beta_1 \wedge \ldots \wedge \alpha_k \approx \beta_k/\ \alpha \approx \beta$ in R such that $p_i \approx q_i$ is obtained from $p_{i_1} \approx q_{i_1}, \ldots, p_{i_k} \approx q_{i_k}$ by an application of r.

(The phrase "application of a rule" is commonly used in the proof-theoretic parlance. Formally, the meaning of the phrase "$p_i \approx q_i$ is obtained from $p_{i_1} \approx q_{i_1}, \ldots, p_{i_k} \approx q_{i_k}$ by an application of r" is that the pair $\langle\{p_{i_1} \approx q_{i_1}, \ldots, p_{i_k} \approx q_{i_k}\}, p_i \approx q_i\rangle$ is an instance of r.)

Let $p \approx q$ be an equation. An R-proof from Σ is called an *R-proof of* $p \approx q$ *from* Σ if $p \approx q$ is the last element of this proof. $p \approx q$ is *R-provable from* Σ if there exists an R-proof of $p \approx q$ from Σ.

For every set $\Sigma \subseteq Eq(\tau)$ we define:

$$C_R^{eq}(\Sigma) := \{p \approx q \in Eq(\tau) : p \approx q \text{ is } R\text{-provable from } \Sigma\}.$$

C_R^{eq} is a structural and finitary consequence operation defined on $Eq(\tau)$ which validates the set of rules R (see below). In particular C_R^{eq} validates Birkhoff's rules for equality *Birkhoff*(τ).

2.3 Equational Logics

$\wp(Eq(\tau))$ is the power set of $Eq(\tau)$, i.e., the family of all subsets of $Eq(\tau)$. A mapping $C^{eq} : \wp(Eq(\tau)) \to \wp(Eq(\tau))$ is a *consequence operation* on $Eq(\tau)$ if it satisfies, for all $X, Y \subseteq Eq(\tau)$:

(Co1) $X \subseteq C^{eq}(X)$ (reflexivity)
(Co2) $C^{eq}(X) \subseteq C^{eq}(Y)$ whenever $X \subseteq Y$ (monotonicity)
(Co3) $C^{eq}(C^{eq}(X)) \subseteq C^{eq}(X)$ (idempotency).

A consequence C^{eq} is *finitary* if for all $X \subseteq Eq(\tau)$:

(Co4) $C^{eq}(X) = \bigcup\{C^{eq}(X_f) : X_f \text{ is a finite subset of } X\}$ (finitariness).

(Note that (Co4) already implies (Co2).)

A consequence C^{eq} is *structural* if for all $X \subseteq Eq(\tau)$:

(Co5) $eC^{eq}(X) \subseteq C^{eq}(eX)$ for every endomorphism $e : \boldsymbol{Te}_\tau \to \boldsymbol{Te}_\tau$ (structurality).

B_τ stands for the consequence operation (on $Eq(\tau)$) determined only by the set of rules *Birkhoff*(τ) in the standard way. B_τ is referred to as *Birkhoff's logic* in the signature τ.

By an *equational logic* we shall understand any structural consequence operation C^{eq} defined on $Eq(\tau)$ which validates Birkhoff's rules for equality *Birkhoff*(τ), and possibly some other rules. This means that $B_\tau(X) \subseteq C^{eq}(X)$, for all $X \subseteq Eq(\tau)$. Birkhoff's logic B_τ is the least equational logic (in the sense of the above inclusions). Thus the class of equational logics in a given signature τ comprises exactly all strengthenings of the logic B_τ.

The terms "equational logic" and "equational deductive system" will be treated as synonyms.

Any set of equations Σ such that $C^{eq}(\Sigma) = \Sigma$ is called a *closed theory of* C^{eq}, or shortly, a *theory of* C^{eq}.

$$Th(C^{eq})$$

is the set of all theories of C^{eq}. Since $Th(C^{eq})$ is a closure system, it forms a complete lattice with respect to inclusion. This lattice is denoted by

$$\boldsymbol{Th}(C^{eq}).$$

Given a class **K** of τ-algebras, we let

$$\mathbf{K}^{\models}$$

denote the *consequence operation* on the set of τ-equations determined by **K**. Thus, for $\{\alpha_i \approx \beta_i : i \in I\} \cup \{\alpha \approx \beta\} \subseteq Eq(\tau)$,

> $\alpha \approx \beta \in \mathbf{K}^{\models}(\{\alpha_i \approx \beta_i : i \in I\}$ if and only if, for every algebra $\boldsymbol{A} \in \mathbf{K}$ and every $h \in Hom(\boldsymbol{Te}_\tau, \boldsymbol{A})$, $h(\alpha) = h(\beta)$ whenever $h(\alpha_i) = h(\beta_i)$ for all $i \in I$.

$\alpha \approx \beta \in \mathbf{K}^{\models}(\{\alpha_i \approx \beta_i : i \in I\}$ is read: $\alpha \approx \beta$ *follows from* $\{\alpha_i \approx \beta_i : i \in I\}$ *relative to* **K**.

The consequence operation $\mathbf{K}^{\models}$ is *structural*, i.e., $\alpha \approx \beta \in \mathbf{K}^{\models}(\{\alpha_i \approx \beta_i : i \in I\})$ implies that $e\alpha \approx e\beta \in \mathbf{K}^{\models}(\{e\alpha_i \approx e\beta_i : i \in I\})$ for all endomorphisms e of the term algebra $\boldsymbol{Te}_\tau$ and all sets $\{\alpha_i \approx \beta_i : i \in I\}$. As $\mathbf{K}^{\models}$ validates Birkhoff's rules, $\mathbf{K}^{\models}$ is an equational logic. Furthermore, if **K** is closed under the formation of ultraproducts, the consequence $\mathbf{K}^{\models}$ is finitary. Note that $\alpha \approx \beta \in \mathbf{K}^{\models}(\emptyset)$ means that the equation $\alpha \approx \beta$ is valid in the class **K**.

Following common practice we suppress parentheses as much as possible and in case of finite sets of equations we usually write

$$\alpha \approx \beta \in \mathbf{K}^{\models}(\alpha_1 \approx \beta_1, \ldots, \alpha_k \approx \beta_k)$$

instead of $\alpha \approx \beta \in \mathbf{K}^{\models}(\{\alpha_1 \approx \beta_1, \ldots, \alpha_k \approx \beta_k\})$.

$\mathbf{K}^{\models}(\{\alpha_i \approx \beta_i : i \in I\})$ is thus the set of all equations $\alpha \approx \beta$ which follow from $\{\alpha_i \approx \beta_i : i \in I\}$ relative to the class $\mathbf{K}$.

A schematic rule $r : \{\alpha_i \approx \beta_i : i \in I\}/\alpha \approx \beta$ is said to be a *rule of the consequence* $\mathbf{K}^{\models}$ if $\alpha \approx \beta \in \mathbf{K}^{\models}(\{\alpha_i \approx \beta_i : i \in I\})$. In this case we also say that r is a *rule of the class* $\mathbf{K}$.

Every equational logic C^{eq} on $Eq(\tau)$ is characterized semantically by some class of algebras, i.e., there exists a class $\mathbf{K}$ of τ-algebras such that $C^{eq} = \mathbf{K}^{\models}$, and there always exists the largest such class $\mathbf{K}$. Furthermore, if C^{eq} is finitary, $\mathbf{K}$ can be assumed to be a quasivariety. These, rather simple observations, form the content of the completeness theorem for equational logics. Consequently, when we are dealing with equational logics, we shall mainly consider consequences $\mathbf{K}^{\models}$ already determined by some fixed class $\mathbf{K}$ of algebras.

Any set of equations Σ such that $\mathbf{K}^{\models}(\Sigma) = \Sigma$ is called a *closed theory of* $\mathbf{K}^{\models}$, or shortly, a *theory of* $\mathbf{K}^{\models}$. A set of equations Σ is a theory of $\mathbf{K}^{\models}$ if and only if Σ is closed with respect to the set of rules of $\mathbf{K}^{\models}$.

According to the adopted notation, $Th(\mathbf{K}^{\models})$ is the set of all theories of $\mathbf{K}^{\models}$. $Th(\mathbf{K}^{\models})$ forms a complete lattice with respect to inclusion. This lattice is denoted by

$$\boldsymbol{Th}(\mathbf{K}^{\models}).$$

For any class $\mathbf{K}$ it is the case that $\mathbf{K}^{\models} = \inf\{\boldsymbol{A}^{\models} : \boldsymbol{A} \in \mathbf{K}\}$, which means that $\mathbf{K}^{\models}(\Sigma) = \bigcap\{\boldsymbol{A}^{\models}(\Sigma) : \boldsymbol{A} \in \mathbf{K}\}$, for any set of equations Σ. (Here $\boldsymbol{A}^{\models}$ is, in accordance with the definition, $\{\boldsymbol{A}\}^{\models}$.) Equivalently, $Th(\mathbf{K}^{\models})$ is the closure system generated by $\bigcup\{Th(\boldsymbol{A}^{\models}) : \boldsymbol{A} \in \mathbf{K}\}$.

There is an obvious translation of $\mathbf{K}^{\models}$ into the language of quasi-identities over $\boldsymbol{Te}_\tau$:

$\alpha \approx \beta \in \mathbf{K}^{\models}(\alpha_1 \approx \beta_1, \ldots, \alpha_k \approx \beta_k)$ if and only if the implication $\alpha_1 \approx \beta_1 \wedge \ldots \wedge \alpha_k \approx \beta_k \to \alpha \approx \beta$ is valid in $\mathbf{K}$.

This observation enables one to express the properties of the consequence operation $\mathbf{K}^{\models}$ on finite sets in terms of quasi-equations valid in $\mathbf{K}$. The definitions given below are formulated in terms of standard rules; but they can be easily reformulated in terms of quasi-identities. The key is in assigning to each standard rule

$$r : \quad \alpha_1 \approx \beta_1 \wedge \ldots \wedge \alpha_k \approx \beta_k/\alpha \approx \beta$$

the quasi-identity

$$(r) : \quad \alpha_1 \approx \beta_1 \wedge \ldots \wedge \alpha_k \approx \beta_k \to \alpha \approx \beta.$$

Thus r is a rule of $\mathbf{K}^{\models}$ if and only if (r) is universally valid in $\mathbf{K}$.

Although quasi-identities and standard rules of inference are interdefinable concepts, in purely syntactic contexts, when one works with consequence relations defined on equations, the notion of a standard rule of inference often appears to be convenient and useful.

For every quasivariety **Q**, the logic $\mathbf{Q}^{\models}$ is characterized proof-theoretically by a set R of standard rules. That is, for every **Q** there is a set R of standard rules including *Birkhoff*(τ) such that

$$\mathbf{Q}^{\models} = C_R^{eq}. \tag{1}$$

(As R one may take the set consisting of all standard rules of $\mathbf{Q}^{\models}$.) If (1) holds, then R is called an *inferential base* for $\mathbf{Q}^{\models}$.

Since the notions of a standard rule of $\mathbf{Q}^{\models}$ and of a quasi-identity of **Q** are interdefinable, the fact that a set of rules R is an inferential base of $\mathbf{Q}^{\models}$ is equivalent to the statement that the set of quasi-equations $\{(r) : r \in R\}$ corresponding to the rules of R forms an axiomatization of **Q**.

In what follows we shall interchangeably speak of inferential bases for $\mathbf{Q}^{\models}$ and axiomatic bases for **Q**. (The latter bases consist of sets of quasi-identities.)

2.4 Relative Congruences

Let τ be a signature. Let R be a binary relation defined on a τ-algebra $\boldsymbol{A}$. R is *closed under a schematic rule* $r : \{\alpha_i \approx \beta_i : i \in I\}/\alpha \approx \beta$ if, for any $\underline{a} \in A^k$, R contains the pair $\langle \alpha^{\boldsymbol{A}}(\underline{a}), \beta^{\boldsymbol{A}}(\underline{a})\rangle$ whenever it contains the pairs $\langle \alpha_i^{\boldsymbol{A}}(\underline{a}), \beta_i^{\boldsymbol{A}}(\underline{a})\rangle$, for all $i \in I$. (Here k is the length of a sequence $\underline{x} = x_1, x_2, \ldots$ which includes every variable occurring in one of the terms of the equations $\{\alpha_i \approx \beta_i : i \in I\}$ and $\alpha \approx \beta$. k may be infinite. In this case we assume that $k = \omega$.)

If the rule r is standard, $r = \alpha_1 \approx \beta_1, \ldots, \alpha_k \approx \beta_k/\alpha \approx \beta$, then we shall interchangeably use the phrases 'R is closed under r' and 'R is closed under the quasi-equation $\alpha_1 \approx \beta_1 \wedge \ldots \wedge \alpha_k \approx \beta_k \to \alpha \approx \beta$'.

A binary relation R on a τ-algebra $\boldsymbol{A}$ is a *congruence relation* (of $\boldsymbol{A}$) if R is closed under Birkhoff's rules *Birkhoff*(τ).

If Φ is a congruence of $\boldsymbol{A}$ and $a \in A$, then a/Φ is the equivalence class of a with respect to Φ. $\boldsymbol{A}/\Phi$ is the quotient algebra whose elements are equivalence classes a/Φ, $a \in A$.

If Φ is a congruence of $\boldsymbol{A}$, then Φ is closed under the rule $\alpha_1 \approx \beta_1, \ldots,$ $\alpha_k \approx \beta_k/\alpha \approx \beta$ if and only if the quotient algebra $\boldsymbol{A}/\Phi$ validates the quasi-identity $(\forall \underline{x})(\alpha_1 \approx \beta_1 \wedge \ldots \wedge \alpha_k \approx \beta_k \to \alpha \approx \beta)$.

Note. Each equation may be identified with a pair of terms. We may therefore identify a set of equations Σ with a set of pairs of terms; Σ is thus a binary relation on the term algebra $\boldsymbol{Te}_\tau$. Consequently, the fact that Σ is closed with respect to a rule r is an instance of the above general definition. In particular, Σ is closed with respect to Birkhoff's rules *Birkhoff*(τ) if and only if the set of ordered pairs $\{\langle\alpha,\beta\rangle : \alpha \approx \beta \in \Sigma\}$ is a congruence of the term algebra $\boldsymbol{Te}_\tau$. □

If $\boldsymbol{A}$ is an algebra, then $Con(\boldsymbol{A})$ is the set of congruences of $\boldsymbol{A}$. $Con(\boldsymbol{A})$ forms an algebraic lattice, denoted by $\boldsymbol{Con}(\boldsymbol{A})$. If $\Phi, \Psi \in Con(\boldsymbol{A})$, then $\Phi + \Psi$ marks their join in $\boldsymbol{Con}(\boldsymbol{A})$. The lattice meet of Φ, Ψ in $\boldsymbol{Con}(\boldsymbol{A})$ coincides with the intersection $\Phi \cap \Psi$. If $X \subseteq A^2$, $\Theta^A(X)$ denotes the least congruence of $\boldsymbol{A}$ that contains X.

Let $\mathbf{Q}$ be a quasivariety of τ-algebras and $\boldsymbol{A}$ a τ-algebra, not necessarily in $\mathbf{Q}$. A congruence Φ on $\boldsymbol{A}$ is called a $\mathbf{Q}$-*congruence* if $\boldsymbol{A}/\Phi \in \mathbf{Q}$. The set of $\mathbf{Q}$-congruences is denoted by $Con_{\mathbf{Q}}(\boldsymbol{A})$. Thus $Con_{\mathbf{Q}}(\boldsymbol{A}) = \{\Phi \in Con(\boldsymbol{A}) : \boldsymbol{A}/\Phi \in \mathbf{Q}\}$. If $\mathbf{Q}$ is not a variety, the elements of $Con_{\mathbf{Q}}(\boldsymbol{A})$ are also called *relative* congruence. $Con_{\mathbf{Q}}(\boldsymbol{A})$ contains the universal congruence $\mathbf{1}_A := A^2$ and it contains the smallest $\mathbf{Q}$-congruence being the intersection of all $\mathbf{Q}$-congruences of $\boldsymbol{A}$. This smallest $\mathbf{Q}$-congruence is the identity congruence $\mathbf{0}_A$ (= diagonal relation on A) if and only if $\boldsymbol{A} \in \mathbf{Q}$.

It is easy to see that if Γ is an axiomatic base for $\mathbf{Q}$, then Φ is a $\mathbf{Q}$-congruence if and only if Φ is closed under every quasi-identity from the base. It follows from this observation that $Con_{\mathbf{Q}}(\boldsymbol{A})$ is closed under arbitrary intersections and the union of directed sets; in other words, $Con_{\mathbf{Q}}(\boldsymbol{A})$ is a finitary closure system on A^2. (This also follows from the fact that $\mathbf{Q}$ is closed under subdirect products and ultraproducts.) $Con_{\mathbf{Q}}(\boldsymbol{A})$ therefore forms the universe of an algebraic lattice $\boldsymbol{Con}_{\mathbf{Q}}(\boldsymbol{A})$ called the *lattice of* $\mathbf{Q}$-*congruences*.

If $\Phi, \Psi \in Con_{\mathbf{Q}}(\boldsymbol{A})$, then $\Phi +_{\mathbf{Q}} \Psi$ denotes their join in $\boldsymbol{Con}_{\mathbf{Q}}(\boldsymbol{A})$. $\Phi +_{\mathbf{Q}} \Psi$ is generally larger than $\Phi + \Psi$. The lattice meet of Φ, Ψ in $\boldsymbol{Con}_{\mathbf{Q}}(\boldsymbol{A})$ coincides with their intersection $\Phi \cap \Psi$.

If $\mathbf{V}$ is a variety, and $\boldsymbol{A}$ is an algebra of type τ, then $Con_{\mathbf{V}}(\boldsymbol{A})$ forms a principal filter in the lattice $\boldsymbol{Con}(\boldsymbol{A})$ of all congruences of $\boldsymbol{A}$. But if $\boldsymbol{A}$ is in $\mathbf{V}$, then $Con_{\mathbf{V}}(\boldsymbol{A})$ coincides with $Con(\boldsymbol{A})$.

For any $X \subseteq A^2$, $\Theta^A_{\mathbf{Q}}(X)$ denotes the least $\mathbf{Q}$-congruence of $\boldsymbol{A}$ that contains X. Thus

$$\Theta^A_{\mathbf{Q}}(X) = \bigcap\{\Phi \in Con_{\mathbf{Q}}(\boldsymbol{A}) : X \subseteq \Phi\}.$$

The congruence $\Theta^A(X)$ is a subset of $\Theta^A_{\mathbf{Q}}(X)$.

Id$(\mathbf{Q})$ denotes the set of all identities valid in $\mathbf{Q}$.

The following characterization of $\Theta^A_{\mathbf{Q}}(X)$ proves convenient in applications:

Theorem 2.1. *Let* $\mathbf{Q}$ *be a quasivariety of algebras of type* τ *and* Γ *a set of quasi-identities which are not identities such that* $\mathbf{Q} = \mathbf{Mod}(Id(\mathbf{Q}) \cup \Gamma)$. *Then for any algebra* $\boldsymbol{A}$, *any set* $X \subseteq A^2$ *and any* $a, b \in A$,

$a \equiv b(\Theta^A_{\mathbf{Q}}(X))$ *if and only if there exists a finite sequence*

$$\langle a_1, b_1\rangle, \ldots, \langle a_n, b_n\rangle \qquad (*)$$

of elements of A^2 *such that* $\langle a_n, b_n \rangle = \langle a, b \rangle$ *and, for every* i, $1 \leqslant i \leqslant n$, *either* $\langle a_i, b_i \rangle \in X$ *or* $a_i = b_i$ *or there exist a set* $J \subseteq \{1, \ldots, i-1\}$, *a quasi-equation* $r_1(\underline{x}) \approx s_1(\underline{x}) \wedge \ldots \wedge r_m(\underline{x}) \approx s_m(\underline{x}) \rightarrow r(\underline{x}) \approx s(\underline{x}) \in \Gamma \cup Birkhoff(\tau)$, *where* $\underline{x} = x_1, \ldots, x_p$, *and a sequence* $\underline{c} = c_1, \ldots, c_p$ *of elements of A such that*

$$\{\langle r_k(\underline{c}), s_k(\underline{c}) \rangle : 1 \leqslant k \leqslant m\} = \{\langle a_j, b_j \rangle : j \in J\} \ \ \textit{and} \ \ \langle r(\underline{c}), s(\underline{c}) \rangle = \langle a_i, b_i \rangle.$$

Proof. See Czelakowski and Dziobiak (1996) or Gorbunov (1984). See also Czelakowski (2001), Lemma Q.2.1. □

The sequence $(*)$ is called a **Q**-*generating sequence* of the pair $\langle a, b \rangle$ from the set X.

Proposition 2.2. *Let* **Q** *be a quasivariety. Suppose that*

$$\alpha \approx \beta \in \mathbf{Q}^{\models}(\{\alpha_i \approx \beta_i : i \in I\})$$

for some set of equations $\{\alpha_i \approx \beta_i : i \in I\}$ *and an equation* $\alpha \approx \beta$. *Let* $\boldsymbol{A}$ *be a* τ*-algebra and* $h : \boldsymbol{Te}_\tau \rightarrow \boldsymbol{A}$ *a homomorphism. Then*

$$\langle h(\alpha), h(\beta) \rangle \in \Theta^{\boldsymbol{A}}_{\mathbf{Q}}(\{\langle h(\alpha_i), h(\beta_i) \rangle : i \in I\}).$$

Proof. Put $\Phi := \Theta^{\boldsymbol{A}}_{\mathbf{Q}}(\{\langle h(\alpha_i), h(\beta_i) \rangle : i \in I\})$. Let g be the composition of h and of the canonical homomorphism from $\boldsymbol{A}$ to the **Q**-algebra $\boldsymbol{A}/\Phi$. As g validates the equations $\alpha_i \approx \beta_i$, $i \in I$, it follows that $g(\alpha) = g(\beta)$. So $\langle h(\alpha), h(\beta) \rangle \in \Theta^{\boldsymbol{A}}_{\mathbf{Q}}(\{\langle h(\alpha_i), h(\beta_i) \rangle : i \in I\})$. □

2.5 Free Algebras

Let K be a class of τ-algebras. $\boldsymbol{F}_{\mathbf{K}}(\omega)$ denotes the free algebra in **K** freely generated by a countably infinite set of generators. $\boldsymbol{F}_{\mathbf{K}}(\omega)$ is also free in the variety $\boldsymbol{Va}(\mathbf{K})$. $\boldsymbol{F}_{\mathbf{K}}(\omega)$ need not belong to **K**, but $\boldsymbol{F}_{\mathbf{K}}(\omega)$ is in $\boldsymbol{Qv}(\mathbf{K})$. We therefore have that $\boldsymbol{F}_{\mathbf{K}}(\omega) = \boldsymbol{F}_{\boldsymbol{Qv}(\mathbf{K})}(\omega) = \boldsymbol{F}_{\boldsymbol{Va}(\mathbf{K})}(\omega)$.

Let $\boldsymbol{\Omega}_0$ be the congruence relation defined on the term algebra $\boldsymbol{Te}_\tau$ as follows: for any terms α, β,

$$\alpha \equiv \beta \ (\mathrm{mod}\ \boldsymbol{\Omega}_0) \quad \Leftrightarrow_{df} \quad \alpha \approx \beta \in \mathbf{K}^{\models}(\emptyset) \qquad (\Leftrightarrow \alpha \approx \beta \text{ is valid in } \mathbf{K}).$$

The equivalence class of any term t with respect to $\boldsymbol{\Omega}_0$ is denoted by $[t]$. (Thus $\alpha \approx \beta$ is valid in **K** if and only if $[\alpha] = [\beta]$.)

Proposition 2.3. *The quotient algebra* $\boldsymbol{Te}_\tau/\boldsymbol{\Omega}_0$ *is free in the class* **K** *and hence in the variety* $\boldsymbol{Va}(\mathbf{K})$. *Moreover* $\{[x] : x \in Var\}$ *is the set of free generators of* $\boldsymbol{Te}_\tau/\boldsymbol{\Omega}_0$.

Proof. The congruence $\boldsymbol{\Omega}_0$ is invariant, i.e., for any terms α, β, if $\alpha \equiv \beta (\text{mod } \boldsymbol{\Omega}_0)$, then $e\alpha \approx e\beta$ (mod $\boldsymbol{\Omega}_0$), for all endomorphisms e of the term algebra $\boldsymbol{Te}_\tau$. Hence $\boldsymbol{Te}_\tau/\boldsymbol{\Omega}_0$ is free in **K**. □

A class **K** is *trivial* if it contains only one-element algebras; otherwise, **K** is *non-trivial.*

We shall identify $\boldsymbol{F}_{\mathbf{K}}(\omega)$ with $\boldsymbol{Te}_\tau/\boldsymbol{\Omega}_0$. Since the congruence $\boldsymbol{\Omega}_0$ does not paste together different variables (unless **K** is trivial), the free generators of $\boldsymbol{F}_{\mathbf{K}}(\omega)$ are often identified with individual variables.

Proposition 2.4. *Let* **Q** *be a quasivariety of algebras. For any set* Γ *of equations and any equation* $\alpha \approx \beta$,

$$\alpha \approx \beta \in \mathbf{Q}^{\models}(\Gamma) \quad \textit{if and only if} \quad \langle[\alpha],[\beta]\rangle \in \Theta^{\boldsymbol{F}}_{\mathbf{Q}}(\{\langle[s],[t]\rangle : s \approx t \in \Gamma\}),$$

where $\boldsymbol{F} := \boldsymbol{F}_{\mathbf{Q}}(\omega)$.

The proof is easy and is omitted. □

Let **Q** be a quasivariety and let $\boldsymbol{\Omega}$ be the mapping which to each (closed) theory Σ of the consequence operation $\mathbf{Q}^{\models}$ assigns the congruence

$$\boldsymbol{\Omega}(\Sigma) := \{\langle\alpha,\beta\rangle : \alpha \approx \beta \in \Sigma\}$$

on the algebra of terms $\boldsymbol{Te}_\tau$. ($\boldsymbol{\Omega}_0$ thus coincides with $\boldsymbol{\Omega}(\mathbf{Q}^{\models}(\emptyset))$.)

In turn, let $\boldsymbol{\Xi}$ be the mapping which to each (closed) theory Σ of the consequence operation $\mathbf{Q}^{\models}$ assigns the set of pairs

$$\boldsymbol{\Xi}(\Sigma) := \{\langle[\alpha],[\beta]\rangle : \alpha \approx \beta \in \Sigma\}$$

of the free algebra $\boldsymbol{F}_{\mathbf{Q}}(\omega)$. $\boldsymbol{\Xi}(\Sigma)$ is a congruence of $\boldsymbol{F}_{\mathbf{Q}}(\omega)$. $\boldsymbol{\Xi}(\Sigma)$ is equal to the quotient congruence $\boldsymbol{\Omega}(\Sigma)/\boldsymbol{\Omega}_0$.

Proposition 2.5. *Let* **Q** *be a quasivariety. The mapping* $\boldsymbol{\Xi}$ *establishes an isomorphism between the lattice of closed theories of* $\mathbf{Q}^{\models}$ *and the congruence lattice* $\boldsymbol{Con}_{\mathbf{Q}}(\boldsymbol{F}_{\mathbf{Q}}(\omega))$.

Proof. Straightforward. □

Proposition 2.6. *Let* $\boldsymbol{F} := \boldsymbol{F}_{\mathbf{Q}}(X)$ *be a free algebra in a non-trivial quasivariety* **Q** *with the set* X *of free generators and* $Z \subseteq X^2$. *Then* $\Theta^{\boldsymbol{F}}_{\mathbf{Q}}(Z) = \Theta^{\boldsymbol{F}}(Z)$, *i.e., the least congruence in* $\boldsymbol{F}$ *containing* X *is a* **Q**-*congruence.*

Proof. $\Theta^{\boldsymbol{F}}(Z)$ is equal to $\Theta^{\boldsymbol{F}}(R(Z))$, where $R(Z)$ is the least equivalence relation on X that includes Z. Let Y be a set of selectors for the abstraction classes of $R(Z)$. Thus, for every $x \in X$ there is a unique $y \in Y$ such that $x\, R(Z)\, y$. The quotient algebra $\boldsymbol{F}/\Theta^{\boldsymbol{F}}(Z)$ is isomorphic with the free algebra $\boldsymbol{F}_{\mathbf{Q}}(Y)$. This fact follows from the observation that the mapping $h_0 : X \to Y$ defined by:

$$h_0(x) := \text{ the unique } y \in Y \text{ such that } x\, R(Z)\, y,$$

extends to a homomorphism h from $\boldsymbol{F}$ to $\boldsymbol{F}_{\mathbf{Q}}(Y)$ and $\ker(h) = \Theta^{\boldsymbol{F}}(R(Z))$. Since $\boldsymbol{F}_{\mathbf{Q}}(Y)$ belongs to $\mathbf{Q}$, $\Theta^{\boldsymbol{F}}(Z)$ is a $\mathbf{Q}$-congruence. (See also, e.g., Czelakowski 2001, Chapter Q, Lemma Q.2.3.) □

It follows from the above proposition that for any free generators x and y of $\boldsymbol{F}$, $\Theta(x, y)$ is a $\mathbf{Q}$-congruence of $\boldsymbol{F}$. But it is not true that every congruence $\Phi \subseteq \Theta(x, y)$ is a $\mathbf{Q}$-congruence. The following example is due to Keith Kearnes[2]. Let $\mathbf{Q}$ be the quasivariety of torsion-free Abelian groups. $\mathbf{Q}$ is known to be relatively congruence-modular. The congruence $\Phi := \Theta(2x, 2y)$ is a proper subset of $\Theta(x, y)$, and it is not a $\mathbf{Q}$-congruence. To see this, it is enough to note that $x - y$ is a nonzero torsion element of $\boldsymbol{F}/\Phi$, so $\boldsymbol{F}/\Phi$ is not in $\mathbf{Q}$.

2.6 More on Congruences

Let A and B be sets and $h : A \to B$ a function. If Y is a subset of B^2, then $h^{-1}(Y) := \{\langle a, b\rangle \in A^2 : \langle ha, hb\rangle \in Y\}$. Similarly, if X is a subset of A^2, then $h(X) := \{\langle ha, hb\rangle \in B^2 : \langle a, b\rangle \in X\}$.

If $h : \boldsymbol{A} \to \boldsymbol{B}$ is a homomorphism between algebras $\boldsymbol{A}$ and $\boldsymbol{B}$, then

$$\ker(h) := \{\langle a, b\rangle \in A^2 : ha = hb\}.$$

$\ker(h)$ is a congruence of $\boldsymbol{A}$. It is clear that $\ker(h) = h^{-1}(\mathbf{0}_{\boldsymbol{B}})$.

Proposition 2.7. (The correspondence property). *Let $h : \mathbf{A} \to \mathbf{B}$ be a homomorphism between arbitrary algebras $\mathbf{A}$ and $\mathbf{B}$. If $\Phi \in Con(\mathbf{A})$ and* $\ker(h) \subseteq \Phi$, *then $h^{-1}h(\Phi) = \Phi$.*

Proof. ($\supseteq$). Suppose $\langle a, b\rangle \in \Phi$. Then $\langle ha, hb\rangle \in h(\Phi)$. It follows that $\langle a, b\rangle \in h^{-1}h(\Phi)$.

($\subseteq$). Assume $\langle a, b\rangle \in h^{-1}h(\Phi)$. Then $\langle ha, hb\rangle \in h(\Phi)$. It follows that there are $x, y \in A$ such that $\langle ha, hb\rangle = \langle hx, hy\rangle$ and $\langle x, y\rangle \in \Phi$. As $ha = hx$, $hb = hy$, we get that $\langle a, x\rangle$, $\langle b, y\rangle \in \ker(h) \subseteq \Phi$. Hence $\langle x, y\rangle$, $\langle a, x\rangle$, $\langle b, y\rangle \in \Phi$. This gives that $\langle a, b\rangle \in \Phi$. □

Note. Let $\mathbf{Q}$ be a quasivariety. Let $h : \boldsymbol{A} \to \boldsymbol{B}$ be a homomorphism, where $\boldsymbol{A}$ and $\boldsymbol{B}$ are arbitrary algebras. If $\Phi \in Con_{\mathbf{Q}}(\boldsymbol{B})$, then $h^{-1}(\Phi)$ is a $\mathbf{Q}$-congruence on $\boldsymbol{A}$, i.e., $h^{-1}(\Phi) \in Con_{\mathbf{Q}}(\boldsymbol{A})$. This follows from the fact that the relation $h^{-1}(\Phi)$ is closed under the rules of $\mathbf{Q}^{eq\models}$. Indeed, let $r : \alpha_1 \approx \beta_1, \ldots, \alpha_n \approx \beta_n/\alpha \approx \beta$ be a rule of $\mathbf{Q}^{eq\models}$ and let $g : \boldsymbol{Te}_\tau \to \boldsymbol{A}$ be a homomorphism such that $\langle g\alpha_i, g\beta_i\rangle \in h^{-1}(\Phi)$ for $i = 1, \ldots, n$. Hence $\langle hg\alpha_i, hg\beta_i\rangle \in \Theta$ for $i = 1, \ldots, n$. As $hg : \boldsymbol{Te}_\tau \to \boldsymbol{B}$ is a homomorphism and Θ, being a $\mathbf{Q}$-congruence, is closed with respect to r, we get that $\langle hg\alpha, hg\beta\rangle \in \Phi$. Hence $\langle g\alpha, g\beta\rangle \in h^{-1}(\Phi)$.

---

[2]Personal correspondence.

In particular, if $\boldsymbol{B} \in \mathbf{Q}$, then $\ker(h)$ $(= h^{-1}(\mathbf{0}_B))$ is a $\mathbf{Q}$-congruence on $\boldsymbol{A}$. □

Corollary 2.8. *Let $\mathbf{Q}$ be a quasivariety of algebras of a signature τ. Let $h : \boldsymbol{A} \to \boldsymbol{B}$ be a homomorphism between arbitrary τ-algebras and let $\Phi \in Con(\boldsymbol{A})$ be a congruence such that* $\ker(h) \subseteq \Phi$.

(a) *If h is surjective and $\Phi \in Con_{\mathbf{Q}}(\boldsymbol{A})$, then $h(\Phi) \in Con_{\mathbf{Q}}(\boldsymbol{B})$.*
(b) *If $h(\Phi) \in Con_{\mathbf{Q}}(\boldsymbol{B})$, then $\Phi \in Con_{\mathbf{Q}}(\boldsymbol{A})$.*
(c) *If h is surjective, then $\Phi \in Con_{\mathbf{Q}}(\boldsymbol{A})$ if and only if $h(\Phi) \in Con_{\mathbf{Q}}(\boldsymbol{B})$.*

Proof. As $\ker(h) \subseteq \Phi$, we have that $h^{-1}h(\Phi) = \Phi$, by the correspondence property. It follows that the algebra $\boldsymbol{A}/\Phi$ is embeddable into $\boldsymbol{B}/h(\Phi)$. (The embedding is established by the mapping ϕ which to each equivalence class $a/\Phi \in \boldsymbol{A}/\Phi$ assigns the equivalence class $ha/h(\Phi)$, $a \in A$.)

If $h(\Phi) \in Con_{\mathbf{Q}}(\boldsymbol{B})$, then $\boldsymbol{B}/h(\Phi) \in \mathbf{Q}$. It follows that $\boldsymbol{A}/\Phi \in \mathbf{Q}$, because it is isomorphic with a subalgebra of the $\mathbf{Q}$-algebra $\boldsymbol{B}/h(\Phi)$. This proves (b).

If h is surjective, then the above mapping is an isomorphism between $\boldsymbol{A}/\Phi$ and $\boldsymbol{B}/h(\Phi)$. If $\Phi \in Con_{\mathbf{Q}}(\boldsymbol{A})$, then $\boldsymbol{A}/\Phi$ belongs to $\mathbf{Q}$, and hence $\boldsymbol{B}/h(\Phi)$ belongs to $\mathbf{Q}$ as well. Hence $h(\Phi) \in Con_{\mathbf{Q}}(\boldsymbol{B})$. This proves (a).

(c) follows from (a) and (b). □

Proposition 2.9. *Let $\mathbf{Q}$ be a quasivariety of algebras of a signature τ. Let $h : \boldsymbol{A} \to \boldsymbol{B}$ be a homomorphism between arbitrary τ-algebras. Then for every set $X \subseteq A^2$,*

$$h(\Theta^{\boldsymbol{A}}_{\mathbf{Q}}(X)) \subseteq \Theta^{\boldsymbol{B}}_{\mathbf{Q}}(h(X)).$$

Proof. As $\boldsymbol{A}/h^{-1}(\Theta^{\boldsymbol{B}}_{\mathbf{Q}}(h(X)))$ is isomorphic with a subalgebra of $\boldsymbol{B}/(\Theta^{\boldsymbol{B}}_{\mathbf{Q}}(h(X))) \in \mathbf{Q}$, it follows that $\boldsymbol{A}/h^{-1}(\Theta^{\boldsymbol{B}}_{\mathbf{Q}}(h(X))) \in \mathbf{Q}$. Hence $h^{-1}(\Theta^{\boldsymbol{B}}_{\mathbf{Q}}(h(X)))$ is a $\mathbf{Q}$-congruence on $\boldsymbol{A}$. Since $X \subseteq h^{-1}(\Theta^{\boldsymbol{B}}_{\mathbf{Q}}(h(X)))$, we therefore get that $\Theta^{\boldsymbol{A}}_{\mathbf{Q}}(X) \subseteq h^{-1}(\Theta^{\boldsymbol{B}}_{\mathbf{Q}}(h(X)))$. Consequently, $h(\Theta^{\boldsymbol{A}}_{\mathbf{Q}}(X) \subseteq \Theta^{\boldsymbol{B}}_{\mathbf{Q}}(h(X))$.

(An alternative proof of the above inclusion is based on Theorem 2.1. For let $\langle a, b\rangle \in \Theta^{\boldsymbol{A}}_{\mathbf{Q}}(X)$ and let $\langle a_1, b_1\rangle, \ldots, \langle a_n, b_n\rangle$ be a $\mathbf{Q}$-generating sequence of $\langle a, b\rangle$ from X in $\boldsymbol{A}$. Then $\langle ha_1, hb_1\rangle, \ldots, \langle ha_n, hb_n\rangle$ is a $\mathbf{Q}$-generating sequence of $\langle ha, hb\rangle$ from $h(X)$ in $\boldsymbol{B}$. Hence $\langle a, b\rangle \in \Theta^{\boldsymbol{B}}_{\mathbf{Q}}(h(X))$.) □

Note. Proposition 2.9 implies that for any set of equations X and any equation $\alpha \approx \beta$, if $\alpha \approx \beta \in \mathbf{Q}^{eq\models}(X)$, then for any τ-algebra $\boldsymbol{A}$ and any homomorphism $h : \boldsymbol{Te}_\tau \to \boldsymbol{A}$ it is the case that $\langle h(\alpha), h(\beta)\rangle \in \Theta^{\boldsymbol{A}}_{\mathbf{Q}}(\{\langle h(\alpha), h(\beta)\rangle : \gamma \approx \delta \in X\})$. □

Let $\mathbf{Q}$ be a quasivariety of τ-algebras, $\boldsymbol{A}, \boldsymbol{B}$ arbitrary τ-algebras, and $h : \boldsymbol{A} \to \boldsymbol{B}$ a homomorphism. We define:

$$\ker_{\mathbf{Q}}(h) := h^{-1}(\Theta^{\boldsymbol{B}}_{\mathbf{Q}}(\mathbf{0}_B)),$$

i.e., $\ker_{\mathbf{Q}}(h)$ is the h-preimage of the least $\mathbf{Q}$-congruence of $\boldsymbol{B}$. As $\boldsymbol{A}/\ker_{\mathbf{Q}}(h)$ is isomorphic with a subalgebra of $\boldsymbol{B}/\Theta^{\boldsymbol{B}}_{\mathbf{Q}}(\mathbf{0}_B) \in \mathbf{Q}$, $\ker_{\mathbf{Q}}(h)$ is a $\mathbf{Q}$-congruence on $\boldsymbol{A}$. If $\boldsymbol{B} \in \mathbf{Q}$, then $\ker_{\mathbf{Q}}(h) = \ker(h)$, because $\Theta^{\boldsymbol{B}}_{\mathbf{Q}}(\mathbf{0}_B) = \mathbf{0}_B$.

Proposition 2.10. *Let* $\mathbf{Q}$ *be a quasivariety of* τ*-algebras,* $\boldsymbol{A}$*,* $\boldsymbol{B}$ *arbitrary* τ*-algebras, and* $h : \boldsymbol{A} \to \boldsymbol{B}$ *a surjective homomorphism. Then for any set* $X \subseteq A^2$,

$$h(\Theta_{\mathbf{Q}}^{\boldsymbol{A}}(X) +_{\mathbf{Q}} \ker_{\mathbf{Q}}(h)) = \Theta_{\mathbf{Q}}^{\boldsymbol{B}}(h(X)). \qquad (*)$$

Note. If $\boldsymbol{B} \in \mathbf{Q}$, then $\ker_{\mathbf{Q}}(h) = \ker(h)$. Hence $(*)$ implies that

$$h(\Theta_{\mathbf{Q}}^{\boldsymbol{A}}(X) +_{\mathbf{Q}} \ker(h)) = \Theta_{\mathbf{Q}}^{\boldsymbol{B}}(h(X)). \quad \square$$

Proof. Since $\ker_{\mathbf{Q}}(h)$ is a $\mathbf{Q}$-congruence on $\boldsymbol{A}$, therefore $\Phi := \Theta_{\mathbf{Q}}^{\boldsymbol{A}}(X) +_{\mathbf{Q}} \ker_{\mathbf{Q}}(h)$ is a well-defined $\mathbf{Q}$-congruence on $\boldsymbol{A}$. As $\ker(h) \subseteq \mathit{ker}_{\mathbf{Q}}(h) \in \mathit{Con}_{\mathbf{Q}}(\boldsymbol{A})$, Corollary 2.8.(a) implies that $h(\Phi)$ is a $\mathbf{Q}$-congruence on $\boldsymbol{B}$. Since $X \subseteq \Phi$, we get that $h(\Theta_{\mathbf{Q}}^{\boldsymbol{A}}(X) +_{\mathbf{Q}} \mathit{ker}_{\mathbf{Q}}(h)) = h(\Phi) \supset \Theta_{\mathbf{Q}}^{\boldsymbol{B}}(h(X))$.

On the other hand, as $h(\Theta_{\mathbf{Q}}^{\boldsymbol{A}}(X) \subseteq \Theta_{\mathbf{Q}}^{\boldsymbol{B}}(h(X))$ and $h(\ker_{\mathbf{Q}}(h)) = \Theta_{\mathbf{Q}}^{\boldsymbol{B}}(\mathbf{0}_{\boldsymbol{B}})$, we get that $h(\Theta_{\mathbf{Q}}^{\boldsymbol{A}}(X) +_{\mathbf{Q}} \ker_{\mathbf{Q}}(h)) = h(\Theta_{\mathbf{Q}}^{\boldsymbol{A}}(X \cup \ker_{\mathbf{Q}}(h)) \subseteq \Theta_{\mathbf{Q}}^{\boldsymbol{B}}(h(X) \cup h(\ker_{\mathbf{Q}}(h))) = \Theta_{\mathbf{Q}}^{\boldsymbol{B}}(h(X))$, by Proposition 2.9. $\square$

Corollary 2.11. *Let* $\mathbf{Q}$ *be a quasivariety,* $\boldsymbol{A}$*,* $\boldsymbol{B}$ *be arbitrary* τ*-algebras, and* $h: \boldsymbol{A} \to \boldsymbol{B}$ *a surjective homomorphism. Then for any set* $X \subseteq A^2$,

$$h^{-1}(\Theta_{\mathbf{Q}}^{\boldsymbol{B}}(hX)) = \ker_{\mathbf{Q}}(h) +_{\mathbf{Q}} \Theta_{\mathbf{Q}}^{\boldsymbol{A}}(X). \quad \square \qquad (**)$$

Proof. Since $\boldsymbol{B} \in \mathbf{Q}$, $\ker(h) = \ker_{\mathbf{Q}}(h)$. $\square$

Corollary 2.12. *Let* $\mathbf{Q}$ *be a quasivariety, let* $\boldsymbol{A}$*,* $\boldsymbol{B}$ *be arbitrary* τ*-algebras, and* $h : \boldsymbol{A} \to \boldsymbol{B}$ *a surjective homomorphism. Then for any* $\Phi, \Psi \in \mathit{Con}_{\mathbf{Q}}(\boldsymbol{B})$,

$$h^{-1}(\Phi +_{\mathbf{Q}} \Psi) = h^{-1}(\Phi) +_{\mathbf{Q}} h^{-1}(\Psi).$$

Proof. As h is "onto", $hh^{-1}(\Phi) = \Phi$ and $hh^{-1}(\Psi) = \Psi$. Moreover $\ker_{\mathbf{Q}}(h) \subseteq h^{-1}(\Phi)$ and $\ker_{\mathbf{Q}}(h) \subseteq h^{-1}(\Psi)$. Applying Corollary 2.11 we get:

$$h^{-1}(\Phi +_{\mathbf{Q}} \Psi) = h^{-1}(\Theta_{\mathbf{Q}}^{\boldsymbol{B}}(\Phi \cup \Psi)) = h^{-1}(\Theta_{\mathbf{Q}}^{\boldsymbol{B}}(hh^{-1}(\Phi) \cup hh^{-1}(\Psi))) =$$

$$h^{-1}(\Theta_{\mathbf{Q}}^{\boldsymbol{B}}(h(h^{-1}(\Phi) \cup h^{-1}(\Psi)))) = \ker_{\mathbf{Q}}(h) +_{\mathbf{Q}} \Theta_{\mathbf{Q}}^{\boldsymbol{A}}(h^{-1}(\Phi) \cup h^{-1}(\Psi)) =$$

$$\ker_{\mathbf{Q}}(h) +_{\mathbf{Q}} \Theta_{\mathbf{Q}}^{\boldsymbol{A}}(h^{-1}(\Phi)) +_{\mathbf{Q}} \Theta_{\mathbf{Q}}^{\boldsymbol{A}}(h^{-1}(\Psi)) = \ker_{\mathbf{Q}}(h) +_{\mathbf{Q}} h^{-1}(\Phi) +_{\mathbf{Q}} h^{-1}(\Psi) =$$

$$h^{-1}(\Phi) +_{\mathbf{Q}} h^{-1}(\Psi). \quad \square$$

Let $\boldsymbol{F}$ be a non-trivial infinitely generated free algebra and let $h : \boldsymbol{F} \to \boldsymbol{F}$ be an epimorphism. Let x and y be arbitrary free generators of $\boldsymbol{F}$. Assume $\boldsymbol{F}$ is free in a quasivariety $\mathbf{Q}$. We know that $\Theta^{\boldsymbol{F}}(x, y)$ is a $\mathbf{Q}$-congruence of $\boldsymbol{F}$. The kernel $\ker(h)$ is also a $\mathbf{Q}$-congruence of $\boldsymbol{F}$. Question: Is $\ker(h) + \Theta^{\boldsymbol{F}}(x, y)$ a $\mathbf{Q}$-congruence of $\boldsymbol{F}$? ("+" stands for the least upper bound in the lattice of congruences of $\boldsymbol{F}$.) If hx and hy are free generators, then indeed $\ker(h) + \Theta^{\boldsymbol{F}}(x, y)$ is a $\mathbf{Q}$-congruence. But if hx and hy are not free generators, the answer may be: No! Here is a simple example provided by Keith Kearnes.

Let $\mathbf{Q}$ be the quasivariety in the language with only two constant symbols 0, 1 (and no other operations) that is axiomatized by the quasi-identity $0 \approx 1 \rightarrow x \approx y$. Let $\boldsymbol{F}$ be the free algebra in $\mathbf{Q}$ with free generators $\{x_0, x_1, \ldots\}$. Then $F = \{0, 1, x_0, x_1, \ldots\}$.

Let $h : F \rightarrow F$ be the function which x_0 and 0 maps to 0, x_1 and 1 maps to 1, and x_{n+1} maps to x_n for all $n \geqslant 1$. h is an epimorphism and $\ker(h) = \{0, x_0\}^2 \cup \{1, x_1\}^2 \cup \mathbf{0}_F$.

The join $\ker(h) + \Theta^F(x_0, x_1)$ equals $\{0, 1, x_0, x_1\}^2 \cup \mathbf{0}_F$, i.e., it relates the four elements $0, 1, x_0, x_1$ and relates no other pair of distinct elements. But the $\mathbf{Q}$-join $\ker(h) +_{\mathbf{Q}} \Theta^F(x_0, x_1)$ is $F \times F$. Hence $\ker(h) + \Theta^F(x_0, x_1)$ is not a $\mathbf{Q}$-congruence.

Other useful facts and aspects of the theory of quasivarieties can be found, e.g., in Pigozzi (1988), Czelakowski and Dziobiak (1996), Czelakowski (2001) and Kearnes and McKenzie (1992).

2.7 Properties of Equational Theories

The theory of the equationally defined commutator to a large extent uses syntactical tools derived from the properties of the equational consequences associated with quasivarieties. In this subsection we shall present some of these properties.

The facts presented in the above section on relative congruences have their counterparts for the theories of $\mathbf{Q}^{\models}$, where $\mathbf{Q}$ is a quasivariety of a signature τ.

Let $e : Te_\tau \rightarrow Te_\tau$ be a function. If X is a set of equations, then

$$e(X) := \{ep \approx eq : p \approx q \in X\}.$$

$$e^{-1}(X) := \{p \approx q \in Eq(\tau) : ep \approx eq \in X\}.$$

We shall mark the theory $\mathbf{Q}^{\models}(\Sigma_1 \cup \Sigma_2)$ as

$$\Sigma_1 +_{\mathbf{Q}} \Sigma_2,$$

thus using the notation applied to $\mathbf{Q}$-congruences. (Σ_1 and Σ_2 are arbitrary theories of $\mathbf{Q}^{\models}$.)

If $e : \boldsymbol{Te}_\tau \rightarrow \boldsymbol{Te}_\tau$ is an endomorphism and Σ is a theory of $\mathbf{Q}^{\models}$, then $e^{-1}(\Sigma)$ is a theory of $\mathbf{Q}^{\models}$ as well. This follows from the fact that the set of equations $e^{-1}(\Sigma)$ is closed with respect to the rules of $\mathbf{Q}^{\models}$ (cf. Note following Proposition 2.7).

Some care is needed when one wants to define the kernel of an endomorphism e of the term algebra. The set $\{p \approx q \in Eq(\tau)$: the term ep is identical with $eq\}$ is a theory of Birkhoff's consequence B_τ. Not much can be said about the properties of this set when one wants to connect it with a non-trivial quasivariety. We therefore adopt the following definition.

If $e : \boldsymbol{Te}_\tau \to \boldsymbol{Te}_\tau$ is an endomorphism, then

$$\ker_{\mathbf{Q}}(e) := e^{-1}(\mathbf{Q}^{\models}(\emptyset)).$$

$\ker_{\mathbf{Q}}(e)$ is called the *kernel* of the endomorphism e *relative to* $\mathbf{Q}$. Thus $p \approx q \in \ker_{\mathbf{Q}}(e)$ if and only if the equation $ep \approx eq$ is valid in $\mathbf{Q}$. One may directly verify that $\ker_{\mathbf{Q}}(e)$ is closed with respect to the rules of $\mathbf{Q}^{\models}$ and hence it is a theory of $\mathbf{Q}^{\models}$. We obviously have that $\{p \approx q \in Eq(\tau) : ep = eq\} \subseteq \ker_{\mathbf{Q}}(e)$.

The consequence operation $\boldsymbol{Va}(\mathbf{Q})^{\models}$ is weaker than $\mathbf{Q}^{\models}$ but both operations agree on $\emptyset$, that is, $\mathbf{Q}^{\models}(\emptyset) = \boldsymbol{Va}(\mathbf{Q})^{\models}(\emptyset)$. From the purely inferential viewpoint, $\boldsymbol{Va}(\mathbf{Q})^{\models}$ is the consequence operation determined by the set of all $\mathbf{Q}$-valid equations and the rules of inference of the Birkhoff's logic B_τ. In other words, $\boldsymbol{Va}(\mathbf{Q})^{\models}$ is an axiomatic strengthening of Birkhoff's logic B_τ (in the signature of $\mathbf{Q}$).

As $\mathbf{Q}^{\models}(\emptyset) = \boldsymbol{Va}(\mathbf{Q})^{\models}(\emptyset)$, it follows that

$$\ker_{\mathbf{Q}}(e) = \ker_{\boldsymbol{Va}(\mathbf{Q})}(e).$$

Proposition 2.13. (The correspondence property for equational theories). *Let* $\mathbf{Q}$ *be a quasivariety of* τ*-algebras and* $e : \boldsymbol{Te}_\tau \to \boldsymbol{Te}_\tau$ *an endomorphism. If* Σ *is a theory of* $\mathbf{Q}^{\models}$ *and* $\ker_{\mathbf{Q}}(e) \subseteq \Sigma$, *then* $e^{-1}e(\Sigma) = \Sigma$. □

Corollary 2.14. *Let* $\mathbf{Q}$ *be a quasivariety of* τ*-algebras and* $e : \boldsymbol{Te}_\tau \to \boldsymbol{Te}_\tau$ *an endomorphism. Let* Σ *be a theory of Birkhoff's logic* B_τ *in* $\boldsymbol{Te}_\tau$, *i.e.,* $\Sigma \in Th(B_\tau)$, *such that* $\ker_{\mathbf{Q}}(e) \subseteq \Sigma$.

(a) *If* e *is surjective and* $\Sigma \in Th(\mathbf{Q}^{\models})$, *then* $e(\Sigma) \in Th(\mathbf{Q}^{\models})$.
(b) *If* $e(\Sigma) \in Th(\mathbf{Q}^{\models})$, *then* $\Sigma \in Th(\mathbf{Q}^{\models})$.
(c) *If* e *is surjective, then* $\Sigma \in Th(\mathbf{Q}^{\models})$ *if and only if* $e(\Sigma) \in Th(\mathbf{Q}^{\models})$. □

Proposition 2.15. *Let* $\mathbf{Q}$ *be a quasivariety of* τ*-algebras and* $e : \boldsymbol{Te}_\tau \to \boldsymbol{Te}_\tau$ *an epimorphism. Then*

$$e(\mathbf{Q}^{\models}(X) +_{\mathbf{Q}} \ker_{\mathbf{Q}}(e)) = \mathbf{Q}^{\models}(e(X))$$

for any set of equations X. □

Corollary 2.16. *Let* $\mathbf{Q}$ *be a quasivariety of* τ*-algebras and let* $e : \boldsymbol{Te}_\tau \to \boldsymbol{Te}_\tau$ *be an epimorphism. Then for any set of equations* X,

$$e^{-1}(\mathbf{Q}^{\models}(eX)) = \ker_{\mathbf{Q}}(e) +_{\mathbf{Q}} \mathbf{Q}^{\models}(X). \quad \square$$

Corollary 2.17. *Let* $\mathbf{Q}$ *be a quasivariety of* τ*-algebras and* $e : \boldsymbol{Te}_\tau \to \boldsymbol{Te}_\tau$ *an epimorphism. Then for any theories* Σ_1 *and* Σ_2 *of* $\mathbf{Q}^{\models}$,

$$e^{-1}(\Sigma_1 +_{\mathbf{Q}} \Sigma_2) = e^{-1}(\Sigma_1) +_{\mathbf{Q}} e^{-1}(\Sigma_2)$$

in the term algebra $\boldsymbol{Te}_\tau$. □

Proposition 2.18. *Let X be a set of equations of variables, i.e., $X = \{x_i \approx y_i : i \in I\}$, where x_i, y_i ($i \in I$) are individual variables. Then $\boldsymbol{Va}(\mathbf{Q})^{\models}(X)$ is a theory of $\mathbf{Q}^{\models}$ for any non-trivial quasivariety $\mathbf{Q}$.* □

$\boldsymbol{Va}(\mathbf{Q})^{\models}(X)$ is a theory of the consequence $\boldsymbol{Va}(\mathbf{Q})^{\models}$. The proposition states that $\boldsymbol{Va}(\mathbf{Q})^{\models}(X)$ is a closed theory of the stronger consequence operation $\mathbf{Q}^{\models}$.

The proofs of the above facts are easy modifications of the corresponding results from the preceding subsection.

2.8 Epimorphisms and Isomorphic Embeddings of the Lattice of Theories

Let $\mathbf{Q}$ be a quasivariety of τ-algebras and let $e : \boldsymbol{Te}_\tau \to \boldsymbol{Te}_\tau$ be an endomorphism. We define the function $\boldsymbol{f}_e : Th(\mathbf{Q}^{\models}) \to Th(\mathbf{Q}^{\models})$ by

$$\boldsymbol{f}_e(\Sigma) := e^{-1}(\Sigma) \qquad \text{for all } \Sigma \in Th(\mathbf{Q}^{\models}).$$

As $e^{-1}(\Sigma)$, the e-preimage of Σ, is a closed theory, $\boldsymbol{f}_e$ is well-defined.

The following fact is immediate:

Lemma 2.19. *For any epimorphism $e : \boldsymbol{Te}_\tau \to \boldsymbol{Te}_\tau$, the function $\boldsymbol{f}_e$ is an isomorphic embedding of the lattice $\boldsymbol{Th}(\mathbf{Q}^{\models})$ into $\boldsymbol{Th}(\mathbf{Q}^{\models})$.*

Proof. Let Σ_1 and Σ_2 be theories of $\mathbf{Q}^{\models}$. Corollary 2.17 yields that

$$\boldsymbol{f}_e(\Sigma_1 +_{\mathbf{Q}} \Sigma_2) = e^{-1}(\Sigma_1 +_{\mathbf{Q}} \Sigma_2) = e^{-1}(\Sigma_1) +_{\mathbf{Q}} e^{-1}(\Sigma_2) = \boldsymbol{f}_e(\Sigma_1) +_{\mathbf{Q}} \boldsymbol{f}_e(\Sigma_2),$$

$$\boldsymbol{f}_e(\Sigma_1 \cap \Sigma_2) = e^{-1}(\Sigma_1 \cap \Sigma_2) = e^{-1}(\Sigma_1) \cap e^{-1}(\Sigma_2) = \boldsymbol{f}_e(\Sigma_1) \cap \boldsymbol{f}_e(\Sigma_2).$$

The verification that $\boldsymbol{f}_e$ is one-to-one is also straightforward. □

We define: $Th^e(\mathbf{Q}^{\models}) := \{\Sigma \in Th(\mathbf{Q}^{\models}) : \ker_{\mathbf{Q}}(e) \subseteq \Sigma\}$. The set $Th^e(\mathbf{Q}^{\models})$ forms a sublattice $\boldsymbol{Th}^e(\mathbf{Q}^{\models})$ of $\boldsymbol{Th}(\mathbf{Q}^{\models})$. The theory $\ker_{\mathbf{Q}}(e)$ is the least element of $\boldsymbol{Th}^e(\mathbf{Q}^{\models})$.

Corollary 2.20. *For any epimorphism $e : \boldsymbol{Te}_\tau \to \boldsymbol{Te}_\tau$, the function $\boldsymbol{f}_e$ is an isomorphism between the lattices $\boldsymbol{Th}(\mathbf{Q}^{\models})$ and $\boldsymbol{Th}^e(\mathbf{Q}^{\models})$.*

Proof. It suffices to check that $\boldsymbol{f}_e$ is a surjection from $Th(\mathbf{Q}^{\models})$ onto $Th^e(\mathbf{Q}^{\models})$. Suppose $Y \in Th^e(\mathbf{Q}^{\models})$. We claim that $Y = \boldsymbol{f}_e(X)$ for some $X \in Th(\mathbf{Q}^{\models})$. We put: $X := \mathbf{Q}^{\models}(eY)$. Then, by Corollary 2.16, $\boldsymbol{f}_e(X) = e^{-1}(X) = e^{-1}(\mathbf{Q}^{\models}(eY)) = \ker_{\mathbf{Q}}(e) +_{\mathbf{Q}} \mathbf{Q}^{\models}(Y) = \ker_{\mathbf{Q}}(e) +_{\mathbf{Q}} Y = Y$. □

It follows from the correspondence property (Proposition 2.13) that the function $g_e : Th^e(\mathbf{Q}^{\models}) \to Th(\mathbf{Q}^{\models})$ given by

$$g_e(\Sigma) := e(\Sigma) \quad (= \{e(\sigma) : \sigma \in \Sigma\}), \qquad \text{for all } \Sigma \in Th^e(\mathbf{Q}^{\models}),$$

is the inverse of the isomorphism $\boldsymbol{f}_e$.

The function $\boldsymbol{k}_e : Th(\mathbf{Q}^{\models}) \to Th^e(\mathbf{Q}^{\models})$ given by

$$\boldsymbol{k}_e(\Sigma) := \ker_{\mathbf{Q}}(e) +_{\mathbf{Q}} \Sigma, \qquad \Sigma \in Th(\mathbf{Q}^{\models}),$$

is a *retraction*, that is $\boldsymbol{k}_e$ is a surjection and $\boldsymbol{k}_e$ is the identity map on the sublattice $\boldsymbol{Th}^e(\mathbf{Q}^{\models})$. $\boldsymbol{k}_e$ need not be a lattice homomorphism: $\boldsymbol{k}_e$ preserves joins but does not preserve meets of theories.

2.9 The Kernels of Epimorphisms

Let $\mathbf{Q}$ be a quasivariety of τ-algebras. *Var* is the (countably infinite) set of individual variables of $\boldsymbol{Te}_\tau$.

We know that for every endomorphism $e : \boldsymbol{Te}_\tau \to \boldsymbol{Te}_\tau$

$$\ker_{\mathbf{Q}}(e) := e^{-1}(\mathbf{Q}^{\models}(\emptyset))$$

is a closed theory of $\mathbf{Q}^{\models}$. As $\mathbf{Q}^{\models}(\emptyset) = \boldsymbol{Va}(\mathbf{Q})^{\models}(\emptyset)$, $\ker_{\mathbf{Q}}(e)$ is also a theory of $\boldsymbol{Va}(\mathbf{Q})^{\models}(\emptyset)$. ($\boldsymbol{Va}(\mathbf{Q})^{\models}$ is the consequence operation determined by the set of all $\mathbf{Q}$-valid equations and the rules of inference of Birkhoff's logic B_τ.)

Let $e : \boldsymbol{Te}_\tau \to \boldsymbol{Te}_\tau$ be an endomorphism. We define:

$$V_e := \{x \in Var : e(x) \in Var\}.$$

Thus $V_e := e^{-1}(Var)$.

From now on we assume $e : \boldsymbol{Te}_\tau \to \boldsymbol{Te}_\tau$ is a fixed epimorphism, i.e., a surjective endomorphism. Then the set V_e is infinite and e surjectively maps V_e onto *Var*. Moreover, e assigns a compound term to each variable $x \in Var \setminus V_e$.

$Te_\tau(V_e)$ is the set of all terms of Te_τ in the variables V_e and $\boldsymbol{Te}_\tau(V_e)$ is the corresponding subalgebra of $\boldsymbol{Te}_\tau$.

For each variable $x \in Var$ we mark the term $e(x)$ as t_x. If $x \in V_e$, then t_x is a variable, not necessarily in V_e. It is clear that then there is a variable $s_x \in V_e$ such that $e(s_x) = t_x$ $(= e(x))$, viz., $s_x := x$. If $x \in Var \setminus V_e$, the term t_x is compound, and we write $t_x = t_x(\underline{y})$, where $\underline{y} = y_1, \dots, y_n$ is the list of variables occurring in t_x. There are (different) variables $\underline{x} = x_1, \dots, x_n$ in V_e such that $\underline{y} = e(\underline{x})$, that is, $y_1 = e(x_1), \dots, y_n = e(x_n)$. Then $e(x) = t_x = t_x(\underline{y}) = t_x(e(\underline{x})) = e(t_x(\underline{x}))$. Hence the term $e(x)$ is identical with $e(t_x(\underline{x}))$. Let us denote the term $t_x(\underline{x})$ by s_x. We thus get:

Fact 1. *For every variable $x \in Var$ there is a term $s_x \in Te_\tau(V_e)$ such that the term $e(x)$ is identical with $e(s_x)$. If $x \in V_e$, then s_x is the variable x. If $x \in Var \setminus V_e$, the term s_x is compound.* □

To simplify notation, we shall omit parentheses in the symbols like '$e(x)$'.

e surjectively maps V_e onto Var but it may glue some variables of V_e. We then put:

$$A_0 := \{x \approx s_x : x \in Var \setminus V_e\} \cup \{x \approx y : x, y \in V_e \text{ and } ex = ey\} \cup$$
$$\{x \approx y : x, y \in Var \setminus V_e \text{ and } ex = ey\},$$

where, for each $x \in Var \setminus V_e$, s_x is an *arbitrary* but fixed term in $Te_\tau(V_e)$ such that $ex = es_x$.

The following observation, which seems to have not been considered in the literature, provides a canonical characterization of sets of equations generating the kernel of an arbitrary epimorphism of the term algebra. This characterization will be applied in Chapters 5 and 6.

Theorem 2.21. $\ker_{\mathbf{Q}}(e) = \mathbf{Va}(\mathbf{Q})^{\models}(A_0)$.

Notes. 1. If e injectively maps V_e onto Var, the set $\{x \approx y : x, y \in V_e \text{ and } ex = ey\}$, being a component of A_0, reduces to $\{x \approx x : x \in V_e\}$ and therefore it may be disregarded in the above definition. Similarly, if e does not paste together different variables in $Var \setminus V_e$, the set $\{x \approx y : x, y \in Var \setminus V_e \text{ and } ex = ey\}$ may be omitted too.

2. The theorem implies that $\ker_{\mathbf{Q}}(e) = \mathbf{Q}^{\models}(A_0)$, because $\ker_{\mathbf{Q}}(e)$ is a theory of $\mathbf{Q}^{\models}$. □

Proof. To simplify notation we put $C := \mathbf{Q}^{\models}$. We also put: $C_0 := \mathbf{Va}(\mathbf{Q})^{\models}$. The definitions of A_0 and $\ker_{\mathbf{Q}}(e)$ immediately give that $A_0 \subseteq \ker_{\mathbf{Q}}(e)$. Hence $C_0(A_0) \subseteq \ker_{\mathbf{Q}}(e)$. We shall show that $\ker_{\mathbf{Q}}(e) \subseteq C_0(A_0)$.

Claim 1. *If $x, y \in Var \setminus V_e$ and ex is identical with ey, then $s_x \approx s_y \in C_0(A_0)$.*

Proof (of the claim). We have: $x \approx s_x, y \approx s_y \in A_0$. Moreover $x \approx y \in A_0$, because $ex = ey$. It follows that $s_x \approx s_y \in C_0(A_0)$. □

Claim 2. *For each term $t \in Te_\tau$ there is a term $s_t \in Te_\tau(V_e)$ such that $t \approx s_t \in C_0(A_0)$. Moreover, et is identical with es_t.*

Proof. We use induction on the complexity of terms. Assume t is a variable x. If $x \in V_e$, then s_x is identical with x. Hence trivially $x \approx s_x \in C_0(A_0)$. If $x \in Var \setminus V_e$, we take the compound term $s_x \in Te_\tau(V_e)$ corresponding to x and defined as above. Then $ex = es_x$ and $x \approx s_x \in A_0$, by the definition of A_0. Hence $x \approx s_x \in C_0(A_0)$.

Let t be a compound term $F(t_1 \dots t_k)$, where F is a k-ary function symbol, and assume the thesis holds for the terms $t_1, \dots, t_k$. There are terms $s_{t_1}, \dots, s_{t_k} \in Te_\tau(V_e)$ such that $t_1 \approx s_{t_1}, \dots, t_k \approx s_{t_k} \in C_0(A_0)$ and $et_1 = es_{t_1}, \dots, et_k = es_{t_k}$. We put $s_t := F(s_{t_1} \dots s_{t_k})$. We have: $s_t \in Te_\tau(V_e)$, $t \approx s_t \in C_0(t_1 \approx s_{t_1}, \approx, t_k \approx s_{t_k}) \in C_0(A_0)$ by functionality rules, and $et = es_t$. □

It should be noted that if t is in $Te_\tau(V_e)$, then it follows from the above definitions that t is identical with s_t.

Claim 3. *Let r and s be terms in $Te_\tau(V_e)$ such that $er \approx es \in C_0(\emptyset)$. Then $r \approx s \in C_0(A_0)$.*

Proof. Write $r = r(\underline{x})$, $s = s(\underline{x})$, where $\underline{x} = x_1, \ldots, x_n$. As r and s are in $Te_\tau(V_e)$, the substitution e merely replaces the variables of $\underline{x}$ by a block of other variables $\underline{y} = ex_1, \ldots, ex_n$ (but not necessarily in a one-to-one way). As $er \approx es$ is identical with $r(\underline{y}) \approx s(\underline{y})$, we see that $r(\underline{y}) \approx s(\underline{y}) \in C_0(\emptyset)$ implies that $r(\underline{x}) \approx s(\underline{x}) \in C_0(\{x_i \approx x_j : 1 \leqslant i < j \leqslant n \text{ and } ex_i = ex_j\})$. Since $\{x_i \approx x_j : 1 \leqslant i < j \leqslant n \text{ and } ex_i = ex_j\} \in A_0$, it follows that $r \approx s \in C_0(A_0)$. □

Claim 4. *Let p and q be arbitrary terms in Te_τ such that $ep \approx eq \in C_0(\emptyset)$. Then $p \approx q \in C_0(A_0)$.*

Proof. According to Claim 2, there are terms $s_p, s_q \in Te_\tau(V_e)$ such that $p \approx s_p$, $q \approx s_q \in C_0(A_0)$ and $ep = es_p$, $eq = es_q$. As $ep \approx eq \in C_0(\emptyset)$, we trivially get that $es_p \approx es_q \in C_0(\emptyset)$. Hence, by Claim 3, $s_p \approx s_q \in C_0(A_0)$. This fact together with $p \approx s_p, q \approx s_q \in C_0(A_0)$ yields that $p \approx q \in C_0(A_0)$. □

From the above claim the theorem follows. □

Example. Let $p(\underline{y})$ and $q(\underline{y})$ be two different compound terms in variables $\underline{y}$ and x, y be different variables. Let $e : \boldsymbol{Te}_\tau \to \boldsymbol{Te}_\tau$ be an epimorphism such that $ex = p$, $ey = q$ and e bijectively maps $Var \setminus \{x, y\}$ onto Var. Then $V_e = Var \setminus \{x, y\}$. Let $\underline{x}$ be the set of variables of V_e such that $e\underline{x} = \underline{y}$.

Let $\mathbf{Q}$ be an arbitrary quasivariety. It follows from the above theorem that the set

$$A_0 := \{x \approx p(\underline{x}), y \approx q(\underline{x})\}$$

generates the kernel $\ker_{\mathbf{Q}}(e)$, that is,

$$\ker_{\mathbf{Q}}(e) = \boldsymbol{Va}(\mathbf{Q})^{\models}(A_0).$$

Note. In reference to Theorem 2.21 we also note the following distributivity law involving the kernels of epimorphisms.

The following notions will be used in the chapters devoted to the equationally defined commutator. A *set of equations of variables* is any set $\{x_i \approx y_i : i \in I\}$, where x_i and y_i are variables for $i \in I$ and the variables occurring in the equations $x_i \approx y_i$ $(i \in I)$ are all pairwise different. Two sets X and Y of equations of variables are *separated* if the equations of X and Y do not share a common variable.

Theorem 2.22. *Let $\mathbf{Q}$ be a quasivariety and $e : \boldsymbol{Te}_\tau \to \boldsymbol{Te}_\tau$ an epimorphism. Define the set V_e as above. Let X and Y be separated sets of equations of variables from V_e. If e is injective on the set of variables occurring in $X \cup Y$, then*

$$(\ker_{\mathbf{Q}}(e) +_{\mathbf{Q}} \mathbf{Q}^{\models}(X)) \cap (\ker_{\mathbf{Q}}(e) +_{\mathbf{Q}} \mathbf{Q}^{\models}(Y)) = \ker_{\mathbf{Q}}(e) +_{\mathbf{Q}} \mathbf{Q}^{\models}(X) \cap \mathbf{Q}^{\models}(Y).$$

The above theorem can be proved rather easily by applying Theorem 2.21 and working with the kernel $\ker_{\mathbf{Q}}(e)$ (cf. also Lemma 5.2.10). Another proof of the theorem is presented in Section 3.3 (see Theorem 3.3.7). □

The thesis of Theorem 2.22 holds for any relatively congruence-distributive quasivariety **Q** (without any restrictions imposed on X, Y and e).

Chapter 3
Commutator Equations and the Equationally-Defined Commutator

3.1 Commutator Equations and the Equationally-Defined Commutator of Congruences

If $\underline{s} = s_1, \ldots, s_m$, and $\underline{t} = t_1, \ldots, t_m$ are sequences of terms (both sequences of the same length m) and X is a set of equations then

$$\underline{s} \approx \underline{t} \in X$$

abbreviates the fact that $s_i \approx t_i \in X$ for $i = 1, \ldots, m$.)

Let m and n be positive integers and let $\underline{x} = x_1, \ldots, x_m$, $\underline{y} = y_1, \ldots, y_m$, $\underline{z} = z_1, \ldots, z_n$, $\underline{w} = w_1, \ldots, w_n$, and $\underline{u} = u_1, \ldots, u_k$ be sequences of *pairwise distinct* individual variables. The lengths of the strings $\underline{x}$ and $\underline{y}$ are equal, $|\underline{x}| = |\underline{y}| = m$ and, similarly, $|\underline{z}| = |\underline{w}| = n$, $|\underline{u}| = k$.

Let

$$\alpha(\underline{x}, \underline{y}, \underline{z}, \underline{w}, \underline{u}) := \alpha(x_1, \ldots, x_m, y_1, \ldots, y_m, z_1, \ldots, z_n, w_1, \ldots, w_n, u_1, \ldots, u_k)$$

and

$$\beta(\underline{x}, \underline{y}, \underline{z}, \underline{w}, \underline{u}) := \beta(x_1, \ldots, x_m, y_1, \ldots, y_m, z_1, \ldots, z_n, w_1, \ldots, w_n, u_1, \ldots, u_k)$$

be terms in Te_τ built up with at most the variables $\underline{x} = x_1, \ldots, x_m$, $\underline{y} = y_1, \ldots, y_m$, $\underline{z} = z_1, \ldots, z_n$, $\underline{w} = w_1, \ldots, w_n$, and $\underline{u} = u_1, \ldots, u_k$.

Definition 3.1.1. Let $\mathbf{K}$ be a class of algebras. $\alpha(\underline{x}, \underline{y}, \underline{z}, \underline{w}, \underline{u}) \approx \beta(\underline{x}, \underline{y}, \underline{z}, \underline{w}, \underline{u})$ is called a *commutator equation for* $\mathbf{K}$ *in the variables* $\underline{x}, \underline{y}$, and $\underline{z}, \underline{w}$ if the following quasi-equations are valid in $\mathbf{K}$:

$$x_1 \approx y_1 \wedge \ldots \wedge x_m \approx y_m \rightarrow \alpha(\underline{x}, \underline{y}, \underline{z}, \underline{w}, \underline{u}) \approx \beta(\underline{x}, \underline{y}, \underline{z}, \underline{w}, \underline{u})$$

$$z_1 \approx w_1 \wedge \ldots \wedge z_n \approx w_n \rightarrow \alpha(\underline{x}, \underline{y}, \underline{z}, \underline{w}, \underline{u}) \approx \beta(\underline{x}, \underline{y}, \underline{z}, \underline{w}, \underline{u}).$$

J. Czelakowski, *The Equationally-Defined Commutator*,
DOI 10.1007/978-3-319-21200-5_3

Equivalently, $\alpha(\underline{x}, \underline{y}, \underline{z}, \underline{w}, \underline{u}) \approx \beta(\underline{x}, \underline{y}, \underline{z}, \underline{w}, \underline{u})$ is a commutator equation for **K** in the variables $\underline{x}, \underline{y}$ and $\underline{z}, \underline{w}$ if and only if **K** validates the equations

$$\alpha(\underline{x}, \underline{x}, \underline{z}, \underline{w}, \underline{u}) \approx \beta(\underline{x}, \underline{x}, \underline{z}, \underline{w}, \underline{u}) \quad \text{and} \quad \alpha(\underline{x}, \underline{y}, \underline{z}, \underline{z}, \underline{u}) \approx \beta(\underline{x}, \underline{y}, \underline{z}, \underline{z}, \underline{u}).$$

CoEq(**K**) is the set of all commutator equations for **K**.

A *quaternary commutator equation* for **K** (with parameters) is any commutator equation $\alpha(x, y, z, w, \underline{u}) \approx \beta(x, y, z, w, \underline{u})$ for **K** in the variables x, y and z, w.

An equivalent definition of a commutator equation is formulated in terms of inference rules: $\alpha(\underline{x}, \underline{y}, \underline{z}, \underline{w}, \underline{u}) \approx \beta(\underline{x}, \underline{y}, \underline{z}, \underline{w}, \underline{u})$ is a commutator equation for **K** in the variables $\underline{x}, \underline{y}$ and $\underline{z}, \underline{w}$ if and only if

$$x_1 \approx y_1 \wedge \ldots \wedge x_m \approx y_m / \alpha(\underline{x}, \underline{y}, \underline{z}, \underline{w}, \underline{u}) \approx \beta(\underline{x}, \underline{y}, \underline{z}, \underline{w}, \underline{u})$$

and

$$z_1 \approx w_1 \wedge \ldots \wedge z_n \approx w_n / \alpha(\underline{x}, \underline{y}, \underline{z}, \underline{w}, \underline{u}) \approx \beta(\underline{x}, \underline{y}, \underline{z}, \underline{w}, \underline{u}).$$

are rules of the consequence operation $\mathbf{K}^{\models}$. They are called *absorption rules* for $\alpha \approx \beta$. □

Notes.

1. It follows from the above definition that $\alpha(\underline{x}, \underline{y}, \underline{z}, \underline{w}, \underline{u}) \approx \beta(\underline{x}, \underline{y}, \underline{z}, \underline{w}, \underline{u})$ is a commutator equation for **K** (in the variables $\underline{x}, \underline{y}$ and $\underline{z}, \underline{w}$) if and only if it is a commutator equation (in the variables $\underline{x}, \underline{y}$ and $\underline{z}, \underline{w}$) for the variety ***Va***(**K**) generated by **K**. Consequently, the classes **K**, ***Qv***(**K**) and ***Va***(**K**) possess the same commutator equations.
2. Definition 3.1.1 is reformulated in terms of the consequence operation $\mathbf{K}^{\models}$ as follows: for fixed $m, n \geqslant 1$,

$$\mathbf{K}^{\models}(x_1 \approx y_1, \ldots, x_m \approx y_m) \cap \mathbf{K}^{\models}(z_1 \approx w_1, \ldots, z_n \approx w_n)$$

is the set of all commutator equations in the variables $\underline{x} = x_1, \ldots, x_m$, $\underline{y} = y_1, \ldots, y_m$, $\underline{z} = z_1, \ldots, z_n$, $\underline{w} = w_1, \ldots, w_n$. In particular

$$\mathbf{K}^{\models}(x \approx y) \cap \mathbf{K}^{\models}(z \approx w)$$

is the set of all quaternary commutator equations for **K** (with parameters) in the variables x, y and z and w.

The definition of $\mathbf{K}^{\models}$ directly implies that $\mathbf{K}^{\models}(x_1 \approx y_1, \ldots, x_m \approx y_m) = \boldsymbol{Va}(\mathbf{K})^{\models}(x_1 \approx y_1, \ldots, x_m \approx y_m)$ for any variables $x_1, \ldots, x_m, y_1, \ldots, y_m$ (see Proposition 2.18). It follows that for any $m, n \geqslant 1$,

$$\mathbf{K}^{\models}(x_1 \approx y_1, \ldots, x_m \approx y_m) \cap \mathbf{K}^{\models}(z_1 \approx w_1, \ldots, z_n \approx w_n) =$$

$$\boldsymbol{Va}(\mathbf{K})^{\models}(x_1 \approx y_1, \ldots, x_m \approx y_m) \cap \boldsymbol{Va}(\mathbf{K})^{\models}(z_1 \approx w_1, \ldots, z_n \approx w_n).$$

In particular,

$$\mathbf{K}^{\models}(x \approx y) \cap \mathbf{K}^{\models}(z \approx w) = \boldsymbol{Va}(\mathbf{K})^{\models}(x \approx y) \cap \boldsymbol{Va}(\mathbf{K})^{\models}(z \approx w). \qquad \square$$

The following lemma is a straightforward consequence of the above definition:

Lemma 3.1.2.

(1) *Let* $\alpha := \alpha(\underline{x}, \underline{y}, \underline{z}, \underline{w}, \underline{u})$ *be any term. Then* $\alpha \approx \alpha$ *is a commutator equation for* **K** *in* $\underline{x}, \underline{y}$ *and* $\underline{z}, \underline{w}$.
(2) *More generally, if* $\alpha \approx \beta$ *is an identity of* **K**, *then it is a commutator equation for* **K** *(in whatever variables* $\underline{x}, \underline{y}$ *and* $\underline{z}, \underline{w}$*).*
(3) *If* $\alpha \approx \beta$ *is a commutator equation for* **K** *in* $\underline{x}, \underline{y}$ *and* $\underline{z}, \underline{w}$, *then so is* $\beta \approx \alpha$.
(4) *If* $\alpha \approx \beta$ *is a commutator equation for* **K** *in* $\underline{x}, \underline{y}$ *and* $\underline{z}, \underline{w}$, *and* $\alpha' \approx \beta'$ *is an equation* **K***-deductively equivalent to* $\alpha \approx \beta$, *then* $\alpha' \approx \beta'$ *is a commutator equation for* **K** *in the same variables* $\underline{x}, \underline{y}$ *and* $\underline{z}, \underline{w}$.
(5) *If* $\alpha \approx \beta$ *and* $\beta \approx \gamma$ *are commutator equations in* $\underline{x}, \underline{y}$ *and* $\underline{z}, \underline{w}$, *then* $\alpha \approx \gamma$ *is a commutator equation in* $\underline{x}, \underline{y}$ *and* $\underline{z}, \underline{w}$. $\square$

Lemma 3.1.3. *Let*

$$\alpha_i(\underline{x_i}, \underline{y_i}, \underline{z_i}, \underline{w_i}, \underline{u_i}) \approx \beta_i(\underline{x_i}, \underline{y_i}, \underline{z_i}, \underline{w_i}, \underline{u_i})$$

be a commutator equation for **K** *in* $\underline{x_i}, \underline{y_i}$ *and* $\underline{z_i}, \underline{w_i}$, *for* $i = 1, \ldots, k$. *Let* f *be a* k*-ary operation symbol of* τ *and let* $\underline{x} :=$ *the union of* $\underline{x_1}, \ldots, \underline{x_k}$, $\underline{y} :=$ *the union of* $\underline{y_1}, \ldots, \underline{y_k}$, $\underline{z} :=$ *the union of* $\underline{z_1}, \ldots, \underline{z_k}$, $\underline{w} :=$ *the union of* $\underline{w_1}, \ldots, \underline{w_k}$, $\underline{u} :=$ *the union of* $\underline{u_1}, \ldots, \underline{u_k}$. *[The sets* $\underline{x}, \underline{y}, \underline{z}, \underline{w}$ *and* $\underline{u}$ *are assumed to be pairwise disjoint.] Let*

$$\alpha(\underline{x}, \underline{y}, \underline{z}, \underline{w}, \underline{u}) := f(\alpha_1, \ldots, \alpha_k) \quad \text{and} \quad \beta(\underline{x}, \underline{y}, \underline{z}, \underline{w}, \underline{u}) := f(\beta_1, \ldots, \beta_k).$$

Then

$$\alpha(\underline{x}, \underline{y}, \underline{z}, \underline{w}, \underline{u}) \approx \beta(\underline{x}, \underline{y}, \underline{z}, \underline{w}, \underline{u})$$

is a commutator equation for **K** *in* $\underline{x}, \underline{y}$ *and* $\underline{z}, \underline{w}$.[1]

The above lemma states that the set of commutator equations for **K** has the substitution property.

Proof (of the lemma). In the proof, the symbol "$\mathbf{K} \models \phi$" denotes the validity of a first-order formula ϕ of $\boldsymbol{L}_\tau$ in the class **K**. We will present the proof in the case where

[1] In the above notation there is some ambiguity. $\underline{x}$ is for finite sequences of variables without repetitions. To simplify notation, each such a sequence $\underline{x}$ is identified here with the set of variables which occur in $\underline{x}$.

$k = 2$ and $\underline{x_i} = x_i$, $\underline{y_i} = y_i$, $\underline{z_i} = z_i$, $\underline{w_i} = w_i$ and $\underline{u_i} = \emptyset$, $i = 1, 2$. (The proof of the general case is a straightforward modification of the reasoning presented below.) We then have $\alpha_i = \alpha_i(x_i, y_i, z_i, w_i)$, $\beta_i = \beta_i(x_i, y_i, z_i, w_i)$ for $i = 1, 2$, and

$$\mathbf{K} \models x_1 \approx y_1 \rightarrow \alpha_1(x_1, y_1, z_1, w_1) \approx \beta_1(x_1, y_1, z_1, w_1); \tag{1}$$

$$\mathbf{K} \models z_1 \approx w_1 \rightarrow \alpha_1(x_1, y_1, z_1, w_1) \approx \beta_1(x_1, y_1, z_1, w_1); \tag{2}$$

$$\mathbf{K} \models x_2 \approx y_2 \rightarrow \alpha_2(x_2, y_2, z_2, w_2) \approx \beta_2(x_2, y_2, z_2, w_2); \tag{3}$$

$$\mathbf{K} \models z_2 \approx w_2 \rightarrow \alpha_2(x_2, y_2, z_2, w_2) \approx \beta_2(x_2, y_2, z_2, w_2). \tag{4}$$

We wish to show that

$$\mathbf{K} \models x_1 \approx y_1 \wedge x_2 \approx y_2 \rightarrow f(\alpha_1, \alpha_2) \approx f(\beta_1, \beta_2) \tag{5}$$

and

$$\mathbf{K} \models z_1 \approx w_1 \wedge z_2 \approx w_2 \rightarrow f(\alpha_1, \alpha_2) \approx f(\beta_1, \beta_2). \tag{6}$$

We have:

$$\mathbf{K} \models x_1 \approx y_1 \wedge x_2 \approx y_2 \rightarrow \alpha_1 \approx \beta_1 \qquad \text{by (1);} \tag{7}$$

$$\mathbf{K} \models x_1 \approx y_1 \wedge x_2 \approx y_2 \rightarrow \alpha_2 \approx \beta_2 \qquad \text{by (3).} \tag{8}$$

But $\mathbf{K} \models \alpha_1 \approx \beta_1 \rightarrow f(\alpha_1, \alpha_2) \approx f(\beta_1, \alpha_2)$. Hence, by (7) and the transitivity of $\approx$,

$$\mathbf{K} \models x_1 \approx y_1 \wedge x_2 \approx y_2 \rightarrow f(\alpha_1, \alpha_2) \approx f(\beta_1, \alpha_2). \tag{9}$$

But also $\mathbf{K} \models \alpha_2 \approx \beta_2 \rightarrow f(\beta_1, \alpha_2) \approx f(\beta_1, \beta_2)$. Hence, by (8),

$$\mathbf{K} \models x_1 \approx y_1 \wedge x_2 \approx y_2 \rightarrow f(\beta_1, \alpha_2) \approx f(\beta_1, \beta_2). \tag{10}$$

(9) and (10) imply (5). The implication (6) is proved in a similar manner. □

Let A be a non-empty set. If $X \subseteq A^2$ is a set of pairs of elements of A and $\underline{a} = \langle a_1, \ldots, a_m \rangle$, $\underline{b} = \langle b_1, \ldots, b_m \rangle$ are sequences of elements of A of the same length, we write

$$\underline{a} \equiv \underline{b}\ (X)$$

to indicate that $\langle a_i, b_i \rangle \in X$ for $i = 1, \ldots, m$. We shall also occasionally write $(\underline{a}, \underline{b}) \in X$.

The above notation is extensively used for algebras and congruences. Thus if $\mathbf{A}$ is an algebra, Φ is a congruence of $\mathbf{A}$, and $\underline{a} = \langle a_1, \ldots, a_m \rangle$, $\underline{b} = \langle b_1, \ldots, b_m \rangle$ are sequences of elements of A of the same length, we write

$$\underline{a} \equiv \underline{b}\ (\Phi)$$

to mark that a_i is *congruent to* b_i modulo Φ for $i = 1, \ldots, m$, i.e., $\langle a_i, b_i \rangle \in \Phi$ for $i = 1, \ldots, m$.

Let $\boldsymbol{A} \in \boldsymbol{K}$ and let Φ, Ψ be fixed congruences on $\boldsymbol{A}$. $Com_{\mathbf{K}}(\Phi, \Psi)$ is the binary relation on A defined as follows:

> $\langle f, g \rangle \in Com_{\mathbf{K}}(\Phi, \Psi)$ if and only if there exist a commutator equation $\alpha(\underline{x}, \underline{y}, \underline{z}, \underline{w}, \underline{u}) \approx \beta(\underline{x}, \underline{y}, \underline{z}, \underline{w}, \underline{u})$ for $\mathbf{K}$ (in some variables $\underline{x}, \underline{y}$ and $\underline{z}, \underline{w}$) and strings $\underline{a}, \underline{b}, \underline{c}, \underline{d}, \underline{e}$ of elements of A such that
>
> $$\underline{a} \equiv \underline{b}\ (\Phi),\ \ \underline{c} \equiv \underline{d}\ (\Psi),\ \ \text{and}\ \ f = \alpha(\underline{x}, \underline{y}, \underline{z}, \underline{w}, \underline{u}),\ \ g = \beta(\underline{x}, \underline{y}, \underline{z}, \underline{w}, \underline{u}).$$

Lemma 3.1.4. *The relation $Com_{\mathbf{K}}(\Phi, \Psi)$ is a tolerance, i.e., it is reflexive, symmetric on A and has the substitution property.*

Proof (of the lemma). Reflexivity of $Com_{\mathbf{K}}(\Phi, \Psi)$ follows from the fact that $x \approx x$ is a commutator equation in whatever variables $\underline{x}, \underline{y}$ and $\underline{z}, \underline{w}$. The symmetry of $Com_{\mathbf{K}}(\Phi, \Psi)$ follows from Lemma 3.1.2.(iii). The substitution property is a consequence of Lemma 3.1.3. □

It is an open problem when $Com_{\mathbf{K}}(\Phi, \Psi)$ is a congruence relation of $\boldsymbol{A}$.

Before passing to presentation of other definitions and theorems we adopt a certain notational convention we shall adhere to throughout the paper.

Convention 1. Let $\Delta = \{\alpha_i(\underline{x}, \underline{y}, \underline{z}, \underline{w}, \underline{u}) \approx \beta_i(\underline{x}, \underline{y}, \underline{z}, \underline{w}, \underline{u}) : i \in I\}$ be a set of commutator equations (in the variables $\underline{x}, \underline{y}$ and $\underline{z}, \underline{w}$) for a class $\mathbf{K}$ of signature τ. Let $\boldsymbol{A}$ be an algebra of signature τ and suppose $\underline{a}, \underline{b}, \underline{c}, \underline{d}, \underline{e}$ are sequences of elements of A whose lengths are equal to the length of the strings $\underline{x}, \underline{y}, \underline{z}, \underline{w}, \underline{u}$, respectively. The set of pairs

$$\{\alpha_i(\underline{a}, \underline{b}, \underline{c}, \underline{d}, \underline{e}) \approx \beta_i(\underline{a}, \underline{b}, \underline{c}, \underline{d}, \underline{e}) : i \in I\},$$

which is a subset of A^2, is denoted by $\Delta^{\boldsymbol{A}}(\underline{a}, \underline{b}, \underline{c}, \underline{d}, \underline{e})$, for short. □

According to the above notation we have that

$$Com_{\mathbf{K}}(\Phi, \Psi) = \bigcup\{\Delta^{\boldsymbol{A}}(\underline{a}, \underline{b}, \underline{c}, \underline{d}, \underline{e}) : \underline{a} \equiv \underline{b}\ (\Phi),\ \underline{c} \equiv \underline{d}\ (\Psi),\ \underline{e} \in A^{\omega}\},$$

where $\Delta := CoEq(\mathbf{Q})$ is the set of all commutator equations for $\mathbf{K}$.

To shorten notation, we shall often use the phrase "c.e." as an abbreviation for "commutator equation".

Convention 2. We shall make a further step and we shall write down the set of pairs $\bigcup\{\Delta^{\boldsymbol{A}}(\underline{a}, \underline{b}, \underline{c}, \underline{d}, \underline{e}) : \underline{e} \in A^{\omega}\}$ in a more compact way as

$$(\forall \underline{e})\ \Delta^{\boldsymbol{A}}(\underline{a}, \underline{b}, \underline{c}, \underline{d}, \underline{e}).$$

(The superscript '$\boldsymbol{A}$' is often omitted when $\boldsymbol{A}$ is clear from context.) □

Consequently, we may write the above equality in a more compact form as

$$Com_{\mathbf{K}}(\Phi, \Psi) = \bigcup\{(\forall \underline{e})\ \Delta^{A}(\underline{a}, \underline{b}, \underline{c}, \underline{d}, \underline{e}) : \underline{a} \equiv \underline{b}\ (\Phi),\ \underline{c} \equiv \underline{d}\ (\Psi)\}.$$

Definition 3.1.5. Let $\mathbf{Q}$ be a quasivariety of algebras of signature τ. Let $\boldsymbol{A}$ be an algebra of type τ, and let Φ and Ψ be $\mathbf{Q}$-congruences on $\boldsymbol{A}$. The *equationally defined commutator of* Φ and Ψ on $\boldsymbol{A}$ relative to $\mathbf{Q}$, in symbols

$$[\Phi, \Psi]^{\boldsymbol{A}}_{edc(\mathbf{Q})}$$

is the least $\mathbf{Q}$-congruence on $\boldsymbol{A}$ which contains the following set of pairs:

$$\{\langle \alpha(\underline{a}, \underline{b}, \underline{c}, \underline{d}, \underline{e}), \beta(\underline{a}, \underline{b}, \underline{c}, \underline{d}, \underline{e})\rangle : \alpha(\underline{x}, \underline{y}, \underline{z}, \underline{w}, \underline{u}) \approx \beta(\underline{x}, \underline{y}, \underline{z}, \underline{w}, \underline{u})$$
$$\text{is a c.e. for } \mathbf{Q}, \underline{a} \equiv \underline{b}\ (\Phi),\ \underline{c} \equiv \underline{d}\ (\Psi),\ \text{and } \underline{e} \in A^{\omega}\}.$$

Equivalently, $[\Phi, \Psi]^{\boldsymbol{A}}_{edc(\mathbf{Q})}$ is the least $\mathbf{Q}$-congruence of $\boldsymbol{A}$ which contains the relation $Com_{\mathbf{Q}}(\Phi, \Psi)$, i.e.,

$$[\Phi, \Psi]^{\boldsymbol{A}}_{edc(\mathbf{Q})} := \Theta^{\boldsymbol{A}}_{\mathbf{Q}}(Com_{\mathbf{Q}}(\Phi, \Psi)) =$$
$$\Theta^{\boldsymbol{A}}_{\mathbf{Q}}(\bigcup\{(\forall \underline{e})\ \Delta^{A}(\underline{a}, \underline{b}, \underline{c}, \underline{d}, \underline{e}) : \underline{a} \equiv \underline{b}\ (\Phi),\ \underline{c} \equiv \underline{d}\ (\Psi)\}). \quad \square$$

Comments. The use of the name *equationally defined commutator* is justified by the fact that the above definition is formulated in terms of commutator equations.

The definition of the equationally defined commutator for $\mathbf{Q}$ refers to the set of all commutator equations for $\mathbf{Q}$. But, in fact, the definition of $[\Phi, \Psi]^{\boldsymbol{A}}_{edc(\mathbf{Q})}$ is more parsimonious in the number of involved commutator equations.

Let $X = \{x_n \approx y_n : n \in \mathbb{N}\}$ and $Y = \{z_n \approx w_n : n \in \mathbb{N}\}$ be infinite sets of equations of variables, where all involved variables are pairwise different. In particular, the equations of X and Y do not share a common variable. Then evidently $\mathbf{Q}^{\models}(X) \cap \mathbf{Q}^{\models}(Y)$ is the set of all commutator equations $\alpha(\underline{x}, \underline{y}, \underline{z}, \underline{w}, \underline{u}) \approx \beta(\underline{x}, \underline{y}, \underline{z}, \underline{w}, \underline{u})$ with $\underline{x} \approx \underline{y} \in X$ and $\underline{z} \approx \underline{w} \in Y$. (According to the adopted earlier notation, if $\underline{x} = x_1, \ldots, x_m$, and $\underline{y} = y_1, \ldots, y_m$ sequences of variables, then $\underline{x} \approx \underline{y} \in X$ abbreviates the fact that $x_i \approx y_i \in X$ for $i = 1, \ldots, n$.)

Let Φ and Ψ be any $\mathbf{Q}$-congruences on a τ-algebra $\boldsymbol{A}$. The set $\mathbf{Q}^{\models}(X) \cap \mathbf{Q}^{\models}(Y)$ determines the commutator $[\Phi, \Psi]^{\boldsymbol{A}}_{edc(\mathbf{Q})}$ in the following sense:

$[\Phi, \Psi]^{\boldsymbol{A}}_{edc(\mathbf{Q})}$ is the $\mathbf{Q}$-congruence generated by the set of pairs

$$\{\langle \alpha(\underline{a}, \underline{b}, \underline{c}, \underline{d}, \underline{e}), \beta(\underline{a}, \underline{b}, \underline{c}, \underline{d}, \underline{e})\rangle : \alpha(\underline{x}, \underline{y}, \underline{z}, \underline{w}, \underline{u}) \approx \beta(\underline{x}, \underline{y}, \underline{z}, \underline{w}, \underline{u}) \in \mathbf{Q}^{\models}(X) \cap \mathbf{Q}^{\models}(Y),$$
$$\underline{a} \equiv \underline{b}\ (\Phi),\ \underline{c} \equiv \underline{d}\ (\Psi),\ \text{and } \underline{e} \in A^{<\omega}\}. \quad \text{(a)}$$

This follows from the fact that any commutator equation $\alpha'(\underline{x}', \underline{y}', \underline{z}', \underline{w}', \underline{u}') \approx \beta'(\underline{x}', \underline{y}', \underline{z}', \underline{w}', \underline{u}')$ can be transformed into an equation in $\mathbf{Q}^{\models}(X) \cap \mathbf{Q}^{\models}(Y)$ by way of renaming the variables occurring in $\underline{x}', \underline{y}', \underline{z}', \underline{w}', \underline{u}'$ in a one-to-one way. It is clear that $\alpha' \approx \beta'$ and the transformed equation yield the same pairs when computed on the same sequences of elements of A.

But one may even go further. Let Δ be an arbitrary set of equations such that $\mathbf{Q}^{\models}(\Delta) = \mathbf{Q}^{\models}(X) \cap \mathbf{Q}^{\models}(Y)$. We then obtain that

$$[\Phi, \Psi]^{\boldsymbol{A}}_{edc(\mathbf{Q})} \text{ is the } \mathbf{Q}\text{-congruence generated by the set of pairs}$$
$$\{\langle \alpha(\underline{a}, \underline{b}, \underline{c}, \underline{d}, \underline{e}), \beta(\underline{a}, \underline{b}, \underline{c}, \underline{d}, \underline{e})\rangle : \alpha(\underline{x}, \underline{y}, \underline{z}, \underline{w}, \underline{u}) \approx \beta(\underline{x}, \underline{y}, \underline{z}, \underline{w}, \underline{u}) \in \Delta,$$
$$\underline{a} \equiv \underline{b}\ (\Phi),\ \underline{c} \equiv \underline{d}\ (\Psi), \text{ and } \underline{e} \in A^{<\omega}\}. \quad \text{(b)}$$

The equivalence of (a) and (b) is a direct consequence of Note following Proposition 2.9.

2. After the identification of the free algebra $\boldsymbol{F} := \boldsymbol{F}_{\mathbf{Q}}(\omega)$ with the quotient algebra $\boldsymbol{Te}_{\tau}/\boldsymbol{\Omega}_0$, the carrier of the free algebra $\boldsymbol{F} := \boldsymbol{F}_{\mathbf{Q}}(\omega)$ is equal to $\{[t] : t \in Te_{\tau}\}$ and then $\{[x] : x \in Var\}$ is the set of free generators of $\boldsymbol{F}$.

If Σ is a set of equations, then

$$[\Sigma] := \{\langle [s], [t]\rangle : s \approx t \in \Sigma\}.$$

$[\Sigma]$ is thus a subset of $F \times F$.

In many algebraic contexts we shall operate with congruences defined on the free algebra $\boldsymbol{F}_{\mathbf{Q}}(\omega)$. Since the lattice of $\mathbf{Q}$-congruences on $\boldsymbol{F}_{\mathbf{Q}}(\omega)$ is isomorphic with the lattice of (closed) theories of $\mathbf{Q}^{\models}$ (Proposition 2.4), we shall often identify the set of commutator equations $\mathbf{Q}^{\models}(x_1 \approx y_1, \dots, x_m \approx y_m) \cap \mathbf{Q}^{\models}(z_1 \approx w_1, \dots, z_n \approx w_n)$ with the corresponding $\mathbf{Q}$-congruence

$$\Theta^{\boldsymbol{F}}_{\mathbf{Q}}(\langle [x_1], [y_1]\rangle, \dots, \langle [x_m], [y_m]\rangle) \cap \Theta^{\boldsymbol{F}}_{\mathbf{Q}}(\langle [z_1], [w_1]\rangle, \dots, \langle [z_n], [w_n]\rangle). \quad (*)$$

In view of Proposition 2.6, the last congruence is equal to

$$\Theta^{\boldsymbol{F}}(\langle [x_1], [y_1]\rangle, \dots, \langle [x_m], [y_m]\rangle) \cap \Theta^{\boldsymbol{F}}(\langle [z_1], [w_1]\rangle, \dots, \langle [z_n], [w_n]\rangle).$$

Following common practice, each free generator $[x]$ of $\boldsymbol{F}$ is identified with the individual variable x. We shall therefore write the congruence $(*)$ in a more compact way as

$$\Theta^{\boldsymbol{F}}(\langle x_1, y_1\rangle, \dots, \langle x_m, y_m\rangle) \cap \Theta^{\boldsymbol{F}}(\langle z_1, w_1\rangle, \dots, \langle z_n, w_n\rangle). \quad (**)$$

Let $X = \{\langle x_i, y_i\rangle : i \in I\}$, $Y = \{\langle z_j, w_j\rangle : j \in J\}$ be possibly infinite sets of pairs of free generators of $\boldsymbol{F}$, where the generators occurring in the pairs belonging to each set are pairwise different. We say that X and Y are *separated* if moreover the pairs of X and Y do not have a common generator. For example, if $\mathbf{Q}$ is non-trivial, then the above sets $\{\langle x_1, y_1\rangle, \dots, \langle x_m, y_m\rangle\}$ and $\{\langle z_1, w_1\rangle, \dots, \langle z_n, w_n\rangle\}$ are separated.

We accordingly call the congruence $(**)$ the *commutator congruence* determined by the separated sets $\{\langle x_1, y_1\rangle, \ldots, \langle x_m, y_m\rangle\}$ and $\{\langle z_1, w_1\rangle, \ldots, \langle z_n, w_n\rangle\}$ of pairs of free generators of $\boldsymbol{F}$.

(We suppress parentheses here as much as possible.) □

Convention 3. The notation

$$[\Phi, \Psi]^{\boldsymbol{A}}_{edc(\mathbf{Q})}$$

though fully informative, will be simplified when $\mathbf{Q}$ or $\boldsymbol{A}$ are clear from context. Accordingly, the equationally defined commutator $[\Phi, \Psi]^{\boldsymbol{A}}_{edc(\mathbf{Q})}$ is often simply marked as

$$[\Phi, \Psi]_{edc(\mathbf{Q})}$$

if $\boldsymbol{A}$ is clear from context. We also write

$$[\Phi, \Psi]^{\boldsymbol{A}}$$

when $\mathbf{Q}$ is fixed. We shall even write

$$[\Phi, \Psi]$$

when both $\mathbf{Q}$ and $\boldsymbol{A}$ are fixed.

In what follows we shall uniformly use these simplified symbols unless stated otherwise. □

Theorem 3.1.6. *Let $\mathbf{Q}$ be a quasivariety of algebras of type τ, let $\boldsymbol{A}$ be an algebra of type τ, and $\Phi, \Psi \in Con_{\mathbf{Q}}(\boldsymbol{A})$. Then:*

(i) $[\Phi, \Psi]$ *is a* $\mathbf{Q}$*-congruence on* $\boldsymbol{A}$;
(ii) $[\Phi, \Psi] \subseteq \Phi \cap \Psi$;
(iii) $[\Phi, \Psi] = [\Psi, \Phi]$;
(iv) *The commutator is monotone in both arguments, i.e., if Φ, Φ_1, Φ_2 and Ψ, Ψ_1, Ψ_2 are $\mathbf{Q}$-congruences on $\boldsymbol{A}$, $\Phi_1 \subseteq \Phi_2$, and $\Psi_1 \subseteq \Psi_2$, then $[\Phi_1, \Psi] \subseteq [\Phi_2, \Psi]$ and $[\Phi, \Psi_1] \subseteq [\Phi, \Psi_2]$;*
(v) *The commutator is order-continuous, i.e., if $\{\Phi_i : i \in I\}$ is a directed family of $\mathbf{Q}$-congruences on $\boldsymbol{A}$ and $\Psi \in Con_{\mathbf{Q}}(\boldsymbol{A})$, then*

$$[\bigcup\{\Phi_i : i \in I\}, \Psi] = \bigcup\{[\Phi_i, \Psi] : i \in I\}.$$

(vi) *If $\boldsymbol{B}$ is a subalgebra of $\boldsymbol{A}$, then $[\Phi \cap B^2, \Psi \cap B^2]^{\boldsymbol{B}} \subseteq B^2 \cap [\Phi, \Psi]^{\boldsymbol{A}}$;*
(vii) *If $a, b, c, d \in A$, then $[\Theta^{\boldsymbol{A}}_{\mathbf{Q}}(a, b), \Theta^{\boldsymbol{A}}_{\mathbf{Q}}(c, d)]^{\boldsymbol{A}} = \bigcup\{[\Theta^{\boldsymbol{B}}_{\mathbf{Q}}(a, b), \Theta^{\boldsymbol{B}}_{\mathbf{Q}}(c, d)]^{\boldsymbol{B}} : \boldsymbol{B}$ is a countably generated subalgebra of $\boldsymbol{A}$ and $a, b, c, d \in B\}$;*

(viii) $[\Theta^A_{\mathbf{Q}}(X), \Theta^A_{\mathbf{Q}}(Y)]^A = \bigcup\{[\Theta^B_{\mathbf{Q}}(X'), \Theta^B_{\mathbf{Q}}(Y')]^B$: *X' and Y' are countable subsets of X and Y, respectively, and $\boldsymbol{B}$ is a countably generated subalgebra of $\boldsymbol{A}$ which includes $X' \cup Y'$*}.

Note. (viii) is also formulated in the following equivalent form:

(viii)* $[\Theta^A_{\mathbf{Q}}(X), \Theta^A_{\mathbf{Q}}(Y)]^A = \bigcup\{[\Theta^B_{\mathbf{Q}}(X'), \Theta^B_{\mathbf{Q}}(Y')]^B$: X' and Y' are finite subsets of X and Y, respectively, and $\boldsymbol{B}$ is a finitely generated subalgebra of $\boldsymbol{A}$ which includes $X' \cup Y'$}.

The algebras $\boldsymbol{B}$ mentioned in (viii) or (viii)* need not be generated by $X' \cup Y'$. □

Proof. Let us denote for brevity by Δ the set $CoEq(\mathbf{Q})$ of all commutator equations for $\mathbf{Q}$. We note that if Φ, Ψ are $\mathbf{Q}$-congruences on $\boldsymbol{A}$, then

$$[\Phi, \Psi] = \Theta^A_{\mathbf{Q}}(Com_{\mathbf{Q}}(\Phi, \Psi)) =$$

$$\Theta^A_{\mathbf{Q}}(\bigcup\{(\forall \underline{e})\Delta(\underline{a}, \underline{b}, \underline{c}, \underline{d}, \underline{e}) : \underline{a} \equiv \underline{b}\ (\Phi),\ \underline{c} \equiv \underline{d}\ (\Psi)\}).$$

(i) is a part of the definition of the commutator.
(ii) Let $\alpha(\underline{x}, \underline{y}, \underline{z}, \underline{w}, \underline{u}) \approx \beta(\underline{x}, \underline{y}, \underline{z}, \underline{w}, \underline{u})$ be a commutator equation for $\mathbf{Q}$. Furthermore, let $\underline{a} \equiv \underline{b}\ (\Phi)$, $\underline{c} \equiv \underline{d}\ (\Psi)$ and $\underline{e} \in A^k$. Since $\alpha(\underline{x}, \underline{y}, \underline{z}, \underline{w}, \underline{u}) \approx \beta(\underline{x}, \underline{y}, \underline{z}, \underline{w}, \underline{u}) \in \mathbf{Q}^{\models}(\underline{x} \approx \underline{y})$, $\underline{a} \equiv \underline{b}\ (\Phi)$, and Φ is a $\mathbf{Q}$-congruence, we have that $\alpha(\underline{a}, \underline{b}, \underline{c}, \underline{d}, \underline{e}) \equiv \beta(\underline{a}, \underline{b}, \underline{c}, \underline{d}, \underline{e})\ (\Phi)$. Similarly, since $\alpha(\underline{x}, \underline{y}, \underline{z}, \underline{w}, \underline{u}) \approx \beta(\underline{x}, \underline{y}, \underline{z}, \underline{w}, \underline{u}) \in \mathbf{Q}^{\models}(\underline{z} \approx \underline{w})$, $\underline{c} \equiv \underline{d}\ (\Psi)$ and Ψ is a $\mathbf{Q}$-congruence, we have that $\alpha(\underline{a}, \underline{b}, \underline{c}, \underline{d}, \underline{e}) \equiv \beta(\underline{a}, \underline{b}, \underline{c}, \underline{d}, \underline{e})\ (\Psi)$. It follows that $\alpha(\underline{a}, \underline{b}, \underline{c}, \underline{d}, \underline{e}) \equiv \beta(\underline{a}, \underline{b}, \underline{c}, \underline{d}, \underline{e})\ (\Phi \cap \Psi)$. Hence $[\Phi, \Psi] \subseteq \Phi \cap \Psi$.
(iii) To prove symmetry, it suffices to show that $Com_{\mathbf{Q}}(\Phi, \Psi) = Com_{\mathbf{Q}}(\Psi, \Phi)$ because then $[\Phi, \Psi] = \Theta^A_{\mathbf{Q}}(Com_{\mathbf{Q}}(\Phi, \Psi)) = \Theta^A_{\mathbf{Q}}(Com_{\mathbf{Q}}(\Psi, \Phi)) = [\Psi, \Phi]$.

Assume $\langle a, b\rangle \in Com_{\mathbf{Q}}(\Phi, \Psi)$. Hence

$$a = \alpha(\underline{a}, \underline{b}, \underline{c}, \underline{d}, \underline{e}), \quad b = \beta(\underline{a}, \underline{b}, \underline{c}, \underline{d}, \underline{e})$$

for some commutator equation $\alpha(\underline{x}, \underline{y}, \underline{z}, \underline{w}, \underline{u}) \approx \beta(\underline{x}, \underline{y}, \underline{z}, \underline{w}, \underline{u})$ for $\mathbf{Q}$ (in the variables $\underline{x}$, $\underline{y}$ and $\underline{z}$, $\underline{w}$) and strings $\underline{a}, \underline{b}, \underline{c}, \underline{d}, \underline{e}$ of elements of A of appropriate length such that $\underline{a} \equiv \underline{b}\ (\Phi)$, $\underline{c} \equiv \underline{d}\ (\Psi)$.

We shall rename the variables occurring in the equation $\alpha \approx \beta$. Let $\underline{x}', \underline{y}', \underline{z}', \underline{w}'$ be new sequences of variables such that $|\underline{x}'| = |\underline{z}|$, $|\underline{y}'| = |\underline{w}|$, $|\underline{z}'| = |\underline{x}|$, and $|\underline{w}'| = |\underline{y}|$. Let

$$\alpha'(\underline{x}', \underline{y}', \underline{z}', \underline{w}', \underline{u}) := \alpha(\underline{x}/\underline{z}', \underline{y}/\underline{w}', \underline{z}/\underline{x}', \underline{w}/\underline{y}', \underline{u})$$

and

$$\beta'(\underline{x}', \underline{y}', \underline{z}', \underline{w}', \underline{u}) := \beta(\underline{x}/\underline{z}', \underline{y}/\underline{w}', \underline{z}/\underline{x}', \underline{w}/\underline{y}', \underline{u}).$$

It is clear that $\alpha' \approx \beta'$ is a commutator equation in $\underline{x}', \underline{y}'$ and $\underline{z}', \underline{w}'$. Then

$$\alpha'(\underline{c}, \underline{d}, \underline{a}, \underline{b}, \underline{e}) := \alpha'(\underline{x}'/\underline{c}, \underline{y}'/\underline{d}, \underline{z}'/\underline{a}, \underline{w}'/\underline{b}, \underline{u}/\underline{e}) = \alpha(\underline{a}, \underline{b}, \underline{c}, \underline{d}, \underline{e}) = a$$

and

$$\beta'(\underline{c}, \underline{d}, \underline{a}, \underline{b}, \underline{e}) := \beta'(\underline{x}'/\underline{c}, \underline{y}'/\underline{d}, \underline{z}'/\underline{a}, \underline{w}'/\underline{b}, \underline{u}/\underline{e}) = \beta(\underline{a}, \underline{b}, \underline{c}, \underline{d}, \underline{e}) = b.$$

This shows that $\langle a, b\rangle \in Com_{\mathbf{Q}}(\Psi, \Phi)$. Hence $Com_{\mathbf{Q}}(\Phi, \Psi) \subseteq Com_{Q}(\Psi, \Phi)$. In a similar manner one proves the opposite inclusion.

As (iv) follows from (iii) and (v), we shall prove (v). Let $\{\Phi_i : i \in I\}$ be a directed (in the sense of inclusion) family of **Q**-congruences of $\boldsymbol{A}$. It follows from the definition of the commutator and the fact that every term has a finite length that the family of **Q**-congruences $\{[\Phi_i, \Psi] : i \in I\}$ is directed as well. Since the operator $\Theta^{\boldsymbol{A}}_{\mathbf{Q}}$ is finitary, it follows that both sets $\bigcup_{i\in I} \Phi_i$ and $\bigcup_{i\in I}[\Phi_i, \Psi]$ are **Q**-congruences of $\boldsymbol{A}$. We compute:

$$[\bigcup_{i\in I} \Phi_i, \Psi] =$$

$$\Theta^{\boldsymbol{A}}_{\mathbf{Q}}(\bigcup\{(\forall \underline{e})\Delta(\underline{a}, \underline{b}, \underline{c}, \underline{d}, \underline{e}) : \underline{a} \equiv \underline{b}\ (\bigcup_{i\in I} \Phi_i),\ \underline{c} \equiv \underline{d}\ (\Psi)\}) =$$

$$\Theta^{\boldsymbol{A}}_{\mathbf{Q}}(\bigcup_{i\in I}\bigcup\{(\forall \underline{e})\Delta(\underline{a}, \underline{b}, \underline{c}, \underline{d}, \underline{e}) : \underline{a} \equiv \underline{b}\ (\Phi_i),\ \underline{c} \equiv \underline{d}\ (\Psi)\}) =$$

$$\bigcup_{i\in I}\Theta^{\boldsymbol{A}}_{\mathbf{Q}}(\bigcup\{(\forall \underline{e})\Delta(\underline{a}, \underline{b}, \underline{c}, \underline{d}, \underline{e}) : \underline{a} \equiv \underline{b}\ (\Phi_i),\ \underline{c} \equiv \underline{d}\ (\Psi)\}) =$$

$$\bigcup_{i\in I}[\Phi_i, \Psi].$$

The third equality follows from the fact that the operator $\Theta^{\boldsymbol{A}}_{\mathbf{Q}}(\cdot)$ is finitary and the family of sets $\bigcup\{(\forall \underline{e})\Delta(\underline{a}, \underline{b}, \underline{c}, \underline{d}, \underline{e}) : \underline{a} \equiv \underline{b}\ (\Phi_i),\ \underline{c} \equiv \underline{d}\ (\Psi)\}$, $i \in I$, is directed.

(vi) directly follows from the definition by comparing the equationally defined commutator in $\boldsymbol{B}$ and $\boldsymbol{A}$.

(vii) follows from (viii). We shall give a proof of (viii).

($\supseteq$). Let $X' \subseteq X$, $Y' \subseteq Y$ and let $\boldsymbol{B}$ be a subalgebra of $\boldsymbol{A}$ which includes the set of elements occurring in the pairs of $X' \cup Y'$. As $\Theta^{\boldsymbol{B}}_{\mathbf{Q}}(X') \subseteq \Theta^{\boldsymbol{A}}_{\mathbf{Q}}(X')$ and $\Theta^{\boldsymbol{B}}_{\mathbf{Q}}(Y') \subseteq \Theta^{\boldsymbol{A}}_{\mathbf{Q}}(Y')$, we then have:

$$[\Theta^{\boldsymbol{B}}_{\mathbf{Q}}(X'), \Theta^{\boldsymbol{B}}_{\mathbf{Q}}(Y')]^{\boldsymbol{B}} =$$

$$\Theta^{\boldsymbol{B}}_{\mathbf{Q}}(\bigcup\{(\forall \underline{e})\Delta^{\boldsymbol{B}}(\underline{a}, \underline{b}, \underline{c}, \underline{d}, \underline{e}) : \underline{a} \equiv \underline{b}\ (\Theta^{\boldsymbol{B}}_{\mathbf{Q}}(X')),\ \underline{c} \equiv \underline{d}\ (\Theta^{\boldsymbol{B}}_{\mathbf{Q}}(Y'))\}) \subseteq$$

$$\Theta^{\boldsymbol{A}}_{\mathbf{Q}}(\bigcup\{(\forall \underline{e})\Delta^{\boldsymbol{B}}(\underline{a}, \underline{b}, \underline{c}, \underline{d}, \underline{e}) : \underline{a} \equiv \underline{b}\ (\Theta^{\boldsymbol{B}}_{\mathbf{Q}}(X')),\ \underline{c} \equiv \underline{d}\ (\Theta^{\boldsymbol{B}}_{\mathbf{Q}}(Y'))\}) \subseteq$$

$$\Theta^{\boldsymbol{A}}_{\mathbf{Q}}(\bigcup\{(\forall \underline{e})\Delta^{\boldsymbol{A}}(\underline{a}, \underline{b}, \underline{c}, \underline{d}, \underline{e}) : \underline{a} \equiv \underline{b}\ (\Theta^{\boldsymbol{A}}_{\mathbf{Q}}(X')),\ \underline{c} \equiv \underline{d}\ (\Theta^{\boldsymbol{A}}_{\mathbf{Q}}(Y'))\}) \subseteq$$

$$\Theta^A_Q(\bigcup\{(\forall \underline{e})\Delta^A(\underline{a}, \underline{b}, \underline{c}, \underline{d}, \underline{e}) : \underline{a} \equiv \underline{b}\ (\Theta^A_Q(X)),\ \underline{c} \equiv \underline{d}\ (\Theta^A_Q(Y))\}) =$$
$$[\Theta^A_Q(X), \Theta^A_Q(Y)]^A.$$

From the above facts the $\supseteq$-part of the equality (viii) follows.
(If X' and Y' are finite, $\boldsymbol{B}$ can be taken to be any *finitely generated* subalgebra of $\boldsymbol{A}$ containing the elements occurring in the pairs of $X' \cup Y'$.)

($\subseteq$) Suppose $\langle u, v\rangle \in [\Theta^A_Q(X), \Theta^A_Q(Y)]^A$, i.e.,

$$\langle u, v\rangle \in [\Theta^A_Q(\bigcup\{(\forall \underline{e})\Delta^A(\underline{a}, \underline{b}, \underline{c}, \underline{d}, \underline{e}) : \underline{a} \equiv \underline{b}\ (\Theta^A_Q(X)),\ \underline{c} \equiv \underline{d}\ (\Theta^A_Q(Y))\}). \quad (1)$$

As the operator $\Theta^A_Q(\ \cdot\ )$ is finitary, (1) implies that there exist a finite subset $\Delta_0(\underline{x}, \underline{y}, \underline{z}, \underline{w}, \underline{u})$ of Δ, where $|\underline{x}| = |\underline{y}| = k$, $|\underline{z}| = |\underline{w}| = l$, $|\underline{u}| = r$, and finitely many sequences $\underline{a_i}, \underline{b_i}$ and $\underline{c_i}, \underline{d_i}, \underline{e_i}$ of elements of A, where $i = 1, 2, \ldots, m$, such that

$$|\underline{a_i}| = |\underline{b_i}| = k, \quad |\underline{c_i}| = |\underline{d_i}| = l, \quad |\underline{e_i}| = r,$$

$$\underline{a_i} \equiv \underline{b_i}\ (\Theta^A_Q(X)),\ \ \underline{c_i} \equiv \underline{d_i}\ (\Theta^A_Q(Y)) \ \text{ for } i = 1, 2, \ldots, m \quad (2)$$

and

$$\langle u, v\rangle \in [\Theta^A_Q(\bigcup\{\Delta_0(\underline{a_i}, \underline{b_i}, \underline{c_i}, \underline{d_i}, \underline{e_i}) : i = 1, 2, \ldots, m\}). \quad (3)$$

We may write:

$$\underline{a_i} = (a_{i,1}, \ldots, a_{i,k}), \qquad \underline{b_i} = (b_{i,1}, \ldots, b_{i,k}),$$
$$\underline{c_i} = (c_{i,1}, \ldots, c_{i,l}), \qquad \underline{d_i} = (d_{i,1}, \ldots, d_{i,l}),$$
$$\underline{e_i} = (e_{i,1}, \ldots, e_{i,r}),$$

for $i = 1, 2, \ldots, m$.

Let X' be the set of pairs $\langle a_{i,j}, b_{i,j}\rangle$ for $i = 1, 2, \ldots, m$, $j = 1, 2, \ldots, k$, and let Y' be the set of pairs $\langle c_{i,j}, d_{i,j}\rangle$, for $i = 1, 2, \ldots, m$, $j = 1, 2, \ldots, l$. Furthermore, let $E_f := \{\underline{e_i} : i = 1, 2, \ldots, m\}$.

It follows from (3) that

$$\langle u, v\rangle \in \Theta^A_Q(\bigcup\{\Delta_0(\underline{a}, \underline{b}, \underline{c}, \underline{d}, \underline{e}) : \underline{a} \equiv \underline{b}\ (X'),\ \underline{c} \equiv \underline{d}\ (Y'),\ \underline{e} \in E_f\}). \quad (4)$$

($\underline{a} \equiv \underline{b}$ (X) means that for $\underline{a} = (a_1, \ldots, a_k)$, $\underline{b} = (b_1, \ldots, b_k)$ it is the case that $\langle a_j, b_j\rangle \in X$ for $j = 1, 2, \ldots, k$.)

It is clear that in view of (4) there exists a countably generated subalgebra $\boldsymbol{B}$ of $\boldsymbol{A}$ (in fact, $\boldsymbol{B}$ is finitely generated) with the following properties:

B contains the elements of A occurring in the pairs of X', Y' and in the sequences of E_f. (5)

(Equivalently, B contains the elements of A occurring in $\underline{a_i}$, $\underline{b_i}$, $\underline{c_i}$, $\underline{d_i}$, $\underline{e_i}$ for $i = 1, 2, \ldots, m$. (5) implies that the finite set $\bigcup\{\Delta_0(\underline{a}, \underline{b}, \underline{c}, \underline{d}, \underline{e}) : \underline{a} \equiv \underline{b}\ (X'),\ \underline{c} \equiv \underline{d}\ (Y'),\ \underline{e} \in E_f\}$ is included in B.)

B contains all elements of A that appear in the definition of a $\mathbf{Q}$-generating sequence of the pair $\langle u, v\rangle$ from the set (6)

$$\bigcup\{\Delta_0(\underline{a}, \underline{b}, \underline{c}, \underline{d}, \underline{e}) : \underline{a} \equiv \underline{b}\ (X'),\ \underline{c} \equiv \underline{d}\ (Y'),\ \underline{e} \in E_f\}$$

(see (4) and Theorem 2.1).

Conditions (5) and (6) give that

$$\begin{aligned}\langle u, v\rangle \in\ & \Theta^{\boldsymbol{B}}_{\mathbf{Q}}(\bigcup\{\Delta^{\boldsymbol{B}}_0(\underline{a}, \underline{b}, \underline{c}, \underline{d}, \underline{e}) : \underline{a} \equiv \underline{b}\ (X'),\ \underline{c} \equiv \underline{d}\ (Y'),\ \underline{e} \in E_f\}) \\ \subseteq\ & \Theta^{\boldsymbol{B}}_{\mathbf{Q}}(\bigcup\{(\forall \underline{e})\Delta^{\boldsymbol{B}}_0(\underline{a}, \underline{b}, \underline{c}, \underline{d}, \underline{e}) : \underline{a} \equiv \underline{b}\ (\Theta^{\boldsymbol{B}}_{\mathbf{Q}}(X')),\ \underline{c} \equiv \underline{d}\ (\Theta^{\boldsymbol{B}}_{\mathbf{Q}}(Y'))\}) \\ \subseteq\ & \Theta^{\boldsymbol{B}}_{\mathbf{Q}}(\bigcup\{(\forall \underline{e})\Delta^{\boldsymbol{B}}(\underline{a}, \underline{b}, \underline{c}, \underline{d}, \underline{e}) : \underline{a} \equiv \underline{b}\ (\Theta^{\boldsymbol{B}}_{\mathbf{Q}}(X')),\ \underline{c} \equiv \underline{d}\ (\Theta^{\boldsymbol{B}}_{\mathbf{Q}}(Y'))\}) \\ =\ & [\Theta^{\boldsymbol{B}}_{\mathbf{Q}}(X'), \Theta^{\boldsymbol{B}}_{\mathbf{Q}}(Y')]^{\boldsymbol{B}}.\end{aligned}$$

Hence $\langle u, v\rangle \in [\Theta^{\boldsymbol{B}}_{\mathbf{Q}}(X'), \Theta^{\boldsymbol{B}}_{\mathbf{Q}}(Y')]^{\boldsymbol{B}}$.

This concludes the proof of (viii). The theorem has been proved. □

The property expressed in (v) is referred to as the *order-continuity* of the commutator. In Chapter 5 we shall investigate a stronger property than order-continuity, namely the additivity of the commutator.

Theorem 3.1.7. *Let* $\mathbf{Q}$ *be a quasivariety of* τ*-algebras and* $h : \boldsymbol{A} \to \boldsymbol{B}$ *a homomorphism, where* $\boldsymbol{A}$ *and* $\boldsymbol{B}$ *are arbitrary* τ*-algebras. Then for any sets* $X, Y \subseteq A^2$,

$$h([\Theta^{\boldsymbol{A}}_{\mathbf{Q}}(X), \Theta^{\boldsymbol{A}}_{\mathbf{Q}}(Y)]^{\boldsymbol{A}}) \subseteq [\Theta^{\boldsymbol{B}}_{\mathbf{Q}}(hX), \Theta^{\boldsymbol{B}}_{\mathbf{Q}}(hY)]^{\boldsymbol{B}}.$$

Note. The above theorem states the property of *structurality* of the equationally defined commutator is analogous to the property of structurality studied in metalogic.

Proof. It suffices to show that for any c.e. $\alpha(\underline{x}, \underline{y}, \underline{z}, \underline{w}, \underline{u}) \approx \beta(\underline{x}, \underline{y}, \underline{z}, \underline{w}, \underline{u})$ and any sequences $\underline{a} \equiv \underline{b}\ (\Theta^{\boldsymbol{A}}_{\mathbf{Q}}(X))$, $\underline{c} \equiv \underline{d}\ (\Theta^{\boldsymbol{A}}_{\mathbf{Q}}(Y))$ and $\underline{e}$ of elements of A (of appropriate lengths) it is the case that

$$\langle h\alpha(\underline{a}, \underline{b}, \underline{c}, \underline{d}, \underline{e}), h\beta(\underline{a}, \underline{b}, \underline{c}, \underline{d}, \underline{e})\rangle \in [\Theta^{\boldsymbol{B}}_{\mathbf{Q}}(hX), \Theta^{\boldsymbol{B}}_{\mathbf{Q}}(hY)]^{\boldsymbol{B}}.$$

But $\underline{a} \equiv \underline{b}\ (\Theta^A_{\mathbf{Q}}(X))$, $\underline{c} \equiv \underline{d}\ (\Theta^A_{\mathbf{Q}}(Y))$ imply that $h\underline{a} \equiv h\underline{b}\ (\Theta^A_{\mathbf{Q}}(hX))$, $h\underline{c} \equiv h\underline{d}\ (\Theta^A_{\mathbf{Q}}(hY))$. As

$$\langle h\alpha(\underline{a},\underline{b},\underline{c},\underline{d},\underline{e}), h\beta(\underline{a},\underline{b},\underline{c},\underline{d},\underline{e})\rangle = \langle \alpha(h\underline{a},h\underline{b},h\underline{c},h\underline{d},h\underline{e}), \beta(h\underline{a},h\underline{b},h\underline{c},h\underline{d},h\underline{e})\rangle$$

and the second pair belongs to $[\Theta^{\boldsymbol{B}}_{\mathbf{Q}}(hX), \Theta^{\boldsymbol{B}}_{\mathbf{Q}}(hY)]^{\boldsymbol{B}}$, the theorem follows. □

Theorem 3.1.8. *Let* $\mathbf{Q}$ *be a quasivariety of* τ*-algebras and* $h : \boldsymbol{A} \to \boldsymbol{B}$ *a surjective homomorphism. Then for any* $\Phi, \Psi \in Con_{\mathbf{Q}}(\boldsymbol{A})$,

$$\ker_{\mathbf{Q}}(h) +_{\mathbf{Q}} [\ker_{\mathbf{Q}}(h) +_{\mathbf{Q}} \Phi, \ker_{\mathbf{Q}}(h) +_{\mathbf{Q}} \Psi]^{\boldsymbol{A}} = h^{-1}([\Theta^{\boldsymbol{B}}_{\mathbf{Q}}(h\Phi), \Theta^{\boldsymbol{B}}_{\mathbf{Q}}(h\Psi)]^{\boldsymbol{B}}).$$

($\boldsymbol{A}$ and $\boldsymbol{B}$ are arbitrary τ-algebras; they need not belong to $\mathbf{Q}$.)

A stronger version of the above equality when the equationally defined commutator is additive is discussed in Chapter 5 (cf. Theorem 5.1.1).

Proof. Suppose $\Phi, \Psi \in Con_{\mathbf{Q}}(\boldsymbol{A})$. $\ker_{\mathbf{Q}}(h) = h^{-1}(\Theta^{\boldsymbol{B}}_{\mathbf{Q}}(\mathbf{0}_{\boldsymbol{B}}))$ is a $\mathbf{Q}$-congruence on $\boldsymbol{A}$.

Let $\Phi^* := \ker_{\mathbf{Q}}(h) +_{\mathbf{Q}} \Phi$ and $\Phi^* := \ker_{\mathbf{Q}}(h) +_{\mathbf{Q}} \Psi$. Clearly, Φ^* and Ψ^* are $\mathbf{Q}$-congruences on $\boldsymbol{A}$. But, more importantly, as $\ker(h) \subseteq \ker_{\mathbf{Q}}(h)$, the h-images $h\Phi^*$ and $h\Psi^*$ are $\mathbf{Q}$-congruences on $\boldsymbol{B}$, by Corollary 2.8.(a). Moreover, $h\Phi^* = \Theta^{\boldsymbol{B}}_{\mathbf{Q}}(h\Phi)$ and $h\Psi^* = \Theta^{\boldsymbol{B}}_{\mathbf{Q}}(h\Psi)$. Indeed, $\Theta^{\boldsymbol{B}}_{\mathbf{Q}}(h\Phi) = h(\Theta^{\boldsymbol{A}}_{\mathbf{Q}}(\Phi) +_{\mathbf{Q}} \ker_{\mathbf{Q}}(h)) = h(\Phi +_{\mathbf{Q}} \ker_{\mathbf{Q}}(h)) = h\Phi^*$, by Proposition 2.10. The proof of the equality $h\Psi^* = \Theta^{\boldsymbol{B}}_{\mathbf{Q}}(h\Psi)$ is similar. We also have that $h^{-1}(h\Phi^*) = \Phi^*$ and $h^{-1}(h\Psi^*) = \Psi^*$, by Proposition 2.7.

We define:

$$X := \bigcup_{\alpha\approx\beta\in CoEq(\mathbf{Q})} \{\langle \alpha(\underline{a},\underline{b},\underline{c},\underline{d},\underline{e}), \beta(\underline{a},\underline{b},\underline{c},\underline{d},\underline{e})\rangle :$$
$$\underline{a} \equiv \underline{b}\ (\text{mod } \Phi^*),\ \underline{c} \equiv \underline{d}\ (\text{mod } \Psi^*),\ \underline{e} \in A^{<\omega}\}.$$

Thus $\Theta^{\boldsymbol{A}}_{\mathbf{Q}}(X) = [\Phi^*, \Psi^*]^{\boldsymbol{A}}$, by the definition of the equationally defined commutator.

We have:

$$h^{-1}([\Theta^{\boldsymbol{B}}_{\mathbf{Q}}(h\Phi), \Theta^{\boldsymbol{B}}_{\mathbf{Q}}(h\Psi)]^{\boldsymbol{B}}) = h^{-1}([h\Phi^*, h\Psi^*]^{\boldsymbol{B}}) = \text{(by the definition of the equationally defined commutator)}$$

$$h^{-1}(\Theta^{\boldsymbol{B}}_{\mathbf{Q}}(\bigcup_{\alpha\approx\beta\in CoEq(\mathbf{Q})} \{\langle \alpha(\underline{a}^*,\underline{b}^*,\underline{c}^*,\underline{d}^*,\underline{e}^*), \beta(\underline{a}^*,\underline{b}^*,\underline{c}^*,\underline{d}^*,\underline{e}^*)\rangle :$$
$$\underline{a}^* \equiv \underline{b}^*\ (\text{mod } h\Phi^*),\ \underline{c}^* \equiv \underline{d}^*\ (\text{mod } h\Psi^*),\ \underline{e} \in B^k\})) = \text{(by surjectivity of } h)$$

$$h^{-1}(\Theta^{\boldsymbol{B}}_{\mathbf{Q}}(\bigcup_{\alpha\approx\beta\in CoEq(\mathbf{Q})}\{\langle\alpha(h\underline{a},h\underline{b},h\underline{c},h\underline{d},h\underline{e}),\beta(h\underline{a},h\underline{b},h\underline{c},h\underline{d},h\underline{e})\rangle :$$

$$h\underline{a}\equiv h\underline{b}\ (\mathrm{mod}\ h\Phi^*),\ h\underline{c}\equiv h\underline{d}\ (\mathrm{mod}\ h\Psi^*),\ \underline{e}\in A^{<\omega}\})) =$$

$$h^{-1}(\Theta^{\boldsymbol{B}}_{\mathbf{Q}}(\bigcup_{\alpha\approx\beta\in CoEq(\mathbf{Q})}\{\langle\alpha(h\underline{a},h\underline{b},h\underline{c},h\underline{d},h\underline{e}),\beta(h\underline{a},h\underline{b},h\underline{c},h\underline{d},h\underline{e})\rangle :$$

$$\underline{a}\equiv \underline{b}\ (\mathrm{mod}\ \Phi^*),\ \underline{c}\equiv \underline{d}\ (\mathrm{mod}\ \Psi^*),\ \underline{e}\in A^{<\omega}\})) =$$

$$h^{-1}(\Theta^{\boldsymbol{B}}_{\mathbf{Q}}(hX)) = \text{(by Corollary 2.11) } \Theta^{\boldsymbol{A}}_{\mathbf{Q}}(X) +_{\mathbf{Q}} \ker_{\mathbf{Q}}(h) =$$

$$[\Phi^*,\Psi^*]^{\boldsymbol{A}} +_{\mathbf{Q}} \ker_{\mathbf{Q}}(h) = \ker_{\mathbf{Q}}(h) +_{\mathbf{Q}} [\ker_{\mathbf{Q}}(h) +_{\mathbf{Q}} \Phi, \ker_{\mathbf{Q}}(h) +_{\mathbf{Q}} \Psi]^{\boldsymbol{A}}.$$

It follows that $h^{-1}([\Theta^{\boldsymbol{B}}_{\mathbf{Q}}(h\Phi),\Theta^{\boldsymbol{B}}_{\mathbf{Q}}(h\Psi)]^{\boldsymbol{B}}) = \ker_{\mathbf{Q}}(h) +_{\mathbf{Q}} [\ker_{\mathbf{Q}}(h) +_{\mathbf{Q}} \Phi, \ker_{\mathbf{Q}}(h) +_{\mathbf{Q}} \Psi]^{\boldsymbol{A}}$. □

Corollary 3.1.9. *Let* **Q** *be a quasivariety in a signature* τ. *Let* ***A***, ***B*** *be any* τ*-algebras and* $h : \boldsymbol{A} \to \boldsymbol{B}$ *a surjective homomorphism. Then for any* $\Phi,\Psi \in Con_{\mathbf{Q}}(\boldsymbol{B})$,

$$h^{-1}([\Phi,\Psi)]^{\boldsymbol{B}}) = \ker_{\mathbf{Q}}(h) +_{\mathbf{Q}} [h^{-1}(\Phi), h^{-1}(\Psi)]^{\boldsymbol{A}}.$$

Proof. Since $\ker(h) \subseteq \ker_{\mathbf{Q}}(h)$, we have that $\ker(h) \subseteq h^{-1}(\Phi)$ and $\ker(h) \subseteq h^{-1}(\Psi)$. As h is a surjection, $hh^{-1}(\Phi) = \Phi$ and $hh^{-1}(\Psi) = \Psi$. In view of Note following Proposition 2.7, $h^{-1}(\Phi)$ and $h^{-1}(\Psi)$ are **Q**-congruences on ***A***. Applying Theorem 3.1.8 we get:

$$\ker_{\mathbf{Q}}(h) +_{\mathbf{Q}} [h^{-1}(\Phi), h^{-1}(\Psi)]^{\boldsymbol{A}} =$$

$$\ker_{\mathbf{Q}}(h) +_{\mathbf{Q}} [\ker_{\mathbf{Q}}(h) +_{\mathbf{Q}} h^{-1}(\Phi), \ker_{\mathbf{Q}}(h) +_{\mathbf{Q}} h^{-1}(\Psi)]^{\boldsymbol{A}} =$$

$$h^{-1}([\Theta^{\boldsymbol{B}}_{\mathbf{Q}}(hh^{-1}(\Phi)),\Theta^{\boldsymbol{B}}_{\mathbf{Q}}(hh^{-1}(\Psi))]^{\boldsymbol{B}}) = h^{-1}([\Theta^{\boldsymbol{B}}_{\mathbf{Q}}(\Phi),\Theta^{\boldsymbol{B}}_{\mathbf{Q}}(\Psi)]^{\boldsymbol{B}}) =$$

$$h^{-1}([\Phi,\Psi]^{\boldsymbol{B}}). \qquad \square$$

Note. Corollary 3.1.9 and Theorem 3.1.8 are equivalent conditions. It suffices to check that Corollary 3.1.9 implies this theorem. Indeed, let Φ, Ψ be **Q**-congruences on ***A***. Then $\Theta^{\boldsymbol{B}}_{\mathbf{Q}}(h(\Phi))$ and $\Theta^{\boldsymbol{B}}_{\mathbf{Q}}(h(\Psi))$ are **Q**-congruences on ***B***. Corollary 3.1.9 and Corollary 2.11 then give:

$$h^{-1}([\Theta^{\boldsymbol{B}}_{\mathbf{Q}}(h\Phi),\Theta^{\boldsymbol{B}}_{\mathbf{Q}}(h\Psi)]^{\boldsymbol{B}}) = \ker_{\mathbf{Q}}(h) +_{\mathbf{Q}} [h^{-1}(\Theta^{\boldsymbol{B}}_{\mathbf{Q}}(h\Phi)), h^{-1}(\Theta^{\boldsymbol{B}}_{\mathbf{Q}}(h\Psi))]^{\boldsymbol{A}} =$$

$$\ker_{\mathbf{Q}}(h) +_{\mathbf{Q}} [\ker_{\mathbf{Q}}(h) +_{\mathbf{Q}} \Phi, \ker_{\mathbf{Q}}(h) +_{\mathbf{Q}} \Psi]^{\boldsymbol{A}}. \qquad \square$$

Let **Q** be an arbitrary quasivariety. As ***Va***(**Q**) is also a quasivariety, the definition of the equationally defined commutator in the *sense of* ***Va***(**Q**) is also meaningful for *arbitrary* ***Va***(**Q**)- congruences on any τ-algebra. Thus to each quasivariety **Q** not being a variety, *two* equationally defined congruences on the class of τ-algebras are assigned. The first commutator is

$$[\cdot\,]_{edc(\mathbf{Q})}$$

which we have defined above. $[\cdot\,]_{edc(\mathbf{Q})}$ operates on **Q**-congruences Φ, Ψ on an arbitrary algebra $\boldsymbol{A}$ and the value $[\Phi, \Psi]_{edc(\mathbf{Q})}$ is a **Q**-congruence of $\boldsymbol{A}$. The other commutator is

$$[\cdot\,]_{edc(\boldsymbol{Va}(\mathbf{Q}))}.$$

According to Definition 3.1.5, the commutator $[\cdot\,]_{edc(\boldsymbol{Va}(\mathbf{Q}))}$ assigns to any ***Va***(**Q**)-congruences Φ, Ψ on arbitrary algebra $\boldsymbol{A}$ a certain ***Va***(**Q**)-congruence on $\boldsymbol{A}$. More precisely, let $\boldsymbol{A}$ be an algebra of type τ, and let Φ and Ψ be ***Va***(**Q**)-congruences on $\boldsymbol{A}$. $[\Phi, \Psi]_{edc(\boldsymbol{Va}(\mathbf{Q}))}$ is the least ***Va***(**Q**)-congruence on $\boldsymbol{A}$ which contains the following set of pairs:

$$\{\langle \alpha(\underline{a}, \underline{b}, \underline{c}, \underline{d}, \underline{e}), \beta(\underline{a}, \underline{b}, \underline{c}, \underline{d}, \underline{e})\rangle : \alpha(\underline{x}, \underline{y}, \underline{z}, \underline{w}, \underline{u}) \approx \beta(\underline{x}, \underline{y}, \underline{z}, \underline{w}, \underline{u})$$
$$\text{is a c.e. for } \boldsymbol{Va}(\mathbf{Q}),\ \underline{a} \equiv \underline{b}\ (\Phi),\ \underline{c} \equiv \underline{d}\ (\Psi),\ \text{and } \underline{e} \in A^{<\omega}\}.$$

Equivalently, $[\Phi, \Psi]_{edc(\boldsymbol{Va}(\mathbf{Q}))}$ is the least ***Va***(**Q**)-congruence of $\boldsymbol{A}$ which contains the relation $Com_{\boldsymbol{Va}(\mathbf{Q})}(\Phi, \Psi)$.

All the above theorems also apply to the commutator $[\cdot\,]_{edc(\boldsymbol{Va}(\mathbf{Q}))}$ and to the lattices of ***Va***(**Q**)-congruences.

If $\boldsymbol{A}$ is in ***Va***(**Q**), the family $Con_{\boldsymbol{Va}(\mathbf{Q})}(\boldsymbol{A})$ of ***Va***(**Q**)-congruences coincides with the set of *all* congruences of $\boldsymbol{A}$. In this case the commutator $[\cdot\,]_{edc(\boldsymbol{Va}(\mathbf{Q}))}$ assigns to *arbitrary* congruences Φ, Ψ on $\boldsymbol{A}$ a congruence on $\boldsymbol{A}$.

We have an embarrassment of riches here—there are *two* equationally defined commutators determined by the same commutator equations on any τ-algebra $\boldsymbol{A}$, viz. $[\cdot\,]_{edc(\mathbf{Q})}$ and $[\cdot\,]_{edc(\boldsymbol{Va}(\mathbf{Q}))}$. The difference between them is that $[\cdot\,]_{edc(\mathbf{Q})}$ is defined only for **Q**-congruences of $\boldsymbol{A}$ and its values are **Q**-congruences while $[\cdot\,]_{edc(\boldsymbol{Va}(\mathbf{Q}))}$ is defined for the larger set of ***Va***(**Q**)-congruences of $\boldsymbol{A}$ and its values are ***Va***(**Q**)-congruences of $\boldsymbol{A}$. It is therefore quite natural to ask about interrelations holding between the two commutators. For example, the question arises as to whether $[\cdot\,]_{edc(\mathbf{Q})}$ is the restriction of $[\cdot\,]_{edc(\boldsymbol{Va}(\mathbf{Q}))}$ to **Q**-congruences of $\boldsymbol{A}$; that is, whether $[\Phi, \Psi]_{edc(\boldsymbol{Va}(\mathbf{Q}))}$ is already a **Q**-congruence whenever Φ, Ψ are **Q**-congruences of $\boldsymbol{A}$. This seems to be a difficult problem and it is natural to try to solve it under additional assumptions such as the additivity of the equationally defined commutator for **Q**. (The additivity property is investigated in Chapter 5.)

As $\mathbf{Q}$ and $\mathbf{Va}(\mathbf{Q})$ share the *same* commutator equations, it is possible to compare the properties of $[\Phi, \Psi]_{edc(\mathbf{Va}(\mathbf{Q}))}$ with those of the commutator $[\Phi, \Psi]_{edc(\mathbf{Q})}$ for any congruences $\Phi, \Psi \in Con_{\mathbf{Q}}(\boldsymbol{A})$ on the algebras $\boldsymbol{A}$ of $\mathbf{Va}(\mathbf{Q})$. We shall formulate the following identity in Section 5.3:

For any algebra $\boldsymbol{A} \in \mathbf{Va}(\mathbf{Q})$ and any congruences $\Phi, \Psi \in Con(\boldsymbol{A})$,

$$\Theta^{\boldsymbol{A}}_{\mathbf{Q}}([\Phi, \Psi]_{edc(\mathbf{Va}(\mathbf{Q}))}) = [\Theta^{\boldsymbol{A}}_{\mathbf{Q}}(\Phi), \Theta^{\boldsymbol{A}}_{\mathbf{Q}}(\Psi)]_{edc(\mathbf{Q})},$$

called the *Extension Principle for the Equationally defined Commutator* (see Theorem 5.3.8).

3.1.1 The Standard Commutator

Let $\mathbf{Q}$ be an RCM quasivariety in a signature τ. Suppose that $\boldsymbol{A} \in \mathbf{Q}$ and $\Phi, \Psi \in Con(\boldsymbol{A})$. $\boldsymbol{A}(\Phi)$ denotes the subalgebra of $\boldsymbol{A} \times \boldsymbol{A}$ whose universe is Φ. In turn, $\Delta_{\Phi,\Psi}$ is the congruence on $\boldsymbol{A}(\Phi)$ generated by identifying the pairs $\langle a, a\rangle$ and $\langle b, b\rangle$ whenever $\langle a, b\rangle \in \Psi$.

Definition 3.1.10. The commutator of the congruences Φ, Ψ in the sense of Smith-Hagemann-Herrmann-Kearnes-McKenzie in the algebra $\boldsymbol{A}$, hereafter called the *standard* commutator, in symbols:

$$[\Phi, \Psi]_{st},$$

is the set of all ordered pairs $\langle a, b\rangle \in A \times A$ such that $\langle a, b\rangle$ is congruent to $\langle a, a\rangle$ modulo the $\mathbf{Q}$-congruence $\Theta^{\boldsymbol{A}(\Phi)}_{\mathbf{Q}}(\Delta_{\Phi,\Psi})$. □

According to Lemma 2.6 in Kearnes and McKenzie (1992), $[\Phi, \Psi]_{st}$ is a congruence on $\boldsymbol{A}$. If $\Phi \in Con_{\mathbf{Q}}(\boldsymbol{A})$, then $[\Phi, \Psi]_{st}$ is a $\mathbf{Q}$-congruence. If both $\Phi, \Psi \in Con_{\mathbf{Q}}(\boldsymbol{A})$, then $[\Phi, \Psi]_{st} \subseteq \Phi \cap \Psi$.

The standard commutator in itself is not the main object of study in this book. Its theory for quasivarieties is expounded in Kearnes and McKenzie (1992). But the standard commutator plays a very significant role here as a reference point—in many contexts the equationally defined commutator is compared with the standard one. In fact, we are also interested here with the issue of identity of these two notions.

The above observations give rise to the following three questions:

1. Does the equationally defined commutator coincide with the standard commutator for quasivarieties defined by Kearnes and McKenzie (1992)?
2. When is the equationally defined commutator additive?
3. What new, non-trivial facts can be established with the help of the equationally-defined commutator?

The above questions determine the logical and narrative structure of this book.

In the following subsections of Chapter 3 as well as in Chapter 4 we will answer the first of the above questions. Our purpose is to prove the following fact:

Theorem A. *Let* **Q** *be a relatively congruence-modular quasivariety. The equationally defined commutator for* **Q** *coincides with the commutator for* **Q** *in the sense of Kearnes and McKenzie.* □

The proof we shall give in Chapter 3 takes a circuitous journey through geometrical properties of the equationally defined commutator and various centralization relations but offers a better understanding of the behaviour of this commutator in the context of RCM quasivarieties.

The second question will answered in Chapters 5 and 6. A partial answer to the third question will be provided in the final chapters.

3.2 The Equationally-Defined Commutator of Equational Theories

Let **Q** be a quasivariety of τ-algebras. The equationally defined commutator (in the sense of **Q**) is defined in *any* algebra $\boldsymbol{A}$ similar to the algebras of **Q**. We are mainly interested in examining the properties of the commutator in the algebras of **Q** and, occasionally, in the algebras belonging to the variety $\boldsymbol{Va}(\mathbf{Q})$ generated by **Q**. One important exception is the term algebra $\boldsymbol{Te}_\tau$, which plays a distinguished role in the investigation of logical properties of the commutator, especially for finitely generated quasivarieties. The equationally defined commutator in the term algebra is defined for **Q**-congruences of $\boldsymbol{Te}_\tau$ but it is also defined for (closed) *equational theories* of the consequence operation $\mathbf{Q}^{\models}$. The lattice of theories of $\mathbf{Q}^{\models}$ is isomorphic with the lattice of **Q**-congruences of $\boldsymbol{Te}_\tau$ via the map $\boldsymbol{\Omega}$ which to each theory T of $\mathbf{Q}^{\models}$ assigns the **Q**-congruence $\boldsymbol{\Omega} T$, where $p \equiv q \pmod{\boldsymbol{\Omega} T}$ if and only if $p \approx q \in T$, for all terms p and q. However, in metalogical applications it is often more convenient to work rather with the commutator defined for *equational theories* than the one defined for the **Q**-congruences of $\boldsymbol{Te}_\tau$.

Definition 3.2.1. Suppose T_1 and T_2 are theories of $\mathbf{Q}^{\models}$. Then, by definition,

$$
\begin{aligned}
[T_1, T_2] := \mathbf{Q}^{\models}(\{\alpha(\underline{p}, \underline{q}, \underline{r}, \underline{s}, \underline{t}) \approx \beta(\underline{p}, \underline{q}, \underline{r}, \underline{s}, \underline{t}) :\\
\alpha(\underline{x}, \underline{y}, \underline{z}, \underline{w}, \underline{u}) \approx \beta(\underline{x}, \underline{y}, \underline{z}, \underline{w}, \underline{u}) \text{ is a c.e. for } \mathbf{Q},\\
\underline{p}, \underline{q}, \underline{r}, \underline{s} \text{ are sequences of terms such that } \underline{p} \approx \underline{q} \in T_1,\ \underline{r} \approx \underline{s} \in T_2,\\
\text{and } \underline{t} \text{ is an arbitrary sequence of terms}\}). \quad (1)
\end{aligned}
$$

(If $\underline{p} = p_1, \dots, p_m$ and $\underline{q} = q_1, \dots, q_m$, then "$\underline{p} \approx \underline{q} \in T$" abbreviates "$p_1 \approx q_1 \in T, \dots, p_m \approx q_m \in T$".)

$[T_1, T_2]$ is called the *equationally defined commutator of the theories* T_1 and T_2. □

The equationally defined commutator of theories of $\mathbf{Q}^{\models}$ and the equationally defined commutator defined for $\mathbf{Q}$-congruences of $\boldsymbol{Te}_\tau$ are "isomorphic" objects because the above map $\boldsymbol{\Omega}$, being a lattice isomorphism, also preserves the commutator operation in the sense that for any theories T_1, T_2 of $\mathbf{Q}^{\models}$, $\boldsymbol{\Omega}[T_1, T_2] = [\boldsymbol{\Omega} T_1, \boldsymbol{\Omega} T_2]$ as one can easily check. (On the right side of the above equation the commutator is defined for $\mathbf{Q}$-congruences of $\boldsymbol{Te}_\tau$.)

The following observation directly follows from the above remarks:

Proposition 3.2.2. *Let* $\mathbf{Q}$ *be a quasivariety of algebras of type* τ. *The theory commutator for* $\mathbf{Q}^{\models}$ *has the same properties as the equationally defined commutator for the* $\mathbf{Q}$*-congruences on* τ*-algebras exhibited in Theorem* 3.1.6 □

(One may also produce a direct proof of Proposition 3.2.2 by emulating the proof of Theorem 3.1.6.)

Proposition 3.2.3. *For any positive integers m and n and for any disjoint sequences* $\underline{x}, \underline{y}, \underline{z}, \underline{w}$ *of pairwise different variables, where* $\underline{x} = x_1, \dots, x_m$, $\underline{y} = y_1, \dots, y_m$, *and* $\underline{z} = z_1, \dots, z_n$, $\underline{w} = w_1, \dots, w_n$,

$$[\mathbf{Q}^{\models}(x_1 \approx y_1, \dots, x_m \approx y_m), \mathbf{Q}^{\models}(z_1 \approx w_1, \dots, z_n \approx w_n)] = \qquad (2)$$

$$\mathbf{Q}^{\models}(x_1 \approx y_1, \dots, x_m \approx y_m) \cap \mathbf{Q}^{\models}(z_1 \approx w_1, \dots, z_n \approx w_n).$$

In particular, if x, y, z, w *are different variables, then*

$$[\mathbf{Q}^{\models}(x \approx y), \mathbf{Q}^{\models}(z \approx w)] = \mathbf{Q}^{\models}(x \approx y) \cap \mathbf{Q}^{\models}(z \approx w).$$

Proof. To prove (2), we put: $T_1 := \mathbf{Q}^{\models}(x_1 \approx y_1, \dots, x_m \approx y_m)$ and $T_2 := \mathbf{Q}^{\models}(z_1 \approx w_1, \dots, z_n \approx w_n)$. In view of Proposition 3.2.2, we have that $[T_1, T_2] \subseteq T_1 \cap T_2$.

(We also give a direct proof of the inclusion $[T_1, T_2] \subseteq T_1 \cap T_2$. To prove it, suppose $\alpha \approx \beta$ is an arbitrary commutator equation for $\mathbf{Q}$ in the variables $\underline{x} = x_1, \dots, x_k$, $\underline{y} = y_1, \dots, y_k$, and $\underline{z} = z_1, \dots, z_l$, $\underline{w} = w_1, \dots, w_l$ and $\underline{u}$. Let $\underline{p} = p_1, \dots, p_k$ and $\underline{q} = q_1, \dots, q_k$ and $\underline{r} = r_1, \dots, r_l$ and $\underline{s} = s_1, \dots, s_l$ be sequences of terms such that $\underline{p} \approx \underline{q} \in T_1$, $\underline{r} \approx \underline{s} \in T_2$, and let $\underline{t}$ be an arbitrary sequence of terms of the length of $\underline{u}$. As $\alpha(\underline{x}, \underline{y}, \underline{z}, \underline{w}, \underline{u}) \approx \beta(\underline{x}, \underline{y}, \underline{z}, \underline{w}, \underline{u}) \in \mathbf{Q}^{\models}(\underline{x} \approx \underline{y})$, structurality gives that $\alpha(\underline{p}, \underline{q}, \underline{r}, \underline{s}, \underline{t}) \approx \beta(\underline{p}, \underline{q}, \underline{r}, \underline{s}, \underline{t}) \in \mathbf{Q}^{\models}(\underline{p} \approx \underline{q})$. Similarly $\alpha(\underline{p}, \underline{q}, \underline{r}, \underline{s}, \underline{t}) \approx \beta(\underline{p}, \underline{q}, \underline{r}, \underline{s}, \underline{t}) \in \mathbf{Q}^{\models}(\underline{r} \approx \underline{s})$. It follows that $\alpha(\underline{p}, \underline{q}, \underline{r}, \underline{s}, \underline{t}) \approx \beta(\underline{p}, \underline{q}, \underline{r}, \underline{s}, \underline{t}) \in \mathbf{Q}^{\models}(\underline{p} \approx \underline{q}) \cap \mathbf{Q}^{\models}(\underline{r} \approx \underline{s}) \subseteq T_1 \cap T_2$.)

"$\supseteq$". Since $T_1 \cap T_2$ is the set of commutator equations in the variables $\underline{x}, \underline{y}$ and $\underline{z}, \underline{w}$ (and hence a subset of the set of all commutator equations) and $\underline{x} \approx \underline{y} \in T_1$, $\underline{z} \approx \underline{w} \in T_2$, definition (1) gives that

$$[T_1, T_2] \supseteq$$

$$\mathbf{Q}^{\models}(\{\alpha(\underline{x}, \underline{y}, \underline{z}, \underline{w}, \underline{t}) \approx \beta(\underline{x}, \underline{y}, \underline{z}, \underline{w}, \underline{t}) : \alpha(\underline{x}, \underline{y}, \underline{z}, \underline{w}, \underline{u}) \approx \beta(\underline{x}, \underline{y}, \underline{z}, \underline{w}, \underline{u}) \in T_1 \cap T_2$$

$$\text{and } \underline{t} \text{ is an arbitrary sequence of terms}\}) \supseteq$$

$$\mathbf{Q}^{\models}(\{\alpha(\underline{x}, \underline{y}, \underline{z}, \underline{w}, \underline{u}) \approx \beta(\underline{x}, \underline{y}, \underline{z}, \underline{w}, \underline{u}) \colon \alpha(\underline{x}, \underline{y}, \underline{z}, \underline{w}, \underline{u}) \approx \beta(\underline{x}, \underline{y}, \underline{z}, \underline{w}, \underline{u}) \in T_1 \cap T_2\}) = T_1 \cap T_2.$$

So (2) holds. □

Theorem 3.1.7 implies:

Corollary 3.2.4. *Let* $e : \boldsymbol{Te}_\tau \to \boldsymbol{Te}_\tau$ *be an endomorphism and* X, Y*—sets of equations of* $Eq(\tau)$*. Then*

$$e([\mathbf{Q}^{\models}(X), \mathbf{Q}^{\models}(Y)]) \subseteq [\mathbf{Q}^{\models}(eX), \mathbf{Q}^{\models}(eY)]. \quad \square$$

The property of the equationally defined commutator in the lattice $\boldsymbol{Th}(\mathbf{Q}^{\models})$ encapsulated in Corollary 3.2.4 is referred to as the *structurality* of the theory-defined commutator.

Theorem 3.1.8 and Corollary 3.1.9 yield:

Corollary 3.2.5. *Let* $e : \boldsymbol{Te}_\tau \to \boldsymbol{Te}_\tau$ *be an epimorphism. For any sets of equations* X *and* Y *of* $Eq(\tau)$*,*

$$\ker_{\mathbf{Q}}(e) +_{\mathbf{Q}} [\ker_{\mathbf{Q}}(e) +_{\mathbf{Q}} \mathbf{Q}^{\models}(X), \ker_{\mathbf{Q}}(e) +_{\mathbf{Q}} \mathbf{Q}^{\models}(Y)]$$
$$= e^{-1}([\mathbf{Q}^{\models}(eX), \mathbf{Q}^{\models}(eY)]);$$

equivalently, for any X *and* Y*,*

$$e^{-1}([\mathbf{Q}^{\models}(X), \mathbf{Q}^{\models}(Y)]) = \ker_{\mathbf{Q}}(e) +_{\mathbf{Q}} [e^{-1}(\mathbf{Q}^{\models}(X)), e^{-1}(\mathbf{Q}^{\models}(Y))]. \quad \square$$

One may look at the above corollary from the perspective of the isomorphism $\boldsymbol{f}_e : \boldsymbol{Th}(\mathbf{Q}^{\models}) \to \boldsymbol{Th}^e(\mathbf{Q}^{\models})$, where $\boldsymbol{f}_e(\Sigma) := e^{-1}(\Sigma)$ for all $\Sigma \in Th(\mathbf{Q}^{\models})$ (see Corollary 2.20). The second equation of Corollary 3.2.5 states that for any closed theories Σ_1 and Σ_2 of $\mathbf{Q}^{\models}$:

$$e^{-1}([\Sigma_1, \Sigma_2]) = \ker_{\mathbf{Q}}(e) +_{\mathbf{Q}} [e^{-1}(\Sigma_1), e^{-1}(\Sigma_2)], \qquad (*)$$

that is,

$$\boldsymbol{f}_e([\Sigma_1, \Sigma_2]) = \ker_{\mathbf{Q}}(e) +_{\mathbf{Q}} [\boldsymbol{f}_e(\Sigma_1), \boldsymbol{f}_e(\Sigma_2)].$$

3.3 More on Epimorphisms and the Equationally-Defined Commutator

Var is the (infinite) set of individual variables of $\boldsymbol{Te}_\tau$. *Var* absolutely freely generates the term algebra $\boldsymbol{Te}_\tau$.

Let $e : \boldsymbol{Te}_\tau \to \boldsymbol{Te}_\tau$ be an epimorphism, i.e., a surjective endomorphism. Then for every $x \in Var$,

$$\{t \in Te_\tau : et \text{ is identical with } x\}$$

is a non-empty set of variables. We recall that

$$V_e := \{x \in Var : e(x) \in Var\}.$$

Thus $V_e := e^{-1}(Var)$. The set V_e is infinite and e surjectively maps V_e onto Var. (e may glue together some variables belonging to V_e.) Moreover, e assigns a compound term to each variable $x \in Var \setminus V_e$.

For each $x \in Var$ we select a variable $x' \in \{t \in Te_\tau : et = x\}$. This choice function is denoted by g. Thus $gx = x'$ for each $x \in Var$. Let $V := g[Var] \subseteq V_e$ be the set of so selected variables. (The Principle of Countable Choice AC_ω, a weaker version of the Axiom of Choice, is used here to show the existence of g.) g is a bijection from Var onto V and $e(g(x)) = x$ for all $x \in Var$. V is an infinite subset of V_e and the restriction of e to V is a bijection from V onto Var. It is the inverse of g. (The set $Var \setminus V$ may be non-empty.) Let T be the set of terms generated by V. T forms a subalgebra $\boldsymbol{T}$ of the term algebra $\boldsymbol{Te}_\tau$.

Let f be the restriction of e to T,

$$f := e\lceil T.$$

f is an isomorphism from $\boldsymbol{T}$ onto $\boldsymbol{Te}_\tau$,

$$f : \boldsymbol{T} \cong \boldsymbol{Te}_\tau.$$

It follows that the extension of g onto the terms of T, which is denoted by the same letter g, is the inverse of f, $g := f^{-1}$. Thus g is an isomorphism from $\boldsymbol{Te}_\tau$ onto $\boldsymbol{T}$.

$$g : \boldsymbol{Te}_\tau \cong \boldsymbol{T}.$$

We therefore have that

$$eg(t) = t$$

for all terms $t \in Te_\tau$.

If $t = t(x_1, \ldots, x_n)$ is a term of $\boldsymbol{Te}_\tau$, then $gt = t(x_1/gx_1, \ldots, x_n/gx_n)$ is in T. (We shall mark the last term as $t(gx_1, \ldots, gx_n)$ for brevity.) Moreover gt is a variant of t, which means that $t(gx_1, \ldots, gx_n)$ results from $t(x_1, \ldots, x_n)$ by an application of a one-to-one substitution of variables for variables, viz. the substitution $x_1/gx_1, \ldots, x_n/gx_n$ that assigns the variable gx_i to the variable x_i for $i = 1, \ldots, n$.

The composition

$$k := g \circ e$$

assigns to each term in Te_τ a term in T. k is a *retraction* from $\boldsymbol{Te}_\tau$ onto $\boldsymbol{T}$, that is, k is a surjective homomorphism from $\boldsymbol{Te}_\tau$ onto $\boldsymbol{T}$ being the identity map on the subalgebra $\boldsymbol{T}$. k is an idempotent operation: $k \circ k = k$.

Similarly to e, the retraction k surjectively maps the set V_e onto V. Indeed, if $x \in V_e$, then $e(x)$ is a variable, say y. Hence, by the definition of g, $g(y)$ is a variable in V. So $k(x) = (g \circ e)(x) \in V$. Now, if $x' \in V$, then $x' = k(x')$, because k is the identity map on V. As $V \subseteq V_e$, this proves the claim. If $x \in Var \setminus V_e$, then $e(x)$ is a compound term, hence $k(x)$ is compound as well. Thus k maps exactly the variables in V_e onto V.

Let $\mathbf{Q}$ be a quasivariety of τ-algebras. To simplify notation, we put

$$C := \mathbf{Q}^{\models}$$

and

$$C_0 := \boldsymbol{Va}(\mathbf{Q})^{\models}.$$

C_0 is thus the consequence operation determined by the set of all $\mathbf{Q}$-valid equations and the rules of inference of Birkhoff's logic B_τ.

We know that $\ker_{\mathbf{Q}}(e) := e^{-1}(C(\emptyset))$ is a closed theory of C. (This fact directly follows from structurality of C.) As $C(\emptyset) = C_0(\emptyset)$, $\ker_{\mathbf{Q}}(e)$ is also a theory of C_0.

Since $C(\emptyset) = C_0(\emptyset)$, we immediately get that $\ker_{\mathbf{Q}}(e) = \ker_{\boldsymbol{Va}(\mathbf{Q})}(e)$.

Lemma 3.3.1. *Let* $\mathbf{Q}$ *be a quasivariety of* τ*-algebras. and let* $e : \boldsymbol{Te}_\tau \to \boldsymbol{Te}_\tau$ *be an epimorphism. Define* $\boldsymbol{T}$ *and the retraction* $k : \boldsymbol{Te}_\tau \to \boldsymbol{T}$ *as above. Then* $\ker_{\mathbf{Q}}(k) = \ker_{\mathbf{Q}}(e)$.

Proof. Let $p, q \in Te_\tau$. Then

$$\begin{array}{ll}
p \approx q \in \ker_{\mathbf{Q}}(k) & \Leftrightarrow \\
kp \approx kq \in C(\emptyset) & \Leftrightarrow \\
g(ep) \approx g(eq) \in C(\emptyset) & \Leftrightarrow \\
ep \approx eq \in C(\emptyset) & \Leftrightarrow \\
p \approx q \in \ker_{\mathbf{Q}}(e). &
\end{array}$$

(The third equivalence follows from the fact that $C(\emptyset) = \mathbf{Q}^{\models}(\emptyset)$ is an invariant set of equations and g is an isomorphism from $\boldsymbol{Te}_\tau$ onto $\boldsymbol{T}$. In other words, the difference between the equation $g(ep) \approx g(eq)$ and the equation $ep \approx eq$ is such that $g(ep) \approx g(eq)$ results from $ep \approx eq$ by renaming the variables occurring $ep \approx eq$ in a one-to-one way. Consequently, $g(ep) \approx g(eq) \in C(\emptyset)$ if and only if $ep \approx eq \in C(\emptyset)$.) Hence $\ker_{\mathbf{Q}}(k) = \ker_{\mathbf{Q}}(e)$. □

Lemma 3.3.2. *For every variable $x \in Var$, $x \approx kx \in \ker_{\mathbf{Q}}(k)$.*

Proof. The lemma directly follows from the fact that k is an idempotent operation. □

Corollary 3.3.3. *For every term $t \in Te_\tau$, $t \approx kt \in \ker_{\mathbf{Q}}(k)$.*

Proof. Use induction on the complexity of terms and the above lemma. □

Corollary 3.3.4. *Let X be any set of equations of $Eq(\tau)$. Then*

$$\ker_{\mathbf{Q}}(k) +_{\mathbf{Q}} C(X) = \ker_{\mathbf{Q}}(k) +_{\mathbf{Q}} C(kX).$$

Here $\ker_{\mathbf{Q}}(k) +_{\mathbf{Q}} C(Z)$ stands for $C(\ker_{\mathbf{Q}}(k) \cup C(Z))$ $(= C(\ker_{\mathbf{Q}}(k) \cup Z))$.

Proof. Use Corollary 3.3.3. □

We shall prove the following theorem:

Theorem 3.3.5. *Let $\mathbf{Q}$ be a quasivariety of τ-algebras and $e : \boldsymbol{Te}_\tau \to \boldsymbol{Te}_\tau$ an epimorphism. Given a choice function $g(x) \in \{x' \in Var : ex' = x\}$, for all $x \in Var$, define the term algebra $\boldsymbol{T}$ as above. Let X and Y be any sets of equations of $Eq(T)$. Then*

$$\begin{aligned}&\ker_{\mathbf{Q}}(e) +_{\mathbf{Q}} [\ker_{\mathbf{Q}}(e) +_{\mathbf{Q}} \mathbf{Q}^{\models}(X), \ker_{\mathbf{Q}}(e) +_{\mathbf{Q}} \mathbf{Q}^{\models}(Y)]\\ &= \ker_{\mathbf{Q}}(e) +_{\mathbf{Q}} [\mathbf{Q}^{\models}(X), \mathbf{Q}^{\models}(Y)]\end{aligned}$$

in the term algebra $\boldsymbol{Te}_\tau$.

The above theorem strengthens Corollary 3.2.5 but at the same time it restricts its scope to the equations of terms from the set T.

Proof. Throughout the proof we put: $C := \mathbf{Q}^{\models}$. As $\ker_{\mathbf{Q}}(e) = \ker_{\mathbf{Q}}(k)$, it suffices to prove the following lemma:

Lemma 3.3.6. *Let X and Y be any set of equations of $Eq(T)$. Then*

$$\ker_{\mathbf{Q}}(k) +_{\mathbf{Q}} [\ker_{\mathbf{Q}}(k) +_{\mathbf{Q}} C(X), \ker_{\mathbf{Q}}(k) +_{\mathbf{Q}} C(Y)] = \ker_{\mathbf{Q}}(k) +_{\mathbf{Q}} [C(X), C(Y)] \qquad (*)$$

in the term algebra $\boldsymbol{Te}_\tau$.

Proof (of the lemma). Since the theory on the RHS of $(*)$ is included in the theory on the LHS, we only need to prove the inclusion

$$[\ker_{\mathbf{Q}}(k) +_{\mathbf{Q}} C(X), \ker_{\mathbf{Q}}(k) +_{\mathbf{Q}} C(Y)] \subseteq \ker_{\mathbf{Q}}(k) +_{\mathbf{Q}} [C(X), C(Y)].$$

Let C' be the restriction of C to $Eq(T)$.

Claim 1. $[\ker_{\mathbf{Q}}(k) +_{\mathbf{Q}} C(X), \ker_{\mathbf{Q}}(k) +_{\mathbf{Q}} C(Y)] = k^{-1}([C'(X), C'(Y)]_T)$.

Proof (of the claim). It directly follows from Theorem 3.1.8, when applied to the algebras $\boldsymbol{Te}_\tau$ and $\boldsymbol{T}$, that

$$[\ker_{\mathbf{Q}}(k) +_{\mathbf{Q}} C(Z), \ker_{\mathbf{Q}}(k) +_{\mathbf{Q}} C(W)] = k^{-1}([C'(kZ), C'(kW)]_{\boldsymbol{T}}),$$

for all sets of equations Z, W in $Eq(\tau)$. As X and Y are sets of equations of $Eq(T)$ and k is the identity map on the subalgebra $\boldsymbol{T}$, the claim follows. □

Claim 2. $k^{-1}([C'(X), C'(Y)]_{\boldsymbol{T}}) \subseteq \ker_{\mathbf{Q}}(k) +_{\mathbf{Q}} [C(X), C(Y)]$.

Proof (of the claim). The definition of the equationally defined commutator and the fact that k is the identity map when restricted to $\boldsymbol{T}$ give:

$$k^{-1}([C'(X), C'(Y)]_{\boldsymbol{T}}) =$$

$$k^{-1}(C'(\bigcup_{\alpha\approx\beta\in CoEq(\mathbf{Q})} \{\alpha(\underline{p},\underline{q},\underline{r},\underline{s},\underline{t}) \approx \beta(\underline{p},\underline{q},\underline{r},\underline{s},\underline{t}) : \underline{p},\underline{q},\underline{r},\underline{s} \text{ are sequences}$$
$$\text{of terms of } T \text{ such that } \underline{p} \approx \underline{q} \in C'(X),\ \underline{r} \approx \underline{s} \in C'(Y)$$
$$\text{and } \underline{t} \text{ is an arbitrary sequence of terms of } T\})) =$$

$$k^{-1}(C'(k(\bigcup_{\alpha\approx\beta\in CoEq(\mathbf{Q})} \{\alpha(\underline{p},\underline{q},\underline{r},\underline{s},\underline{t}) \approx \beta(\underline{p},\underline{q},\underline{r},\underline{s},\underline{t}) : \underline{p},\underline{q},\underline{r},\underline{s} \text{ are sequences}$$
$$\text{of terms of } T \text{ such that } \underline{p} \approx \underline{q} \in C'(X),\ \underline{r} \approx \underline{s} \in C'(Y),$$
$$\underline{t} \text{ is an arbitrary sequence of terms of } T\}))) =$$

(by Corollary 2.11 applied to the algebras $\boldsymbol{Te}_\tau$ and $\boldsymbol{T}$)

$$\ker_{\mathbf{Q}}(k) +_{\mathbf{Q}} C(\bigcup_{\alpha\approx\beta\in CoEq(\mathbf{Q})} \{\alpha(\underline{p},\underline{q},\underline{r},\underline{s},\underline{t}) \approx \beta(\underline{p},\underline{q},\underline{r},\underline{s},\underline{t}) : \underline{p},\underline{q},\underline{r},\underline{s} \text{ are}$$
$$\text{sequences of terms of } T \text{ such that } \underline{p} \approx \underline{q} \in C'(X),\ \underline{r} \approx \underline{s} \in C'(Y),$$
$$\underline{t} \text{ is an arbitrary sequence of terms of } T\}) \subseteq$$

$$\ker_{\mathbf{Q}}(k) +_{\mathbf{Q}} C(\bigcup_{\alpha\approx\beta\in CoEq(\mathbf{Q})} \{\alpha(\underline{p},\underline{q},\underline{r},\underline{s},\underline{t}) \approx \beta(\underline{p},\underline{q},\underline{r},\underline{s},\underline{t}) : \underline{p},\underline{q},\underline{r},\underline{s} \text{ are}$$
$$\text{sequences of terms of } T \text{ such that } \underline{p} \approx \underline{q} \in C(X),\ \underline{r} \approx \underline{s} \in C(Y),$$
$$\underline{t} \text{ is an arbitrary sequence of terms of } T\}) \subseteq$$

$$\ker_{\mathbf{Q}}(k) +_{\mathbf{Q}} C(\bigcup_{\alpha\approx\beta\in CoEq(\mathbf{Q})} \{\alpha(\underline{p},\underline{q},\underline{r},\underline{s},\underline{t}) \approx \beta(\underline{p},\underline{q},\underline{r},\underline{s},\underline{t}) : \underline{p},\underline{q},\underline{r},\underline{s} \text{ are}$$
$$\text{sequences of terms of } Te_\tau \text{ such that } \underline{p} \approx \underline{q} \in C(X),\ \underline{r} \approx \underline{s} \in C(Y),$$
$$\underline{t} \text{ is an arbitrary sequence of terms of } Te_\tau\}) =$$

$$\ker_{\mathbf{Q}} +_{\mathbf{Q}}[C(X), C(Y)].$$

The last inclusion is due to the fact that $T \subseteq Te_\tau$ and the next to last one follows from the fact that $C'(Z) \subseteq C(Z)$ for any set $Z \subseteq Eq(T)$. □

From the claims the lemma follows. □

The proof of the theorem is concluded. □

Applying the above reasoning we also prove:

Theorem 3.3.7. (This is Theorem 2.22 repeated.) *Let* $\mathbf{Q}$ *be a quasivariety of* τ*-algebras and* $e : \boldsymbol{Te}_\tau \to \boldsymbol{Te}_\tau$ *an epimorphism. Define the set* V_e *as above. Then for any separated sets X and Y of equations of variables from* V_e *such that e is injective on* $Var(X \cup Y)$ *it is the case that*

$$(\ker_{\mathbf{Q}}(e) +_{\mathbf{Q}} \mathbf{Q}^{\models}(X)) \cap (\ker_{\mathbf{Q}}(e) +_{\mathbf{Q}} \mathbf{Q}^{\models}(Y)) = \ker_{\mathbf{Q}}(e) +_{\mathbf{Q}} \mathbf{Q}^{\models}(X) \cap \mathbf{Q}^{\models}(Y).$$

Proof. Let X and Y be separated equations of variables from V_e. The inclusion

$$(\ker_{\mathbf{Q}}(e) +_{\mathbf{Q}} C(X)) \cap (\ker_{\mathbf{Q}}(e) +_{\mathbf{Q}} C(Y)) \subseteq \ker_{\mathbf{Q}}(e) +_{\mathbf{Q}} C(X) \cap C(Y). \quad (1)$$

is a non-trivial part of the theorem. Moreover, it suffices to prove this inclusion only for finite sets X and Y. Let $X = \{\underline{x} \approx \underline{y}\}$, $Y = \{\underline{z} \approx \underline{w}\}$, where $\{\underline{x} \approx \underline{y}\} = \{x_1 \approx y_1, \ldots, x_m \approx y_m\}$, $\{\underline{z} \approx \underline{w}\} = \{z_1 \approx w_1, \ldots, z_n \approx w_n\}$.

We assume e is injective on $Var(X \cup Y)$. Let $V_0 := Var(X \cup Y) = \{x_1, \ldots, x_m\} \cup \{y_1, \ldots, y_m\} \cup \{z_1, \ldots, z_n\} \cup \{w_1, \ldots, w_n\}$.

Given the epimorphism $e : \boldsymbol{Te}_\tau \to \boldsymbol{Te}_\tau$, we define $\boldsymbol{T}$, the mappings g, f and the retraction $k : \boldsymbol{Te}_\tau \to \boldsymbol{T}$ as above. The carrier of $\boldsymbol{T}$ contains exactly the terms in the variables from an infinite set V of variables, viz. $V = g(Var)$. As e is one-to-one on the set V_0 we may also assume that the selector g is so defined that $V_0 \subseteq V$.

Let C' be the restriction of C to the term algebra $\boldsymbol{T}$, i.e., $C'(\Sigma) = C(\Sigma) \cap Eq(\boldsymbol{T})$, where $Eq(\boldsymbol{T})$ is the set of all equations of terms of $\boldsymbol{T}$. We select a set $\Delta(\underline{x}, \underline{y}, \underline{z}, \underline{w}, \underline{u})$ of equations in $Eq(\boldsymbol{T})$ such that

$$C'(\underline{x} \approx \underline{y}) \cap C'(\underline{z} \approx \underline{w}) = C'(\Delta(\underline{x}, \underline{y}, \underline{z}, \underline{w}, \underline{u})).$$

Then

$$\Delta(\underline{x}, \underline{y}, \underline{z}, \underline{w}, \underline{u}) \subseteq C(\underline{x} \approx \underline{y}) \cap C(\underline{z} \approx \underline{w}), \quad (2)$$

because $\Delta(\underline{x}, \underline{y}, \underline{z}, \underline{w}, \underline{u}) \subseteq C'(\underline{x} \approx \underline{y}) \cap C'(\underline{z} \approx \underline{w}) \subseteq C(\underline{x} \approx \underline{y}) \cap C(\underline{z} \approx \underline{w})$.

To prove (1), assume $p \approx q \in \ker_{\mathbf{Q}}(e) +_{\mathbf{Q}} C(\underline{x} \approx \underline{y}) = C(\underline{x} \approx \underline{y} \cup \ker_{\mathbf{Q}}(e))$ and $p \approx q \in \ker_{\mathbf{Q}}(e) +_{\mathbf{Q}} C(\underline{z} \approx \underline{w}) = C(\underline{z} \approx \underline{w} \cup \ker_{\mathbf{Q}}(e))$.

Lemma 3.3.1, structurality and the fact that k is the identity map on $\boldsymbol{T}$ give:

$$kp \approx kq \in C'(k\underline{x} \approx k\underline{y} \cup k(\ker_{\mathbf{Q}}(e))) = C'(\underline{x} \approx \underline{y} \cup k(\ker_{\mathbf{Q}}(k))) = C'(\underline{x} \approx \underline{y})$$

and

$$kp \approx kq \in C'(k\underline{z} \approx k\underline{w} \cup k(\ker_{\mathbf{Q}}(e))) = C'(\underline{z} \approx \underline{w} \cup k(\ker_{\mathbf{Q}}(k))) = C'(\underline{z} \approx \underline{w}).$$

Hence

$$kp \approx kq \in C'(\underline{x} \approx \underline{y}) \cap C'(\underline{z} \approx \underline{w}) = C'(\Delta(\underline{x}, \underline{y}, \underline{z}, \underline{w}, \underline{u})). \tag{3}$$

As k is the identity map on the variables $\underline{x}, \underline{y}, \underline{z}, \underline{w}$ and $\underline{u}$, we also have:

$$C'(\Delta(\underline{x}, \underline{y}, \underline{z}, \underline{w}, \underline{u})) = C'(\Delta(k\underline{x}, k\underline{y}, k\underline{z}, k\underline{w}, k\underline{u})) = \\ C'(k(\Delta(\underline{x}, \underline{y}, \underline{z}, \underline{w}, \underline{u}))) \subseteq C(k(\Delta(\underline{x}, \underline{y}, \underline{z}, \underline{w}, \underline{u}))).$$

This and (3) give that

$$kp \approx kq \in C(k(\Delta(\underline{x}, \underline{y}, \underline{z}, \underline{w}, \underline{u}))),$$

i.e.,

$$p \approx q \in k^{-1}C(k(\Delta(\underline{x}, \underline{y}, \underline{z}, \underline{w}, \underline{u}))). \tag{4}$$

As $k : \boldsymbol{Te}_\tau \to \boldsymbol{T}$ is surjective, applying Proposition 2.5 and Corollary 2.16 to (4) we obtain:

$$p \approx q \in \ker_{\mathbf{Q}}(k) +_{\mathbf{Q}} C(\Delta(\underline{x}, \underline{y}, \underline{z}, \underline{w}, \underline{u})) \subseteq \ker_{\mathbf{Q}}(e) +_{\mathbf{Q}} C(\underline{x} \approx \underline{y}) \cap C(\underline{z} \approx \underline{w}),$$

by (2). So (1) holds. This proves the theorem. □

Note. A theorem analogous (and equivalent) to Theorem 3.3.5 is formulated below in terms of congruences generated by separated sets of pairs of free generators in the free algebra $\boldsymbol{F}$.

Let $\boldsymbol{B}$ be a subalgebra of an algebra $\boldsymbol{A}$. A *retraction* of $\boldsymbol{A}$ onto $\boldsymbol{B}$ is a surjective epimorphism $k : \boldsymbol{A} \to \boldsymbol{B}$ being the identity map on the subalgebra $\boldsymbol{B}$, i.e., $k(b) = b$ for all $b \in B$. $\boldsymbol{B}$ is then called a *retract* of $\boldsymbol{A}$.

Every free subalgebra $\boldsymbol{G}$ of a free algebra $\boldsymbol{F}$ is a retract of $\boldsymbol{F}$. ($\boldsymbol{G}$ is generated by a subset of the set of free generators of $\boldsymbol{F}$.)

The following facts are reformulations for free algebras of the above results established for the consequence $\mathbf{Q}^{\models}$.

Retraction Lemma. *Let* $\mathbf{F}$ *be a free algebra and* $h : \boldsymbol{F} \to \boldsymbol{F}$ *an epimorphism. There exists a free subalgebra* $\mathbf{G}$ *of* $\mathbf{F}$*, generated by a subset of the set of free generators of* $\mathbf{F}$*, and a retraction* $k : \boldsymbol{F} \to \mathbf{G}$ *such that* $\ker(k) = \ker(h)$*. Moreover, if* $\mathbf{F}$ *has an infinite set of free generators, then so does* $\mathbf{G}$*.*

The above lemma trivializes if $\boldsymbol{F}$ has only finitely many generators because h is then an automorphism of $\boldsymbol{F}$. Consequently, $\mathbf{G} = \boldsymbol{F}$ and k is the identity mapping.

Theorem 1.3.5*. *Let* $\mathbf{Q}$ *be a quasivariety and* $h : \boldsymbol{F} \to \boldsymbol{F}$ *an epimorphism of the free algebra* $\boldsymbol{F} = \mathbf{F}_{\mathbf{Q}}(\omega)$. *Let* $\mathbf{G}$ *be the free subalgebra of* $\mathbf{F}$ *defined as in the above lemma with the infinite set* V *of free generators. Let* X *and* Y *be subsets of* $G \times G$. *Then*

$$\ker(h) +_{\mathbf{Q}} [\Theta^{F}(X) +_{\mathbf{Q}} \ker(h), \Theta^{F}(Y) +_{\mathbf{Q}} \ker(h)]^{F} = \ker(h) +_{\mathbf{Q}} \Theta^{F}(X) \cap \Theta^{F}(Y). \quad \Box \quad (1)$$

(As $\boldsymbol{F} \in \mathbf{Q}$, $\ker(h)$ is $\mathbf{Q}$-congruence of $\boldsymbol{F}$.)

In a similar manner one may paraphrase Theorem 3.3.7.

3.4 The Relative Shifting Property and the Commutator

In this section logical aspects of the geometrical approach to the theory of quasivarieties are examined. The shifting and cube properties play a central role in this approach. These properties were defined by H.P. Gumm (1983) for varieties of algebras and then extended to quasivarieties by Kearnes and McKenzie (1992). For a variety, the validity of the shifting property is equivalent to congruence-modularity, a result proved by Gumm (1983).

Let $\mathbf{Q}$ be a quasivariety of algebras with signature τ. The *relative shifting property of* $\mathbf{Q}$ (see, e.g., Kearnes and McKenzie 1992) is the following statement:

> *Suppose that* $\mathbf{A} \in \mathbf{Q}$, $\Phi, \Psi, \Xi \in Con_{\mathbf{Q}}(\mathbf{A})$, *and that* $a, b, c, d \in A$ *satisfy* $\langle a, b\rangle, \langle c, d\rangle \in \Phi$, $\langle a, c\rangle, \langle b, d\rangle \in \Psi$, *and* $\langle a, b\rangle \in \Xi$. *Then* $\langle c, d\rangle \in \Xi +_{\mathbf{Q}}(\Phi \cap \Psi)$.

This statement is expressed pictorially in Fig. 3.1:

In diagrams like this, lines are assumed to be labeled by any label appearing on a parallel line.

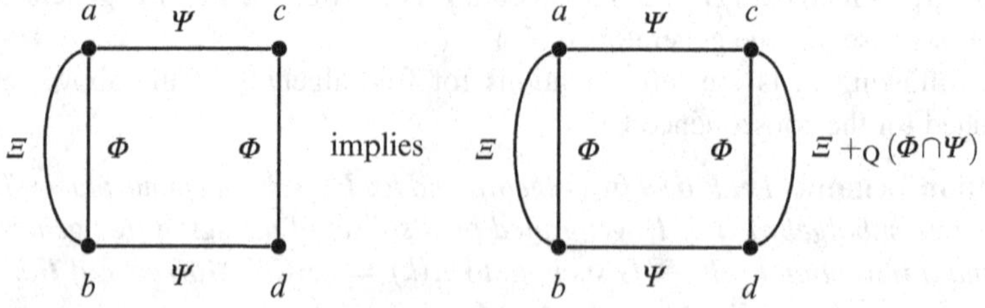

Fig. 3.1 The relative shifting property.

The adjective "relative" is usually dropped when we speak of the relative shifting property for *varieties* of algebras.

The following theorem is a combination of two results due to A. Day (1969), and recalled as the equivalence of conditions (1) and (3) below, and H.P. Gumm (1983), respectively. It shows that the shifting property for varieties is equivalent to congruence-modularity

Theorem 3.4.1. *For any variety* **V** *of algebras the following conditions are equivalent:*

(1) **V** *is congruence-modular;*
(2) **V** *has the shifting property;*
(3) *For some n there are terms* $m_0(x, y, z, w), \dots, m_n(x, y, z, w)$ *such that* **V** *satisfies*

(i) $m_0(x, y, z, w) \approx z, \quad m_n(x, y, z, w) \approx w,$
(ii) $m_i(x, x, z, z) \approx z, \quad i \leqslant n,$
(iii) $m_i(x, y, x, y) \approx m_{i+1}(x, y, x, y)$ *for all even* $i < n$,
(iv) $m_i(x, x, z, w) \approx m_{i+1}(x, x, z, w)$ *for all odd* $i < n$.

Proof. See, e.g., Freese and McKenzie (1987) or Czelakowski (2001). □

Notes 3.4.2.
(1). The terms $m_0(x, y, z, w), \dots, m_n(x, y, z, w)$ satisfying condition (3) of Theorem 3.4.1 are called *Day terms.*
(2). Equations (i)–(iv) depart from the standard formulation of Day equations for congruence-modular varieties given, e.g., in Freese and McKenzie (1987), Theorem 2.2. But the above equations are equivalent to the standard ones (up to a permutation of variables). Indeed, taking the permutation σ of $\{x, y, z, w\}$ given by $\sigma(x) := y$, $\sigma(y) := z$, $\sigma(z) := x$, $\sigma(w) := w$, it is not difficult to check that the terms $\sigma m_0(x, y, z, w), \dots, \sigma m_n(x, y, z, w)$ satisfy the equations given in Theorem 2.2 of Freese and McKenzie (1987).

The reason for displaying Day equations in the form (i)–(iv) is explained in Section 3.5.

(3). The implication (1) ⇒ (2) also holds for relatively congruence-modular *quasivarieties* of algebras (RCM quasivarieties, for short). Consequently,

Every RCM quasivariety has the relative shifting property.

Proposition 6.1.7 supplies a short proof of the above fact. The relative shifting property *need not* imply relative congruence-modularity unless a quasivariety **Q** is a variety.

(4). A quasivariety **Q** satisfies the *Extension Principle* if for every algebra $\boldsymbol{A} \in \mathbf{Q}$ the operator $\Theta_{\mathbf{Q}}(\,\cdot\,)$ is homomorphism from the lattice $\boldsymbol{Con}(\boldsymbol{A})$ to the lattice $\boldsymbol{Con}_{\mathbf{Q}}(\boldsymbol{A})$. Equivalently, **Q** satisfies the Extension Principle if for every algebra $\boldsymbol{A} \in \mathbf{Q}$ and for every pair of congruences $\Phi, \Psi \in Con(\boldsymbol{A})$, it is the case that $\Theta_{\mathbf{Q}}(\Phi \cap \Psi) = \Theta_{\mathbf{Q}}(\Phi) \cap \Theta_{\mathbf{Q}}(\Psi)$.

Kearnes and McKenzie (1992) have proved a surprising result:

A quasivariety **Q** *is RCM if and only if it has the relative shifting property and satisfies the Extension Principle.* □

We shall return to the Extension Principle in Chapter 6.

3.5 Day Implication Systems and the Relative Shifting Property

We investigate quasivarieties **Q** such that the associated equational logic $\mathbf{Q}^{\models}$ is endowed with a Day implication system. A Day implication $\Rightarrow_D$ for $\mathbf{Q}^{\models}$ is a finite set of quaternary equations which collectively possess the detachment property relative to the equational logic $\mathbf{Q}^{\models}$ and furthermore satisfy two other natural conditions. A Day implication is the notion extracted from the well-known Mal'cev-style characterization of congruence-modular varieties due to Day (1969) and recalled in Theorem 3.4.1 above. The fact that a variety has a Day implication is equivalent to congruence-modularity.

In this section we show that for quasivarieties of algebras, the geometrical properties discovered by Gumm are naturally linked with the presence of Day implication: the relative shifting property for **Q** is equivalent to the existence of a Day implication system for $\mathbf{Q}^{\models}$. Furthermore, Day implication provides a simple syntactic characterization of the cube property. These observations show the *logical* perspective of Gumm's approach and shed new light on the results proved by Kearnes and McKenzie (1992).

We shall give an intrinsic characterization of the relative shifting property for a quasivariety **Q** in terms of properties of the equational logic $\mathbf{Q}^{\models}$ associated with **Q**.

The relative shifting property is defined in terms of relative congruences of a quasivariety. But the definition of the relative shifting property also makes sense for (closed) theories of equational consequence operations.

Let **Q** be a quasivariety. The *relative shifting property for the theories of* $\mathbf{Q}^{\models}$ says that for any theories $X, Y, Z \in Th(\mathbf{Q}^{\models})$ and any terms $\alpha, \beta, \gamma, \delta$, the conditions $\alpha \approx \beta$, $\gamma \approx \delta \in X$, $\alpha \approx \gamma, \beta \approx \delta \in Y$ and $\alpha \approx \beta \in Z$ imply that $\gamma \approx \delta \in \mathbf{Q}^{\models}(Z \cup (X \cap Y))$.

A finite set of equations in four variables

$$\{p_i(x,y,z,w) \approx q_i(x,y,z,w) : i \in I\}, \tag{Impl}$$

more suggestively denoted by $x \approx y \Rightarrow z \approx w$ or simply by $\Rightarrow$, is called an *implication system* for the equational logic $\mathbf{Q}^{\models}$ if it satisfies two conditions:

$$z \approx w \in \mathbf{Q}^{\models}(x \approx y, x \approx y \Rightarrow z \approx w), \tag{iD1}$$

i.e., $\Rightarrow$ has the detachment property relative to $\mathbf{Q}^{\models}$, and

$$x \approx y \Rightarrow x \approx y \in \mathbf{Q}^{\models}(\emptyset), \tag{iD2}$$

i.e., $\Rightarrow$ has the identity property relative to $\mathbf{Q}^{\models}$.

If $\Rightarrow$ additionally satisfies

$$x \approx x \Rightarrow y \approx y \in \mathbf{Q}(\emptyset), \tag{iD3}$$

then it is called a *Day implication system* for $\mathbf{Q}^{\models}$, or simply a *Day implication*.

(iD2) thus states that $\{p_i(x,y,x,y) \approx q_i(x,y,x,y) : i \in I\} \subseteq \mathbf{Q}^{\models}(\emptyset)$, i.e., $\mathbf{Q}$ validates the equations

$$p_i(x,y,x,y) \approx q_i(x,y,x,y), \qquad i \in I.$$

Similarly, (iD3) states that $\{p_i(x,x,y,y) \approx q_i(x,x,y,y) : i \in I\} \subseteq \mathbf{Q}^{\models}(\emptyset)$, i.e., the equations

$$p_i(x,x,y,y) \approx q_i(x,x,y,y), \qquad i \in I,$$

are valid in $\mathbf{Q}$.

(iD3) is equivalent to the fact that $x \approx y, z \approx w / x \approx y \Rightarrow z \approx w$ is a set of rules of $\mathbf{Q}^{\models}$.

Any Day implication will be more suggestively marked by $\Rightarrow_D$.

The following theorem relates the shifting property to the syntactic notion of a Day implication system:

Theorem 3.5.1. *For any quasivariety* $\mathbf{Q}$ *of* τ*-algebras the following conditions are equivalent:*

(A) *The relative shifting property holds for* $\mathbf{Q}$;
(B) *The relative shifting property holds for the* $\mathbf{Q}$*-congruences of the free algebra* $\boldsymbol{F}_{\mathbf{Q}}(\omega)$;
(C) *The relative shifting property holds for the equational theories of* $\mathbf{Q}^{\models}$;
(D) *The consequence operation* $\mathbf{Q}^{\models}$ *has a finite Day implication system* $\Rightarrow_D$;

Proof. We first prove the following fact:

Lemma 3.5.2. *For any quasivariety* $\mathbf{Q}$*, the following conditions are equivalent:*

(i) *The relative shifting property for the theories of* $\mathbf{Q}^{\models}$.
(ii) *For any (equivalently, for some) different variables* x, y, z, w,

$$z \approx w \in \mathbf{Q}^{\models}(\{x \approx y\} \cup \mathbf{Q}^{\models}(x \approx y, z \approx w) \cap \mathbf{Q}^{\models}(x \approx z, y \approx w)).$$

(iii) *The consequence operation* $\mathbf{Q}^{\models}$ *has a finite Day implication system* $\Rightarrow_D$.

Proof (of the lemma). (i) $\Rightarrow$ (ii). This is trivial.

(ii) $\Rightarrow$ (iii). Assume (ii) for some x, y, z, w. (ii) implies that there exists a finite set of equations $\Sigma(x,y,z,w,\underline{u})$ such that $z \approx w \in \mathbf{Q}^{\models}(\{x \approx y\} \cup \Sigma(x,y,z,w,\underline{u}))$ and $\Sigma(x,y,z,w,\underline{u}) \subseteq \mathbf{Q}^{\models}(x \approx y, z \approx w)$, $\Sigma(x,y,z,w,\underline{u}) \subseteq \mathbf{Q}^{\models}(x \approx z, y \approx w)$. Taking a substitution e being the identity map on x, y, z, w and which assigns the variable x to each variable of $\underline{u}$, we have that $z \approx w \in \mathbf{Q}^{\models}(\{x \approx y\} \cup \Sigma(x, y, z, w, e\underline{u}))$ and $\Sigma(x, y, z, w, e\underline{u}) \subseteq \mathbf{Q}^{\models}(x \approx y, z \approx w)$, $\Sigma(x, y, z, w, e\underline{u}) \subseteq \mathbf{Q}^{\models}(x \approx z, y \approx w)$, by the structurality of $\mathbf{Q}^{\models}$. Putting

$$x \approx y \Rightarrow_D z \approx w := \Sigma(x, y, z, w, e\underline{u}),$$

we see that $z \approx w \in \mathbf{Q}^{\models}(x \approx y, x \approx y \Rightarrow_D z \approx w)$, $x \approx y \Rightarrow_D z \approx w \subseteq \mathbf{Q}^{\models}(x \approx y, z \approx w)$, and $x \approx y \Rightarrow_D z \approx w \subseteq \mathbf{Q}^{\models}(x \approx z, y \approx w)$. Applying structurality to the last two inclusions we get that $x \approx x \Rightarrow_D y \approx y \subseteq \mathbf{Q}^{\models}(x \approx x, y \approx y) = \mathbf{Q}^{\models}(\emptyset)$ and $x \approx y \Rightarrow_D x \approx y \subseteq \mathbf{Q}^{\models}(x \approx x, y \approx y) = \mathbf{Q}^{\models}(\emptyset)$, respectively. Thus $\Rightarrow_D$ satisfies (iD1), (iD3), and (iD2). This proves (iii).

(iii) $\Rightarrow$ (i). Suppose that for $X, Y, Z \in Th(\mathbf{Q}^{\models})$ and terms $\alpha, \beta, \gamma, \delta$, it is the case that $\alpha \approx \beta, \gamma \approx \delta \in X$, $\alpha \approx \gamma, \beta \approx \delta \in Y$ and $\alpha \approx \beta \in Z$. We shall show that $\gamma \approx \delta \in \mathbf{Q}^{\models}(Z \cup (X \cap Y))$. Let $x \approx y \Rightarrow_D z \approx w$ be a Day implication for $\mathbf{Q}^{\models}$.

Claim. *$p(\alpha, \beta, \gamma, \delta) \approx q(\alpha, \beta, \gamma, \delta) \in X \cap Y$, for any equation $p \approx q$ in $\Rightarrow_D$.*

Proof (of the claim). Let $p \approx q$ be in $\Rightarrow_D$. Since $\alpha \approx \beta, \gamma \approx \delta \in X$, we have that $p(\alpha, \beta, \gamma, \delta) \approx p(\alpha, \alpha, \gamma, \gamma) \in X$ and $q(\alpha, \beta, \gamma, \delta) \approx q(\alpha, \alpha, \gamma, \gamma) \in X$. But by (iD3), $p(\alpha, \alpha, \gamma, \gamma) \approx q(\alpha, \alpha, \gamma, \gamma) \in \mathbf{Q}^{\models}(\emptyset)$. It follows that

$$p(\alpha, \beta, \gamma, \delta) \approx q(\alpha, \beta, \gamma, \delta) \in X. \tag{a}$$

Furthermore, as $\alpha \approx \gamma$, $\beta \approx \delta \in Y$, we also have that $p(\alpha, \beta, \gamma, \delta) \approx p(\alpha, \beta, \alpha, \beta) \in X$ and $q(\alpha, \beta, \gamma, \delta) \approx q(\alpha, \beta, \alpha, \beta) \in Y$. But by (iD2), $p(\alpha, \beta, \alpha, \beta) \approx q(\alpha, \beta, \alpha, \beta) \in \mathbf{Q}^{\models}(\emptyset)$. Consequently,

$$p(\alpha, \beta, \gamma, \delta) \approx q(\alpha, \beta, \gamma, \delta) \in Y. \tag{b}$$

The claim follows from (a) and (b). □

(iD1), the claim and $\alpha \approx \beta \in Z$ imply that

$$\gamma \approx \delta \in \mathbf{Q}^{\models}(\{\alpha \approx \beta\} \cup \alpha \approx \beta \Rightarrow_D \gamma \approx \delta) \subseteq \mathbf{Q}^{\models}(Z \cup (X \cap Y)). \quad \square$$

We now pass to the proof of the theorem.

Implication (A) $\Rightarrow$ (B) is trivial. The equivalence of (B) and (C) follows from Proposition 2.5. Implication (C) $\Rightarrow$ (D) follows from the above lemma. The implication (D) $\Rightarrow$ (A) is proved by a straightforward modification of the proof of implication (iii) $\Rightarrow$ (i) of Lemma 3.5.2. (See also Czelakowski 2001, Theorem Q.10.4.) □

Notes.

1. Since for every Day implication system $\Rightarrow_D$ for a quasivariety $\mathbf{Q}$, $x \approx y, z \approx w / x \approx y \Rightarrow_D z \approx w$ is a set of rules of $\mathbf{Q}^{\models}$, it follows that for any terms $\alpha, \beta, \gamma, \delta$,

$$\alpha \approx \beta, \gamma \approx \delta \in \mathbf{Q}^{\models}(\emptyset) \ \text{ implies } \ \alpha \approx \beta \Rightarrow_D \gamma \approx \delta \in \mathbf{Q}^{\models}(\emptyset).$$

(This fact also follows from the proof of condition (a) of the above claim.)

2. Condition (ii) of Lemma 3.5.2 is equivalent to the following identity:

$$\mathbf{Q}^{\models}(\{x \approx y\} \cup \mathbf{Q}^{\models}(x \approx y, z \approx w) \cap \mathbf{Q}^{\models}(x \approx z, y \approx w)) =$$
$$\mathbf{Q}^{\models}(\{z \approx w\} \cup \mathbf{Q}^{\models}(x \approx y, z \approx w) \cap \mathbf{Q}^{\models}(x \approx z, y \approx w)). \quad \text{(ii)}^*$$

Indeed, by applying the permutation σ of the individual variables, where $\sigma(x) = z$, $\sigma(y) = w$, $\sigma(z) = x$, $\sigma(w) = y$ and σ does not move the remaining variables, to inclusion (ii) we see that

$$x \approx y \in \mathbf{Q}^{\models}(\{z \approx w\} \cup \mathbf{Q}^{\models}(z \approx w, x \approx y) \cap \mathbf{Q}^{\models}(z \approx x, w \approx y)) =$$
$$\mathbf{Q}^{\models}(\{z \approx w\} \cup \mathbf{Q}^{\models}(x \approx y, z \approx w) \cap \mathbf{Q}^{\models}(x \approx z, y \approx w)).$$

This together with (ii) gives (ii)*. Trivially, (ii)* implies (ii). □

3. Let $\Pi(x, y, z, w, \underline{u})$ be a set of equations (possibly with parameters $\underline{u}$) such that

$$\mathbf{Q}^{\models}(\Pi(x, y, z, w, \underline{u})) = \mathbf{Q}^{\models}(x \approx y, z \approx w) \cap \mathbf{Q}^{\models}(x \approx z, y \approx w). \quad \text{(eq)}$$

Marking the set $\Pi(x, y, z, w, \underline{u})$ as $x \approx y \Leftrightarrow_{\underline{u}} z \approx w$ (note the occurrence of parameters), we see that condition (ii) of Lemma 3.5.2 implies that the following are rules for $\mathbf{Q}^{\models}$:

$$x \approx y, x \approx y \Leftrightarrow_{\underline{u}} z \approx w / z \approx w \quad \text{and} \quad z \approx w, x \approx y \Leftrightarrow_{\underline{u}} z \approx w / x \approx y. \quad \text{(eD1)}$$

$x \approx y \Leftrightarrow_{\underline{u}} z \approx w$ may be called a parameterized *equivalence system* for the consequence $\mathbf{Q}^{\models}$; (eD1) are detachment rules for the equivalence $x \approx y \Leftrightarrow_{\underline{u}} z \approx w$. (eq) implies that $\mathbf{Q}^{\models}(x \approx y \Leftrightarrow_{\underline{u}} z \approx w) = \mathbf{Q}^{\models}(z \approx w \Leftrightarrow_{\underline{u}} x \approx y)$ which means that

$$x \approx y \Leftrightarrow_{\underline{u}} z \approx w / z \approx w \Leftrightarrow_{\underline{u}} x \approx y$$

and

$$z \approx w \Leftrightarrow_{\underline{u}} x \approx y / x \approx y \Leftrightarrow_{\underline{u}} z \approx w$$

are sets of rules of $\mathbf{Q}^{\models}$. These are commutativity rules for the equivalence system $x \approx y \Leftrightarrow_{\underline{u}} z \approx w$. Moreover, $\Leftrightarrow_{\underline{u}}$ retains the characteristic properties of any Day implication system, viz.,

$$x \approx y \Leftrightarrow_{\underline{u}} x \approx y \subseteq \mathbf{Q}^{\models}(\emptyset) \quad \text{(eD2)}$$

and

$$x \approx x \Leftrightarrow_{\underline{u}} y \approx y \subseteq \mathbf{Q}^{\models}(\emptyset). \quad \text{(eD3)}$$

(eD3) is equivalent to the fact that

$$x \approx y, z \approx w / x \approx y \Leftrightarrow_{\underline{u}} z \approx w \tag{eD3$_G$}$$

is a set of rules of $\mathbf{Q}^{\models}$.

By an analogy to propositional logic, the rules (eD3)$_G$ may be collectively called *Gödel rules* for $\mathbf{Q}^{\models}$.

Let $x \approx y \Leftrightarrow z \approx w$ be the set of equations which results from $x \approx y \Leftrightarrow_{\underline{u}} z \approx w$ by the uniform substitution of the variable x for each parameter of $\underline{u}$. (The equations $x \approx y \Leftrightarrow z \approx w$ therefore do not involve parameters.) It follows from structurality that $x \approx y \Leftrightarrow z \approx w$ shares properties (eD1)–(eD3) and (eD3)$_G$ with $x \approx y \Leftrightarrow_{\underline{u}} z \approx w$. But (eq) need not hold for $x \approx y \Leftrightarrow z \approx w$, that is, $\mathbf{Q}^{\models}(x \approx y \Leftrightarrow z \approx w)$ may be a proper subset of $\mathbf{Q}^{\models}(x \approx y, z \approx w) \cap \mathbf{Q}^{\models}(x \approx z, y \approx w)$.

The set $x \approx y \Leftrightarrow z \approx w$ may be infinite, but from the definition a Day implication system $\Rightarrow_D$ for $\mathbf{Q}^{\models}(\emptyset)$ we get that

$$x \approx y \Rightarrow_D z \approx w \subseteq \mathbf{Q}^{\models}(x \approx y \Leftrightarrow z \approx w).$$

(eD1)–(eD3) give an equivalent characterization of the relative shifting property. □

The above result provides Mal'cev-type conditions which characterize the relative shifting property. Theorem 3.5.1 is essentially due to Kearnes and McKenzie (1992, Theorem 2.1). They do not use the term "Day implication system", but their syntactic characterization of the relative shifting property is, after making suitable rearrangements of variables, equivalent to the above three properties (iD1)–(iD3) that define Day implications.

Corollary 3.5.3. *Let* $\mathbf{Q}'$ *and* $\mathbf{Q}$ *be quasivarieties of* τ*-algebras such that* $\mathbf{Q}' \subseteq \mathbf{Q}$. *If the relative shifting property holds for* $\mathbf{Q}$, *then it also holds for* $\mathbf{Q}'$.

Proof. Assume $\mathbf{Q}^{\models}$ has a Day implication system $\Rightarrow_D$. By the inclusion $\mathbf{Q}' \subseteq \mathbf{Q}$, $\Rightarrow_D$ is also a Day implication system for $\mathbf{Q}'^{\models}$. □

Note 3.5.4. Let $\mathbf{V}$ be a congruence-modular variety and let $m_0(x, y, z, w), \ldots, m_n(x, y, z, w)$ be Day terms for $\mathbf{V}$. Define:

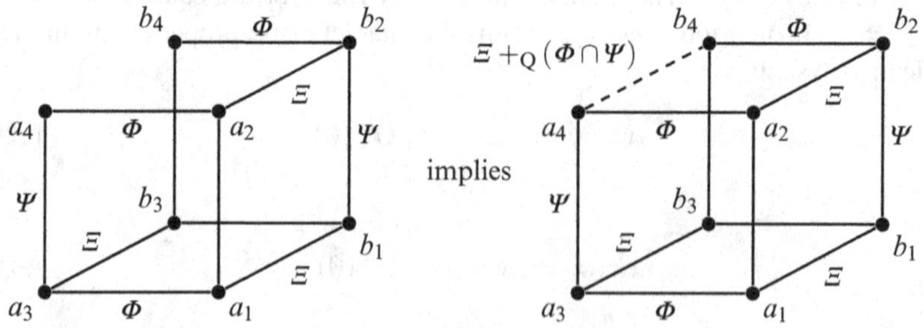

Fig. 3.2 The relative cube property

$$x \approx y \Rightarrow_D z \approx w := \{m_i(x,y,z,w) \approx m_{i+1}(x,y,z,w) : i < n,\ i \text{ even}\}.$$

We claim that $x \approx y \Rightarrow_D z \approx w$ is a Day implication system for $\mathbf{V}^{\models}$.

We apply conditions (i)–(iv) of Theorem 3.4.1.(3). As the equations $m_i(x,y,x,y) \approx m_{i+1}(x,y,x,y)$, $i < n$, i even, are valid in $\mathbf{V}$ (by (iii), we see that $x \approx y \Rightarrow_D z \approx w$ satisfies (iD2). But by (ii) the equations $m_i(x,x,y,y) \approx m_{i+1}(x,x,y,y)$, $i < n$, also hold in $\mathbf{V}$. Hence the system $x \approx y \Rightarrow_D z \approx w$ satisfies (iD3). Finally, suppose that for any algebra $\boldsymbol{A} \in \mathbf{V}$ and $a,c,d \in \boldsymbol{A}$ it is the case that $m_i(a,a,c,d) = m_{i+1}(a,a,c,d)$, $i < n$, i even. In view of condition (iv) of Theorem 3.4.1.(3), it follows that $m_i(a,a,c,d) = m_{i+1}(a,a,c,d)$, for all $i < n$. Consequently, $c = m_0(a,a,c,d) = m_n(a,a,c,d) = d$. Hence $c = d$. This shows that $x \approx y \Rightarrow_D z \approx w$ satisfies (iD1).

This implication system actually satisfies a stronger system of equations than (iD3), viz.

$$m_i(x,x,y,y) \approx y \approx m_{i+1}(x,x,y,y), \qquad i < n, \quad i \text{ even} \tag{iD4}$$

(see also Theorem 4.2.2). □

3.6 The Relative Cube Property

Let $\mathbf{Q}$ be a quasivariety of algebras of a given signature τ. The *relative cube property of* $\mathbf{Q}$ is the following statement:

> *Suppose that* $\boldsymbol{A} \in \mathbf{Q}$*, that* $\Phi, \Psi, \Xi \in Con_{\mathbf{Q}}(\boldsymbol{A})$*, and that* $a_1, a_2, a_3, a_4, b_1, b_2, b_3, b_4 \in A$ *satisfy* $\langle a_1, a_3\rangle, \langle a_2, a_4\rangle, \langle b_1, b_3\rangle, \langle b_2, b_4\rangle \in \Phi$, $\langle a_1, a_2\rangle, \langle a_3, a_4\rangle, \langle b_1, b_2\rangle, \langle b_3, b_4\rangle \in \Psi$*, and* $\langle a_1, b_1\rangle, \langle a_2, b_2\rangle, \langle a_3, b_3\rangle \in \Xi$*. Then* $\langle a_4, b_4\rangle \in \Xi +_{\mathbf{Q}} \Phi \cap \Psi$.

This statement is expressed pictorially in Fig. 3.2:

The following theorem is essentially due to Kearnes and McKenzie (1992) (modulo a permutation of the variables in condition (B) below):

Theorem 3.6.1. *For any quasivariety* $\mathbf{Q}$ *of* τ*-algebras the following conditions are equivalent:*

(A) *The relative cube property holds for* $\mathbf{Q}$*;*
(B) *There exists a finite set of equations* $\Sigma_c(x_1, x_2, x_3, x_4, y_1, y_2, y_3, y_4)$ *in eight variables such that*

  $(\alpha)^*$ $\Sigma_c(x_1, x_2, x_1, x_2, y_1, y_2, y_1, y_2) \subseteq \mathbf{Q}^{\models}(\emptyset)$
  $(\beta)^*$ $\Sigma_c(x_1, x_1, x_3, x_3, y_1, y_1, y_3, y_3) \subseteq \mathbf{Q}^{\models}(\emptyset)$

  and

  $(\delta)^*$ $x_4 \approx y_4 \in \mathbf{Q}^{\models}(\Sigma_c(x_1, x_2, x_3, x_4, x_1, x_2, x_3, y_4))$.

Proof. (B) $\Rightarrow$ (A). Let $\mathbf{A} \in \mathbf{Q}$, $\Phi, \Psi, \Xi \in Con_{\mathbf{Q}}(\mathbf{A})$ and $a_1, a_2, a_3, a_4, b_1, b_2, b_3, b_4 \in A$ so that the hypothesis of the relative cube property is satisfied. Let $\Sigma_c(x_1, x_2, x_3, x_4, y_1, y_2, y_3, y_4)$ be the set of equations supplied by (B). We show that

For every equation $p \approx q \in \Sigma_c$,
$p(a_1, a_2, a_3, a_4, a_1, a_2, a_3, b_4) \equiv q(a_1, a_2, a_3, a_4, a_1, a_2, a_3, b_4) \pmod{\Phi \cap \Psi +_{\mathbf{Q}} \Xi}$.

Let $p \approx q \in \Sigma_c$. We have:

$$\begin{aligned} p(a_1, a_2, a_3, a_4, b_1, b_2, b_3, b_4) &\equiv_{\Phi} p(a_1, a_2, a_1, a_2, b_1, b_2, b_1, b_2) = \text{(by } (\alpha)^*) \\ q(a_1, a_2, a_1, a_2, b_1, b_2, b_1, b_2) &\equiv_{\Phi} q(a_1, a_2, a_3, a_4, b_1, b_2, b_3, b_4) \end{aligned}$$

and

$$\begin{aligned} p(a_1, a_2, a_3, a_4, b_1, b_2, b_3, b_4) &\equiv_{\Psi} p(a_1, a_1, a_3, a_3, b_1, b_1, b_3, b_3) = \text{(by } (\beta)^*) \\ q(a_1, a_1, a_3, a_3, b_1, b_1, b_3, b_3) &\equiv_{\Psi} q(a_1, a_2, a_3, a_4, b_1, b_2, b_3, b_4). \end{aligned}$$

Thus

$$p(a_1, a_2, a_3, a_4, b_1, b_2, b_3, b_4) \equiv_{\Phi \cap \Psi} q(a_1, a_2, a_3, a_4, b_1, b_2, b_3, b_4). \tag{2}$$

Furthermore

$$p(a_1, a_2, a_3, a_4, a_1, a_2, a_3, b_4) \equiv_{\Xi} p(a_1, a_2, a_3, a_4, b_1, b_2, b_3, b_4) \tag{3}$$

and

$$p(a_1, a_2, a_3, a_4, a_1, a_2, a_3, b_4) \equiv_{\Xi} p(a_1, a_2, a_3, a_4, b_1, b_2, b_3, b_4). \tag{4}$$

From (2), (3), and (4) the condition (1) follows.

Applying now $(\delta)^*$ to (1) we see that $a_4 \equiv b_4 \pmod{\Phi \cap \Psi +_{\mathbf{Q}} \Xi}$. So (A) holds.

(A) $\Rightarrow$ (B). Let $\boldsymbol{F}$ be the free algebra in $\mathbf{Q}$, freely generated by x_1, x_2, x_3, x_4, y_1, y_2, y_3, y_4. We define:

$\Phi :=$ the congruence of $\boldsymbol{F}$ generated by $\{\langle x_1, x_3\rangle, \langle x_2, x_4\rangle, \langle y_1, y_3\rangle, \langle y_2, y_4\rangle\}$,
$\Psi :=$ the congruence of $\boldsymbol{F}$ generated by $\{\langle x_1, x_2\rangle, \langle x_3, x_4\rangle, \langle y_1, y_2\rangle, \langle y_3, y_4\rangle\}$,
$\Xi :=$ the congruence of $\boldsymbol{F}$ generated by $\{\langle x_1, y_1\rangle, \langle x_2, y_2\rangle, \langle x_3, y_3\rangle\}$.

Φ, Ψ and Ξ are **Q**-congruences because every congruence relation of $\boldsymbol{F}$ generated by some equivalence relation on a set of free generators is a **Q**-congruence (see Proposition 2.6). Applying the relative cube property of **Q** to this situation, we have that

$$\langle x_4, y_4 \rangle \in \Xi +_{\mathbf{Q}} \Phi \cap \Psi.$$

Since the operator $\Theta_{\mathbf{Q}}$ is finitary, there exists a finite set T of ordered pairs contained in $\Phi \cap \Psi$ such that $\langle x_4, y_4 \rangle$ belongs to $\Theta_{\mathbf{Q}}(\{\langle x_1, y_1 \rangle, \langle x_2, y_2 \rangle, \langle x_3, y_3 \rangle\} \cup T)$. We can write

$$T = \{\langle p^{\boldsymbol{F}}(x_1, x_2, x_3, x_4, y_1, y_2, y_3, y_4), q^{\boldsymbol{F}}(x_1, x_2, x_3, x_4, y_1, y_2, y_3, y_4) \rangle : \langle p, q \rangle \in \Sigma_c\}$$

for a finite set Σ_c of equations.

Conditions $(\alpha)^*$ and $(\beta)^*$ hold because T is a subset of $\Phi \cap \Psi$. Condition $(\delta)^*$ holds because $\langle x_4, y_4 \rangle$ is in the **Q**-congruence of $\boldsymbol{F}$ generated by T together with the pairs $\langle x_1, y_1 \rangle, \langle x_2, y_2 \rangle, \langle x_3, y_3 \rangle$ (see Proposition 2.5). □

3.6.1 *The Relative Shifting Property and the Relative Cube Property are Equivalent Properties*

Theorem 3.6.2. *Let* **Q** *be a quasivariety with the relative shifting property. Then there exists a finite set of equations in eight variables* $\Sigma_c(x_1, x_2, x_3, x_4, y_1, y_2, y_3, y_4)$ *such that the sets of equations*

$$\Sigma_c(x_1, x_2, x_1, x_2, y_1, y_2, y_1, y_2) \qquad (\alpha)^*$$

$$\Sigma_c(x_1, x_1, x_3, x_3, y_1, y_1, y_3, y_3) \qquad (\beta)^*$$

$$\Sigma_c(x_1, x_2, x_3, x_4, x_1, x_2, x_3, x_4) \qquad (\gamma)^*$$

are valid in **Q**. *Furthermore*

$$x_4 \approx y_4 \in \mathbf{Q}^{\models}(\Sigma_c(x_1, x_2, x_3, x_4, x_1, x_2, x_3, y_4)). \qquad (\delta)^*$$

Consequently, any quasivariety with the relative shifting property has the relative cube property.

(Note that the set Σ_c additionally satisfies condition $(\gamma)^*$ which is not mentioned in the characterization of the relative cube property.)

Proof. Let $x \approx y \Rightarrow_D z \approx w$ be a Day implication system for $\mathbf{Q}^{\models}$. We define the set of equations:

$$\Sigma_c(x_1, x_2, x_3, x_4, y_1, y_2, y_3, y_4) :=$$

$$\bigcup\{p(x_1,y_1,x_2,y_2) \approx q(x_1,y_1,x_2,y_2) \Rightarrow_D p(x_3,y_3,x_4,y_4) \approx q(x_3,y_3,x_4,y_4) : \\ p \approx q \in x \approx y \Rightarrow_D z \approx w\}.$$

Σ_c is obtained from $x \approx y \Rightarrow_D z \approx w$ by the uniform substitution of the term $p(x_1,y_1,x_2,y_2)$ for x, the term $q(x_1,y_1,x_2,y_2)$ for y, $p(x_3,y_3,x_4,y_4)$ for z and $q(x_3,y_3,x_4,y_4)$ for w in each equation of $x \approx y \Rightarrow_D z \approx w$. The set Σ_c is also written down in a more compact form as

$$(x_1 \approx y_1 \Rightarrow_D x_2 \approx y_2) \Rightarrow_D (x_3 \approx y_3 \Rightarrow_D x_4 \approx y_4).$$

Lemma 3.6.3. *The following sets of equations are valid in* **Q***:*

$$(x_1 \approx y_1 \Rightarrow_D x_2 \approx y_2) \Rightarrow_D (x_1 \approx y_1 \Rightarrow_D x_2 \approx y_2), \tag{α}$$

$$(x_1 \approx y_1 \Rightarrow_D x_1 \approx y_1) \Rightarrow_D (x_3 \approx y_3 \Rightarrow_D x_3 \approx y_3), \tag{β}$$

$$(x_1 \approx x_1 \Rightarrow_D x_2 \approx x_2) \Rightarrow_D (x_3 \approx x_3 \Rightarrow_D x_4 \approx x_4). \tag{γ}$$

Moreover,

$$x_4 \approx y_4 \in \mathbf{Q}^{\models}((x_1 \approx x_1 \Rightarrow_D x_2 \approx x_2) \Rightarrow_D (x_3 \approx x_3 \Rightarrow_D x_4 \approx y_4)). \tag{δ}$$

Proof (of the lemma). The set of equations defined in (α) is equal to

$$\bigcup\{p(x_1,y_1,x_2,y_2) \approx q(x_1,y_1,x_2,y_2) \Rightarrow_D p(x_1,y_1,x_2,y_2) \approx q(x_1,y_1,x_2,y_2) : \\ p \approx q \in x \approx y \Rightarrow_D z \approx w\}.$$

The above set is contained in the union of the following sets of equations

$$s \approx t \Rightarrow_D s \approx t, \tag{1}$$

with s, t ranging over arbitrary terms. The equations of (1) are **Q**-valid by (iD2). Hence the equations of (α) are **Q**-valid as well.

The set of equations defined in (β) is equal to

$$\bigcup\{p(x_1,y_1,x_1,y_1) \approx q(x_1,y_1,x_1,y_1) \Rightarrow_D p(x_3,y_3,x_3,y_3) \approx q(x_3,y_3,x_3,y_3) : \\ p \approx q \in x \approx y \Rightarrow_D z \approx w\}.$$

But, by (iD2), the equations

$$p(x_1,y_1,x_1,y_1) \approx q(x_1,y_1,x_1,y_1)$$

and

$$p(x_3, y_3, x_3, y_3) \approx q(x_3, y_3, x_3, y_3)$$

are **Q**-valid, for all $p \approx q \in x \approx y \Rightarrow_D z \approx w$. The set of equations (β) is therefore equivalent (on the basis of $\mathbf{Q}^{\models}$) to the set

$$\bigcup\{p(x_1, y_1, x_1, y_1) \approx p(x_1, y_1, x_1, y_1) \Rightarrow {}_D p(x_3, y_3, x_3, y_3) \approx p(x_3, y_3, x_3, y_3) : \\ p \approx q \in x \approx y \Rightarrow_D z \approx w\}.$$

The last set is contained in the union of the sets of equations

$$s \approx s \Rightarrow_D t \approx t, \tag{2}$$

where s, t are arbitrary terms. The equations of (2) are **Q**-valid by (iD3). Hence the equations of (β) are **Q**-valid as well.

The set of equations defined in (γ) is equal to

$$\bigcup\{p(x_1, x_1, x_2, x_2) \approx q(x_1, x_1, x_2, x_2) \Rightarrow_D p(x_3, x_3, x_4, x_4) \approx q(x_3, x_3, x_4, x_4) : \\ p \approx q \in x \approx y \Rightarrow_D z \approx w\}.$$

But, by (iD3), the equations

$$p(x_1, x_1, x_2, x_2) \approx q(x_1, x_1, x_2, x_2) \quad \text{and} \quad p(x_3, x_3, x_4, x_4) \approx q(x_3, x_3, x_4, x_4)$$

are **Q**-valid, for all $p \approx q \in x \approx y \Rightarrow_D z \approx w$. Hence the set of equations (γ) is equivalent (on the basis of $\mathbf{Q}^{\models}$) to the set

$$\bigcup\{p(x_1, x_1, x_2, x_2) \approx p(x_1, x_1, x_2, x_2) \Rightarrow {}_D p(x_3, x_3, x_4, x_4) \approx p(x_3, x_3, x_4, x_4) : \\ p \approx q \in x \approx y \Rightarrow_D z \approx w\}.$$

This set is contained in the union of all sets (2). Since the latter set is **Q**-valid, the former is **Q**-valid as well. This shows that the equations of (γ) are valid in **Q**.

We now show the validity of (δ). By virtue of (iD3) we have that the set of equations $x_1 \approx x_1 \Rightarrow_D x_2 \approx x_2$ is **Q**-valid, i.e.,

$$x_1 \approx x_1 \Rightarrow_D x_2 \approx x_2 \in \mathbf{Q}^{\models}(\emptyset). \tag{3}$$

Using (iD1) and (3) we obtain

$$x_3 \approx x_3 \Rightarrow_D x_4 \approx y_4 \subseteq \mathbf{Q}^{\models}((x_1 \approx x_1 \Rightarrow_D x_2 \approx x_2) \\ \Rightarrow_D (x_3 \approx x_3 \Rightarrow_D x_4 \approx y_4)). \tag{4}$$

Since $x_3 \approx x_3$ is trivially $\mathbf{Q}$-valid, applying again (iD1) we derive from (4) that

$$x_4 \approx y_4 \in \mathbf{Q}^{\models}((x_1 \approx x_1 \Rightarrow_D x_2 \approx x_2) \Rightarrow_D (x_3 \approx x_3 \Rightarrow_D x_4 \approx y_4)).$$

So (δ) holds. □

The theorem follows from Lemma 3.6.3. □

The proof of Theorem 3.6.2 shows that the cube property is a consequence of some simple iterations of Day's implication (Lemma 3.6.3). The above considerations show the logical side of this genuinely geometric property.

The converse of Theorem 3.6.2 is also true; that is, the relative cube property implies the relative shifting property under the additional assumption that the set of equations Σ_c supplied by Theorem 3.6.1, which characterize the relative cube property, apart of conditions $(\alpha)^*$, $(\beta)^*$ and $(\delta)^*$ satisfies condition $(\gamma)^*$ as well. We have:

Theorem 3.6.4. *Let* $\mathbf{Q}$ *be a quasivariety and suppose there exists a finite set of equations* Σ_c *in eight variables which satisfies* $(\alpha)^*$, $(\beta)^*$, $(\gamma)^*$ *and* $(\delta)^*$. *Define* $x \approx y \Rightarrow_D z \approx w := \Sigma_c(x, x, x, z, y, y, y, w)$. *The system* $x \approx y \Rightarrow_D z \approx w$ *is a Day implication for* $\mathbf{Q}$. *Consequently,* $\mathbf{Q}$ *has the relative shifting property.*

Proof. We shall check that $\Rightarrow_D$ satisfies conditions (iD1)–(iD3) with respect to $\mathbf{Q}^{\models}$.

$x \approx y \Rightarrow_D x \approx y$ is equal to $\Sigma_c(x, x, x, x, y, y, y, y)$. The last set of equations is $\mathbf{Q}$-valid by $(\alpha)^*$ or $(\beta)^*$. So $\Rightarrow_D$ satisfies (iD2).

$x \approx x \Rightarrow_D y \approx y$ is equal to $\Sigma_c(x, x, x, y, x, x, x, y)$. The last set of equations is $\mathbf{Q}$-valid by $(\gamma)^*$. So $\Rightarrow_D$ satisfies (iD3).

Finally, we observe that $\Rightarrow_D$ satisfies (iD1) by $(\delta)^*$. □

Corollary 3.6.5. *A quasivariety* $\mathbf{Q}$ *has the relative shifting property if and only if it has the relative cube property.* □

Chapter 4
Centralization Relations

4.1 Four Centralization Relations for Quasivarieties

This section is devoted to the study of various forms of the ternary relation of centralization holding on congruences, viz. Φ centralizes Ψ modulo Ξ, where Φ, Ψ, Ξ are congruences on an algebra. We show the *logical* dimensions of this relation by linking these relations with Day implication systems and commutator equations.

Let $\boldsymbol{A}$ be a $\mathbf{Q}$-algebra. Let Φ, Ψ, Ξ be congruences of $\boldsymbol{A}$. We write:

$$Z_{4,\mathrm{com}}(\Phi, \Psi; \Xi)$$

(Φ *centralizes* Ψ modulo Ξ in the sense of *quaternary commutator equations*). By definition,

$Z_{4,\mathrm{com}}(\Phi, \Psi; \Xi) \quad \Leftrightarrow_{df} \quad$ for any quadruple a, b, c, d of elements of A, the conditions

$$a \equiv b(\mathrm{mod}\ \Phi) \text{ and } c \equiv d(\mathrm{mod}\ \Psi) \quad \text{imply} \quad p(a, b, c, d, \underline{e}) \equiv q(a, b, c, d, \underline{e})(\mathrm{mod}\ \Xi)$$

for any quaternary commutator equation $p(x, y, z, w, \underline{u}) \approx q(x, y, z, w, \underline{u})$ for $\mathbf{Q}$ and any sequence $\underline{e}$ of elements of A of the length of $\underline{u}$.

The ternary relation $Z_{4,\mathrm{com}}$ on congruences is called the *centralization relation* in the sense of *quaternary commutator equations* for $\mathbf{Q}$.

We also write:

$$Z_{\mathrm{com}}(\Phi, \Psi; \Xi)$$

J. Czelakowski, *The Equationally-Defined Commutator*,
DOI 10.1007/978-3-319-21200-5_4

(Φ *centralizes* Ψ modulo Ξ in the sense of *arbitrary commutator equations*). By definition,

> $Z_{\text{com}}(\Phi, \Psi; \Xi)$ $\Leftrightarrow_{df}$ for any $m, n \geqslant 1$, any m-tuples $\underline{a}, \underline{b}$, any n-tuples $\underline{c}, \underline{d}$ of elements of A,

$$\underline{a} \equiv \underline{b}\,(\text{mod } \Phi) \text{ and } \underline{c} \equiv \underline{d}\,(\text{mod } \Psi) \text{ imply that } p(\underline{a}, \underline{b}, \underline{c}, \underline{d}, \underline{e}) \equiv q(\underline{a}, \underline{b}, \underline{c}, \underline{d}, \underline{e})(\text{mod } \Xi)$$

> for any commutator equation $p(\underline{x}, \underline{y}, \underline{z}, \underline{w}, \underline{u}) \approx q(\underline{x}, \underline{y}, \underline{z}, \underline{w}, \underline{u})$ for **Q** with $|\underline{x}| = |\underline{y}| = m$, $|\underline{z}| = |\underline{w}| = n$, and any sequence $\underline{e}$ of elements of A of the length of $\underline{u}$.

The ternary relation Z_{com} is called the *centralization relation* in the sense of *arbitrary commutator equations* for **Q**.

We have: $Z_{4,\text{com}}(\Phi, \Psi; \Xi) \Leftrightarrow Z_{4,\text{com}}(\Psi, \Phi; \Xi)$ and $Z_{\text{com}}(\Phi, \Psi; \Xi) \Leftrightarrow Z_{\text{com}}(\Psi, \Phi; \Xi)$. It is also obvious that $Z_{\text{com}}(\Phi, \Psi; \Xi)$ implies $Z_{4,\text{com}}(\Phi, \Psi; \Xi)$.

Proposition 4.1.1. *Let* **Q** *be an arbitrary quasivariety and* **A** *an algebra in* **Q**. *Then, for any* **Q**-*congruences* Φ, Ψ, Ξ_i $(i \in I)$ *of* **A**:

(i) *if* $Z_{\text{com}}(\Phi, \Psi; \Xi_i)$ *for all* $i \in I$, *then* $Z_{\text{com}}(\Phi, \Psi; \bigcap_{i \in I} \Xi_i)$.
(ii) $[\Phi, \Psi]^{\mathbf{A}} = \bigcap\{\Xi \in Con_{\mathbf{Q}}(\mathbf{A}) : Z_{\text{com}}(\Phi, \Psi; \Xi)\}$.

Proof. Immediate. □

We need two more definitions. We write

$$Z_{2,2}(\Phi, \Psi; \Xi)$$

(Φ *centralizes* Ψ modulo Ξ in the sense of the *classical two-binary term* condition). We put:

> $Z_{2,2}(\Phi, \Psi; \Xi)$ $\Leftrightarrow_{df}$ for any pairs $\langle a, b\rangle \in \Phi$, $\langle c, d\rangle \in \Psi$, any two terms $f(x, y, \underline{u})$, $g(x, y, \underline{u})$, and any sequence $\underline{e}$ of elements of A of the length of $\underline{u}$, the conditions
>
> $$f(a, c, \underline{e}) \equiv g(a, c, \underline{e})(\text{mod } \Xi),$$
> $$f(a, d, \underline{e}) \equiv g(a, d, \underline{e})(\text{mod } \Xi),$$
> $$f(b, c, \underline{e}) \equiv g(b, c, \underline{e})(\text{mod } \Xi)$$
>
> imply
>
> $$f(b, d, \underline{e}) \equiv g(b, d, \underline{e})(\text{mod } \Xi).$$

The ternary relation $Z_{2,2}$ is called the *centralization relation* in the sense of the *classical two-binary term condition.*

Note 1. $Z_{2,2}(\Phi, \Psi; \Xi)$ is equivalently expresses in terms of binary polynomial operation of $\mathbf{A}$: for all pairs $\langle a, b\rangle \in \Phi$, $\langle c, d\rangle \in \Psi$, for all polynomial operations $f(x, y)$, $g(x, y)$ of $\mathbf{A}$, the conditions

$$f(a, c) \equiv g(a, c)(\text{mod } \Xi),$$

$$f(a, d) \equiv g(a, d)(\text{mod } \Xi),$$

$$f(b, c) \equiv g(b, c)(\text{mod } \Xi)$$

imply

$$f(b, d) \equiv g(b, d)(\text{mod } \Xi). \quad \square$$

We also define

$$.Z_2(\Phi, \Psi; \Xi)$$

(Φ *centralizes* Ψ modulo Ξ in the sense of the *two-term condition*). We put:

> $Z_2(\Phi, \Psi; \Xi) \quad \Leftrightarrow_{df}$ for any $m, n \geqslant 1$, any m-tuples $\underline{a}, \underline{b}$, any n-tuples $\underline{c}, \underline{d}$ of elements of A such that $\underline{a} \equiv \underline{b}(\text{mod } \Phi)$ and $\underline{c} \equiv \underline{d}(\text{mod } \Psi)$, any terms $f(\underline{x}, \underline{y}, \underline{v})$, $g(\underline{x}, \underline{y}, \underline{v})$, where $|\underline{x}| = m$, $|\underline{y}| = n$, and any sequence $\underline{e}$ of elements of A of the length of $\underline{v}$, the conditions
>
> $$f(\underline{a}, \underline{c}, \underline{e}) \equiv g(\underline{a}, \underline{c}, \underline{e})(\text{mod } \Xi)$$
>
> $$f(\underline{a}, \underline{d}, \underline{e}) \equiv g(\underline{a}, \underline{d}, \underline{e})(\text{mod } \Xi)$$
>
> $$f(\underline{b}, \underline{c}, \underline{e}) \equiv g(\underline{b}, \underline{c}, \underline{e})(\text{mod } \Xi)$$
>
> imply
>
> $$f(\underline{b}, \underline{d}, \underline{e}) \equiv g(\underline{b}, \underline{d}, \underline{e})(\text{mod } \Xi).$$

The relation Z_2 is called the *centralization relation* in the sense of the *two-term condition*.

Note 2. We have:

$$Z_{2,2}(\Phi, \Psi; \Xi) \Leftrightarrow Z_{2,2}(\Psi, \Phi; \Xi) \quad \text{and} \quad Z_2(\Phi, \Psi; \Xi) \Leftrightarrow Z_2(\Psi, \Phi; \Xi).$$

For the sake of completeness, we prove the second equivalence. We assume Φ, Ψ, $\Xi \in Con_{\mathbf{Q}}(\mathbf{A})$ so that $Z_2(\Phi, \Psi; \Xi)$. To show that $Z_2(\Psi, \Phi; \Xi)$, suppose $f(\underline{x}, \underline{y}, \underline{v})$, $g(\underline{x}, \underline{y}, \underline{v})$ are arbitrary terms, where $|\underline{x}| = m \geqslant 1$, $|\underline{y}| = n \geqslant 1$, and let $\underline{a}, \underline{b}$ be m-tuples and $\underline{c}, \underline{d}$—$n$-tuples of elements of A such that $\underline{a} \equiv \underline{b}$ (mod Ψ), $\underline{c} \equiv \underline{d}$ (mod Φ), and let $\underline{e}$ be a sequence of elements of A of the length of $\underline{v}$, so that the conditions

$$f(\underline{a},\underline{c},\underline{e}) \equiv g(\underline{a},\underline{c},\underline{e})(\bmod\ \Xi) \tag{a}$$
$$f(\underline{a},\underline{d},\underline{e}) \equiv g(\underline{a},\underline{d},\underline{e})(\bmod\ \Xi)$$
$$f(\underline{b},\underline{c},\underline{e}) \equiv g(\underline{b},\underline{c},\underline{e})(\bmod\ \Xi)$$

hold. We claim that

$$f(\underline{b},\underline{d},\underline{e}) \equiv g(\underline{b},\underline{d},\underline{e})(\bmod\ \Xi).$$

We select two sequences of new and different variables $\underline{x}'$ and $\underline{y}'$ such that $|\underline{x}'| = |\underline{y}|$ and $|\underline{y}'| = |\underline{x}|$ and put

$$f' = f'(\underline{x}',\underline{y}',\underline{u}) := f(\underline{x}/\underline{y}',\underline{y}/\underline{x}',\underline{u}), \qquad g' = g'(\underline{x}',\underline{y}',\underline{u}) := g(\underline{x}/\underline{y}',\underline{y}/\underline{x}',\underline{u}).$$

We then have:

$$f'(\underline{c},\underline{a},\underline{e}) = f'(\underline{x}'/\underline{c},\underline{y}'/\underline{a},\underline{u}/\underline{e}) = f(\underline{x}/\underline{a},\underline{y}/\underline{c},\underline{u}/\underline{e}) = f(\underline{a},\underline{c},\underline{e}); \tag{b}$$
$$g'(\underline{c},\underline{a},\underline{e}) = g'(\underline{x}'/\underline{c},\underline{y}'/\underline{a},\underline{u}/\underline{e}) = g(\underline{x}/\underline{a},\underline{y}/\underline{c},\underline{u}/\underline{e}) = g(\underline{a},\underline{c},\underline{e});$$
$$f'(\underline{d},\underline{a},\underline{e}) = f'(\underline{x}'/\underline{d},\underline{y}'/\underline{a},\underline{u}/\underline{e}) = f(\underline{x}/\underline{a},\underline{y}/\underline{d},\underline{u}/\underline{e}) = f(\underline{a},\underline{d},\underline{e});$$
$$g'(\underline{d},\underline{a},\underline{e}) = g'(\underline{x}'/\underline{d},\underline{y}'/\underline{a},\underline{u}/\underline{e}) = g(\underline{x}/\underline{a},\underline{y}/\underline{d},\underline{u}/\underline{e}) = g(\underline{a},\underline{d},\underline{e});$$
$$f'(\underline{c},\underline{b},\underline{e}) = f'(\underline{x}'/\underline{c},\underline{y}'/\underline{b},\underline{u}/\underline{e}) = f(\underline{x}/\underline{b},\underline{y}/\underline{c},\underline{u}/\underline{e}) = f(\underline{b},\underline{c},\underline{e});$$
$$g'(\underline{c},\underline{b},\underline{e}) = g'(\underline{x}'/\underline{x},\underline{y}'/\underline{b},\underline{u}/\underline{e}) = g(\underline{x}/\underline{b},\underline{y}/\underline{c},\underline{u}/\underline{e}) = g(\underline{b},\underline{c},\underline{e});$$
$$f'(\underline{d},\underline{b},\underline{e}) = f'(\underline{x}'/\underline{d},\underline{y}'/\underline{b},\underline{u}/\underline{e}) = f(\underline{x}/\underline{b},\underline{y}/\underline{d},\underline{u}/\underline{e}) = f(\underline{b},\underline{d},\underline{e}).$$
$$g'(\underline{d},\underline{b},\underline{e}) = g'(\underline{x}'/\underline{d},\underline{y}'/\underline{b},\underline{u}/\underline{e}) = g(\underline{x}/\underline{b},\underline{y}/\underline{d},\underline{u}/\underline{e}) = g(\underline{b},\underline{d},\underline{e}).$$

(a) and the first six identities of (b) give:

$$f'(\underline{c},\underline{a},\underline{e}) \equiv g'(\underline{c},\underline{a},\underline{e})(\bmod\ \Xi) \tag{c}$$
$$f'(\underline{c},\underline{b},\underline{e}) \equiv g'(\underline{c},\underline{b},\underline{e})(\bmod\ \Xi)$$
$$f'(\underline{d},\underline{a},\underline{e}) \equiv g'(\underline{d},\underline{a},\underline{e})(\bmod\ \Xi).$$

As $\underline{c} \equiv \underline{d}\ (\bmod\ \Phi)$ and $\underline{a} \equiv \underline{b}\ (\bmod\ \Psi)$, the assumption $Z_2(\Phi, \Psi; \Xi)$ and (c) yield that

$$f'(\underline{d},\underline{b},\underline{e}) \equiv g'(\underline{d},\underline{b},\underline{e})(\bmod\ \Xi).$$

Hence

$$f(\underline{b},\underline{d},\underline{e}) \equiv g(\underline{b},\underline{d},\underline{e})(\bmod\ \Xi),$$

by the last two identities of (b). So $Z_2(\Psi, \Phi; \Xi)$ holds.

By a symmetric argument one proves the reverse implication: $Z_2(\Psi, \Phi; \Xi) \Rightarrow Z_2(\Phi, \Psi; \Xi)$. □

The following theorem is a crucial result of this section. It shows that for every quasivariety with the relative shifting property all the four centralization relations coincide.

Theorem 4.1.2. *Let* $\mathbf{Q}$ *be a quasivariety with the relative shifting property. Then for any algebra* $\mathbf{A} \in \mathbf{Q}$ *and any* $\mathbf{Q}$*-congruences* Φ, Ψ, Ξ *of* $\mathbf{A}$,

$$Z_{4,\mathrm{com}}(\Phi, \Psi; \Xi) \;\Leftrightarrow\; Z_{\mathrm{com}}(\Phi, \Psi; \Xi) \;\Leftrightarrow\; Z_{2,2}(\Phi, \Psi; \Xi) \;\Leftrightarrow\; Z_2(\Phi, \Psi; \Xi).$$

Proof. The implications "$Z_{\mathrm{com}}(\Phi, \Psi; \Xi)$ implies $Z_{4,\mathrm{com}}(\Phi, \Psi; \Xi)$" and "$Z_2(\Phi, \Psi; \Xi)$ implies $Z_{2,2}(\Phi, \Psi; \Xi)$" are trivial. The proof of the remaining implications is based on several simple lemmas.

Lemma 4.1.3. *Let* $\mathbf{Q}$ *be any quasivariety.* $Z_2(\Phi, \Psi; \Xi)$ *implies* $Z_{\mathrm{com}}(\Phi, \Psi; \Xi)$, *for any algebra* $\mathbf{A} \in \mathbf{Q}$ *and any* $\mathbf{Q}$*-congruences* Φ, Ψ, Ξ *of* $\mathbf{A}$.

Proof. Fix $\mathbf{A} \in \mathbf{Q}$ and $\mathbf{Q}$-congruences Φ, Ψ, Ξ of $\mathbf{A}$. Assume $Z_2(\Phi, \Psi; \Xi)$. Suppose that $\underline{a} \equiv \underline{b} \pmod{\Phi}$, $\underline{c} \equiv \underline{d} \pmod{\Psi}$ for sequences $\underline{a}, \underline{b}, \underline{c}, \underline{d}$ of elements of A, where $|\underline{a}| = |\underline{b}| = m$, $|\underline{c}| = |\underline{d}| = n$, and let $p(\underline{x}, \underline{y}, \underline{z}, \underline{w}, \underline{u}) \approx q(\underline{x}, \underline{y}, \underline{z}, \underline{w}, \underline{u})$ be any commutator equation for $\mathbf{Q}$ with $|\underline{x}| = |\underline{y}| = m$, $|\underline{z}| = |\underline{w}| = n$. Let $\underline{e}$ be a sequence of elements of A whose length is $|\underline{u}|$. We must show that $p(\underline{a}, \underline{b}, \underline{c}, \underline{d}) \equiv q(\underline{a}, \underline{b}, \underline{c}, \underline{d}) \pmod{\Xi}$.

We consider the following terms:

$$f(\underline{x}, \underline{y}, \underline{v}) := p(\underline{v}_1, \underline{x}, \underline{v}_2, \underline{y}, \underline{u}), \qquad g(\underline{x}, \underline{y}, \underline{v}) := q(\underline{v}_1, \underline{x}, \underline{v}_2, \underline{y}, \underline{u}),$$

where $\underline{v}_1$ and $\underline{v}_2$ are sequences of distinct variables not occurring on the list $\underline{x} + \underline{y} + \underline{z} + \underline{w} + \underline{u}$ such that $|\underline{v}_1| = m$, $|\underline{v}_2| = n$, and $\underline{v} := \underline{v}_1 + \underline{v}_2 + \underline{u}$ (see the footnote on p. 29). The variables of $\underline{v}$ are thus treated as parametric variables in f and g. We define the list of elements of A: $\underline{e}' := \underline{a} + \underline{c} + \underline{e}$. By the fact that $p \approx q$ is a commutator equation for $\mathbf{Q}$, we have that $f(\underline{a}, \underline{c}, \underline{e}') = p(\underline{a}, \underline{a}, \underline{c}, \underline{c}, \underline{e}) = q(\underline{a}, \underline{a}, \underline{c}, \underline{c}, \underline{e}) = g(\underline{a}, \underline{c}, \underline{e}')$. Hence evidently,

$$f(\underline{a}, \underline{c}, \underline{e}') \equiv g(\underline{a}, \underline{c}, \underline{e}') \pmod{\Xi}. \tag{1}$$

Furthermore $f(\underline{a}, \underline{d}, \underline{e}') = p(\underline{a}, \underline{a}, \underline{c}, \underline{d}, \underline{e}) = q(\underline{a}, \underline{a}, \underline{c}, \underline{d}, \underline{e}) = g(\underline{a}, \underline{d}, \underline{e}')$. Hence trivially

$$f(\underline{a}, \underline{d}, \underline{e}') \equiv g(\underline{a}, \underline{d}, \underline{e}') \pmod{\Xi}. \tag{2}$$

We also have $f(\underline{b}, \underline{c}, \underline{e}') = p(\underline{a}, \underline{b}, \underline{c}, \underline{c}, \underline{e}) = q(\underline{a}, \underline{b}, \underline{c}, \underline{c}, \underline{e}) = g(\underline{b}, \underline{c}, \underline{e}')$. Hence

$$f(\underline{b}, \underline{c}, \underline{e}') \equiv g(\underline{b}, \underline{c}, \underline{e}') \pmod{\Xi}. \tag{3}$$

Then (1), (2), (3), $\underline{a} \equiv \underline{b}$ (mod Φ), $\underline{c} \equiv \underline{d}$ (mod Ψ) and $Z_2(\Phi, \Psi; \Xi)$ imply

$$f(\underline{b}, \underline{d}, \underline{e}') \equiv g(\underline{b}, \underline{d}, \underline{e}')(\text{mod } \Xi)$$

which means that

$$p(\underline{a}, \underline{b}, \underline{c}, \underline{d}) \equiv q(\underline{a}, \underline{b}, \underline{c}, \underline{d})(\text{mod } \Xi).$$

So $Z_{\text{com}}(\Phi, \Psi; \Xi)$. □

As a particular case of the above proof, we obtain:

Lemma 4.1.4. *Let* **Q** *be any quasivariety.* $Z_{2,2}(\Phi, \Psi; \Xi)$ *implies* $Z_{4,\text{com}}(\Phi, \Psi; \Xi)$, *for any algebra* $\mathbf{A} \in \mathbf{Q}$ *and any* **Q**-*congruences* Φ, Ψ, Ξ *of* **A**. □

Lemma 4.1.5. *Let* **Q** *be any quasivariety with the relative shifting property. Then* $Z_{\text{com}}(\Phi, \Psi; \Xi)$ *implies* $Z_2(\Phi, \Psi; \Xi)$, *for every algebra* $\mathbf{A} \in \mathbf{Q}$ *and any* **Q**-*congruences* Φ, Ψ, Ξ *of* **A**.

Proof. By Theorem 3.6.2, **Q** has the relative cube property. Let Σ_c be the set of pairs of terms in 8 variables supplied by Theorem 3.6.1, and let $(\alpha)^*$, $(\beta)^*$, and $(\delta)^*$ be the equations and the quasi-equations provided by that theorem.

Let $\mathbf{A} \in \mathbf{Q}$ and assume $Z_{\text{com}}(\Phi, \Psi; \Xi)$, $\underline{a} \equiv \underline{b}$ (mod Φ), $\underline{c} \equiv \underline{d}$ (mod Ψ) for tuples $\underline{a}, \underline{b}, \underline{c}, \underline{d}$ of elements of A, where $|\underline{a}| = |\underline{b}| = m$, $|\underline{c}| = |\underline{d}| = n$. Let $f(\underline{x}, \underline{y}, \underline{v})$, $g(\underline{x}, \underline{y}, \underline{v})$ be terms and let $\underline{e}$ be a sequence of elements of A of the length of $\underline{v}$. Furthermore, let us assume that

$$f(\underline{a}, \underline{c}, \underline{e}) \equiv g(\underline{a}, \underline{c}, \underline{e})(\text{mod } \Xi), \quad f(\underline{a}, \underline{d}, \underline{e}) \equiv g(\underline{a}, \underline{d}, \underline{e})(\text{mod } \Xi), \quad \text{and} \qquad (*)$$
$$f(\underline{b}, \underline{c}, \underline{e}) \equiv g(\underline{b}, \underline{c}, \underline{e})(\text{mod } \Xi).$$

We shall show that $f(\underline{b}, \underline{d}, \underline{e}) \equiv g(\underline{b}, \underline{d}, \underline{e})(\text{mod } \Xi)$.

For each pair $\langle p, q\rangle \in \Sigma_c$ define:

$$p'(\underline{x}, \underline{y}, \underline{z}, \underline{w}, \underline{v}) :=$$
$$p(f(\underline{x}, \underline{z}, \underline{v}), f(\underline{x}, \underline{w}, \underline{v}), f(\underline{y}, \underline{z}, \underline{v}), f(\underline{y}, \underline{w}, \underline{v}), g(\underline{x}, \underline{z}, \underline{v}), g(\underline{x}, \underline{w}, \underline{v}), g(\underline{y}, \underline{z}, \underline{v}), g(\underline{y}, \underline{w}, \underline{v})),$$
$$q'(\underline{x}, \underline{y}, \underline{z}, \underline{w}, \underline{v}) :=$$
$$q(f(\underline{x}, \underline{z}, \underline{v}), f(\underline{x}, \underline{w}, \underline{v}), f(\underline{y}, \underline{z}, \underline{v}), f(\underline{y}, \underline{w}, \underline{v}), g(\underline{x}, \underline{z}, \underline{v}), g(\underline{x}, \underline{w}, \underline{v}), g(\underline{y}, \underline{z}, \underline{v}), g(\underline{y}, \underline{w}, \underline{v})),$$

Claim. $p' \approx q'$ *is a commutator equation for* **Q** *in* $\underline{x}, \underline{y}$ *and* $\underline{z}, \underline{w}$.

Proof (of the claim). To show that $p'(\underline{x}, \underline{x}, \underline{z}, \underline{w}, \underline{v}) \approx q'(\underline{x}, \underline{x}, \underline{z}, \underline{w}, \underline{v})$ holds in **Q**, observe that for any **A** in **Q** and any sequences $\underline{a}, \underline{b}, \underline{c}, \underline{d}, \underline{e}$ of elements of A:

$$p'(\underline{a},\underline{a},\underline{c},\underline{d},\underline{e}) =$$
$$p(f(\underline{a},\underline{c},\underline{e}), f(\underline{a},\underline{d},\underline{e}), f(\underline{a},\underline{c},\underline{e}), f(\underline{a},\underline{d},\underline{e}), g(\underline{a},\underline{c},\underline{e}), g(\underline{a},\underline{d},\underline{e}), g(\underline{a},\underline{c},\underline{e}), g(\underline{a},\underline{d},\underline{e})),$$
$$= \text{(by } (\alpha)^*)$$
$$q(f(\underline{a},\underline{c},\underline{e}), f(\underline{a},\underline{d},\underline{e}), f(\underline{a},\underline{c},\underline{e}), f(\underline{a},\underline{d},\underline{e}), g(\underline{a},\underline{c},\underline{e}), g(\underline{a},\underline{d},\underline{e}), g(\underline{a},\underline{c},\underline{e}), g(\underline{a},\underline{d},\underline{e})),$$
$$= q'(\underline{a},\underline{a},\underline{c},\underline{d},\underline{e}).$$

To show that $p'(\underline{x},\underline{y},\underline{z},\underline{z},\underline{v}) \approx q'(\underline{x},\underline{y},\underline{z},\underline{z},\underline{v})$ holds in $\mathbf{Q}$, observe that in $\boldsymbol{A}$:

$$p'(\underline{a},\underline{b},\underline{c},\underline{c},\underline{e}) =$$
$$p(f(\underline{a},\underline{c},\underline{d}), f(\underline{a},\underline{c},\underline{d}), f(\underline{b},\underline{c},\underline{e}), f(\underline{b},\underline{c},\underline{e}), g(\underline{a},\underline{c},\underline{e}), g(\underline{a},\underline{c},\underline{e}), g(\underline{b},\underline{c},\underline{e}), g(\underline{b},\underline{c},\underline{e})),$$
$$= \text{(by } (\beta)^*)$$
$$q(f(\underline{a},\underline{c},\underline{e}), f(\underline{a},\underline{c},\underline{e}), f(\underline{b},\underline{c},\underline{e}), f(\underline{b},\underline{c},\underline{e}), g(\underline{a},\underline{c},\underline{e}), g(\underline{a},\underline{c},\underline{e}), g(\underline{b},\underline{c},\underline{e}), g(\underline{b},\underline{c},\underline{e})),$$
$$= q'(\underline{a},\underline{b},\underline{c},\underline{c},\underline{e}).$$

This proves the claim. □

Since $\underline{a} \equiv \underline{b}$ (mod Φ) and $\underline{c} \equiv \underline{d}$ (mod Ψ), the above claim and the fact that $Z_{\mathrm{com}}(\Phi, \Psi; \Xi)$ holds imply

$$p'(\underline{a},\underline{b},\underline{c},\underline{d},\underline{e}) \equiv q'(\underline{a},\underline{b},\underline{c},\underline{d},\underline{e}) (\mathrm{mod}\ \Xi). \tag{1}$$

We put:

$$a := f(\underline{b},\underline{d},\underline{e}),$$
$$e_1 := f(\underline{a},\underline{c},\underline{e}), \qquad e_2 := f(\underline{a},\underline{d},\underline{e}), \qquad e_3 := f(\underline{b},\underline{c},\underline{e}),$$
$$b := g(\underline{b},\underline{d},\underline{e}),$$
$$e_1' := g(\underline{a},\underline{c},\underline{e}), \qquad e_2' := g(\underline{a},\underline{d},\underline{e}), \qquad e_3' := g(\underline{b},\underline{c},\underline{e}).$$

(1) thus states that

$$p(e_1, e_2, e_3, a, e_1', e_2', e_3', b) \equiv q(e_1, e_2, e_3, a, e_1', e_2', e_3', b) (\mathrm{mod}\ \Xi) \tag{2}$$

for all $\langle p, q\rangle \in \Sigma_c$. But $(*)$ means that

$$e_1 \equiv e_1' \ (\mathrm{mod}\ \Xi), \quad e_2 \equiv e_2' \ (\mathrm{mod}\ \Xi), \quad e_3 \equiv e_3' \ (\mathrm{mod}\ \Xi).$$

This together with (2) gives that

$$p(e_1, e_2, e_3, a, e_1, e_2, e_3, b) \equiv q(e_1, e_2, e_3, a, e_1, e_2, e_3, b) (\mathrm{mod}\ \Xi),$$

for all $\langle p, q\rangle \in \Sigma_c$. Now applying $(\delta)^*$, we obtain that $a \equiv b$ (mod Ξ). Thus $f(\underline{b},\underline{d},\underline{e}) \equiv g(\underline{b},\underline{d},\underline{e})$(mod Ξ) as required. So $Z_2(\Phi, \Psi; \Xi)$ holds. □

Repeating the above proof for binary terms with parameters $f(x, z, \underline{v})$, $g(x, z, \underline{v})$, we obtain:

Lemma 4.1.6. *Let* **Q** *be any quasivariety with the relative shifting property. Then* $Z_{4,\mathrm{com}}(\Phi, \Psi; \Xi)$ *implies* $Z_{2,2}(\Phi, \Psi; \Xi)$, *for every algebra* $\mathbf{A} \in \mathbf{Q}$ *and any* **Q***-congruences* Φ, Ψ, Ξ *of* **A**. □

The proof of the next lemma is more involved:

Lemma 4.1.7. *Let* **Q** *be a quasivariety of algebras with the relative shifting property. Let* Φ, Ψ, *and* Ξ *be* **Q***-congruences of an algebra* $\mathbf{A} \in \mathbf{Q}$. *Then* $Z_{4,\mathrm{com}}(\Phi, \Psi; \Xi)$ *implies* $Z_{\mathrm{com}}(\Phi, \Psi; \Xi)$.

Proof. Let $\mathbf{A} \in \mathbf{Q}$ and assume that $Z_{4,\mathrm{com}}(\Phi, \Psi; \Xi)$. We apply Theorem 3.5.1. We prove the following statement $P(m, n)$:

$P(m, n)$ For any commutator equation $\alpha(\underline{x}, \underline{y}, \underline{z}, \underline{w}, \underline{u}) \approx \beta(\underline{x}, \underline{y}, \underline{z}, \underline{w}, \underline{u})$ for **Q** with $|\underline{x}| = |\underline{y}| = m$ and $|\underline{z}| = |\underline{w}| = n$, and for any four sequences $\underline{a}, \underline{b}, \underline{c}, \underline{d}$ elements of A such that $|\underline{a}| = |\underline{b}| = m$ and $|\underline{c}| = |\underline{d}| = n$, the conditions

$$\underline{a} \equiv \underline{b} \pmod{\Phi} \quad \text{and} \quad \underline{c} \equiv \underline{d} \pmod{\Psi} \quad \text{imply}$$

$$\alpha(\underline{a}, \underline{b}, \underline{c}, \underline{d}, \underline{e}) \approx \beta(\underline{a}, \underline{b}, \underline{c}, \underline{d}, \underline{e}) (\mathrm{mod}\ \Xi) \text{ for any sequence } \underline{e} \text{ of elements of } A \text{ of the length of } \underline{u}.$$

Lemma 4.1.7 will be proved once we show $P(m, n)$ for all possible values of m and n. $P(m, n)$ is proved by induction on m for $n = 1$, and then by induction on n.

Sublemma 1. $P(m, 1)$, *for all* $m \geqslant 1$.

Proof (of the sublemma). The case $m = 1$ is covered by the assumption $Z_{4,\mathrm{com}}(\Phi, \Psi; \Xi)$. Now suppose the sublemma is true for a certain m. Let $\underline{a} = a_1, \ldots, a_m, a_{m+1}$ and $\underline{b} = b_1, \ldots, b_m, b_{m+1}$ be $(m + 1)$-tuples, let c, d be elements such that $\underline{a} \equiv \underline{b} \pmod{\Phi}$ and $c \equiv d \pmod{\Psi}$, and let $\alpha(\underline{x}, \underline{y}, z, w, \underline{u}) \approx \beta(\underline{x}, \underline{y}, z, w, \underline{u})$ be a commutator equation for **Q** with $\underline{x} = x_1, \ldots, x_m, x_{m+1}$, $\underline{y} = y_1, \ldots, y_m, y_{m+1}$. Let $\underline{e}$ be sequence of elements of A of the length of $\underline{u}$. We need to show that $\alpha(\underline{a}, \underline{b}, c, d, \underline{e}) \equiv \beta(\underline{a}, \underline{b}, c, d, \underline{e}) \pmod{\Xi}$. We prove it in two stages.

Put $\underline{x}_0 := x_1, \ldots, x_m$, $\underline{y}_0 := y_1, \ldots, y_m$. Let $\underline{v}$ be a sequence of variables (of length $m + 1$) not occurring in $\underline{x}, \underline{y}, z, w, \underline{u}$, and let v be a single variable not occurring in $\underline{x}, \underline{y}, z, w, \underline{u}, \underline{v}$. We define the list $\underline{v}' := \underline{u} + \underline{v} + v$. Consider the Day implication $x \approx y \Rightarrow_D z \approx w$ for the consequence operation $\mathbf{Q}^{\models}$. For each equation $p(x, y, z, w) \approx q(x, y, z, w)$ belonging to $\Rightarrow_D$ we define the following two terms:

$$p'(\underline{x}_0, \underline{y}_0, z, w, \underline{v}') :=$$

$$p(\alpha(\underline{v}, \underline{x}_0 v, z, w, \underline{u}), \beta(\underline{v}, \underline{x}_0 v, z, w, \underline{u}), \alpha(\underline{v}, \underline{y}_0 v, z, w, \underline{u}), \beta(\underline{v}, \underline{y}_0 v, z, w, \underline{u})),$$

$$q'(\underline{x}_0, \underline{y}_0, z, w, \underline{v}') :=$$

$$q(\alpha(\underline{v}, \underline{x}_0 v, z, w, \underline{u}), \beta(\underline{v}, \underline{x}_0 v, z, w, \underline{u}), \alpha(\underline{v}, \underline{y}_0 v, z, w, \underline{u}), \beta(\underline{v}, \underline{y}_0 v, z, w, \underline{u})).$$

Claim 1. *$p'(\underline{x}_0, \underline{y}_0, z, w, \underline{v}') \approx q'(\underline{x}_0, \underline{y}_0, z, w, \underline{v}')$ is a commutator equation for $\mathbf{Q}$ in the variables $\underline{x}_0, \underline{y}_0$ and z, w.*

Proof (of the claim). The equation $p'(\underline{x}_0, \underline{x}_0, z, w, \underline{v}') \approx q'(\underline{x}_0, \underline{x}_0, z, w, \underline{v}')$ coincides with

$$p(\alpha(\underline{v}, \underline{x}_0 v, z, w, \underline{u}), \beta(\underline{v}, \underline{x}_0 v, z, w, \underline{u}), \alpha(\underline{v}, \underline{x}_0 v, z, w, \underline{u}), \beta(\underline{v}, \underline{x}_0 v, z, w, \underline{u})) \approx$$
$$q(\alpha(\underline{v}, \underline{x}_0 v, z, w, \underline{u}), \beta(\underline{v}, \underline{x}_0 v, z, w, \underline{u}), \alpha(\underline{v}, \underline{x}_0 v, z, w, \underline{u}), \beta(\underline{v}, \underline{x}_0 v, z, w, \underline{u})).$$

The last equation is $\mathbf{Q}$-valid by (iD2) since it is of the form $p(s, t, s, t) \approx q(s, t, s, t)$.
The equation $p'(\underline{x}_0, \underline{y}_0, z, z, \underline{v}') \approx q'(\underline{x}_0, \underline{y}_0, z, z, \underline{v}')$ coincides with

$$p(\alpha(\underline{v}, \underline{x}_0 v, z, z, \underline{u}), \beta(\underline{v}, \underline{x}_0 v, z, z, \underline{u}), \alpha(\underline{v}, \underline{y}_0 v, z, z, \underline{u}), \beta(\underline{v}, \underline{y}_0 v, z, z, \underline{u})) \approx \quad (1)$$
$$q(\alpha(\underline{v}, \underline{x}_0 v, z, z, \underline{u}), \beta(\underline{v}, \underline{x}_0 v, z, z, \underline{u}), \alpha(\underline{v}, \underline{y}_0 v, z, z, \underline{u}), \beta(\underline{v}, \underline{y}_0 v, z, z, \underline{u})).$$

Since the equations $\alpha(\underline{v}, \underline{x}_0 v, z, z, \underline{u}) \approx \beta(\underline{v}, \underline{x}_0 v, z, z, \underline{u})$ and $\alpha(\underline{v}, \underline{y}_0 v, z, z, \underline{u}) \approx \beta(\underline{v}, \underline{y}_0 v, z, z, \underline{u})$ are $\mathbf{Q}$-valid because $\alpha \approx \beta$ is a commutator equation for $\mathbf{Q}$, we see that the equation (1) is $\mathbf{Q}$-valid if and only if the equation

$$p(\alpha(\underline{v}, \underline{x}_0 v, z, z, \underline{u}), \alpha(\underline{v}, \underline{x}_0 v, z, z, \underline{u}), \alpha(\underline{v}, \underline{y}_0 v, z, z, \underline{u}), \alpha(\underline{v}, \underline{y}_0 v, z, z, \underline{u})) \approx$$
$$q(\alpha(\underline{v}, \underline{x}_0 v, z, z, \underline{u}), \alpha(\underline{v}, \underline{x}_0 v, z, z, \underline{u}), \alpha(\underline{v}, \underline{y}_0 v, z, z, \underline{u}), \alpha(\underline{v}, \underline{y}_0 v, z, z, \underline{u})).$$

is $\mathbf{Q}$-valid. But the last equation is $\mathbf{Q}$-valid by (iD3) since it is of the form $p(s, s, t, t) \approx q(s, s, t, t)$. □

If f, g, k, l are polynomials over an algebra $\mathbf{A}$, then we write $\langle f, g\rangle \Rightarrow_D \langle k, l\rangle$ to denote the set of pairs $\{\langle p(f, g, k, l), q(f, g, k, l)\rangle : p \approx q \in x \approx y \Rightarrow_D z \approx w\}$.

Let $\underline{a}_0 := a_1, \ldots, a_m$, $\underline{b}_0 := b_1, \ldots, b_m$ and $\underline{e}' := \underline{e} + \underline{a} + a_{m+1}$. By the above claim and the inductive hypothesis (in the proof of Sublemma 1), we obtain that

$$p'(\underline{a}_0, \underline{b}_0, c, d, \underline{e}') \equiv q'(\underline{a}_0, \underline{b}_0, c, d, \underline{e}') \pmod{\Xi}.$$

This means that

$$p(\alpha(\underline{a}, \underline{a}_0 a_{m+1}, c, d, \underline{e}), \beta(\underline{a}, \underline{a}_0 a_{m+1}, c, d, \underline{e}), \alpha(\underline{a}, \underline{b}_0 a_{m+1}, c, d, \underline{e}), \beta(\underline{a}, \underline{b}_0 a_{m+1}, c, d, \underline{e})) \equiv_{\Xi}$$
$$q(\alpha(\underline{a}, \underline{a}_0 a_{m+1}, c, d, \underline{e}), \beta(\underline{a}, \underline{a}_0 a_{m+1}, c, d, \underline{e}), \alpha(\underline{a}, \underline{b}_0 a_{m+1}, c, d, \underline{e}), \beta(\underline{a}, \underline{b}_0 a_{m+1}, c, d, \underline{e})),$$

that is,

$$p(\alpha(\underline{a},\underline{a},c,d,\underline{e}),\beta(\underline{a},\underline{a},c,d,\underline{e}),\alpha(\underline{a},\underline{b}_0a_{m+1},c,d,\underline{e}),\beta(\underline{a},\underline{b}_0a_{m+1},c,d,\underline{e}))\equiv_{\Xi} \tag{2}$$
$$q(\alpha(\underline{a},\underline{a},c,d,\underline{e}),\beta(\underline{a},\underline{a},c,d,\underline{e}),\alpha(\underline{a},\underline{b}_0a_{m+1},c,d,\underline{e}),\beta(\underline{a},\underline{b}_0a_{m+1},c,d,\underline{e})).$$

Since (2) holds for an arbitrary member $p \approx q$ of $\Rightarrow_D$, we may write

$$\langle\alpha(\underline{a},\underline{a},c,d,\underline{e}),\beta(\underline{a},\underline{a},c,d,\underline{e})\rangle$$
$$\Rightarrow_D \langle\alpha(\underline{a},\underline{b}_0a_{m+1},c,d,\underline{e}),\beta(\underline{a},\underline{b}_0a_{m+1},c,d,\underline{e})\rangle \subseteq \Xi.$$

But evidently $\langle\alpha(\underline{a},\underline{a},c,d,\underline{e}),\beta(\underline{a},\underline{a},c,d,\underline{e})\rangle \in \Xi$, again by the fact that $\alpha \approx \beta$ is a commutator equation for **Q**. Hence, applying (iD1) to the above situation, we conclude that

$$\langle\alpha(\underline{a},\underline{b}_0a_{m+1},c,d,\underline{e}),\beta(\underline{a},\underline{b}_0a_{m+1},c,d,\underline{e})\rangle \in \Xi. \tag{3}$$

In the second stage, for each equation $p(x,y,z,w) \approx q(x,y,z,w)$ belonging to $\Rightarrow_D$, we define the following two terms:

$$p'(x,y,z,w,\underline{v}'){:=}p(\alpha(\underline{v},\underline{v}_0x,z,w,\underline{u}),\beta(\underline{v},\underline{v}_0x,z,w,\underline{u}),\alpha(\underline{v},\underline{v}_0y,z,w,\underline{u}),\beta(\underline{v},\underline{v}_0y,z,w,\underline{u})),$$
$$q'(x,y,z,w,\underline{v}'){:=}p(\alpha(\underline{v},\underline{v}_0x,z,w,\underline{u}),\beta(\underline{v},\underline{v}_0x,z,w,\underline{u}),\alpha(\underline{v},\underline{v}_0y,z,w,\underline{u}),\beta(\underline{v},\underline{v}_0y,z,w,\underline{u})),$$

where:

$\underline{v}$ is a sequence of variables (of length $m+1$) not occurring in $\underline{x},\underline{y},z,w,\underline{u}$,
$\underline{v}_0$ is a sequence of variables (of length m) not occurring in $\underline{x},\underline{y},z,w,\underline{u},\underline{v}$, and
$\underline{v}' := \underline{u}+\underline{v}+\underline{v}_0$.

Claim 2. $p'(x,y,z,w,\underline{v}') \approx q'(x,y,z,w,\underline{v}')$ *is a quaternary commutator equation for* **Q**

Proof (of the claim). The equation $p'(x,x,z,w,\underline{v}') \approx q'(x,x,z,w,\underline{v}')$ coincides with

$$p(\alpha(\underline{v},\underline{v}_0x,z,w,\underline{u}),\beta(\underline{v},\underline{v}_0x,z,w,\underline{u}),\alpha(\underline{v},\underline{v}_0x,z,w,\underline{u}),\beta(\underline{v},\underline{v}_0x,z,w,\underline{u}))\approx$$
$$q(\alpha(\underline{v},\underline{v}_0x,z,w,\underline{u}),\beta(\underline{v},\underline{v}_0x,z,w,\underline{u}),\alpha(\underline{v},\underline{v}_0x,z,w,\underline{u}),\beta(\underline{v},\underline{v}_0x,z,w,\underline{u})).$$

It is **Q**-valid by (iD2) since it is of the form $p(s,t,s,t) \approx q(s,t,s,t)$.

The equation $p'(x,y,z,z,\underline{v}') \approx q'(x,y,z,z,\underline{v}')$ coincides with

$$p(\alpha(\underline{v},\underline{v}_0x,z,z,\underline{u}),\beta(\underline{v},\underline{v}_0x,z,z,\underline{u}),\alpha(\underline{v},\underline{v}_0y,z,z,\underline{u}),\beta(\underline{v},\underline{v}_0y,z,z,\underline{u}))\approx \tag{4}$$
$$q(\alpha(\underline{v},\underline{v}_0x,z,z,\underline{u}),\beta(\underline{v},\underline{v}_0x,z,z,\underline{u}),\alpha(\underline{v},\underline{v}_0y,z,z,\underline{u}),\beta(\underline{v},\underline{v}_0y,z,z,\underline{u})).$$

Since the equations $\alpha(\underline{v}, \underline{v}_0 x, z, z, \underline{u}) \approx \beta(\underline{v}, \underline{v}_0 x, z, z, \underline{u})$ and $\alpha(\underline{v}, \underline{v}_0 y, z, z, \underline{u}) \approx \beta(\underline{v}, \underline{v}_0 y, z, z, \underline{u})$ are **Q**-valid because $\alpha \approx \beta$ is a commutator equation for **Q**, we see that the **Q**-validity of (4) is equivalent to the **Q**-validity of the equation

$$p(\alpha(\underline{v}, \underline{v}_0 x, z, z, \underline{u}), \alpha(\underline{v}, \underline{v}_0 x, z, z, \underline{u}), \alpha(\underline{v}, \underline{v}_0 y, z, z, \underline{u}), \alpha(\underline{v}, \underline{v}_0 y, z, z, \underline{u})) \approx$$
$$q(\alpha(\underline{v}, \underline{v}_0 x, z, z, \underline{u}), \alpha(\underline{v}, \underline{v}_0 x, z, z, \underline{u}), \alpha(\underline{v}, \underline{v}_0 y, z, z, \underline{u}), \alpha(\underline{v}, \underline{v}_0 y, z, z, \underline{u})).$$

But the last equation is **Q**-valid by (iD3) since it is of the form $p(s, s, t, t) \approx q(s, s, t, t)$. □

By the above claim and the inductive hypothesis (in the proof of Sublemma 1), we obtain that

$$p'(a_{m+1}, b_{m+1}, c, d, \underline{e}') \equiv q'(a_{m+1}, b_{m+1}, c, d, \underline{e}') \ (\mathrm{mod}\ \Xi),$$

where $\underline{e}'$ is the list $\underline{e} + \underline{a} + \underline{b}_0$. This means that

$$p(\alpha(\underline{a}, \underline{b}_0 a_{m+1}, c, d, \underline{e}), \beta(\underline{a}, \underline{b}_0 a_{m+1}, c, d, \underline{e}), \alpha(\underline{a}, \underline{b}_0 b_{m+1}, c, d, \underline{e}), \beta(\underline{a}, \underline{b}_0 b_{m+1}, c, d, \underline{e})) \equiv_{\Xi}$$
$$q(\alpha(\underline{a}, \underline{b}_0 a_{m+1}, c, d, \underline{e}), \beta(\underline{a}, \underline{b}_0 a_{m+1}, c, d, \underline{e}), \alpha(\underline{a}, \underline{b}_0 b_{m+1}, c, d, \underline{e}), \beta(\underline{a}, \underline{b}_0 b_{m+1}, c, d, \underline{e}))$$

i.e.,

$$p(\alpha(\underline{a}, \underline{b}_0 a_{m+1}, c, d, \underline{e}), \beta(\underline{a}, \underline{b}_0 a_{m+1}, c, d, \underline{e}), \alpha(\underline{a}, \underline{b}, c, d, \underline{e}), \beta(\underline{a}, \underline{b}, c, d, \underline{e})) \tag{5}$$
$$\equiv_{\Xi} q(\alpha(\underline{a}, \underline{b}_0 a_{m+1}, c, d, \underline{e}), \beta(\underline{a}, \underline{b}_0 a_{m+1}, c, d, \underline{e}), \alpha(\underline{a}, \underline{b}, c, d, \underline{e}), \beta(\underline{a}, \underline{b}, c, d, \underline{e})).$$

Since (5) holds for an arbitrary member $p \approx q$ of $\Rightarrow_D$, we may write

$$\langle \alpha(\underline{a}, \underline{b}_0 a_{m+1}, c, d, \underline{e}), \beta(\underline{a}, \underline{b}_0 a_{m+1}, c, d, \underline{e}) \rangle$$
$$\Rightarrow_D \langle \alpha(\underline{a}, \underline{b}, c, d, \underline{e}), \beta(\alpha(\underline{a}, \underline{b}, c, d, \underline{e}) \rangle \subseteq \Xi.$$

Now using (3) and applying (iD1) to the above situation, we infer that $\langle \alpha(\underline{a}, \underline{b}, c, d, \underline{e}), \beta(\alpha(\underline{a}, \underline{b}, c, d, \underline{e}) \rangle \in \Xi$.

The inductive proof of $P(m, 1)$ for any positive value of m is complete. Sublemma 1 has been proved. □

Sublemma 2. *Let m be an arbitrary but fixed positive integer. Then $P(m, n)$ holds, for all $n \geqslant 1$.*

Proof (of the sublemma). $P(m, 1)$ holds by Sublemma 1. Now suppose the sublemma is true for a certain n. Let $\underline{a} = a_1, \ldots, a_m$ and $\underline{b} = b_1, \ldots, b_m$ be m-tuples, $\underline{c} = c_1, \ldots, c_n, c_{n+1}$ and $\underline{d} = d_1, \ldots, d_n, d_{n+1}$ be $(n + 1)$-tuples of elements of

A such that $\underline{a} \equiv \underline{b} \pmod{\Phi}$ and $\underline{c} \equiv \underline{d} \pmod{\Psi}$, and let $\alpha(\underline{x}, \underline{y}, \underline{z}, \underline{w}, \underline{u}) \approx \beta(\underline{x}, \underline{y}, \underline{z}, \underline{w}, \underline{u})$ be a commutator equation for $\mathbf{Q}$ with $\underline{x} = x_1, \ldots, x_m$, $\underline{y} = y_1, \ldots, y_m$, $\underline{z} = z_1, \ldots, z_n, z_{n+1}$ and $\underline{w} = w_1, \ldots, w_n, w_{n+1}$. Let $\underline{e}$ be sequence of elements of A of the length of $\underline{u}$. We need to show that $\alpha(\underline{a}, \underline{b}, \underline{c}, \underline{d}, \underline{e}) \equiv \beta(\underline{a}, \underline{b}, \underline{c}, \underline{d}, \underline{e}) \pmod{\Xi}$. We prove it in two stages.

Put $\underline{z}_0 := z_1, \ldots, z_n$, $\underline{w}_0 := w_1, \ldots, w_n$. Let $\underline{v}$ be a sequence of variables (of length $n+1$) not occurring in $\underline{x}, \underline{y}, \underline{z}, \underline{w}, \underline{u}$, and let v be a single variable not occurring in $\underline{x}, \underline{y}, \underline{z}, \underline{w}, \underline{u}, \underline{v}$. Put $\underline{v}' := \underline{u} + \underline{v} + v$. Consider the Day implication $x \approx y \Rightarrow_D z \approx w$ for the consequence $\mathbf{Q}^{\models}$. For each equation $p(x, y, z, w) \approx q(x, y, z, w)$ belonging to $\Rightarrow_D$ we define the following two terms:

$$p'(\underline{x}, \underline{y}, \underline{z}_0, \underline{w}_0, \underline{v}') :=$$
$$p(\alpha(\underline{x}, \underline{y}, \underline{v}, \underline{z}_0 v, \underline{u}), \beta(\alpha(\underline{x}, \underline{y}, \underline{v}, \underline{z}_0 v, \underline{u}), \alpha(\underline{x}, \underline{y}, \underline{v}, \underline{w}_0 v, \underline{u}), \beta(\underline{x}, \underline{y}, \underline{v}, \underline{w}_0 v, \underline{u})),$$
$$q'(\underline{x}, \underline{y}, \underline{z}_0, \underline{w}_0, \underline{v}') :=$$
$$q(\alpha(\underline{x}, \underline{y}, \underline{v}, \underline{z}_0 v, \underline{u}), \beta(\alpha(\underline{x}, \underline{y}, \underline{v}, \underline{z}_0 v, \underline{u}), \alpha(\underline{x}, \underline{y}, \underline{v}, \underline{w}_0 v, \underline{u}), \beta(\underline{x}, \underline{y}, \underline{v}, \underline{w}_0 v, \underline{u})).$$

Claim 3. *$p'(\underline{x}, \underline{y}, \underline{z}_0, \underline{w}_0, \underline{v}') \approx q'(\underline{x}, \underline{y}, \underline{z}_0, \underline{w}_0, \underline{v}')$ is a commutator equation for $\mathbf{Q}$ in the variables $\underline{x}, \underline{y}$ and $\underline{z}_0, \underline{w}_0$.*

The proof of the claim is left as an exercise to the reader. □

By the above claim and the inductive hypothesis (in the proof of Sublemma 2), we obtain that

$$p'(\underline{a}, \underline{b}, \underline{c}_0, \underline{d}_0, \underline{e}') \equiv q'(\underline{a}, \underline{b}, \underline{c}_0, \underline{d}_0, \underline{e}') \pmod{\Xi},$$

where $\underline{e}' := \underline{e} + \underline{c} + c_{n+1}$. This means that

$$p(\alpha(\underline{a}, \underline{b}, \underline{c}, \underline{c}_0 c_{n+1}, \underline{e}), \beta(\underline{a}, \underline{b}, \underline{c}, \underline{c}_0 c_{n+1}, \underline{e}), \alpha(\underline{a}, \underline{b}, \underline{c}, \underline{d}_0 c_{n+1}, \underline{e}), \beta(\underline{a}, \underline{b}, \underline{c}, \underline{d}_0 c_{n+1}, \underline{e})) \equiv_{\Xi}$$
$$q(\alpha(\underline{a}, \underline{b}, \underline{c}, \underline{c}_0 c_{n+1}, \underline{e}), \beta(\underline{a}, \underline{b}, \underline{c}, \underline{c}_0 c_{n+1}, \underline{e}), \alpha(\underline{a}, \underline{b}, \underline{c}, \underline{d}_0 c_{n+1}, \underline{e}), \beta(\underline{a}, \underline{b}, \underline{c}, \underline{d}_0 c_{n+1}, \underline{e})),$$

i.e.,

$$p(\alpha(\underline{a}, \underline{b}, \underline{c}, \underline{c}, \underline{e}), \beta(\underline{a}, \underline{b}, \underline{c}, \underline{c}, \underline{e}), \alpha(\underline{a}, \underline{b}, \underline{c}, \underline{d}_0 c_{n+1}, \underline{e}), \beta(\underline{a}, \underline{b}, \underline{c}, \underline{d}_0 c_{n+1}, \underline{e})) \equiv_{\Xi} \quad (6)$$
$$q(\alpha(\underline{a}, \underline{b}, \underline{c}, \underline{c}, \underline{e}), \beta(\underline{a}, \underline{b}, \underline{c}, \underline{c}, \underline{e}), \alpha(\underline{a}, \underline{b}, \underline{c}, \underline{d}_0 c_{n+1}, \underline{e}), \beta(\underline{a}, \underline{b}, \underline{c}, \underline{d}_0 c_{n+1}, \underline{e})).$$

Since (6) holds for an arbitrary member $p \approx q$ of $\Rightarrow_D$, we may write

$$\langle \alpha(\underline{a}, \underline{b}, \underline{c}, \underline{c}, \underline{e}), \beta(\underline{a}, \underline{b}, \underline{c}, \underline{c}, \underline{e}) \rangle$$
$$\Rightarrow_D \langle \alpha(\underline{a}, \underline{b}, \underline{c}, \underline{d}_0 c_{n+1}, \underline{e}), \beta(\underline{a}, \underline{b}, \underline{c}, \underline{d}_0 c_{n+1}, \underline{e}) \rangle \subseteq \Xi.$$

But evidently $\langle \alpha(\underline{a}, \underline{b}, \underline{c}, \underline{c}, \underline{e}), \beta(\underline{a}, \underline{b}, \underline{c}, \underline{c}, \underline{e})\rangle \in \Xi$, again by the fact that $\alpha \approx \beta$ is a commutator equation for **Q**. Hence, applying (iD1) to the above situation, we obtain that

$$\langle \alpha(\underline{a}, \underline{b}, \underline{c}, \underline{d}_0 c_{n+1}, \underline{e}), \beta(\underline{a}, \underline{b}, \underline{c}, \underline{d}_0 c_{n+1}, \underline{e})\rangle \in \Xi. \tag{7}$$

In the second stage, for each equation $p(x, y, z, w) \approx q(x, y, z, w)$ belonging to $\Rightarrow_D$, we define the following two terms:

$$p'(\underline{x}, \underline{y}, z, w, \underline{v}') :=$$

$$p(\alpha(\underline{x}, \underline{y}, \underline{v}, \underline{v}_0 z, \underline{u}), \beta(\underline{x}, \underline{y}, \underline{v}, \underline{v}_0 z, \underline{u}), \alpha(\underline{x}, \underline{y}, \underline{v}, \underline{v}_0 w, \underline{u}), \beta(\underline{x}, \underline{y}, \underline{v}, \underline{v}_0 w, \underline{u})),$$

$$q'(\underline{x}, \underline{y}, z, w, \underline{v}') :=$$

$$q(\alpha(\underline{x}, \underline{y}, \underline{v}, \underline{v}_0 z, \underline{u}), \beta(\underline{x}, \underline{y}, \underline{v}, \underline{v}_0 z, \underline{u}), \alpha(\underline{x}, \underline{y}, \underline{v}, \underline{v}_0 w, \underline{u}), \beta(\underline{x}, \underline{y}, \underline{v}, \underline{v}_0 w, \underline{u})).$$

where:

$\underline{v}$ is a sequence of variables (of length $n + 1$) not occurring in $\underline{x}, \underline{y}, z, w, \underline{u}$,
$\underline{v}_0$ is a sequence of variables (of length n) not occurring in $\underline{x}, \underline{y}, z, w, \underline{u}, \underline{v}$,
$\underline{v}' := \underline{u} + \underline{v} + \underline{v}_0$.

Claim 4. $p'(\underline{x}, \underline{y}, z, w, \underline{v}') \approx q'(\underline{x}, \underline{y}, z, w, \underline{v}')$ *is a commutator equation for* **Q** *in the variables* $\underline{x}, \underline{y}$ *and* z, w.

The proof of the claim is left as an exercise. □

By the above claim and the inductive hypothesis (in the proof of Sublemma 2), we obtain that

$$p'(\underline{a}, \underline{b}, c_{m+1}, d_{m+1}, \underline{e}') \equiv q'(\underline{a}, \underline{b}, c_{m+1}, d_{m+1}, \underline{e}') \pmod{\Xi},$$

where $\underline{e}' = \underline{e} + \underline{c} + \underline{d}_0$. This means that

$$p(\alpha(\underline{a}, \underline{b}, \underline{c}, \underline{d}_0 c_{n+1}, \underline{e}), \beta(\underline{a}, \underline{b}, \underline{c}, \underline{d}_0 c_{n+1}, \underline{e}), \alpha(\underline{a}, \underline{b}, \underline{c}, \underline{d}_0 d_{n+1}, \underline{e}), \beta(\underline{a}, \underline{b}, \underline{c}, \underline{d}_0 d_{n+1}, \underline{e})) \equiv_{\Xi}$$
$$q(\alpha(\underline{a}, \underline{b}, \underline{c}, \underline{d}_0 c_{n+1}, \underline{e}), \beta(\underline{a}, \underline{b}, \underline{c}, \underline{d}_0 c_{n+1}, \underline{e}), \alpha(\underline{a}, \underline{b}, \underline{c}, \underline{d}_0 d_{n+1}, \underline{e}), \beta(\underline{a}, \underline{b}, \underline{c}, \underline{d}_0 d_{n+1}, \underline{e})),$$

i.e.,

$$p(\alpha(\underline{a}, \underline{b}, \underline{c}, \underline{d}_0 c_{n+1}, \underline{e}), \beta(\underline{a}, \underline{b}, \underline{c}, \underline{d}_0 c_{n+1}, \underline{e}), \alpha(\underline{a}, \underline{b}, \underline{c}, \underline{d}, \underline{e}), \beta(\underline{a}, \underline{b}, \underline{c}, \underline{d}, \underline{e})) \equiv_{\Xi} \tag{8}$$
$$q(\alpha(\underline{a}, \underline{b}, \underline{c}, \underline{d}_0 c_{n+1}, \underline{e}), \beta(\underline{a}, \underline{b}, \underline{c}, \underline{d}_0 c_{n+1}, \underline{e}), \alpha(\underline{a}, \underline{b}, \underline{c}, \underline{d}, \underline{e}), \beta(\underline{a}, \underline{b}, \underline{c}, \underline{d}, \underline{e})).$$

Since (8) holds for an arbitrary member $p \approx q$ of $\Rightarrow_D$, we may write

$$\langle \alpha(\underline{a}, \underline{b}, \underline{c}, \underline{d}_0 c_{n+1}, \underline{e}), \beta(\underline{a}, \underline{b}, \underline{c}, \underline{d}_0 c_{n+1}, \underline{e})\rangle$$
$$\Rightarrow_D \langle \alpha(\underline{a}, \underline{b}, \underline{c}, \underline{d}, \underline{e}), \beta(\underline{a}, \underline{b}, \underline{c}, \underline{d}, \underline{e})\rangle \subseteq \Xi$$

Now using (7) and applying (iD1) to the above situation, we obtain that $\langle \alpha(\underline{a}, \underline{b}, \underline{c}, \underline{d}, \underline{e}), \beta(\underline{a}, \underline{b}, \underline{c}, \underline{d}, \underline{e})\rangle \in \Xi$.

The inductive proof that $P(m, n)$ holds for any positive value of n is complete. Sublemma 2 has been proved. □

This concludes the proof of Lemma 4.1.7. □

The theorem is an easy consequence of the above lemmas. In virtue of Lemmas 4.1.3 and 4.1.5, $Z_2(\Phi, \Psi; \Xi)$ is equivalent to $Z_{com}(\Phi, \Psi; \Xi)$. In turn $Z_{2,2}(\Phi, \Psi; \Xi)$ is equivalent to $Z_{4,com}(\Phi, \Psi; \Xi)$ by Lemmas 4.1.4 and 4.1.6. $Z_{com}(\Phi, \Psi; \Xi)$ is equivalent to $Z_{4,com}(\Phi, \Psi; \Xi)$ by Lemma 4.1.7. Thus all the four conditions are equivalent. □

Note. Conditions $Z_2(\Phi, \Psi; \Xi)$ and $Z_{2,2}(\Phi, \Psi; \Xi)$ were essentially formulated by Kearnes and McKenzie (1992) in terms of polynomials of algebras. They proved that these conditions are equivalent for any algebra belonging to a relatively congruence-modular quasivariety. The conditions $Z_{com}(\Phi, \Psi; \Xi)$ and $Z_{4,com}(\Phi, \Psi; \Xi)$ seem to have not been considered in the literature. □

Corollary 4.1.8. *Let* **Q** *be a quasivariety with the relative shifting property. Then for any algebra* **A** $\in$ **Q** *and any* **Q**-*congruences* Φ, Ψ *of* **A**,

$$[\Phi, \Psi] = \bigcap\{\Xi \in Con_{\mathbf{Q}}(\mathbf{A}) : Z(\Phi, \Psi; \Xi)\},$$

where Z is an arbitrary but fixed centralization relation from the above list of four centralization relations. □

The following observation is useful:

Corollary 4.1.9. *Let* **Q** *be a quasivariety with the relative shifting property. Then for any algebra* **A** $\in$ **Q** *and any* **Q**-*congruences* Φ, Ψ *of* **A**,

$$[\Phi, \Psi] = \Theta_{\mathbf{Q}}(\{\langle \alpha(a, b, c, d, \underline{e}), \beta(a, b, c, d, \underline{e})\rangle : \alpha(x, y, z, w, \underline{u}) \approx \beta(x, y, z, w, \underline{u})$$

is a quaternary commutator equation **Q**, $a \equiv b\ (\Phi),\ c \equiv d\ (\Psi),$ *and* $\underline{e} \in A^{<\omega}\})$.

Proof. Take the centralization relation $Z_{4,com}$ and apply the above corollary. □

On the right-hand side of the above corollary one may even take a subset of the set of all quaternary commutator equations in x, y and z, w. E.g. one may take the quaternary equations belonging to any set Δ such that $\mathbf{Q}^{\models}(\Delta) = \mathbf{Q}^{\models}(x \approx y) \cap \mathbf{Q}^{\models}(z \approx w)$ (see Theorem 4.3.9) and even smaller sets (Theorem 4.1.11 below).

Theorem 4.1.2 implies:

Corollary 4.1.10. *Let* **Q** *be a quasivariety with the relative shifting property. Then for any algebra* **A** $\in$ **Q** *and all* $\Phi, \Psi \in Con_{\mathbf{Q}}(\mathbf{A})$*:*

$$[\Phi, \Psi]^{\mathbf{A}} = \sup\{[\Theta_{\mathbf{Q}}(a, b), \Theta_{\mathbf{Q}}(c, d)]^{\mathbf{A}} : a \equiv b\ (\Phi),\ c \equiv d\ (\Psi)\}.$$

(The supremum on the right-hand side of the above formula is taken in the lattice $\boldsymbol{Con}_{\mathbf{Q}}(\mathbf{A})$.)

Proof. Let Δ be the set of all quaternary commutator equations for $\mathbf{Q}$. Suppose $\mathbf{A} \in \mathbf{Q}$ and $\Phi, \Psi \in Con_{\mathbf{Q}}(\mathbf{A})$. It is easy to see that

$$\Theta_{\mathbf{Q}}(\bigcup\{(\forall \underline{e})\Delta(a, b, c, d, \underline{e}) : a \equiv b\ (\Phi),\ c \equiv d\ (\Psi)\}).$$

is the least $\mathbf{Q}$-congruence Ξ of $\mathbf{A}$ for which $Z_{4,\mathrm{com}}(\Phi, \Psi; \Xi)$ holds. Hence

$$[\Phi, \Psi]^{\mathbf{A}} = \Theta_{\mathbf{Q}}(\bigcup\{(\forall \underline{e})\Delta(a, b, c, d, \underline{e}) : a \equiv b\ (\Phi),\ c \equiv d\ (\Psi)\})$$

by Corollary 4.1.8.

Let $a, b, c, d \in A$. By the above remark, the least $\mathbf{Q}$-congruence Ξ of $\mathbf{A}$ for which $Z_{4,\mathrm{com}}(\Theta_{\mathbf{Q}}(a, b), \Theta_{\mathbf{Q}}(c, d); \Xi)$ holds coincides with

$$\Theta_{\mathbf{Q}}(\bigcup\{(\forall \underline{e})\Delta(a', b', c', d', \underline{e}) : a' \equiv b'\ (\mathrm{mod}\ \Theta_{\mathbf{Q}}(a, b)),\ c' \equiv d'\ (\mathrm{mod}\ \Theta_{\mathbf{Q}}(c, d))\}).$$

But $[\Theta_{\mathbf{Q}}(a, b), \Theta_{\mathbf{Q}}(c, d)]^{\mathbf{A}}$ is the least $\mathbf{Q}$-congruence Ξ of $\mathbf{A}$ for which $Z_{4,\mathrm{com}}$ $(\Theta_{\mathbf{Q}}(a, b), \Theta_{\mathbf{Q}}(c, d); \Xi)$ holds, again by Corollary 4.1.8. It follows that

$$\Theta_{\mathbf{Q}}(\bigcup\{(\forall \underline{e})\Delta(a, b, c, d, \underline{e}) : a \equiv b\ (\Phi),\ c \equiv d\ (\Psi)\}) =$$

$$\sup\{\Theta_{\mathbf{Q}}(\bigcup\{(\forall \underline{e}))\Delta(a', b', c', d', \underline{e}) : a' \equiv b'\ (\Theta_{\mathbf{Q}}(a, b)),\ c' \equiv d'\ (\Theta_{\mathbf{Q}}(c, d))\}) :$$
$$a \equiv b\ (\Phi),\ c \equiv d\ (\Psi)\}) =$$
$$\sup\{[\Theta_{\mathbf{Q}}(a, b), \Theta_{\mathbf{Q}}(c, d)]^{\mathbf{A}} : a \equiv b\ (\Phi),\ c \equiv d\ (\Psi)\}.$$

□

Let $\mathbf{Q}$ be a quasivariety with the relative shifting property. Let $x \approx y \Rightarrow_D z \approx w$ be a Day implication system for $\mathbf{Q}$. By Theorem 3.6.2, $\mathbf{Q}$ has the relative cube property. Let $\Sigma_c(x_1, x_2, x_3, x_4, y_1, y_2, y_3, y_4)$ be the set of equations in 8 variables supplied by Theorem 3.6.1 and let $(\alpha)^*$, $(\beta)^*$, and $(\delta)^*$ be the equations and the rule of inference of $\mathbf{Q}^{\models}$ provided by that theorem.

For each equation $p \approx q \in \Sigma_c$, for any m-tuple $\underline{t} = t_1, \dots, t_m$ of terms, and any two terms $f(x, y, \underline{u})$, $g(x, y, \underline{u})$, we define the following terms:

$$p' := p(f(x, z, \underline{t}), f(x, w, \underline{t}), f(y, z, \underline{t}), f(y, w, \underline{t}), g(x, z, \underline{t}), g(x, w, \underline{t}), g(y, z, \underline{t}), g(y, w, \underline{t})),$$
$$q' := q(f(x, z, \underline{t}), f(x, w, \underline{t}), f(y, z, \underline{t}), f(y, w, \underline{t}), g(x, z, \underline{t}), g(x, w, \underline{t}), g(y, z, \underline{t}), g(y, w, \underline{t}))$$

(see the proof of Lemma 4.1.5).

Let $\Delta_c(x, y, z, w, \underline{u})$ be the set of all equations of the form $p' \approx q'$ with p' and q' defined as above, f and g ranging over $(m + 2)$-ary terms (for all m), and $\underline{t}$ ranging over strings of terms. The set Δ_c is infinite.

In a more compact form, Δ_c is the union of the following finite sets of equations

$$\Sigma_c(f(x, z, \underline{t}), f(x, w, \underline{t}), f(y, z, \underline{t}), f(y, w, \underline{t}), g(x, z, \underline{t}), g(x, w, \underline{t}), g(y, z, \underline{t}), g(y, w, \underline{t}))$$

obtained from the equations of $\Sigma_c(x_1, x_2, x_3, x_4, y_1, y_2, y_3, y_4)$ by the uniform substitution of the terms $f(x, z, \underline{t})$, $f(x, w, \underline{t})$, $f(y, z, \underline{t})$, $f(y, w, \underline{t})$, $g(x, z, \underline{t})$, $g(x, w, \underline{t})$, $g(y, z, \underline{t})$, $g(y, w, \underline{t})$ for the variables $x_1, x_2, x_3, x_4, y_1, y_2, y_3, y_4$, respectively.

Applying the notation adopted in the proof of Theorem 3.6.2 we may write:

$$\Delta_c(x, y, z, w, u) :=$$

$$\bigcup\{(f(x, z, \underline{t}) \approx g(x, z, \underline{t}) \Rightarrow_D f(x, w, \underline{t}) \approx g(x, w, \underline{t}))$$

$$\Rightarrow_D (f(y, z, \underline{t}) \approx g(y, z, \underline{t}) \Rightarrow_D f(y, w, \underline{t}) \approx g(y, w, \underline{t})) : f \text{ and } g \text{ range}$$

$$\text{over } (m + 2)\text{-ary terms (for all } m) \text{ and } \underline{t} \text{ ranges over strings of terms}\}.$$

Note. The set $\Delta_c(x, y, z, w, \underline{u})$ has the following absolute invariance property: it is the *same* in any quasivariety $\mathbf{Q}$ for which $x \approx y \Rightarrow_D z \approx w$ is a Day implication system. This fact directly follows from the definitions of $\Sigma_c(x_1, x_2, x_3, x_4, y_1, y_2, y_3, y_4)$ and Δ_c. □

Theorem 4.1.11. *Let* $\mathbf{Q}$ *be a quasivariety with the relative shifting property. Let* $x \approx y \Rightarrow_D z \approx w$ *be a Day implication system for* $\mathbf{Q}$. *Define the set of equations* Δ_c *as above.* Δ_c *is a set of quaternary commutator equations for* $\mathbf{Q}$ *such that for every algebra* $\mathbf{A} \in \mathbf{Q}$ *and any* $\Phi, \Psi \in Con_{\mathbf{Q}}(\mathbf{A})$:

$$[\Phi, \Psi]^A = \Theta_{\mathbf{Q}}(\bigcup\{(\forall \underline{e})\Delta_c(a, b, c, d, \underline{e}) : a \equiv b\ (\Phi),\ c \equiv d\ (\Psi)\}).$$

Δ_c is called the set of *quaternary commutator equations for* $\mathbf{Q}$ *supplied by the (relative) cube property and the centralization relation in the sense of the classical two-binary term condition.*

Proof. The fact that Δ_c is indeed a set of quaternary commutator equations for $\mathbf{Q}$ follows from the claim being a part of the proof of Lemma 4.1.5.

To prove the above equality, we shall make use of the centralization relations $Z_{4,\mathrm{com}}$ and $Z_{2,2}$ (defined in the algebras of $\mathbf{Q}$). Suppose $\mathbf{A} \in \mathbf{Q}$ and $\Phi, \Psi \in Con_{\mathbf{Q}}(\mathbf{A})$. We let Ξ_0 denote the congruence $\Theta_{\mathbf{Q}}(\bigcup\{(\forall \underline{e})\ \Delta_c(a, b, c, d, \underline{e}) : a \equiv b\ (\Phi), c \equiv d\ (\Psi)\})$.

Since $\Delta_c(x, y, z, w, \underline{u})$ is a set of quaternary commutator equations for $\mathbf{Q}$, it immediately follows that Ξ_0 is included in any $\mathbf{Q}$-congruence Ξ of $\mathbf{A}$ such that $Z_{4,\mathrm{com}}(\Phi, \Psi; \Xi)$ (see the definition of $Z_{4,\mathrm{com}}$). It follows, by Theorem 4.1.2, that Ξ_0 is included in any $\mathbf{Q}$-congruence Ξ of $\mathbf{A}$ such that $Z_{2,2}(\Phi, \Psi; \Xi)$. To prove the theorem, it therefore suffices to show that

$$Z_{2,2}(\Phi, \Psi; \Xi_0). \tag{1}$$

To show (1), assume $f(x, y, \underline{u})$, $g(x, y, \underline{u})$ are arbitrary terms, where $\underline{u} = u_1, \ldots, u_m$, $a \equiv b\ (\Phi)$, $c \equiv d\ (\Psi)$ and let $\underline{e} = e_1, \ldots, e_m$ be a sequence of terms. Suppose that

$$\text{the pairs } \langle f(a,c,\underline{e}), g(a,c,\underline{e})\rangle,\ \langle f(a,d,\underline{e}), g(a,d,\underline{e})\rangle, \qquad (2)$$
$$\langle f(b,c,\underline{e}), g(b,c,\underline{e})\rangle \text{ belong to } \Xi_0.$$

We must show that $\langle f(b,d,\underline{e}), g(b,d,\underline{e})\rangle \in \Xi_0$.
But by the definitions of Δ_c and Ξ_0 we have that

$$\Sigma_c(f(a,c,\underline{e}), f(a,d,\underline{e}), f(b,c,\underline{e}), f(b,d,\underline{e}),$$
$$g(a,c,\underline{e}), g(a,d,\underline{e}), g(b,c,\underline{e}), g(b,d,\underline{e})) \subseteq \Xi_0. \qquad (3)$$

(2) and (3) imply

$$\Sigma_c(f(a,c,\underline{e}), f(a,d,\underline{e}), f(b,c,\underline{e}), f(b,d,\underline{e}),$$
$$f(a,c,\underline{e}), f(a,d,\underline{e}), f(b,c,\underline{e}), g(b,d,\underline{e})) \subseteq \Xi_0. \qquad (4)$$

But by condition $(\delta)^*$ (see Theorem 3.6.2) we have that

$$\langle f(b,d,\underline{e}), g(b,d,\underline{e})\rangle \in \Theta_{\mathbf{Q}}(\Sigma_c(f(a,c,\underline{e}), f(a,d,\underline{e}), f(b,c,\underline{e}), f(b,d,\underline{e}),$$
$$f(a,c,\underline{e}), f(a,d,\underline{e}), f(b,c,\underline{e}), g(b,d,\underline{e}))) \qquad (5)$$

(4) and (5) imply that

$$\langle f(b,d,\underline{e}), g(b,d,\underline{e}) \in \Xi_0.$$

So (1) holds. The theorem has been proved. □

It follows from Theorem 4.1.2 that the definition of the equationally defined commutator, adopted in this paper, is equivalent to the definition provided by Kearnes and McKenzie (1992) for relatively congruence-modular quasivarieties. Theorem 4.1.2 is a main tool which, in the presence of the facts established by Kearnes and McKenzie (1992), enables us to derive the crucial Theorem A announced in the beginning of this section:

Theorem 4.1.12. (This is Theorem A repeated.) *Let* $\mathbf{Q}$ *be a relatively congruence-modular quasivariety. The equationally defined commutator for* $\mathbf{Q}$ *coincides with the commutator for* $\mathbf{Q}$ *in the sense of Kearnes and McKenzie.*

Proof. Let $\mathbf{Q}$ be an RCM quasivariety of algebras. Let $\boldsymbol{A} \in \mathbf{Q}$. Then, for any $\mathbf{Q}$-congruences Φ, Ψ of $\boldsymbol{A}$, the commutator of Φ and Ψ in the sense of Kearnes and McKenzie coincides with the $\mathbf{Q}$-congruence $\bigcap\{\Xi \in Con_{\mathbf{Q}}(\boldsymbol{A}) : Z_{2,2}(\Phi, \Psi; \Xi)\}$ (Theorem 2.13(3) of Kearnes and McKenzie 1992). But $\mathbf{Q}$ has the relative shifting property. It follows from Theorem 4.1.2 and Proposition 4.1.1 that

$$\bigcap\{\Xi \in Con_{\mathbf{Q}}(\boldsymbol{A}) : Z_{2,2}(\Phi, \Psi; \Xi)\} = \bigcap\{\Xi \in Con_{\mathbf{Q}}(\boldsymbol{A}) : Z_{\mathrm{com}}(\Phi, \Psi; \Xi)\} = [\Phi, \Psi]^{\boldsymbol{A}}.$$

From the last two equalities the thesis follows. □

4.2 Centralization Relations in the Sense of One-Term Conditions

4.2.1 *Three More Centralization Relations*

We shall examine three more centralization relations. Let $\boldsymbol{A}$ be an arbitrary algebra and let Φ, Ψ, Ξ be congruences of $\boldsymbol{A}$. We write:

$$Z_1(\Phi, \Psi; \Xi)$$

(Φ *centralizes* Ψ modulo Ξ in the sense of the classical *one-term* condition). By definition,

$Z_1(\Phi, \Psi; \Xi)$ $\Leftrightarrow_{df}$ for any positive integers m, n, for any term $t(\underline{x}, \underline{y}, \underline{u})$ with $\underline{x} = x_1, \ldots, x_m$, $\underline{y} = y_1, \ldots, y_n$, for any two m-tuples $\underline{a}$, $\underline{b}$ of elements of A and any two n-tuples $\underline{c}$, $\underline{d}$ of elements of A such that $\underline{a} \equiv \underline{b} \pmod{\Phi}$ and $\underline{c} \equiv \underline{d} \pmod{\Psi}$,

$$t(\underline{a}, \underline{c}, \underline{e}) \equiv t(\underline{a}, \underline{d}, \underline{e}) \pmod{\Xi} \quad \text{implies} \quad t(\underline{b}, \underline{c}, \underline{e}) \equiv t(\underline{b}, \underline{d}, \underline{e}) \pmod{\Xi},$$

for any sequence $\underline{e}$ of elements of A of the length of $\underline{u}$.

We also define

$$Z_{0.5}(\Phi, \Psi; \Xi),$$

a relaxation of the above condition $Z_1(\Phi, \Psi; \Xi)$. We put:

$Z_{0.5}(\Phi, \Psi; \Xi)$ $\Leftrightarrow_{df}$ for any positive integer n, for every term $t(x, \underline{y}, \underline{u})$ with $\underline{y} = y_1, \ldots, y_n$, for every pair a, b and any two n-tuples $\underline{c}, \underline{d}$ of elements of A such that $a \equiv b \pmod{\Phi}$ and $\underline{c} \equiv \underline{d} \pmod{\Psi}$,

$$t(a, \underline{c}, \underline{e}) \equiv t(a, \underline{d}, \underline{e}) \pmod{\Xi} \quad \text{implies} \quad t(b, \underline{c}, \underline{e}) \equiv t(b, \underline{d}, \underline{e}) \pmod{\Xi},$$

for any sequence $\underline{e}$ of elements of A of the length of $\underline{u}$.

While the centralization relation $Z_{0.5}$ only plays an auxiliary role in our considerations, the subsequent centralization relation Z_0 we define plays a significant role in the theory of congruence-modular varieties (see, e.g., McKenzie 1987). $Z_0(\Phi, \Psi; \Xi)$ is a relaxation of the above condition $Z_{0.5}(\Phi, \Psi; \Xi)$. We put:

$Z_0(\Phi, \Psi; \Xi)$ $\Leftrightarrow_{df}$ for every term $t(x, y, \underline{u})$, for every quadruple a, b, c, d of elements of A such that $a \equiv b \pmod{\Phi}$ and $c \equiv d \pmod{\Psi}$, and for any sequence $\underline{e}$ of elements of A of the length of $\underline{u}$,

$$t(a, c, \underline{e}) \equiv t(a, d, \underline{e}) \pmod{\Xi} \quad \text{implies} \quad t(b, c, \underline{e}) \equiv t(b, d, \underline{e}) \pmod{\Xi}.$$

Trivially, $Z_1(\Phi, \Psi; \Xi)$ implies $Z_{0.5}(\Phi, \Psi; \Xi)$ and $Z_{0.5}(\Phi, \Psi; \Xi)$ implies $Z_0(\Phi, \Psi; \Xi)$.

The next theorem establishes the equivalence of the above three centralization relations for a wide class of quasivarieties.

Theorem 4.2.1. *Let* $\mathbf{Q}$ *be a quasivariety with the relative shifting property. Then, for every algebra* $\mathbf{A} \in \mathbf{Q}$ *and any congruences* $\Phi, \Psi, \Xi \in Con_{\mathbf{Q}}(\mathbf{A})$,

(a) $Z_1(\Phi, \Psi; \Xi) \;\Leftrightarrow\; Z_{0.5}(\Phi, \Psi; \Xi) \;\Leftrightarrow\; Z_0(\Phi, \Psi; \Xi)$.

(b) $Z_{com}(\Phi, \Psi; \Xi)$ *implies* $Z_1(\Phi, \Psi; \Xi)$, *i.e., the fact that* Φ *centralizes* Ψ *modulo* Ξ *in the sense of arbitrary commutator equations implies that* Φ *centralizes* Ψ *modulo* Ξ *in the sense of the one-term condition.*

Proof. Suppose $\mathbf{A} \in \mathbf{Q}$ and $\Phi, \Psi, \Xi \in Con_{\mathbf{Q}}(\mathbf{A})$. To show (a), we first prove the following lemma.

Lemma 1. $Z_1(\Phi, \Psi; \Xi) \;\Leftrightarrow\; Z_{0.5}(\Phi, \Psi; \Xi)$.

Proof (of the lemma). Evidently $Z_1(\Phi, \Psi; \Xi)$ implies $Z_{0.5}(\Phi, \Psi; \Xi)$. The proof the reverse implication is similar to the proof of Lemma 4.1.6. We apply Theorem 3.5.1. Assuming $Z_{0.5}(\Phi, \Psi; \Xi)$, we prove the following sentence $(\forall m)\, T(m)$, where $T(m)$ is formulated in terms of polynomials over $\mathbf{A}$:

$T(m)$ For any polynomial $t(\underline{x}, \underline{y})$ over $\mathbf{A}$ with $|\underline{x}| = m$, and for any sequences $\underline{a}$, $\underline{b}$, $\underline{c}$, $\underline{d}$ elements of A such that $|\underline{a}| = |\underline{b}| = m$, $|\underline{c}| = |\underline{d}| = |\underline{y}|$ and $\underline{a} \equiv \underline{b} \pmod{\Phi}$ and $\underline{c} \equiv \underline{d} \pmod{\Psi}$, the condition $t(\underline{a}, \underline{c}) \equiv t(\underline{a}, \underline{d}) \pmod{\Xi}$ implies $t(\underline{b}, \underline{c}) \equiv t(\underline{b}, \underline{d}) \pmod{\Xi}$.

(a) will be proved once we show $T(m)$ for all possible values of m. $T(m)$ is proved by induction on m.

Sublemma. $T(m)$, *for all* $m \geqslant 1$.

Proof (of the sublemma). The case $m = 1$ is covered by the assumption $Z_{0.5}(\Phi, \Psi; \Xi)$. Now suppose the sublemma is true for a certain m. Let $\underline{a} = a_1, \dots, a_m, a_{m+1}$ and $\underline{b} = b_1, \dots, b_m, b_{m+1}$ be $(m+1)$-tuples, $\underline{c}$ and $\underline{d}$ arbitrary sequences of elements of the same length such that $\underline{a} \equiv \underline{b} \pmod{\Phi}$ and $\underline{c} \equiv \underline{d} \pmod{\Psi}$, and let $t(\underline{x}, \underline{y})$ be a polynomial over $\mathbf{A}$ with $\underline{x} = x_1, \dots, x_m, x_{m+1}$ and $|\underline{y}| = |\underline{c}| = |\underline{d}|$. Assume that $t(\underline{a}, \underline{c}) \equiv t(\underline{a}, \underline{d}) \pmod{\Xi}$. We need to show that $t(\underline{b}, \underline{c}) \equiv t(\underline{b}, \underline{d}) \pmod{\Xi}$. We prove it in two stages.

Put $\underline{x}_0 := x_1, \dots, x_m$ and $\underline{a}_0 := a_1, \dots, a_m$, $\underline{b}_0 := b_1, \dots, b_m$. For each equation $p(x, y, z, w) \approx q(x, y, z, w)$ belonging to the Day implication $\Rightarrow_D$ for the consequence $\mathbf{Q}^{\models}$, we define the following two polynomials over $\mathbf{A}$:

$$p'(\underline{x}_0, \underline{y}) := p(t(\underline{a}, \underline{y}), t(\underline{a}, \underline{d}), t(\underline{x}_0 a_{m+1}, \underline{y}), t(\underline{x}_0 a_{m+1}, \underline{d})),$$

$$q'(\underline{x}_0, \underline{y}) := q(t(\underline{a}, \underline{y}), t(\underline{a}, \underline{d}), t(\underline{x}_0 a_{m+1}, \underline{y}), t(\underline{x}_0 a_{m+1}, \underline{d})).$$

Claim 1. *Let* $p \approx q$ *be in* $\Rightarrow_D$. *Then* $p'(\underline{a}_0, \underline{c}) \equiv p'(\underline{a}_0, \underline{d}) \pmod{\Xi}$ *and* $q'(\underline{a}_0, \underline{c}) \equiv q'(\underline{a}_0, \underline{d}) \pmod{\Xi}$.

Proof (of the claim). We have:

$$p'(\underline{a}_0, \underline{c}) = p(t(\underline{a}, \underline{c}), t(\underline{a}, \underline{d}), t(\underline{a}_0 a_{m+1}, \underline{c}), t(\underline{a}_0 a_{m+1}, \underline{d})) =$$
$$p(t(\underline{a}, \underline{c}), t(\underline{a}, \underline{d}), t(\underline{a}, \underline{c}), t(\underline{a}, \underline{d})) \equiv_{\Xi} \text{ (by the assumption)}$$
$$p(t(\underline{a}, \underline{d}), t(\underline{a}, \underline{d}), t(\underline{a}, \underline{d}), t(\underline{a}, \underline{d})) = p'(\underline{a}_0, \underline{d}).$$

The second condition is similarly proved. □

It follows from the claim and the inductive hypothesis that

$$p'(\underline{b}_0, \underline{c}) \equiv p'(\underline{b}_0, \underline{d}) \pmod{\Xi} \quad \text{and} \quad q'(\underline{b}_0, \underline{c}) \equiv q'(\underline{b}_0, \underline{d}) \pmod{\Xi},$$

for any $p \approx q$ in $\Rightarrow_D$, which means that

$$p(t(\underline{a}, \underline{c}), t(\underline{a}, \underline{d}), t(\underline{b}_0 a_{m+1}, \underline{c}), t(\underline{a}_0 a_{m+1}, \underline{d})) \equiv_{\Xi}$$
$$p(t(\underline{a}, \underline{d}), t(\underline{a}, \underline{d}), t(\underline{b}_0 a_{m+1}, \underline{d}), t(\underline{b}_0 a_{m+1}, \underline{d})) \quad (1)$$

and

$$q(t(\underline{a}, \underline{c}), t(\underline{a}, \underline{d}), t(\underline{b}_0 a_{m+1}, \underline{c}), t(\underline{a}_0 a_{m+1}, \underline{d})) \equiv_{\Xi}$$
$$q(t(\underline{a}, \underline{d}), t(\underline{a}, \underline{d}), t(\underline{b}_0 a_{m+1}, \underline{d}), t(\underline{b}_0 a_{m+1}, \underline{d})) \quad (2)$$

for all $p \approx q$ in $\Rightarrow_D$. But evidently, by (iD3), the elements standing on the right-hand sides of (1) and (2) are identical, i.e.,

$$p(t(\underline{a}, \underline{d}), t(\underline{a}, \underline{d}), t(\underline{b}_0 a_{m+1}, \underline{d}), t(\underline{b}_0 a_{m+1}, \underline{d})) =$$
$$q(t(\underline{a}, \underline{d}), t(\underline{a}, \underline{d}), t(\underline{b}_0 a_{m+1}, \underline{d}), t(\underline{b}_0 a_{m+1}, \underline{d})).$$

Consequently, the elements standing on the left-hand sides of (1) and (2) must be Ξ-congruent, i.e.,

$$p(t(\underline{a}, \underline{c}), t(\underline{a}, \underline{d}), t(\underline{b}_0 a_{m+1}, \underline{c}), t(\underline{b}_0 a_{m+1}, \underline{d})) \equiv_{\Xi}$$
$$q(t(\underline{a}, \underline{c}), t(\underline{a}, \underline{d}), t(\underline{b}_0 a_{m+1}, \underline{c}), t(\underline{b}_0 a_{m+1}, \underline{d}))$$

for all $p \approx q$ in $\Rightarrow_D$. This shows that $\langle t(\underline{a}, \underline{c}), t(\underline{a}, \underline{d})\rangle \Rightarrow_D \langle t(\underline{b}_0 a_{m+1}, \underline{c}), t(\underline{b}_0 a_{m+1}, \underline{d})\rangle \subseteq \Xi$. Since, by the assumption, $t(\underline{a}, \underline{c}) \equiv_{\Xi} t(\underline{a}, \underline{d})$, we thus obtain that

$$t(\underline{b}_0 a_{m+1}, \underline{c}) \equiv_{\Xi} t(\underline{b}_0 a_{m+1}, \underline{d}), \quad (3)$$

by (iD1).

In the second stage, for each equation $p(x,y,z,w) \approx q(x,y,z,w)$ belonging to $\Rightarrow_D$, we define the following two polynomials:

$$p'(x,\underline{y}) := p(t(\underline{a},\underline{y}), t(\underline{a},\underline{d}), t(\underline{b}_0 x,\underline{y}), t(\underline{b}_0 x,\underline{d})),$$

$$q'(x,\underline{y}) := q(t(\underline{a},\underline{y}), t(\underline{a},\underline{d}), t(\underline{b}_0 x,\underline{y}), t(\underline{b}_0 x,\underline{d})).$$

Claim 2. *Let $p \approx q$ be in $\Rightarrow_D$. Then*

$$p'(a_{m+1},\underline{c}) \equiv_\Xi p'(a_{m+1},\underline{d}) \quad \textit{and} \quad q'(a_{m+1},\underline{c}) \equiv_\Xi q'(a_{m+1},\underline{d}).$$

Proof (of the claim). By (3) and the assumption we have:

$$p'(a_{m+1},\underline{c}) = p(t(\underline{a},\underline{c}), t(\underline{a},\underline{d}), t(\underline{b}_0 a_{m+1},\underline{c}), t(\underline{b}_0 a_{m+1},\underline{d})) \equiv_\Xi$$
$$p(t(\underline{a},\underline{d}), t(\underline{a},\underline{d}), t(\underline{b}_0 a_{m+1},\underline{d}), t(\underline{b}_0 a_{m+1},\underline{d})) = p'(a_{m+1},\underline{d}).$$

The second condition is similarly proved. □

It follows from the claim and the inductive hypothesis that

$$p'(b_{m+1},\underline{c}) \equiv_\Xi p'(b_{m+1},\underline{d}) \quad \text{and} \quad q'(b_{m+1},\underline{c}) \equiv_\Xi q'(b_{m+1},\underline{d}),$$

for all $p \approx q$ in $\Rightarrow_D$, which means that

$$p(t(\underline{a},\underline{c}), t(\underline{a},\underline{d}), t(\underline{b},\underline{c}), t(\underline{b},\underline{d})) \equiv_\Xi p(t(\underline{a},\underline{d}), t(\underline{a},\underline{d}), t(\underline{b},\underline{d}), t(\underline{b},\underline{d})) \tag{4}$$

and

$$q(t(\underline{a},\underline{c}), t(\underline{a},\underline{d}), t(\underline{b},\underline{c}), t(\underline{b},\underline{d})) \equiv_\Xi q(t(\underline{a},\underline{d}), t(\underline{a},\underline{d}), t(\underline{b},\underline{d}), t(\underline{b},\underline{d})), \tag{5}$$

for all $p \approx q$ in $\Rightarrow_D$. But the elements standing on the right-hand sides of (4) and (5) are equal by (iD3). Hence the elements standing on the left-hand sides of (5) and (5) are Ξ-congruent, i.e.,

$$p(t(\underline{a},\underline{c}), t(\underline{a},\underline{d}), t(\underline{b},\underline{c}), t(\underline{b},\underline{d})) \equiv_\Xi q(t(\underline{a},\underline{c}), t(\underline{a},\underline{d}), t(\underline{b},\underline{c}), t(\underline{b},\underline{d})),$$

for all $p \approx q$ in $\Rightarrow_D$. This shows that $\langle t(\underline{a},\underline{c}), t(\underline{a},\underline{d})\rangle \Rightarrow_D \langle t(\underline{b},\underline{c}), t(\underline{b},\underline{d})\rangle \subseteq \Xi$. But $t(\underline{a},\underline{c}) \equiv_\Xi t(\underline{a},\underline{d})$, again by the assumption. Hence, using (iD1), we infer from the above that $t(\underline{b},\underline{c}) \equiv_\Xi t(\underline{b},\underline{d})$. This proves the sublemma and at the same time concludes the proof of Lemma 1. □

Lemma 2. $Z_{0.5}(\Phi,\Psi;\Xi)$ *if and only if* $Z_0(\Phi,\Psi;\Xi)$.

Proof (of the lemma). Evidently $Z_{0.5}(\Phi, \Psi; \Xi)$ implies $Z_0(\Phi, \Psi; \Xi)$. To prove the reverse implication, we assume $Z_0(\Phi, \Psi; \Xi)$ and we prove the following sentence $(\forall m)\, T(m)$, where $T(m)$ is formulated in terms of polynomials over $\mathbf{A}$:

$T(m)$ For any polynomial $t(x, \underline{y})$ over $\mathbf{A}$ with $|\underline{y}| = m$, and for any $a, b \in A$, for any sequences $\underline{c}, \underline{d}$ elements of A such that $|\underline{c}| = |\underline{d}| = |\underline{y}|$, $a \equiv b \pmod{\Phi}$ and $\underline{c} \equiv \underline{d} \pmod{\Psi}$, the condition $t(a, \underline{c}) \equiv t(a, \underline{d}) \pmod{\Xi}$ implies $t(b, \underline{c}) \equiv t(b, \underline{d}) \pmod{\Xi}$.

The lemma will be proved once we show $T(m)$ for all positive values of m. $T(m)$ is proved by induction on m.

Sublemma. *$T(m)$, for all $m \geqslant 1$.*

Proof (of the sublemma). The case $m = 1$ is covered by the assumption $Z_0(\Phi, \Psi; \Xi)$. Now suppose the sublemma is true for a certain m. Let $\underline{c} = c_1, \ldots, c_m, c_{m+1}$ and $\underline{d} = d_1, \ldots, d_m, d_{m+1}$ be $(m+1)$-tuples of elements of A such that $a \equiv b \pmod{\Phi}$ and $\underline{c} \equiv \underline{d} \pmod{\Psi}$, and let $t(x, \underline{y})$ be a polynomial of $\mathbf{A}$ with $\underline{y} = y_1, \ldots, y_m, y_{m+1}$. We assume that $t(a, \underline{c}) \equiv t(a, \underline{d}) \pmod{\Xi}$. We need to show that $t(b, \underline{c}) \equiv t(b, \underline{d}) \pmod{\Xi}$.

We define $\underline{y}_0 := y_1, \ldots, y_m$ and $\underline{c}_0 := c_1, \ldots, c_m$, $\underline{d}_0 := d_1, \ldots, d_m$. For each equation $p(x, y, z, w) \approx q(x, y, z, w)$ belonging to the Day implication $\Rightarrow_D$ for the consequence $\mathbf{Q}^{\models}$, we define the following two polynomials:

$$p(x, \underline{y}_0) := p(t(a, \underline{c}), t(a, \underline{d}), t(x, \underline{y}_0 c_{m+1}), t(x, \underline{d})),$$

$$q(x, \underline{y}_0) := q(t(a, \underline{c}), t(a, \underline{d}), t(x, \underline{y}_0 c_{m+1}), t(x, \underline{d})).$$

Claim A. *Let $p \approx q$ be in $\Rightarrow_D$. Then $p'(a, \underline{c}_0) \equiv q'(a, \underline{c}_0) \pmod{\Xi}$.*

Proof (of the claim). We have:

$$p'(a, \underline{c}_0) = p(t(a, \underline{c}), t(a, \underline{d}), t(a, \underline{c}), t(a, \underline{d}))$$

and

$$q'(a, \underline{c}_0) = q(t(a, \underline{c}), t(a, \underline{d}), t(a, \underline{c}), t(a, \underline{d})).$$

As $\mathbf{Q}$ validates (iD2), we have that

$$p(t(a, \underline{c}), t(a, \underline{d}), t(a, \underline{c}), t(a, \underline{d})) = q(t(a, \underline{c}), t(a, \underline{d}), t(a, \underline{c}), t(a, \underline{d})).$$

Hence $p'(a, \underline{c}_0) = q'(a, \underline{c}_0)$ which trivially implies that $p'(a, \underline{c}_0) \equiv_{\Xi} q'(a, \underline{c}_0)$. □

It follows from the claim and the induction hypothesis that $p'(b, \underline{c}_0) \equiv_{\Xi} q'(b, \underline{c}_0)$ for all $p \approx q$ in $\Rightarrow_D$. This means that

$$p(t(a, \underline{c}), t(a, \underline{d}), t(b, \underline{c}), t(b, \underline{d})) \equiv_{\Xi} q(t(a, \underline{c}), t(a, \underline{d}), t(b, \underline{c}), t(b, \underline{d}))$$

for all $p \approx q$ in $\Rightarrow_D$. But, by assumption, $t(a, \underline{c}) \equiv_\Xi t(a, \underline{d}) \pmod{\Xi}$. Applying condition (iD1) to the above situation, we conclude that $t(b, \underline{c}) \equiv t(b, \underline{d}) \pmod{\Xi}$. So $T(m+1)$ holds. This proves the sublemma and at the same time concludes the proof of Lemma 2. □

The part (a) of Theorem 4.2.1 has been proved.

Proof of (b). Assume $Z_{\mathrm{com}}(\Phi, \Psi; \Xi)$ holds. Assume $\underline{a} \equiv \underline{b} \pmod{\Phi}$, $\underline{c} \equiv \underline{d} \pmod{\Psi}$ for some m-tuples $\underline{a}, \underline{b}$ and some n-tuples $\underline{c}, \underline{d}$ of elements of A, and let $t(\underline{a}, \underline{c}, \underline{e}) \equiv t(\underline{a}, \underline{d}, \underline{e}) \pmod{\Xi}$, where $t(\underline{x}, \underline{y}, \underline{u})$ is a term with $|\underline{x}| = m$ and $|\underline{y}| = n$.

Consider the Day implication $x \approx y \Rightarrow_D z \approx w$ for the consequence $\mathbf{Q}^{\models}$. For each equation $p(x, y, z, w) \approx q(x, y, z, w)$ belonging to $\Rightarrow_D$ we define the following two terms:

$$p'(\underline{x}, \underline{y}, \underline{z}, \underline{w}, \underline{u}) := p(t(\underline{x}, \underline{z}, \underline{u}), t(\underline{x}, \underline{w}, \underline{u}), t(\underline{y}, \underline{z}, \underline{u}), t(\underline{y}, \underline{w}, \underline{u})),$$

$$q'(\underline{x}, \underline{y}, \underline{z}, \underline{w}, \underline{u}) := q(t(\underline{x}, \underline{z}, \underline{u}), t(\underline{x}, \underline{w}, \underline{u}), t(\underline{y}, \underline{z}, \underline{u}), t(\underline{y}, \underline{w}, \underline{u})).$$

Claim B. *$p' \approx q'$ is a commutator equation for $\mathbf{Q}$, for all $p \approx q$ belonging to $\Rightarrow_D$.*

Proof (of the claim). Let $p \approx q$ be in $\Rightarrow_D$. We have:

$$p'(\underline{x}, \underline{x}, \underline{z}, \underline{w}, \underline{u}) = p(t(\underline{x}, \underline{z}, \underline{u}), t(\underline{x}, \underline{w}, \underline{u}), t(\underline{x}, \underline{z}, \underline{u}), t(\underline{x}, \underline{w}, \underline{u})) \quad \text{and}$$
$$q'(\underline{x}, \underline{x}, \underline{z}, \underline{w}, \underline{u}) = q(t(\underline{x}, \underline{z}, \underline{u}), t(\underline{x}, \underline{w}, \underline{u}), t(\underline{x}, \underline{z}, \underline{u}), t(\underline{x}, \underline{w}, \underline{u})).$$

Thus the equation $p'(\underline{x}, \underline{x}, \underline{z}, \underline{w}, \underline{u}) \approx q'(\underline{x}, \underline{x}, \underline{z}, \underline{w}, \underline{u})$ is $\mathbf{Q}$-valid by (iD2) since it is of the form $p(s, t, s, t) \approx q(s, t, s, t)$.

Similarly,

$$p'(\underline{x}, \underline{y}, \underline{z}, \underline{z}, \underline{u}) = p(t(\underline{x}, \underline{z}, \underline{u}), t(\underline{x}, \underline{z}, \underline{u}), t(\underline{y}, \underline{z}, \underline{u}), t(\underline{y}, \underline{z}, \underline{u})) \quad \text{and}$$

$$q'(\underline{x}, \underline{y}, \underline{z}, \underline{z}, \underline{u}) = q(t(\underline{x}, \underline{z}, \underline{u}), t(\underline{x}, \underline{z}, \underline{u}), t(\underline{y}, \underline{z}, \underline{u}), t(\underline{y}, \underline{z}, \underline{u})).$$

The equation $p'(\underline{x}, \underline{y}, \underline{z}, \underline{z}, \underline{u}) \approx q'(\underline{x}, \underline{y}, \underline{z}, \underline{z}, \underline{u})$ is thus $\mathbf{Q}$-valid by (iD3) since it is of the form $p(s, s, t, t) \approx q(s, s, t, t)$. □

The assumptions that $\underline{a} \equiv \underline{b} \pmod{\Phi}$, $\underline{c} \equiv \underline{d} \pmod{\Psi}$, $Z_{\mathrm{com}}(\Phi, \Psi; \Xi)$, and the above claim imply that $p'(\underline{a}, \underline{b}, \underline{c}, \underline{d}, \underline{e}) \equiv q'(\underline{a}, \underline{b}, \underline{c}, \underline{d}, \underline{e}) \pmod{\Xi}$, for all $p \approx q$ belonging to $\Rightarrow_D$ and all sequences $\underline{e}$ of elements of A. Thus

$$p(t(\underline{a}, \underline{c}, \underline{e}), t(\underline{a}, \underline{d}, \underline{e}), t(\underline{b}, \underline{c}, \underline{e}), t(\underline{b}, \underline{d}, \underline{e})) \equiv_\Xi q(t(\underline{a}, \underline{c}, \underline{e}), t(\underline{a}, \underline{d}, \underline{e}), t(\underline{b}, \underline{c}, \underline{e}), t(\underline{b}, \underline{d}, \underline{e})),$$

for all equations $p \approx q$ in $\Rightarrow_D$ and all strings $\underline{e}$. Since $t(\underline{a}, \underline{c}, \underline{e}) \equiv t(\underline{a}, \underline{d}, \underline{e}) \bmod \Xi)$, we thus obtain that $t(\underline{b}, \underline{c}, \underline{e}) \equiv t(\underline{b}, \underline{d}, \underline{e}) \pmod{\Xi}$, by (iD1). So $Z_1(\Phi, \Psi; \Xi)$ holds.

This completes the proof of the theorem. □

For *varieties* of algebras the shifting property is equivalent to congruence-modularity. In this case both Theorems 4.1.2 and 4.2.1 can be considerably strengthened and lifted to the form of equivalence:

Theorem 4.2.2. *Let* $\mathbf{V}$ *be a congruence-modular variety. Then, for any algebra* $\mathbf{A} \in \mathbf{V}$*, all the seven centralization relations defined as above coincide.*

Proof. As $\mathbf{V}$ has the shifting property, in virtue of Theorem 4.1.2 the four centralization relations $Z_{4,\mathrm{com}}$, Z_{com}, $Z_{2,2}$, and Z_2 coincide for $\mathbf{V}$. In view of Theorem 4.2.1, $Z_{\mathrm{com}} \subseteq Z_1 = Z_{0.5} = Z_0$ for any algebra $\mathbf{A} \subseteq \mathbf{V}$. Thus, in order to prove the above theorem, it suffices to show that $Z_0 \subseteq Z_{4,\mathrm{com}}$.

Let $x \approx y \Rightarrow_D z \approx w$ be the Day implication system for the consequence $\mathbf{V}^{\models}$ supplied by Day equations for $\mathbf{V}$. We know that this implication system satisfies a stronger system of equations than (iD3), viz. the equations

$$p(x,x,y,y) \approx y \approx q(x,x,y,y), \tag{iD4}$$

for all $p \approx q$ in $x \approx y \Rightarrow_D z \approx w$ (see Note 3.5.4).

Let $\mathbf{A}$ be an algebra in $\mathbf{V}$, and Φ, Ψ, Ξ congruences of $\mathbf{A}$. Assume that $Z_0(\Phi, \Psi; \Xi)$ holds. This means that

(1) for every term $t(x, y, \underline{u})$, for every quadruple a, b, c, d of elements of A such that $a \equiv b \pmod{\Phi}$ and $c \equiv d \pmod{\Psi}$,

$$t(a, c, \underline{e}) \equiv t(a, d, \underline{e}) \pmod{\Xi} \quad \text{implies} \quad t(b, c, \underline{e}) \equiv t(b, d, \underline{e}) \pmod{\Xi},$$

for any sequence $\underline{e}$ of elements of A of the length of $\underline{u}$.

We claim that $Z_{4,\mathrm{com}}(\Phi, \Psi; \Xi)$ holds in $\mathbf{A}$. Let $p(x, y, z, w, \underline{u}) \approx q(x, y, z, w, \underline{u})$ be a quaternary commutator equation for $\mathbf{V}$ and let a, b, c, d be a quadruple of elements of A such that

$$a \equiv b \pmod{\Phi} \quad \text{and} \quad c \equiv d \pmod{\Psi}.$$

We must show that

$$p(a, b, c, d, \underline{e}) \equiv q(a, b, c, d, \underline{e}) \pmod{\Xi}$$

for any sequence $\underline{e}$ of elements of A of the length of $\underline{u}$.

In the proof that follows we shall be working with polynomials over the algebra $\mathbf{A}$ rather than with terms. This will simplify the proof and, as is easy to check, we will not lose generality of the argument.

For each even i define the following two polynomials over $\mathbf{A}$:

$$t_i(x,y,\underline{u}) := m_i(p(a,x,c,y,\underline{u}),q(a,x,c,y,\underline{u}),p(a,x,c,d,\underline{u}),q(a,x,c,d,\underline{u})),$$

$$t_{i+1}(x,y,\underline{u}) := m_{i+1}(p(a,x,c,y,\underline{u}),q(a,x,c,y,\underline{u}),p(a,x,c,d,\underline{u}),q(a,x,c,d,\underline{u})).$$

Let $\underline{e}$ be a sequence of elements of A of the length of $\underline{u}$.

Claim 1. *$t_i(a,c,\underline{e}) = t_i(a,d,\underline{e})$ and $t_{i+1}(a,c,\underline{e}) = t_{i+1}(a,d,\underline{e})$ for any even i.*

Proof (of the claim). Fix even i. We compute:

$$t_i(a,c,\underline{e}) = m_i(p(a,a,c,c,\underline{e}),q(a,a,c,c,\underline{e}),p(a,a,c,d,\underline{e}),q(a,a,c,d,\underline{e})) =$$

$$m_i(p(a,a,c,c,\underline{e}),p(a,a,c,c,\underline{e}),p(a,a,c,d,\underline{e}),p(a,a,c,d,\underline{e})) = p(a,a,c,d,\underline{e}).$$

(The second equality follows from the fact that $p(x,y,z,w,\underline{u}) \approx q(x,y,z,w,\underline{u})$ is a commutator equation for $\mathbf{V}$ and hence $p(a,a,c,c,\underline{e}) = q(a,a,c,c,\underline{e})$ and $p(a,a,c,d,\underline{e}) = q(a,a,c,d,\underline{e})$. The third equality follows from (iD4).)

Similarly,

$$t_i(a,d,\underline{e}) = m_i(p(a,a,c,d,\underline{e}),q(a,a,c,d,\underline{e}),p(a,a,c,d,\underline{e}),q(a,a,c,d,\underline{e})) =$$

$$m_i(p(a,a,c,d,\underline{e}),p(a,a,c,d,\underline{e}),p(a,a,c,d,\underline{e}),p(a,a,c,d,\underline{e})) = p(a,a,c,d,\underline{e}),$$

by the same argument. Consequently, $t_i(a,c,\underline{e}) = t_i(a,d,\underline{e})$.

The proof of the equality $t_{i+1}(a,c,\underline{e}) = t_{i+1}(a,d,\underline{e})$ is fully analogous. □

Claim 2. *$t_i(b,c,\underline{e}) \equiv t_i(b,d,\underline{e}) \pmod{\Xi}$ and $t_{i+1}(b,c,\underline{e}) \equiv t_{i+1}(b,d,\underline{e}) \pmod{\Xi}$ for any even i.*

Proof (of the claim). It trivially follows from Claim 1 that $t_i(a,c,\underline{e}) \equiv t_i(a,d,\underline{e}) \pmod{\Xi}$ and $t_{i+1}(a,c,\underline{e}) \equiv t_{i+1}(a,d,\underline{e}) \pmod{\Xi}$ for an even i. Then use (1). □

But $t_i(b,c,\underline{e}) \equiv t_i(b,d,\underline{e}) \pmod{\Xi}$ means that

$$m_i(p(a,b,c,c,\underline{e}),q(a,b,c,c,\underline{e}),p(a,b,c,d,\underline{e}),q(a,b,c,d,\underline{e})) \equiv_{\Xi} \tag{2}$$

$$m_i(p(a,b,c,d,\underline{e}),q(a,b,c,d,\underline{e}),p(a,b,c,d,\underline{e}),q(a,b,c,d,\underline{e})) \quad \text{for } i \text{ even}.$$

Similarly and $t_{i+1}(a,c,\underline{e}) \equiv t_{i+1}(a,d,\underline{e}) \pmod{\Xi}$ means that

$$m_{i+1}(p(a,b,c,c,\underline{e}),q(a,b,c,c,\underline{e}),p(a,b,c,d,\underline{e}),q(a,b,c,d,\underline{e})) \equiv_{\Xi} \tag{3}$$

$$m_{i+1}(p(a,b,c,d,\underline{e}),q(a,b,c,d,\underline{e}),p(a,b,c,d,\underline{e}),q(a,b,c,d,\underline{e})) \quad \text{for } i \text{ even}.$$

Moreover

$$m_i(p(a,b,c,d,\underline{e}),q(a,b,c,d,\underline{e}),p(a,b,c,d,\underline{e}),q(a,b,c,d,\underline{e})) = \tag{4}$$

$$m_{i+1}(p(a,b,c,d,\underline{e}),q(a,b,c,d,\underline{e}),p(a,b,c,d,\underline{e}),q(a,b,c,d,\underline{e})) \quad \text{for } i \text{ even},$$

because $x \approx y \Rightarrow_D z\ w$ satisfies (iD2), which means that the equations $m_i(x, y, x, y) \approx m_{i+1}(x, y, x, y)$, i even, are valid in $\mathbf{V}$.

It follows from (2), (3), and (4) that

$$m_i(p(a, b, c, c, \underline{e}), q(a, b, c, c, \underline{e}), p(a, b, c, d, \underline{e}), q(a, b, c, d, \underline{e})) \equiv_{\Xi} \tag{5}$$
$$m_{i+1}(p(a, b, c, c, \underline{e}), q(a, b, c, c, \underline{e}), p(a, b, c, d, \underline{e}), q(a, b, c, d, \underline{e})) \quad \text{for } i \text{ even.}$$

But $p(a, b, c, c, \underline{e}) = q(a, b, c, c, \underline{e})$, because $p \equiv q$ is a commutator equation for $\mathbf{V}$. Hence trivially

$$p(a, b, c, c, \underline{e}) \equiv_{\Xi} q(a, b, c, c, \underline{e}). \tag{6}$$

Applying the detachment property (iD1) of $x \approx y \Rightarrow_D z \approx w$ to (5) and (6) we conclude that $p(a, b, c, d, \underline{e}) \equiv_{\Xi} q(a, b, c, d, \underline{e})$, which is what was to be shown.

This completes the proof of Theorem 4.2.2. □

Corollary 4.2.3. *Let* $\mathbf{V}$ *be a congruence-modular variety. Then for every algebra* $\mathbf{A} \in \mathbf{V}$ *and any congruences* Φ, Ψ *of* $\mathbf{A}$,

$$[\Phi, \Psi]^A = \bigcap\{\Xi \in Con(\mathbf{A}) : Z(\Phi, \Psi; \Xi)\},$$

where Z *is an arbitrary but fixed centralization relation from the above list of seven centralization relations.*

Theorem 4.2.2 and Corollary 4.2.3 imply that the equationally defined commutator for congruence-modular varieties coincides with the "standard" one—see Freese and McKenzie (1987) . But this fact also follows from Theorem 4.2.1 and the results presented in Kearnes and McKenzie (1992).

4.3 Generating Sets of Commutator Equations

We first adopt another notational convention.

Convention 4. The notation introduced in Convention 2 is extended to the term algebra $\boldsymbol{Te}_\tau$. Let $\Delta_0(x, y, z, w, \underline{u})$ be a set of quaternary equations with k parameters $\underline{u}$. For any terms $\alpha, \beta, \gamma, \delta$, the set of equations

$$\bigcup\{\Delta_0(\alpha, \beta, \gamma, \delta, \underline{\sigma}) : \underline{\sigma} \in S^k\}$$

is written as

$$(\forall \underline{u})\Delta_0(\alpha, \beta, \gamma, \delta, \underline{u}).$$ □

The four definitions of centralization relations formulated for **Q**-congruences in Section 4.1 are also applicable, after suitable modifications, to the term algebra $\boldsymbol{Te}_\tau$ and to equational theories of the consequence operation $\mathbf{Q}^{\models}$, where **Q** is an arbitrary quasivariety. For example, let X, Y, Z be theories of $\mathbf{Q}^{\models}$. We write:

$$Z_{4,\mathrm{com}}(X, Y; Z)$$

(X *centralizes* Y modulo Z in the sense of *quaternary commutator equations*). By definition,

$Z_{4,\mathrm{com}}(X, Y; Z)$ $\Leftrightarrow_{df}$ for any quadruple $\alpha, \beta, \gamma, \delta$ of terms, the conditions

$$\alpha \approx \beta \in X \text{ and } \gamma \approx \delta \in Y \text{ imply } p(\alpha, \beta, \gamma, \delta, \underline{t}) \approx q(\alpha, \beta, \gamma, \delta, \underline{t}) \in Z$$

for any quaternary commutator equation $p(x, y, z, w, \underline{u}) \approx q(x, y, z, w, \underline{u})$ for **Q** and for any sequence $\underline{t}$ of terms of the length of $\underline{u}$.

In turn, we write:

$$Z_{2,2}(X, Y; Z)$$

(X *centralizes* Y modulo Z in the sense of the *classical two-binary term* condition). By definition,

$Z_{2,2}(X, Y; Z)$ $\Leftrightarrow_{df}$ for any equations $\alpha \approx \beta \in X$ and $\gamma \approx \delta \in Y$, any two terms $f(x, y, \underline{u})$, $g(x, y, \underline{u})$, and any sequence $\underline{t}$ of terms (of the length of $\underline{u}$),

$$f(\alpha, \gamma, \underline{t}) \approx g(\alpha, \gamma, \underline{t}) \in Z,$$
$$f(\alpha, \delta, \underline{t}) \approx g(\alpha, \delta, \underline{t}) \in Z,$$
$$f(\beta, \gamma, \underline{t}) \approx g(\beta, \gamma, \underline{t}) \in Z$$

imply

$$f(\beta, \delta, \underline{t}) \approx g(\beta, \delta, \underline{t}) \in Z.$$

$Z_{2,2}$ is called the *centralization relation* in the sense of the *classical two-binary term condition*.

Keeping in mind the remaining, original definitions of centralization relations for **Q**-congruences, viz. Z_{com}, and Z_2, one can easily reformulate them in terms of equational theories of $\mathbf{Q}^{\models}$. This is a simple task and the details are omitted.

Similar remarks concern the three additional definitions of centralization relations adopted for congruence-modular *varieties* **V**, viz. $Z_1, Z_{0.5}$, and Z_0. They are also redefined so that they are applicable for equational theories of $\mathbf{V}^{\models}$.

The main theorems of the previous paragraph, which show the equivalence of various centralization relations for congruences, continue to hold for the theories of $\mathbf{Q}^{\models}$ if one adopts the above modified definitions of centralization relations for equational theories. The proofs of these new theorems are straightforward adaptations of the proofs presented above for congruences.

We have the following counterparts of Theorems 4.1.2 and 4.2.2 for equational theories:

Theorem 4.3.1. *Let* **Q** *be a quasivariety with the relative shifting property. Then for any theories* X, Y, Z *of* $\mathbf{Q}^{\models}$*:*

$$Z_{4,\mathrm{com}}(X,Y;Z) \Leftrightarrow Z_{\mathrm{com}}(X,Y;Z) \Leftrightarrow Z_{2,2}(X,Y,Z) \Leftrightarrow Z_2(X,Y;Z). \quad (1)$$

Moreover, the universal closure of (1) (*quantified over all theories* X, Y, Z *of* $\mathbf{Q}^{\models}$) *is equivalent to the universal closure of the formula*

$$Z_{4,\mathrm{com}}(\Phi,\Psi;\Xi) \Leftrightarrow Z_{\mathrm{com}}(\Phi,\Psi;\Xi) \Leftrightarrow Z_{2,2}(\Phi,\Psi;\Xi) \Leftrightarrow Z_2(\Phi,\Psi;\Xi) \quad (2)$$

(*quantified over all algebras* $\mathbf{A} \in \mathbf{Q}$ *and all* **Q***-congruences* Φ, Ψ, Ξ *of* $\mathbf{A}$).

Proof. Given theories X, Y, Z of $\mathbf{Q}^{\models}$, one proves (1) by emulating the proof of Theorem 4.1.2 for the theories of $\mathbf{Q}^{\models}$. In the next step one proves that for any two centralization relations Z_1, Z_2 from the above list of four such relations, the condition

$$\text{For any theories } X, Y, Z \text{ of } \mathbf{Q}^{\models},\ Z_1(X,Y;Z) \Leftrightarrow Z_2(X,Y;Z) \quad \text{(a)}$$

is equivalent to

$$\text{For all algebras } \mathbf{A} \in \mathbf{Q} \text{ and all } \mathbf{Q}\text{-congruences } \Phi, \Psi, \Xi \text{ of } \mathbf{A},$$
$$Z_1(\Phi,\Psi;\Xi) \Leftrightarrow Z_2(\Phi,\Psi;\Xi). \quad \text{(b)}$$

The proof of the equivalence of (a) and (b) makes use of the fact that the mapping $\boldsymbol{\Omega}$ given by $\boldsymbol{\Omega}(\Gamma) := \{\langle[\alpha],[\beta]\rangle : \alpha \approx \beta \in \Gamma\}$ is an isomorphism from the lattice of closed theories of the consequence operation $\mathbf{Q}^{\models}$ onto the lattice $\boldsymbol{Con}_{\mathbf{Q}}(\boldsymbol{F})$ (Proposition 2.4).

Assuming (a), one first proves the equivalence (b) for the **Q**-congruences of the free algebra $\boldsymbol{F}_{\mathbf{Q}}(\omega)$, then one extends (b) onto the **Q**-congruences of arbitrary *countably generated* algebras of **Q**, and, finally, onto the set of all **Q**-congruences of an arbitrary algebra of **Q**. We omit the details.

The proof of the reverse implication (b) $\Rightarrow$ (a) is immediate.

Then (1) and (a) $\Leftrightarrow$ (b) give (2). □

Theorem 4.3.2. *Let* **V** *be a congruence-modular variety. Then in the algebra of terms, all the seven centralization relations for equational theories of* $\mathbf{V}^{\models}$ *defined as above coincide. Moreover, the equivalence of* (1) *and* (2) *continue to hold for the remaining three centralization relations defined for varieties.*

The proof of Theorem 4.3.2 is similar. □

We have the following counterparts of Corollaries 4.1.8 and 4.2.3 for equational theories.

Corollary 4.3.3. *Let* $\mathbf{Q}$ *be a quasivariety with a Day implication system. Then for any theories* $X, Y \in Th(\mathbf{Q}^{\models})$,

$$[X, Y] = \bigcap\{Z \in Th(\mathbf{Q}^{\models}) : R(X, Y; Z)\},$$

where R *is an arbitrary but fixed centralization relation among* $Z_{4,\mathrm{com}}$, Z_{com}, $Z_{2,2}$, Z_2 *for the theories of* $\mathbf{Q}^{\models}$.

If $\mathbf{V}$ *is a congruence-modular variety, then*

$$[X, Y] = \bigcap\{Z \in Th(\mathbf{V}^{\models}) : R(X, Y; Z)\},$$

where R *is any centralization relation from the above list of seven centralization relations.* □

We also have:

Proposition 4.3.4. *Let* $\mathbf{Q}$ *be an arbitrary quasivariety. Let* $\Delta_0(x, y, z, w, \underline{u})$ *be a set of quaternary commutator equations for* $\mathbf{Q}$. *The following conditions are equivalent:*

(i) $[X, Y] = \mathbf{Q}^{\models}(\bigcup\{(\forall \underline{u})\ \Delta_0(\alpha, \beta, \gamma, \delta, \underline{u}) : \alpha \approx \beta \in X,\ \gamma \approx \delta \in Y\})$ *for every pair* X, Y *of theories of* $\mathbf{Q}^{\models}$,

(ii) $[\Phi, \Psi]^{\mathbf{A}} = \Theta^{\mathbf{A}}_{\mathbf{Q}}(\bigcup\{(\forall \underline{e})\ \Delta^{\mathbf{A}}_0(a, b, c, d, \underline{e}) : \langle a, b\rangle \in \Phi, \langle c, d\rangle \in \Psi\})$ *for all algebras* $\mathbf{A} \in \mathbf{Q}$ *and all* $\Phi, \Psi \in Con_{\mathbf{Q}}(\mathbf{A})$.

Proof (an outline). (ii) $\Rightarrow$ (i). Let $\boldsymbol{F}$ be the free algebra $\boldsymbol{F}_{\mathbf{Q}}(\omega)$. (ii) implies that

(iii) $[\Phi, \Psi]^{\boldsymbol{F}} = \Theta^{\boldsymbol{F}}_{\mathbf{Q}}(\bigcup\{(\forall \underline{e}) \Delta^{\boldsymbol{F}}_0(a, b, c, d, \underline{e}) : \langle a, b\rangle \in \Phi,\ \langle c, d\rangle \in \Psi\})$ for all $\Phi, \Psi \in Con_{\mathbf{Q}}(\boldsymbol{F})$.

Then (i) holds, by Proposition 2.4.

(i) $\Rightarrow$ (ii). (i) implies (iii). Then, one extends (iii) onto countably generated algebras, and, finally, onto algebras of arbitrary cardinality. This shows (ii). □

In this section the focus is on some syntactic aspects of the theory of the commutator, namely the axiomatization of the equational theory $\mathbf{Q}^{\models}(x \approx y) \cap \mathbf{Q}^{\models}(z \approx w)$ of all quaternary commutator equations for a quasivariety $\mathbf{Q}$.

Proposition 4.3.5. *Let* $\mathbf{Q}$ *be an arbitrary quasivariety and* $\boldsymbol{F} = \boldsymbol{F}_{\mathbf{Q}}(\omega)$ *the* ω*-generated free algebra in* $\mathbf{Q}$. *For any positive integers* m *and* n, *the equationally defined commutator of the congruences* $\Theta^{\boldsymbol{F}}_{\mathbf{Q}}(\langle [x_1], [y_1]\rangle, \ldots, \langle [x_m], [y_m]\rangle)$ *and* $\Theta^{\boldsymbol{F}}_{\mathbf{Q}}(\langle [z_1], [w_1]\rangle, \ldots, \langle [z_n], [w_n]\rangle)$ *in* $\boldsymbol{F}$ *coincides with their meet.*

Proof. We apply Propositions 2.5 and 3.2.3. It suffices to show that

$$\Theta^{F}_{\mathbf{Q}}(\langle[x_1],[y_1]\rangle,\ldots,\langle[x_m],[y_m]\rangle)\cap\Theta^{F}_{\mathbf{Q}}(\langle[z_1],[w_1]\rangle,\ldots,\langle[z_n],[w_n]\rangle)\subseteq$$
$$[\Theta^{F}_{\mathbf{Q}}(\langle[x_1],[y_1]\rangle,\ldots,\langle[x_m],[y_m]\rangle),\Theta^{F}_{\mathbf{Q}}(\langle[z_1],[w_1]\rangle,\ldots,\langle[z_n],[w_n]\rangle)].$$

We shall make use of the mapping $\boldsymbol{\Omega}$ defined as above on the equational theories of $\mathbf{Q}^{eq\models}$.

Let $\Delta := \mathbf{Q}^{\models}(x_1\approx y_1,\ldots,x_m\approx y_m)\cap\mathbf{Q}^{\models}(z_1\approx w_1,\ldots,z_n\approx w_n)$. Δ is the set of *all* commutator equations of $\mathbf{Q}$ in the variables $\underline{x}=x_1,\ldots,x_m$, $\underline{y}=y_1,\ldots,y_m$ and $\underline{z}=z_1,\ldots,z_n$, $\underline{w}=w_1,\ldots,w_n$. Δ is also a closed theory of $\mathbf{Q}^{\models}$. We have:

$$[\Theta^{F}_{\mathbf{Q}}(\langle[x_1],[y_1]\rangle,\ldots,\langle[x_m],[y_m]\rangle),\Theta^{F}_{\mathbf{Q}}(\langle[z_1],[w_1]\rangle,\ldots,\langle[z_n],[w_n]\rangle)]\supseteq$$
$$\Theta^{F}_{\mathbf{Q}}((\forall\underline{e})\ \Delta^{F}([\underline{x}],[\underline{y}],[\underline{z}],[\underline{w}],\underline{e}))=$$
$$\Theta^{F}_{\mathbf{Q}}(\{\langle[\alpha],[\beta]\rangle:\alpha\approx\beta\in\Delta\})=$$
$$\{\langle[\alpha],[\beta]\rangle:\alpha\approx\beta\in\Delta\}=\boldsymbol{\Omega}(\Delta)=$$
$$\boldsymbol{\Omega}(\mathbf{Q}^{\models}(x_1\approx y_1,\ldots,x_m\approx y_m)\cap\mathbf{Q}^{\models}(z_1\approx w_1,\ldots,z_n\approx w_n))=$$
$$\boldsymbol{\Omega}\mathbf{Q}^{\models}(x_1\approx y_1,\ldots,x_m\approx y_m)\cap\boldsymbol{\Omega}\mathbf{Q}^{\models}(z_1\approx w_1,\ldots,z_n\approx w_n)=$$
$$\Theta^{F}_{\mathbf{Q}}(\langle[x_1],[y_1]\rangle,\ldots,\langle[x_m],[y_m]\rangle)\cap\Theta^{F}_{\mathbf{Q}}(\langle[z_1],[w_1]\rangle,\ldots,\langle[z_n],[w_n]\rangle).$$

The reverse inclusion holds by Theorem 3.1.6.(ii). □

Proposition 4.3.6. *Let* $\mathbf{Q}$ *be an arbitrary quasivariety and* $\boldsymbol{F}=\boldsymbol{F}_{\mathbf{Q}}(\omega)$ *the* ω*-generated free algebra in* $\mathbf{Q}$*. Let m and n be positive integers and let* Δ_0 *be a subset of* $\mathbf{Q}^{\models}(x_1\approx y_1,\ldots,x_m\approx y_m)\cap\mathbf{Q}^{\models}(z_1\approx w_1,\ldots,z_n\approx w_n)$*. The following conditions are equivalent:*

(i) $\mathbf{Q}^{\models}(\Delta_0)=\mathbf{Q}^{\models}(x_1\approx y_1,\ldots,x_m\approx y_m)\cap\mathbf{Q}^{\models}(z_1\approx w_1,\ldots,z_n\approx w_n)$.

$$\Theta^{F}_{\mathbf{Q}}(\{\langle[\alpha],[\beta]\rangle:\alpha\approx\beta\in\Delta_0\})=$$

(ii) $\Theta^{F}_{\mathbf{Q}}(\langle[x_1],[y_1]\rangle,\ldots,\langle[x_m],[y_m]\rangle)\cap\Theta^{F}_{\mathbf{Q}}(\langle[z_1],[w_1]\rangle,\ldots,\langle[z_n],[w_n]\rangle)$.

Proof. This immediately follows from Proposition 2.5. □

Before passing to other results we shall introduce the following definition.

Definition 4.3.7. Let $\mathbf{K}$ be a class of algebras. Any set $\Delta(x,y,z,w,\underline{u})$ of equations such that

$$\mathbf{K}^{\models}(x\approx y)\cap\mathbf{K}^{\models}(z\approx w)=\mathbf{K}^{\models}((\forall\underline{u})\ \Delta(x,y,z,w,\underline{u}))$$

is called a *generating set of quaternary commutator equations for* $\mathbf{K}$. □

Note. The variables x, y, z, w are assumed to be different and $\underline{u} = u_1, u_2, \ldots$ is a possibly infinite string of variables (of length k, where $k \leqslant \omega$). The variables of $\underline{u}$ are called *parametric variables*. (We recall that the equations belonging to $\mathbf{K}^{\models}(x \approx y) \cap \mathbf{K}^{\models}(z \approx w)$ are called *quaternary commutator equations* in the variables x, y, z, w with parameters for the class $\mathbf{K}$.) It is clear that a generating set $\Delta(x, y, z, w, \underline{u})$ exists (as $\Delta(x, y, z, w, \underline{u})$ we may take $\mathbf{K}^{\models}(x \approx y) \cap \mathbf{K}^{\models}(z \approx w)$). □

Let $\mathbf{Q}$ be a quasivariety. As $\mathbf{Q}^{\models}(x \approx y) = \boldsymbol{Va}(\mathbf{Q})^{\models}(x \approx y)$ for any variables x and y, it follows that the classes $\mathbf{Q}$ and $\boldsymbol{Va}(\mathbf{Q})$ have the same quaternary commutator equations in the variables x, y, z, w with parameters. (This also directly follows from the definition of a commutator equation for $\mathbf{Q}$.) As the consequence operations $\mathbf{Q}^{\models}$ and $\boldsymbol{Qv}(\mathbf{Q})^{\models}$ coincide on the equations $x \approx y$ and $z \approx w$ it follows that the classes $\mathbf{Q}$ and $\boldsymbol{Qv}(\mathbf{Q})$ have the same generating sets of quaternary commutator equations. We therefore interchangeably use the terms:

"generating set of quaternary commutator equations for $\mathbf{Q}$",
"generating set of quaternary commutator equations for $\mathbf{Q}^{\models}$",
"generating set of quaternary commutator equations for $\boldsymbol{Va}(\mathbf{Q})$" etc.

In particular, if $\mathbf{Q}$ is a quasivariety, then according to the above definition, any set $\Delta_0(x, y, z, w, \underline{u})$ with the property that

$$\mathbf{Q}^{\models}((\forall \underline{u})\Delta_0(x, y, z, w, \underline{u}) = \mathbf{Q}^{\models}(x \approx y) \cap \mathbf{Q}^{\models}(z \approx w)$$

is also called a *generating set of quaternary commutator equations for the (equationally-defined) commutator of* $\mathbf{Q}$. □

Proposition 4.3.8. *Let $\mathbf{Q}$ be a quasivariety and $\Delta_0(x, y, z, w, \underline{u})$ a generating set of quaternary equations for $\mathbf{Q}$. Let $\mathbf{Q}'$ be a quasivariety such that $\mathbf{Q}' \subseteq \mathbf{Q} \subseteq \boldsymbol{Va}(\mathbf{Q}')$. Then $\Delta_0(x, y, z, w, \underline{u})$ is also a generating set of quaternary equations for $\mathbf{Q}'$, i.e.,*

$$\mathbf{Q}'^{\models}((\forall \underline{u})\ \Delta_0(x, y, z, w, \underline{u})) = \mathbf{Q}'^{\models}(x \approx y) \cap \mathbf{Q}'^{\models}(z \approx w).$$

Proof. The inclusion

$$\Delta_0(x, y, z, w, \underline{u}) \subseteq \mathbf{Q}'^{\models}(x \approx y) \cap \mathbf{Q}'^{\models}(z \approx w)$$

follows from the fact that the equations $\Delta_0(x, x, z, w, \underline{u})$ and $\Delta_0(x, y, z, z, \underline{u})$ are valid in $\mathbf{Q}$ and hence in $\mathbf{Q}'$ (because $\boldsymbol{Va}(\mathbf{Q}) = \boldsymbol{Va}(\mathbf{Q}')$). Consequently,

$$\mathbf{Q}'^{\models}((\forall \underline{u})\ \Delta_0(x, y, z, w, \underline{u})) \subseteq \mathbf{Q}'^{\models}(x \approx y) \cap \mathbf{Q}'^{\models}(z \approx w).$$

As $\mathbf{Q}' \subseteq \mathbf{Q}$ and $\boldsymbol{Va}(\mathbf{Q}) = \boldsymbol{Va}(\mathbf{Q}')$, we also have:

$$\mathbf{Q}'^{\models}(x \approx y) \cap \mathbf{Q}'^{\models}(z \approx w) = \boldsymbol{Va}(\mathbf{Q}')^{\models}(x \approx y) \cap \boldsymbol{Va}(\mathbf{Q}')^{\models}(z \approx w) =$$

$$\boldsymbol{Va}(\mathbf{Q})^{\models}(x \approx y) \cap \boldsymbol{Va}(\mathbf{Q})^{\models}(z \approx w) = \mathbf{Q}^{\models}(x \approx y) \cap \mathbf{Q}^{\models}(z \approx w) =$$

$$\mathbf{Q}^{\models}((\forall \underline{u})\ \Delta_0(x, y, z, w, \underline{u})) \subseteq \mathbf{Q}'^{\models}((\forall \underline{u})\ \Delta_0(x, y, z, w, \underline{u})).$$

Hence $\mathbf{Q}'^{\models}(x \approx y) \cap \mathbf{Q}'^{\models}(z \approx w) = \mathbf{Q}'^{\models}((\forall \underline{u})\ \Delta_0(x, y, z, w, \underline{u}))$. □

Let $\mathbf{K}$ be a class of algebras. Putting $\mathbf{Q} := \boldsymbol{Va}(\mathbf{K})$ and $\mathbf{Q}' := \boldsymbol{Qv}(\mathbf{K})$, we have that $\mathbf{Q}' \subseteq \mathbf{Q} = \boldsymbol{Va}(\mathbf{Q}')$. Applying the above proposition, we get that for any set $\Delta_0(x, y, z, w, \underline{u})$ such that $\mathbf{Q}^{\models}((\forall \underline{u})\ \Delta_0(x, y, z, w, \underline{u})) = \mathbf{Q}^{\models}(x \approx y) \cap \mathbf{Q}^{\models}(z \approx w)$, it is also the case that $\mathbf{Q}'^{\models}((\forall \underline{u})\ \Delta_0(x, y, z, w, \underline{u})) = \mathbf{Q}'^{\models}(x \approx y) \cap \mathbf{Q}'^{\models}(z \approx w)$, i.e..

$$\boldsymbol{Va}(\mathbf{K})^{\models}((\forall \underline{u})\ \Delta_0(x, y, z, w, \underline{u})) = \boldsymbol{Va}(\mathbf{K})^{\models}(x \approx y) \cap \boldsymbol{Va}(\mathbf{K})^{\models}(z \approx w)$$

implies that

$$\boldsymbol{Qv}(\mathbf{K})^{\models}((\forall \underline{u})\ \Delta_0(x, y, z, w, \underline{u})) = \boldsymbol{Qv}(\mathbf{K})^{\models}(x \approx y) \cap \boldsymbol{Qv}(\mathbf{K})^{\models}(z \approx w).$$

As $\boldsymbol{Va}(\mathbf{K})^{\models}(x \approx y) = \boldsymbol{Qv}(\mathbf{K})^{\models}(x \approx y)$ and $\boldsymbol{Va}(\mathbf{K})^{\models}(z \approx w) = \boldsymbol{Qv}(\mathbf{K})^{\models}(z \approx w)$, we thus see that $\boldsymbol{Va}(\mathbf{K})^{\models}((\forall \underline{u})\ \Delta_0(x, y, z, w, \underline{u}))$ is a theory of the consequence $\boldsymbol{Qv}(\mathbf{K})^{\models}$.

The equationally defined commutator is characterized for quasivarieties with the relative shifting property by generating sets in the following way:

Theorem 4.3.9. *Let* $\mathbf{Q}$ *be a quasivariety with the relative shifting property and* $\Delta_0(x, y, z, w, \underline{u})$ *a generating set of quaternary commutator equations for* $\mathbf{Q}$. *Then*

$$[\Phi, \Psi]^{\mathbf{A}} = \Theta^{\mathbf{A}}_{\mathbf{Q}}\Big(\bigcup\{(\forall \underline{e})\ \Delta^{\mathbf{A}}_0(a, b, c, d, \underline{e}) : \langle a, b\rangle \in \Phi, \langle c, d\rangle \in \Psi\}\Big)$$

for all algebras $\mathbf{A} \in \mathbf{Q}$ *and all* $\Phi, \Psi \in Con_{\mathbf{Q}}(\mathbf{A})$.

Proof. In view of Proposition 4.3.4 it suffices to show that

$$[X, Y] = \mathbf{Q}^{\models}\Big(\bigcup\{(\forall \underline{u})\ \Delta_0(\alpha, \beta, \gamma, \delta, \underline{u}) : \alpha \approx \beta \in X, \gamma \approx \delta \in Y\}\Big) \quad (1)$$

for every pair X, Y of theories of $\mathbf{Q}^{\models}$.

Fix theories X and Y of $\mathbf{Q}^{\models}$. Suppose $\alpha \approx \beta \in X$, $\gamma \approx \delta \in Y$. As Δ_0 is a generating set, we have that

$$p(x, y, z, w, \underline{v}) \approx q(x, y, z, w, \underline{v}) \in \mathbf{Q}^{\models}((\forall \underline{u})\ \Delta_0(x, y, z, w, \underline{u})),$$

for every quaternary commutator equation $p(x, y, z, w, \underline{v}) \approx q(x, y, z, w, \underline{v})$ for $\mathbf{Q}$. It follows, by structurality, that

$$p(\alpha, \beta, \gamma, \delta, \underline{t}) \approx q(\alpha, \beta, \gamma, \delta, \underline{t}) \in \mathbf{Q}^{\models}((\forall \underline{u})\ \Delta_0(\alpha, \beta, \gamma, \delta, \underline{u})), \quad (2)$$

for any sequence $\underline{t}$ of terms of the length of $\underline{v}$.

Applying Theorem 4.3.1, Corollary 4.3.3, and the centralization relation $Z_{4,\mathrm{com}}$ for theories, we see that (2) yields

$$[X, Y] =$$
$$\mathbf{Q}^{\models}(\{p(\alpha, \beta, \gamma, \delta, \underline{t}) \approx q(\alpha, \beta, \gamma, \delta, \underline{t}) : \alpha \approx \beta \in X,\ \gamma \approx \delta \in Y,\ p \approx q \text{ is}$$
$$\text{a quaternary commutator equation for } \mathbf{Q} \text{ and } \underline{t} \text{ a sequence of terms}\}) \subseteq$$
$$\mathbf{Q}^{\models}(\bigcup\{(\forall \underline{u})\ \Delta_0(\alpha, \beta, \gamma, \delta, \underline{u})) : \alpha \approx \beta \in X,\ \gamma \approx \delta \in Y\}).$$

But the last inclusion is reversible, because $\Delta_0(x, y, z, w, \underline{u})$ is a set of quaternary commutator equations. So (1) holds.

This proves the theorem. □

Proposition 4.3.10. *Let* $\mathbf{Q}$ *be a quasivariety with the relative shifting property. Let* $\Delta_c(x, y, z, w, \underline{u})$ *be the set of quaternary commutator equations for* $\mathbf{Q}$ *supplied by the relative cube property and the centralization relation in the sense of the classical two-binary term condition (as in Theorem* 4.1.11*). Then*

$$\bigcup\{(\forall \underline{u})\ \Delta_c(p, q, r, s, \underline{u}) : p \approx q \in \mathbf{Q}^{\models}(x \approx y),\ r \approx s \in \mathbf{Q}^{\models}(z \approx w)\}$$

is a generating set of quaternary commutator equations for $\mathbf{Q}$.

Proof. In view of Theorem 4.1.11 and Proposition 4.3.4,

$$\mathbf{Q}^{\models}(\bigcup\{(\forall \underline{u})\ \Delta_c(p, q, r, s, \underline{u}) : p \approx q \in \mathbf{Q}^{\models}(x \approx y),\ r \approx s \in \mathbf{Q}^{\models}(z \approx w)\}) =$$
$$[\mathbf{Q}^{\models}(x \approx y), \mathbf{Q}^{\models}(z \approx w)] = \mathbf{Q}^{\models}(x \approx y) \cap \mathbf{Q}^{\models}(z \approx w). \qquad \square$$

Chapter 5
Additivity of the Equationally-Defined Commutator

In this chapter we are concerned with the problem of additivity of the equationally defined commutator.

5.1 Conditions (C1) and (C2)

Let τ be an algebraic signature. Let $\mathbf{Q}$ be a quasivariety of τ-algebras and let $\boldsymbol{A}$ be an arbitrary τ-algebra, not necessarily in $\mathbf{Q}$. The equationally defined commutator of $\mathbf{Q}$ is *additive on* $\boldsymbol{A}$ if for any $\Phi_1, \Phi_2, \Psi \in Con_{\mathbf{Q}}(\boldsymbol{A})$:

(C1) $$[\Phi_1 +_{\mathbf{Q}} \Phi_2, \Psi] = [\Phi_1, \Psi] +_{\mathbf{Q}} [\Phi_2, \Psi].$$

(C1), quantified over all algebras of $\mathbf{Q}$, is referred to as the *additivity property* of the commutator. (C1) implies that

$(\mathrm{C1})_{fin}$ $$[\Phi_1 +_{\mathbf{Q}} \ldots +_{\mathbf{Q}} \Phi_n, \Psi] = [\Phi_1, \Psi] +_{\mathbf{Q}} \ldots +_{\mathbf{Q}} [\Phi_n, \Psi].$$

for any finite set $\{\Phi_1, \ldots, \Phi_n\}$ of $\mathbf{Q}$-congruences on $\boldsymbol{A}$ and any $\Psi \in Con_{\mathbf{Q}}(\boldsymbol{A})$.

(C1) also implies that for any non-empty set $\{\Phi_i : i \in I\}$ of $\mathbf{Q}$-congruences of $\boldsymbol{A}$ and for any $\Psi \in Con_{\mathbf{Q}}(\boldsymbol{A})$,

$(\mathrm{C1})_{\infty}$ $$[\sup{}_{\mathbf{Q}}\{\Phi_i : i \in I\}, \Psi] = \sup{}_{\mathbf{Q}}\{[\Phi_i, \Psi] : i \in I\}.$$

in the lattice $\boldsymbol{Con}_{\mathbf{Q}}(\boldsymbol{A})$; equivalently

$$[\Theta^{\boldsymbol{A}}_{\mathbf{Q}}(\bigcup\{\Phi_i : i \in I\}), \Psi] = \Theta^{\boldsymbol{A}}_{\mathbf{Q}}(\bigcup\{[\Phi_i, \Psi] : i \in I\}).$$

J. Czelakowski, *The Equationally-Defined Commutator*,
DOI 10.1007/978-3-319-21200-5_5

If $\{\Phi_i : i \in I\}$ is a *directed* family, then $(\mathrm{C1})_\infty$ holds in virtue of Theorem 3.1.6.(v). Additivity thus postulates that $(\mathrm{C1})_\infty$ holds for an *arbitrary* non-empty family of $\mathbf{Q}$-congruences.

$(\mathrm{C1})_\infty$ follows from $(\mathrm{C1})_{fin}$ and the definition of the equationally defined commutator. The inclusion $(\supseteq)$ in $(\mathrm{C1})_\infty$ always holds by the monotonicity of the commutator. To show the reverse inclusion, suppose that $\langle a, b\rangle \in [\Theta^{\boldsymbol{A}}_{\mathbf{Q}}(\bigcup\{\Phi_i : i \in I\}), \Psi]$. It follows from Definition 3.1.5 that there is a finite subset $I_f \subseteq I$ such that $\langle a, b\rangle \in [\Theta^{\boldsymbol{A}}_{\mathbf{Q}}(\bigcup\{\Phi_i : i \in I_f\}), \Psi]$. $(\mathrm{C1})_{fin}$ then gives that $\langle a, b\rangle \in \Theta^{\boldsymbol{A}}_{\mathbf{Q}}(\bigcup\{[\Phi_i, \Psi] : i \in I_f\})$. Hence $\langle a, b\rangle \in \Theta^{\boldsymbol{A}}_{\mathbf{Q}}(\bigcup\{[\Phi_i, \Psi] : i \in I\})$. So $(\mathrm{C1})_\infty$ holds.

The equationally defined commutator of $\mathbf{Q}$ is *additive on* $\mathbf{Q}$ if it is additive on every algebra $\boldsymbol{A} \in \mathbf{Q}$. We then say that the equationally defined commutator satisfies (C1) on $\mathbf{Q}$.

We need one more property of the commutator:

(C2) If $h : \boldsymbol{A} \to \boldsymbol{B}$ is a surjective homomorphism between $\mathbf{Q}$-algebras and $\Phi, \Psi \in Con_{\mathbf{Q}}(\boldsymbol{A})$, then

$$\ker_{\mathbf{Q}}(h) +_{\mathbf{Q}} [\Phi, \Psi]^{\boldsymbol{A}} = h^{-1}([\Theta^{\boldsymbol{B}}_{\mathbf{Q}}(h\Phi), \Theta^{\boldsymbol{B}}_{\mathbf{Q}}(h\Psi)]^{\boldsymbol{B}}).$$

(C2), quantified over all algebras $\boldsymbol{A}, \boldsymbol{B}$ of $\mathbf{Q}$, is referred to as the *correspondence property* of the commutator. The above conditions are extensively applied in the commutator theory—see Freese and McKenzie (1987) or Kearnes and McKenzie (1992).

(C2) is stronger than the statement of Theorem 3.1.8.

It turns out that for any quasivariety $\mathbf{Q}$, (C1) implies (C2). More precisely:

Theorem 5.1.1. *For any quasivariety* $\mathbf{Q}$*, if the equationally defined commutator is additive on* $\mathbf{Q}$*, then it has the correspondence property.*

Proof. Let $h : \boldsymbol{A} \to \boldsymbol{B}$ be a surjective homomorphism between τ-algebras, and $\Phi, \Psi \in Con_{\mathbf{Q}}(\boldsymbol{A})$. According to Theorem 3.1.8,

$$\ker_{\mathbf{Q}}(h) +_{\mathbf{Q}} [\ker_{\mathbf{Q}}(h) +_{\mathbf{Q}} \Phi, \ker_{\mathbf{Q}}(h) +_{\mathbf{Q}} \Psi]^{\boldsymbol{A}} = h^{-1}([\Theta^{\boldsymbol{B}}_{\mathbf{Q}}(h\Phi), \Theta^{\boldsymbol{B}}_{\mathbf{Q}}(h\Psi)]^{\boldsymbol{B}})$$

Thus, to prove (C2), it suffices to show that

$$\ker_{\mathbf{Q}}(h) +_{\mathbf{Q}} [\ker_{\mathbf{Q}}(h) +_{\mathbf{Q}} \Phi, \ker_{\mathbf{Q}}(h) +_{\mathbf{Q}} \Psi]^{\boldsymbol{A}} = \ker_{\mathbf{Q}}(h) +_{\mathbf{Q}} [\Phi, \Psi]^{\boldsymbol{A}}. \qquad (*)$$

As $\ker_{\mathbf{Q}}(h) \in Con_{\mathbf{Q}}(\boldsymbol{A})$, (C1) implies that $\ker_{\mathbf{Q}}(h) +_{\mathbf{Q}} [\ker_{\mathbf{Q}}(h) +_{\mathbf{Q}} \Phi, \ker_{\mathbf{Q}}(h) +_{\mathbf{Q}} \Psi]^{\boldsymbol{A}} = \ker_{\mathbf{Q}}(h) +_{\mathbf{Q}} [\ker_{\mathbf{Q}}(h), \ker_{\mathbf{Q}}(h)]^{\boldsymbol{A}} +_{\mathbf{Q}} [\ker_{\mathbf{Q}}(h), \Psi]^{\boldsymbol{A}} +_{\mathbf{Q}} [\Phi, \ker_{\mathbf{Q}}(h)]^{\boldsymbol{A}} +_{\mathbf{Q}} [\Phi, \Psi]^{\boldsymbol{A}} = \ker_{\mathbf{Q}}(h) +_{\mathbf{Q}} [\Phi, \Psi]^{\boldsymbol{A}}$, because $[\ker_{\mathbf{Q}}(h), \ker_{\mathbf{Q}}(h)]^{\boldsymbol{A}} \subseteq \ker_{\mathbf{Q}}(h)$, $[\ker_{\mathbf{Q}}(h), \Psi]^{\boldsymbol{A}} \subseteq \ker_{\mathbf{Q}}(h)$, and $[\Phi, \ker_{\mathbf{Q}}(h)]^{\boldsymbol{A}} \subseteq \ker_{\mathbf{Q}}(h)$. This proves $(*)$.

It follows that $h^{-1}([\Theta^{\boldsymbol{B}}_{\mathbf{Q}}(h\Phi), \Theta^{\boldsymbol{B}}_{\mathbf{Q}}(h\Psi)]^{\boldsymbol{B}}) = \ker_{\mathbf{Q}}(h) +_{\mathbf{Q}} [\Phi, \Psi]^{\boldsymbol{A}}$. So (C2) holds. □

The following theorem, being one of the crucial results of this section, characterizes the additivity property of the equationally defined commutator in terms of quaternary commutator equations.

Theorem 5.1.2. *Let* $\mathbf{Q}$ *be a quasivariety of algebras. The following conditions are equivalent:*

(1) *The equationally defined commutator for* $\mathbf{Q}$ *is additive, i.e., it satisfies* (C1) *in any* τ*-algebra* $\boldsymbol{A}$*;*
(2) *There exists a set* $\Delta_0(x, y, z, w, \underline{u})$ *of quaternary commutator equations for* $\mathbf{Q}$ *in* x, y *and* z, w *(possibly with k parameters* $\underline{u} = u_1, u_2, \ldots, k \leqslant \omega$*) such that for every* τ*-algebra* $\boldsymbol{A}$ *and for every pair of sets* $X, Y \subseteq A^2$*,*

$$[\Theta^{\boldsymbol{A}}_{\mathbf{Q}}(X), \Theta^{\boldsymbol{A}}_{\mathbf{Q}}(Y)] = \Theta^{\boldsymbol{A}}_{\mathbf{Q}}(\bigcup\{(\forall \underline{e})\ \Delta^{\boldsymbol{A}}_0(a, b, c, d, \underline{e}) : \langle a, b\rangle \in X, \langle c, d\rangle \in Y\}).$$

Note that condition (2) is stronger than Theorem 4.3.9—see Example 4 in Section 6.4.

Notes. (a). We apply here the convention adopted in Chapter 3. We may rewrite the above equality in a more developed form as

$$[\Theta^{\boldsymbol{A}}_{\mathbf{Q}}(X), \Theta^{\boldsymbol{A}}_{\mathbf{Q}}(Y)]^{\boldsymbol{A}} = \Theta^{\boldsymbol{A}}_{\mathbf{Q}}(\{\langle \alpha(a, b, c, d, \underline{e}), \beta(a, b, c, d, \underline{e})\rangle :$$
$$\alpha \approx \beta \in \Delta_0,\ \langle a, b\rangle \in X,\ \langle c, d\rangle \in Y,\ \underline{e} \in A^k\}). \quad \square$$

(b). Condition (2) immediately implies that

$$[\Theta^{\boldsymbol{A}}_{\mathbf{Q}}(a, b), \Theta^{\boldsymbol{A}}_{\mathbf{Q}}(c, d)]^{\boldsymbol{A}} = \Theta^{\boldsymbol{A}}_{\mathbf{Q}}((\forall \underline{e})\ \Delta^{\boldsymbol{A}}_0(a, b, c, d, \underline{e}))$$

for any algebra $\boldsymbol{A}$ and any $a, b, c, d \in A$. This in turn implies that $\Delta_0(x, y, z, w, \underline{u})$ is a generating set for the equational commutator of $\mathbf{Q}$. $\square$

Proof. (1) $\Rightarrow$ (2). Assume (1). In view of Theorem 5.1.1, the equationally defined commutator for $\mathbf{Q}$ also satisfies (C2).

Let $\Delta_0 = \Delta_0(x, y, z, w, \underline{u})$ be an arbitrary set of quaternary commutator equations for $\mathbf{Q}$ such that $\mathbf{Q}^{\models}(\Delta_0) = \mathbf{Q}^{\models}(x \approx y) \cap \mathbf{Q}^{\models}(z \approx w)$. Δ_0 is therefore a generating set for the equational commutator of $\mathbf{Q}$. Let $k \leqslant \omega$ be the length of $\underline{u} = u_1, u_2, \ldots$. In view of Propositions 4.3.4–4.3.5 we have that

$$[\Theta^{\boldsymbol{F}}_{\mathbf{Q}}([x], [y]), \Theta^{\boldsymbol{F}}_{\mathbf{Q}}([z], [w])]^{\boldsymbol{F}} = \Theta^{\boldsymbol{F}}_{\mathbf{Q}}(\Delta^{\boldsymbol{F}}_0([x], [y], [z], [w], [\underline{u}]))$$

in the free algebra $\boldsymbol{F} := \boldsymbol{F}_{\mathbf{Q}}(\omega)$. (Here $[\underline{u}] := [u_1], [u_2], \ldots$.)

We prove the following string of claims.

Claim 1. *Let* $\boldsymbol{B}$ *be any countably generated* τ*-algebra and* $a, b, c, d \in B$*. Then*

$$[\Theta^{\boldsymbol{B}}_{\mathbf{Q}}(a, b), \Theta^{\boldsymbol{B}}_{\mathbf{Q}}(c, d)]^{\boldsymbol{B}} = \Theta^{\boldsymbol{B}}_{\mathbf{Q}}((\forall \underline{e})\ \Delta^{\boldsymbol{B}}_0(a, b, c, d, \underline{e})).$$

Proof (of the claim). Since $\Delta_0(x, y, z, w, \underline{u})$ is a set of commutator equations, the definition of the equationally defined commutator gives:

$$\Theta^{\boldsymbol{B}}_{\mathbf{Q}}(\Delta_0(a, b, c, d, \underline{e})) \subseteq [\Theta^{\boldsymbol{B}}_{\mathbf{Q}}(a, b), \Theta^{\boldsymbol{B}}_{\mathbf{Q}}(c, d)]^{\boldsymbol{B}}, \tag{a}$$

for all sequences $\underline{e} = e_1, e_2, \ldots, e_k$ of elements of B of length k.

To simplify notation we shall identify the free generators $[x], [y], [z], [w]$, and $[\underline{u}] = [u_1], [u_2], \ldots$ of $\boldsymbol{F}$ with x, y, z, w, and $\underline{u} = u_1, u_2, \ldots$, respectively.

Let $\underline{e} \in B^k$ be an arbitrary but fixed sequence of length k. Since $\boldsymbol{B}$ is countably generated, there exists a surjective homomorphism $h : \boldsymbol{F} \to \boldsymbol{B}$ such that $hx = a$, $hy = b$, $hz = c$, $hw = d$. Let $e_j := hu_j$ for $j = 1, \ldots, k$. (h is arbitrarily defined for the remaining free generators.) Then

$$\Theta^{\boldsymbol{B}}_{\mathbf{Q}}(hx, hy) = \Theta^{\boldsymbol{B}}_{\mathbf{Q}}(h(\Theta^{\boldsymbol{F}}_{\mathbf{Q}}(x, y))) \quad \text{and} \quad \Theta^{\boldsymbol{B}}_{\mathbf{Q}}(hz, hw) = \Theta^{\boldsymbol{B}}_{\mathbf{Q}}(h(\Theta^{\boldsymbol{F}}_{\mathbf{Q}}(z, w))).$$

Indeed, putting $X := \Theta^{\boldsymbol{F}}_{\mathbf{Q}}(x, y)$, and applying Proposition 2.9 we obtain that $h(\Theta^{\boldsymbol{F}}_{\mathbf{Q}}(x, y) +_{\mathbf{Q}} \ker_{\mathbf{Q}}(h)) = h(\Theta^{\boldsymbol{F}}_{\mathbf{Q}}(\Theta^{\boldsymbol{F}}_{\mathbf{Q}}(x, y)) +_{\mathbf{Q}} \ker(h)) = \Theta^{\boldsymbol{B}}_{\mathbf{Q}}(h(\Theta^{\boldsymbol{F}}_{\mathbf{Q}}(x, y)))$. But the congruence $h(\Theta^{\boldsymbol{F}}_{\mathbf{Q}}(x, y) +_{\mathbf{Q}} \ker_{\mathbf{Q}}(h))$ is equal to $\Theta^{\boldsymbol{B}}_{\mathbf{Q}}(hx, hy)$, also by Proposition 2.9. Hence $\Theta^{\boldsymbol{B}}_{\mathbf{Q}}(hx, hy) = \Theta^{\boldsymbol{B}}_{\mathbf{Q}}(h(\Theta^{\boldsymbol{F}}_{\mathbf{Q}}(x, y)))$. The proof of the other equality is similar.

We then have:

$$\begin{aligned} h^{-1}([\Theta^{\boldsymbol{B}}_{\mathbf{Q}}(a, b), \Theta^{\boldsymbol{B}}_{\mathbf{Q}}(c, d)]^{\boldsymbol{B}}) &= h^{-1}([\Theta^{\boldsymbol{B}}_{\mathbf{Q}}(hx, hy), \Theta^{\boldsymbol{B}}_{\mathbf{Q}}(hz, hw)]^{\boldsymbol{B}}) = \\ h^{-1}([\Theta^{\boldsymbol{B}}_{\mathbf{Q}}(h(\Theta^{\boldsymbol{F}}_{\mathbf{Q}}(x, y))), \Theta^{\boldsymbol{B}}_{\mathbf{Q}}(h(\Theta^{\boldsymbol{F}}_{\mathbf{Q}}(z, w)))]^{\boldsymbol{B}}) &= \text{(by (C2)} \\ \ker_{\mathbf{Q}}(h) +_{\mathbf{Q}} [\Theta^{\boldsymbol{F}}_{\mathbf{Q}}(x, y), \Theta^{\boldsymbol{F}}_{\mathbf{Q}}(z, w)]^{\boldsymbol{F}} &= \\ \ker_{\mathbf{Q}}(h) +_{\mathbf{Q}} \Theta^{\boldsymbol{F}}_{\mathbf{Q}}(\Delta_0(x, y, z, w, \underline{u})). \end{aligned} \tag{b}$$

On the other hand, Proposition 2.9 also gives:

$$\begin{aligned} h^{-1}(\Theta^{\boldsymbol{B}}_{\mathbf{Q}}(\Delta_0(a, b, c, d, \underline{e}))) &= h^{-1}(\Theta^{\boldsymbol{B}}_{\mathbf{Q}}(\Delta_0(hx, hy, hz, hw, hu))) = \\ h^{-1}(h(\ker_{\mathbf{Q}}(h) +_{\mathbf{Q}} \Theta^{\boldsymbol{F}}_{\mathbf{Q}}(\Delta_0(x, y, z, w, \underline{u})))) &= \\ \ker_{\mathbf{Q}}(h) +_{\mathbf{Q}} \Theta^{\boldsymbol{F}}_{\mathbf{Q}}(\Delta_0(x, y, z, w, \underline{u})). \end{aligned} \tag{c}$$

It follows that the first congruence of (b) is equal to the first congruence of (c), i.e.,

$$h^{-1}([\Theta^{\boldsymbol{B}}_{\mathbf{Q}}(a, b), \Theta^{\boldsymbol{B}}_{\mathbf{Q}}(c, d)]^{\boldsymbol{B}}) = h^{-1}(\Theta^{\boldsymbol{B}}_{\mathbf{Q}}(\Delta_0(a, b, c, d, \underline{e}))).$$

Consequently,

$$[\Theta^{\boldsymbol{B}}_{\mathbf{Q}}(a, b), \Theta^{\boldsymbol{B}}_{\mathbf{Q}}(c, d)]^{\boldsymbol{B}} = \Theta^{\boldsymbol{B}}_{\mathbf{Q}}(\Delta_0(a, b, c, d, \underline{e})). \tag{d}$$

As $\underline{e} \in B^k$ is arbitrary, it follows from (d) that

$$[\Theta^{\boldsymbol{B}}_{\mathbf{Q}}(a,b), \Theta^{\boldsymbol{B}}_{\mathbf{Q}}(c,d)]^{\boldsymbol{B}} = \Theta^{\boldsymbol{B}}_{\mathbf{Q}}((\forall \underline{e})\ \Delta_0(a,b,c,d,\underline{e})).$$

The last equality combined with (a) gives the thesis of the claim. □

Claim 1 continues to hold regardless of the cardinality of $\boldsymbol{B}$:

Claim 2. *Let* $\boldsymbol{A}$ *be an arbitrary* τ*-algebra and* $a, b, c, d \in A$*. Then*

$$[\Theta^{\boldsymbol{A}}_{\mathbf{Q}}(a,b), \Theta^{\boldsymbol{A}}_{\mathbf{Q}}(c,d)]^{\boldsymbol{A}} = \Theta^{\boldsymbol{A}}_{\mathbf{Q}}((\forall \underline{e})\ \Delta^{\boldsymbol{A}}_0(a,b,c,d,\underline{e})).$$

Proof (of the claim). As $\Delta_0(x,y,z,w,\underline{u})$ is a set of commutator equations for $\mathbf{Q}$, the definition of the equational commutator gives the inclusion:

$$[\Theta^{\boldsymbol{A}}_{\mathbf{Q}}(a,b), \Theta^{\boldsymbol{A}}_{\mathbf{Q}}(a,b)]^{\boldsymbol{A}} \supseteq \Theta^{\boldsymbol{A}}_{\mathbf{Q}}((\forall \underline{e})\ \Delta^{\boldsymbol{A}}_0(a,b,c,d,\underline{e})).$$

We shall prove the opposite inclusion. According to Theorem 3.1.6.(vii) and Claim 1 we have:

$$[\Theta^{\boldsymbol{A}}_{\mathbf{Q}}(a,b), \Theta^{\boldsymbol{A}}_{\mathbf{Q}}(c,d)]^{\boldsymbol{A}} =$$

$$\bigcup\{[\Theta^{\boldsymbol{B}}_{\mathbf{Q}}(a,b), \Theta^{\boldsymbol{B}}_{\mathbf{Q}}(c,d)]^{\boldsymbol{B}} : \boldsymbol{B} \text{ is a countably generated subalgebra}$$

$$\text{of } \boldsymbol{A} \text{ and } a,b,c,d \in B\} =$$

$$\bigcup\{\Theta^{\boldsymbol{B}}_{\mathbf{Q}}(\Delta_0(a,b,c,d,\underline{e}))^{\boldsymbol{B}} : \boldsymbol{B}\text{is a countably generated subalgebra}$$

$$\text{of } \boldsymbol{A} \text{ containing } a,b,c,d \in B \text{ and } \underline{e} \in B^k\}.$$

Suppose that a^*, b^* are elements of A and $\langle a^*, b^*\rangle \in [\Theta^{\boldsymbol{A}}_{\mathbf{Q}}(a,b), \Theta^{\boldsymbol{A}}_{\mathbf{Q}}(a,b)]^{\boldsymbol{A}}$. It follows from the above equations that there exists a countably generated subalgebra $\boldsymbol{B} \subseteq \boldsymbol{A}$ containing a,b,c,d and a string $\underline{e} \in B^k$ such that $\langle a^*, b^*\rangle \in \Theta^{\boldsymbol{B}}_{\mathbf{Q}}(\Delta_0(a,b,c,d,\underline{e}))$. But evidently

$$\Theta^{\boldsymbol{B}}_{\mathbf{Q}}(\Delta_0(a,b,c,d,\underline{e})) \subseteq B^2 \cap \Theta^{\boldsymbol{A}}_{\mathbf{Q}}(\Delta_0(a,b,c,d,\underline{e})).$$

Hence $\langle a^*, b^*\rangle \in \Theta^{\boldsymbol{A}}_{\mathbf{Q}}(\Delta_0(a,b,c,d,\underline{e}))$. Consequently,

$$\langle a^*, b^*\rangle \in \Theta^{\boldsymbol{A}}_{\mathbf{Q}}(\bigcup\{\Delta_0(a,b,c,d,\underline{e}) : \underline{e} \in A^k\}).$$

This proves the claim. □

Claim 3. *Let* $\boldsymbol{A}$ *be any* $\mathbf{Q}$*-algebra and* $a_1, \ldots, a_n$*,* $b_1, \ldots, b_n$*,* $a, b, c, d \in A$*. If* $a \equiv b$ $(\Theta^{\boldsymbol{A}}_{\mathbf{Q}}(\langle a_1, b_1\rangle, \ldots, \langle a_n, b_n\rangle)$*, then*

$$(\forall \underline{e})\ \Delta_0(a,b,c,d,\underline{e}) \subseteq \Theta^{\boldsymbol{A}}_{\mathbf{Q}}((\forall \underline{e})\ \Delta_0(a_1,b_1,c,d,\underline{e}) \cup \ldots \cup (\forall \underline{e})\ \Delta_0(a_n,b_n,c,d,\underline{e})).$$

Proof (of the claim). The assumption $a \equiv b\ (\Theta_{\mathbf{Q}}^{A}(\langle a_1, b_1\rangle, \ldots, \langle a_n, b_n\rangle))$ implies that

$$[\Theta_{\mathbf{Q}}^{A}(a,b), \Theta_{\mathbf{Q}}^{A}(c,d)]^{A} \subseteq [\Theta_{\mathbf{Q}}^{A}(\langle a_1, b_1\rangle, \ldots, \langle a_n, b_n\rangle), \Theta_{\mathbf{Q}}^{A}(c,d)]^{A}, \qquad (*)$$

by the monotonicity of the commutator.

We have:

$$\begin{aligned}
\Theta_{\mathbf{Q}}^{A}((\forall \underline{e})\ \Delta_0(a,b,c,d,\underline{e})) &= \text{ (by Claim 2)}\\
[\Theta_{\mathbf{Q}}^{A}(a,b), \Theta_{\mathbf{Q}}^{A}(c,d)]^{A} &\subseteq \text{ (by } (*))\\
[\Theta_{\mathbf{Q}}^{A}(\langle a_1, b_1\rangle, \ldots, \langle a_n, b_n\rangle), \Theta_{\mathbf{Q}}^{A}(c,d)]^{A} &= \text{ (by (C1)}\\
[\Theta_{\mathbf{Q}}^{A}(a_1,b_1), \Theta_{\mathbf{Q}}^{A}(c,d)]^{A} +_{\mathbf{Q}} \ldots +_{\mathbf{Q}} [\Theta_{\mathbf{Q}}^{A}(a_n,b_n), \Theta_{\mathbf{Q}}^{A}(c,d)]^{A} &= \text{ (by Claim 2)}\\
\Theta_{\mathbf{Q}}^{A}((\forall \underline{e})\ \Delta_0(a_1,b_1,c,d,\underline{e})) +_{\mathbf{Q}} \ldots +_{\mathbf{Q}} \Theta_{\mathbf{Q}}^{A}((\forall \underline{e})\ \Delta_0(a_n,b_n,c,d,\underline{e})) &=\\
\Theta_{\mathbf{Q}}^{A}((\forall \underline{e})\ \Delta_0(a_1,b_1,c,d,\underline{e}) \cup \ldots \cup (\forall \underline{e})\ \Delta_0(a_n,b_n,c,d,\underline{e})). &\quad \square
\end{aligned}$$

Claim 4. *Let* $\boldsymbol{A}$ *be any* τ*-algebra and* $X, Y \subseteq A^2$. *Then*

$$[\Theta_{\mathbf{Q}}^{A}(X), \Theta_{\mathbf{Q}}^{A}(Y)]^{A} = \Theta_{\mathbf{Q}}^{A}(\bigcup\{(\forall \underline{e})\ \Delta_0(a,b,c,d,\underline{e}) : \langle a,b\rangle \in X, \langle c,d\rangle \in Y\}).$$

We note that Claim 4 is the same as the thesis of condition (2).

Proof (of the claim). It suffices to show that the congruence on the left side is included in the congruence on the right side.

According to (C1) and Claim 2

$$\begin{aligned}
&[\Theta_{\mathbf{Q}}^{A}(X), \Theta_{\mathbf{Q}}^{A}(Y)]^{A} = &(\alpha)\\
&\sup{}_{\mathbf{Q}}\{[\Theta_{\mathbf{Q}}^{A}(a,b), \Theta_{\mathbf{Q}}^{A}(c,d)]^{A} : \langle a,b\rangle \in \Theta_{\mathbf{Q}}^{A}(X), \langle c,d\rangle \in \Theta_{\mathbf{Q}}^{A}(Y)\} =\\
&\Theta_{\mathbf{Q}}^{A}(\bigcup\{(\forall \underline{e})\ \Delta_0(a,b,c,d,\underline{e}) : \langle a,b\rangle \in \Theta_{\mathbf{Q}}^{A}(X), \langle c,d\rangle \in \Theta_{\mathbf{Q}}^{A}(Y)\}.
\end{aligned}$$

($\sup_{\mathbf{Q}}$ denotes the supremum in the sense of the lattice $\boldsymbol{Con}_{\mathbf{Q}}(\boldsymbol{A})$.)

Assume $\langle a^*, b^*\rangle \in \Theta_{\mathbf{Q}}^{A}(X)$ and $\langle c^*, d^*\rangle \in \Theta_{\mathbf{Q}}^{A}(Y)$. Then there exist finite sets $X^* \subseteq X$ and $Y^* \subseteq Y$ such that $\langle a^*, b^*\rangle \in \Theta_{\mathbf{Q}}^{A}(X^*)$ and $\langle c^*, d^*\rangle \in \Theta_{\mathbf{Q}}^{A}(Y^*)$. According to Claim 3 we then have:

$$(\forall \underline{e})\ \Delta_0(a^*,b^*,c^*,d^*,\underline{e}) \subseteq \Theta_{\mathbf{Q}}^{A}(\bigcup\{(\forall \underline{e})\ \Delta_0(a,b,c^*,d^*,\underline{e}) : \langle a,b\rangle \in X^*\}), \qquad (\beta)$$

because $\langle a^*, b^*\rangle \in \Theta_{\mathbf{Q}}^{A}(X^*)$. But we also have that for every pair $\langle a,b\rangle \in X^*$,

$$(\forall \underline{e})\ \Delta_0(a, b, c^*, d^*, \underline{e}) \subseteq \Theta^{A}_{\mathbf{Q}}(\bigcup\{(\forall \underline{e})\ \Delta_0(a, b, c, d, \underline{e}) : \langle c, d\rangle \in Y^*\}), \qquad (\gamma)$$

because $\langle c^*, d^*\rangle \in \Theta^{A}_{\mathbf{Q}}(Y^*)$.

Conditions (β) and (γ) imply that

$$(\forall \underline{e})\ \Delta_0(a^*, b^*, c^*, d^*, \underline{e}) \subseteq \qquad (\delta)$$

$$\Theta^{A}_{\mathbf{Q}}(\bigcup\{(\forall \underline{e})\ \Delta_0(a, b, c, d, \underline{e}) : \langle a, b\rangle \in X^*, \langle c, d\rangle \in Y^*\}) \subseteq$$

$$\Theta^{A}_{\mathbf{Q}}(\bigcup\{(\forall \underline{e})\ \Delta_0(a, b, c, d, \underline{e}) : \langle a, b\rangle \in X, \langle c, d\rangle \in Y\}).$$

Taking into account (α), (δ) and the fact that $\langle a^*, b^*\rangle$ and $\langle c^*, d^*\rangle$ are arbitrary elements of $\Theta^{A}_{\mathbf{Q}}(X)$ and $\Theta^{A}_{\mathbf{Q}}(Y)$, respectively, we obtain that

$$[\Theta^{A}_{\mathbf{Q}}(X), \Theta^{A}_{\mathbf{Q}}(Y)]^{A} \subseteq \Theta^{A}_{\mathbf{Q}}(\bigcup\{(\forall \underline{e})\ \Delta_0(a, b, c, d, \underline{e}) : \langle a, b\rangle \in X, \langle c, d\rangle \in Y\}).$$

This proves the claim. □

Now, by Claim 4, condition (2) is satisfied for the set $\Delta_0(x, y, z, w, \underline{u})$ of equations defined as above.

(2) ⇒ (1). We assume (2) holds. We show that the equational commutator for **Q** is additive.

Let Φ_i, $i \in I$, be a family of **Q**-congruences on an algebra $\boldsymbol{A} \in \mathbf{Q}$ and $\Psi \in Con_{\mathbf{Q}}(\boldsymbol{A})$. We claim that

$$[\Theta^{A}_{\mathbf{Q}}(\bigcup\{\Phi_i : i \in I\}), \Psi]^{A} = \Theta^{A}_{\mathbf{Q}}(\bigcup\{[\Phi_i, \Psi]^{A} : i \in I\})$$

in the lattice $\boldsymbol{Con}_{\mathbf{Q}}(\boldsymbol{A})$.

We have:

$$[\Theta^{A}_{\mathbf{Q}}(\bigcup\{\Phi_i : i \in I\}), \Psi]^{A} = \text{(by (2))}$$

$$\Theta^{A}_{\mathbf{Q}}(\bigcup\{(\forall \underline{e})\ \Delta_0(a, b, c, d, \underline{e}) : \langle a, b\rangle \in \bigcup\{\Phi_i : i \in I\}, \langle c, d\rangle \in \Psi\}) =$$

$$\Theta^{A}_{\mathbf{Q}}(\bigcup\{\bigcup\{(\forall \underline{e})\ \Delta_0(a, b, c, d, \underline{e}) : \langle a, b\rangle \in \Phi_i, \langle c, d\rangle \in \Psi\} : i \in I\}) =$$

$$\Theta^{A}_{\mathbf{Q}}(\bigcup\{\Theta^{A}_{\mathbf{Q}}(\bigcup\{(\forall \underline{e})\ \Delta_0(a, b, c, d, \underline{e}) : \langle a, b\rangle \in \Phi_i, \langle c, d\rangle \in \Psi\})\} : i \in I\}) =$$

$$\Theta^{A}_{\mathbf{Q}}(\bigcup\{[\Phi_i, \Psi]^{A} : i \in I\}).$$

This concludes the proof of the theorem. □

Corollary 5.1.3. *Let* $\mathbf{Q}$ *be a quasivariety whose equationally defined commutator is additive. Let* $\Delta_0(x, y, z, w, \underline{u})$ *be a generating set of quaternary equations for the equational commutator of* $\mathbf{Q}$*. Then for any* τ*-algebra* $\mathbf{A}$ *and any* $a, b, c, d \in A$,

$$[\Theta_{\mathbf{Q}}(a, b), \Theta_{\mathbf{Q}}(c, d)]^A = \Theta_{\mathbf{Q}}((\forall \underline{e})\ \Delta_0(a, b, c, d, \underline{e})).$$

Proof. This follows from the proof of the implication (1) $\Rightarrow$ (2) above. □

The next observation supplements Theorem 5.1.2.

Theorem 5.1.4. *Let* $\mathbf{Q}$ *be a quasivariety of algebras. The following conditions are equivalent:*

(1) *The equationally defined commutator of* $\mathbf{Q}$ *is additive (on* $\mathbf{Q}$*);*
(2) *There exists a set* $\Delta_0(x, y, z, w, \underline{u})$ *of (parameterized) quaternary commutator equations for* $\mathbf{Q}$ *such that for any* τ*-algebra* $\mathbf{A}$:

(2).(i) *If* $\Phi, \Psi \in Con_{\mathbf{Q}}(\mathbf{A})$*, then*

$$[\Phi, \Psi]^A = \Theta^A_{\mathbf{Q}}(\bigcup\{(\forall \underline{e})\ \Delta_0(a, b, c, d, \underline{e}) : a \equiv b\ (\Phi),\ c \equiv d\ (\Psi)\})$$

and

(2).(ii) *For all* $a_1, \ldots, a_n, b_1, \ldots, b_n, a, b, c, d \in A$*, if*

$$a \equiv b\ (\Theta^A_{\mathbf{Q}}(\langle a_1, b_1\rangle, \ldots, \langle a_n, b_n\rangle),$$

then

$$(\forall \underline{e})\ \Delta_0(a, b, c, d, \underline{e}) \subseteq \Theta^A_{\mathbf{Q}}((\forall \underline{e})\ \Delta_0(a_1, b_1, c, d, \underline{e}) \cup \ldots \cup (\forall \underline{e})\ \Delta_0(a_n, b_n, c, d, \underline{e})).$$

Proof. (1) $\Rightarrow$ (2). Assume (1) holds. Let $\Delta_0(x, y, z, w, \underline{u})$ be a generating set of quaternary commutator equations for $\mathbf{Q}$, i.e., $\mathbf{Q}^{\models}((\forall \underline{u})\ \Delta_0(x, y, z, w, \underline{u})) = \mathbf{Q}^{\models}(x \approx y) \cap \mathbf{Q}^{\models}(z \approx w)$.

Let $\mathbf{A} \in \mathbf{Q}$. To prove (2).(i), assume $\Phi, \Psi \in Con_{\mathbf{Q}}(\mathbf{A})$. Putting $X := \Phi$ and $Y := \Psi$, we immediately infer condition (2).(i) from the above theorem.

As to (2).(ii), we assume that $a \equiv b\,(\Theta^A_{\mathbf{Q}}(\langle a_1, b_1\rangle, \ldots, \langle a_n, b_n\rangle))$ for some $a_1, \ldots, a_n, b_1, \ldots, b_n, a, b, c, d \in A$. Theorem 5.1.2.(2) gives that

$$\Theta^A_{\mathbf{Q}}((\forall \underline{e})\ \Delta_0(a_1, b_1, c, d, \underline{e}) \cup \ldots \cup (\forall \underline{e})\ \Delta_0(a_n, b_n, c, d, \underline{e})) = \qquad (*)$$
$$[\Theta^A_{\mathbf{Q}}(a_1, b_1) +_{\mathbf{Q}} \ldots +_{\mathbf{Q}} \Theta^A_{\mathbf{Q}}(a_n, b_n), \Theta^A_{\mathbf{Q}}(c, d)].$$

As $\Theta^A_{\mathbf{Q}}(a, b) \subseteq \Theta^A_{\mathbf{Q}}(a_1, b_1) +_{\mathbf{Q}} \ldots +_{\mathbf{Q}} \Theta^A_{\mathbf{Q}}(a_n, b_n)$, the monotonicity of the commutator yields that

$$[\Theta^A_{\mathbf{Q}}(a, b), \Theta^A_{\mathbf{Q}}(c, d)] \subseteq [\Theta^A_{\mathbf{Q}}(a_1, b_1) +_{\mathbf{Q}} \ldots +_{\mathbf{Q}} \Theta^A_{\mathbf{Q}}(a_n, b_n), \Theta^A_{\mathbf{Q}}(c, d)].$$

But $[\Theta^A_{\mathbf{Q}}(a,b), \Theta^A_{\mathbf{Q}}(c,d)] = \Theta^A_{\mathbf{Q}}((\forall \underline{e})\ \Delta_0(a,b,c,d,\underline{e}))$. Hence, applying $(*)$, we deduce the thesis of (2).(ii).

(2) $\Rightarrow$ (1). The proof of this implication is based on the following lemma:

Lemma 5.1.5. *Let* $\mathbf{Q}$ *be an arbitrary quasivariety of algebras and* $\Delta_0(x,y,z,w,\underline{u})$ *a set of (parameterized) quaternary commutator equations for* $\mathbf{Q}$*. Suppose* $\Delta_0(x,y,z,w,\underline{u})$ *validates the above condition (2).(ii) in the algebras of* $\mathbf{Q}$*. Then for any* τ*-algebra* $\mathbf{A}$ *and every pair of sets* $X, Y \subseteq A^2$,

$$\Theta^A_{\mathbf{Q}}(\bigcup\{(\forall \underline{e})\ \Delta_0(a,b,c,d,\underline{e}) : a \equiv b\ (\Theta^A_{\mathbf{Q}}(X)),\ c \equiv d\ (\Theta^A_{\mathbf{Q}}(Y))\}) =$$
$$\Theta^A_{\mathbf{Q}}(\bigcup\{(\forall \underline{e})\ \Delta_0(a,b,c,d,\underline{e}) : \langle a,b\rangle \in X,\ \langle c,d\rangle \in Y\}).$$

Proof (of the lemma). Suppose $\mathbf{A} \in \mathbf{Q}$ and $X, Y \subseteq A^2$. Let $\Phi := \Theta^A_{\mathbf{Q}}(X)$ and $\Psi := \Theta^A_{\mathbf{Q}}(Y)$. We trivially have:

$$\Theta^A_{\mathbf{Q}}(\bigcup\{(\forall \underline{e})\ \Delta_0(a,b,c,d,\underline{e}) : a \equiv b\ (\Phi),\ c \equiv d\ (\Psi)\}) \supseteq$$
$$\Theta^A_{\mathbf{Q}}(\bigcup(\forall \underline{e})\ \Delta_0(a,b,c,d,\underline{e}) : \langle a,b\rangle \in X,\ \langle c,d\rangle \in Y\}).$$

We shall show that the above inclusion can be reversed. Suppose

$$\langle a_0, b_0\rangle \in \Theta^A_{\mathbf{Q}}(\bigcup\{(\forall \underline{e})\ \Delta_0(a,b,c,d,\underline{e}) : a \equiv b\ (\Phi),\ c \equiv d\ (\Psi)\}).$$

As the closure operator $\Theta^A_{\mathbf{Q}}(\,\cdot\,)$ is finitary on subsets of A^2, there exist four n-tuples $a_1,\ldots,a_n, b_1,\ldots,b_n, c_1,\ldots,c_n, d_1,\ldots,d_n \in A$ such that $a_i \equiv b_i\ (\Phi)$, $c_i \equiv d_i\ (\Psi)$ for $i = 1,\ldots,n$ and

$$\langle a_0, b_0\rangle \in \Theta^A_{\mathbf{Q}}(\bigcup\{(\forall \underline{e})\ \Delta_0(a_i,b_i,c_i,d_i,\underline{e}) : i = 1,\ldots,n\}). \tag{a}$$

Since the $\mathbf{Q}$-congruences Φ and Ψ are generated by X and Y, respectively, there exist finite subsets $X_f \subseteq X$ and $Y_f \subseteq Y$ such that

$$\langle a_i, b_i\rangle \in \Theta^A_{\mathbf{Q}}(X_f) \qquad \text{for } i = 1,\ldots,n, \tag{b1}$$

and

$$\langle c_i, d_i\rangle \in \Theta^A_{\mathbf{Q}}(Y_f) \qquad \text{for } i = 1,\ldots,n. \tag{b2}$$

But (b1) and (2).(ii) imply that for each i ($i = 1,\ldots,n$)

$$(\forall \underline{e})\ \Delta_0(a_i, b_i, c_i, d_i, \underline{e}) \subseteq \Theta^{A}_{\mathbf{Q}}(\bigcup\{(\forall \underline{e})\ \Delta_0(a, b, c_i, d_i, \underline{e}) : \langle a, b\rangle \in X_f\}). \tag{c}$$

Similarly, by (b2) and (2).(ii) we also have that for each i ($i = 1, \ldots, n$) and each $\langle a, b\rangle \in X_f$,

$$(\forall \underline{e})\ \Delta_0(a, b, c_i, d_i, \underline{e}) \subseteq \Theta^{A}_{\mathbf{Q}}(\bigcup\{(\forall \underline{e})\ \Delta_0(a, b, c, d, e) : \langle c, d\rangle \in Y_f\}). \tag{d}$$

Combining together (c) and (d) we deduce that

$$\bigcup\{(\forall \underline{e})\ \Delta_0(a_i, b_i, c_i, d_i, \underline{e}) : i = 1, \ldots, n\} \subseteq$$
$$\Theta^{A}_{\mathbf{Q}}(\bigcup\{(\forall \underline{e})\ \Delta_0(a, b, c, d, \underline{e}) : \langle a, b\rangle \in X_f, \langle c, d\rangle \in Y_f\}). \tag{e}$$

In the presence of (a), condition (e) gives that

$$\langle a_0, b_0\rangle \in \Theta^{A}_{\mathbf{Q}}(\bigcup\{(\forall \underline{e})\ \Delta_0(a, b, c, d, \underline{e}) : \langle a, b\rangle \in X_f, \langle c, d\rangle \in Y_f\}).$$

Consequently,

$$\langle a_0, b_0\rangle \in \Theta^{A}_{\mathbf{Q}}(\bigcup\{(\forall \underline{e})\ \Delta_0(a, b, c, d, \underline{e}) : \langle a, b\rangle \in X, \langle c, d\rangle \in Y\}).$$

This proves that

$$\Theta^{A}_{\mathbf{Q}}(\bigcup\{(\forall \underline{e})\ \Delta_0(a, b, c, d, \underline{e}) : a \equiv b\ (\Phi),\ c \equiv d\ (\Psi)\}) \subseteq$$
$$\Theta^{A}_{\mathbf{Q}}(\bigcup\{(\forall \underline{e})\ \Delta_0(a, b, c, d, \underline{e}) : \langle a, b\rangle \in X, \langle c, d\rangle \in Y\}. \qquad \square$$

Having established the lemma, we prove the implication (2) ⇒ (1). Assume (2) holds. Suppose $\boldsymbol{A}$ is a τ-algebra and $X, Y \subseteq A^2$. Let $\Phi := \Theta^{A}_{\mathbf{Q}}(X)$ and $\Psi := \Theta^{A}_{\mathbf{Q}}(Y)$. According to condition 2.(i) and Lemma 5.1.5 we have that

$$[\Phi, \Psi]^{A} = \Theta^{A}_{\mathbf{Q}}(\bigcup\{(\forall \underline{e})\ \Delta_0(a, b, c, d, \underline{e}) : a \equiv b\ (\Phi),\ c \equiv d\ (\Psi)\}) =$$
$$\Theta^{A}_{\mathbf{Q}}(\bigcup\{(\forall \underline{e})\ \Delta_0(a, b, c, d, \underline{e}) : \langle a, b\rangle \in X, \langle c, d\rangle \in Y\}).$$

This proves that the equationally defined commutator of $\mathbf{Q}$ satisfies condition (2) of Theorem 5.1.2. It follows that the commutator for $\mathbf{Q}$ is additive. $\square$

Notes. (a). In the above theorem the existence of a Day implication system for the consequence operation $\mathbf{Q}^{\models}$ is *not* assumed. On the other hand, if $\mathbf{Q}^{\models}$ possesses a Day implication, (equivalently, $\mathbf{Q}$ has the relative shifting property), then $\mathbf{Q}$ satisfies condition (2).(i) for some quaternary set of commutator equations $\Delta_0(x, y, z, w, \underline{u})$. As $\Delta_0(x, y, z, w, \underline{u})$ one may take the set Δ_c of quaternary commutator equations supplied by the cube property (see Theorem 4.1.11).

(b). It may happen that (2).(i) is satisfied by some set $\Delta_0(x, y, z, w, \underline{u})$ and the equationally defined commutator for $\mathbf{Q}$ is additive, but implication (2).(ii) does not hold. An appropriate example is provided by the variety of equivalence algebras in Section 6.4 (Example 4). □

Corollary 5.1.6. *Let* $\mathbf{Q}$ *be a quasivariety with the relative shifting property. Suppose that the equationally defined commutator for* $\mathbf{Q}$ *is additive. Let Z be any of the four (equivalent) centralization relations defined as in Theorem* 4.1.2. *Then for any algebra* $\mathbf{A} \in \mathbf{Q}$, *for any sets* $X \subseteq A^2$, $Y \subseteq A^2$ *and any congruence* $\Xi \in Con_{\mathbf{Q}}(\mathbf{A})$:

$$Z(\Theta^{\mathbf{A}}_{\mathbf{Q}}(X), \Theta^{\mathbf{A}}_{\mathbf{Q}}(Y); \Xi) \quad \textit{if and only if}$$

$$Z(\Theta^{\mathbf{A}}_{\mathbf{Q}}(a, b), \Theta^{\mathbf{A}}_{\mathbf{Q}}(c, d); \Xi) \textit{ for all pairs } \langle a, b\rangle \in X, \langle c, d\rangle \in Y.$$

Moreover, if $\mathbf{Q}$ *is a congruence-modular variety, then the above holds for any of the seven equivalent centralization relations defined as in Theorem* 4.2.2.

Proof. Let $\mathbf{A}$, X, Y and Ξ be as above. In view of Theorem 4.1.2 and Proposition 4.1.1, $[\Theta^{\mathbf{A}}_{\mathbf{Q}}(X), \Theta^{\mathbf{A}}_{\mathbf{Q}}(Y)]$ is the least $\mathbf{Q}$-congruence in $\mathbf{A}$ for which $Z(\Theta^{\mathbf{A}}_{\mathbf{Q}}(X), \Theta^{\mathbf{A}}_{\mathbf{Q}}(Y); \Xi)$ holds. Trivially, if $Z(\Theta^{\mathbf{A}}_{\mathbf{Q}}(X), \Theta^{\mathbf{A}}_{\mathbf{Q}}(Y); \Xi_1)$ and $\Xi_1 \subseteq \Xi_2$, then $Z(\Theta^{\mathbf{A}}_{\mathbf{Q}}(X), \Theta^{\mathbf{A}}_{\mathbf{Q}}(Y); \Xi_2)$, for all congruences Ξ_1, Ξ_2 on $\mathbf{A}$. Hence, by the additivity of the commutator, the following conditions are equivalent:

$$Z(\Theta^{\mathbf{A}}_{\mathbf{Q}}(X), \Theta^{\mathbf{A}}_{\mathbf{Q}}(Y); \Xi),$$

$$[\Theta^{\mathbf{A}}_{\mathbf{Q}}(X), \Theta^{\mathbf{A}}_{\mathbf{Q}}(Y)] \subseteq \Xi,$$

$$Z(\Theta^{\mathbf{A}}_{\mathbf{Q}}(a, b), \Theta^{\mathbf{A}}_{\mathbf{Q}}(c, d); \Xi) \quad \text{for all pairs } \langle a, b\rangle \in X, \langle c, d\rangle \in Y.$$

To prove the second statement, apply also Theorem 4.2.2. □

5.2 Additivity of the Equationally-Defined Commutator of Equational Theories

An interest into the equationally defined commutator of *equational theories* is motivated by investigations concerning purely syntactical aspects of this notion.

The equationally defined commutator for $\mathbf{Q}$ is *additive in the lattice* $\boldsymbol{Th}(\mathbf{Q}^{\models})$ if

$$(\mathrm{C1})_{theories} \qquad [X_1 +_{\mathbf{Q}} X_2, Y] = [X_1, Y] +_{\mathbf{Q}} [X_2, Y],$$

for any closed theories X_1, X_2, Y of $\mathbf{Q}^{\models}$.

The above condition is equivalently formulated in the infinite form:

$$[\mathbf{Q}^{\models}(\bigcup_{i \in I} X_i), Y] = \mathbf{Q}^{\models}(\bigcup_{i \in I} [X_i, Y]),$$

for any family X_i, $i \in I$, of closed theories of $\mathbf{Q}^{\models}$ and any closed theory Y.

It is easy to see that (C1) implies:

For any epimorphism $e : \boldsymbol{Te}_{\tau} \to \boldsymbol{Te}_{\tau}$ *and any sets of equation* X *and* Y,

$$(\mathrm{C2})_{theories} \quad \ker_{\mathbf{Q}}(e) +_{\mathbf{Q}} [\mathbf{Q}^{\models}(eX), \mathbf{Q}^{\models}(eY)]] = e^{-1}([\mathbf{Q}^{\models}(eX), \mathbf{Q}^{\models}(eY)]). \qquad \square$$

As

$$e^{-1}([\mathbf{Q}^{\models}(eX), \mathbf{Q}^{\models}(eY)]) = \ker_{\mathbf{Q}}(e) +_{\mathbf{Q}} [\ker_{\mathbf{Q}}(e) +_{\mathbf{Q}} \mathbf{Q}^{\models}(X), \ker_{\mathbf{Q}}(e) +_{\mathbf{Q}} \mathbf{Q}^{\models}(Y)],$$

holds for any quasivariety $\mathbf{Q}$ (see Corollary 3.2.5), condition $(\mathrm{C2})_{theories}$ is equivalently formulated as the equation

$$\ker_{\mathbf{Q}}(e) +_{\mathbf{Q}} [X, Y] = \ker_{\mathbf{Q}}(e) +_{\mathbf{Q}} [\ker_{\mathbf{Q}}(e) +_{\mathbf{Q}} X, \ker_{\mathbf{Q}}(e) +_{\mathbf{Q}} Y],$$

holding for any closed theories X, Y and any epimorphism $e : \boldsymbol{Te}_{\tau} \to \boldsymbol{Te}_{\tau}$.

We recall that, for a given epimorphism e, $\boldsymbol{k}_e : \boldsymbol{Th}(\mathbf{Q}^{\models}) \to \boldsymbol{Th}^e(\mathbf{Q}^{\models})$ is the retraction defined by $\boldsymbol{k}_e(\Sigma) := \ker_{\mathbf{Q}}(e) +_{\mathbf{Q}} \Sigma$, for all $\Sigma \in Th(\mathbf{Q}^{\models})$. The above equation thus states that

$$\boldsymbol{k}_e([X, Y]) = \boldsymbol{k}_e([\boldsymbol{k}_e(X), \boldsymbol{k}_e(Y)]),$$

for any closed theories X, Y.

The following observation supplements Theorem 5.1.2:

Theorem 5.2.1. *Let* $\mathbf{Q}$ *be a quasivariety of algebras. The following conditions are equivalent:*

(1) *The equationally defined commutator for* $\mathbf{Q}$ *is additive, i.e., it satisfies* (C1) *in any algebra* $\mathbf{A} \in \mathbf{Q}$;
(2) *The equationally defined commutator for* $\mathbf{Q}$ *is additive in the lattice* $\boldsymbol{Th}(\mathbf{Q}^{\models})$;
(3) *There exists a set* $\Delta_0(x, y, z, w, \underline{u})$ *of quaternary commutator equations for* $\mathbf{Q}$ *in* x, y *and* z, w *(possibly with* k *parameters* $\underline{u} = u_1, u_2, \ldots, k \leqslant \omega$*) such that*

$$[\mathbf{Q}^{\models}(X), \mathbf{Q}^{\models}(Y)] = \mathbf{Q}^{\models}(\bigcup\{(\forall \underline{u})\ \Delta_0(\alpha, \beta, \gamma, \delta, \underline{u}) : \alpha \approx \beta \in X, \gamma \approx \delta \in Y\}),$$

for any sets of equations X, Y.

Proof. The theorem follows from Theorem 5.1.2 and the following strengthening of Proposition 4.3.4:

Lemma 5.2.2. *Let* $\mathbf{Q}$ *be an arbitrary quasivariety. Let* $\Delta_0(x, y, z, w, \underline{u})$ *be a set of quaternary commutator equations for* $\mathbf{Q}$ *with* $|\underline{u}| = k \leq \omega$. *The following conditions are equivalent:*

(i) $[\mathbf{Q}^{\models}(X), \mathbf{Q}^{\models}(Y)] = \mathbf{Q}^{\models}(\bigcup\{(\forall \underline{u})\ \Delta_0(\alpha, \beta, \gamma, \delta, \underline{u}) : \alpha \approx \beta \in X, \gamma \approx \delta \in Y\})$ *for every pair* X, Y *of sets of equations,*

(ii) $[\Theta^{A}_{\mathbf{Q}}(X), \Theta^{A}_{\mathbf{Q}}(Y)]^{A} = \Theta^{A}_{\mathbf{Q}}(\bigcup\{(\forall \underline{e})\ \Delta^{A}_0(a, b, c, d, \underline{e}) : \langle a, b\rangle \in X, \langle c, d\rangle \in Y\})$ *for all algebras* $\boldsymbol{A} \in \mathbf{Q}$ *and all sets* $X, Y \subseteq A^2$.

Lemma 5.2.2 states that the equational commutator for $\mathbf{Q}$ is additive if and only it is additive in the lattice of equational theories of $\mathbf{Q}^{eq\models}$.

Proof (of the lemma). In view of Proposition 2.5, (i) is equivalent to the fact that

(i)* $[\Theta^{F}_{\mathbf{Q}}(X), \Theta^{F}_{\mathbf{Q}}(Y)]^{F} = \Theta^{F}_{\mathbf{Q}}(\bigcup\{(\forall \underline{e})\ \Delta^{F}_0(a, b, c, d, \underline{e}) : \langle a, b\rangle \in X, \langle c, d\rangle \in Y\})$ *for the free algebra* $\boldsymbol{F} := \boldsymbol{F}_{\mathbf{Q}}(\omega)$ *and all sets* $X, Y \subseteq F^2$.

By repeating the (1) ⇒ (2)-part of the proof of Theorem 5.1.2, we get that (i)* implies (ii). Hence (i) implies (ii). The reverse implication is also immediate, because (ii) implies (i)* and hence (i). □

We now pass to the proof of Theorem 5.2.1. (1) implies (2), because (2) is equivalent to the additivity of the equational commutator on the $\mathbf{Q}$-congruences of the free algebra $\boldsymbol{F}_{\mathbf{Q}}(\omega)$. (2) implies (3) by modifying the proof of the (2) ⇒ (1) part of Theorem 5.1.2. In turn (1) is equivalent to condition (ii) of Lemma 5.2.2. (This is the content of Theorem 5.1.2.) Hence, by Lemma 5.2.2, we get the equivalence of (3) with (1). □

The following observation is also useful.

Theorem 5.2.3. *Let* $\mathbf{Q}$ *be a quasivariety of algebras. The following conditions are equivalent:*

(1) *The equationally defined commutator of* $\mathbf{Q}$ *is additive (on* $\mathbf{Q}$*);*
(2) *There exists a set* Δ_0 *of quaternary commutator equations for* $\mathbf{Q}$ *(possibly with parameters* $\underline{u}$*) such that for every pair* X, Y *of theories of* $\mathbf{Q}^{\models}$ *it is the case that*

(2).(i) $[X, Y] = \mathbf{Q}^{\models}(\bigcup\{(\forall \underline{u})\ \Delta_0(\alpha, \beta, \gamma, \delta, \underline{u}) : \alpha \approx \beta \in X, \gamma \approx \delta \in Y\})$ *and*

(2).(ii) *For any terms* $\alpha_1, \ldots, \alpha_n$, $\beta_1, \ldots, \beta_n$, α, β *and any variables* z *and* w *not occurring in these terms, if*

$$\alpha \approx \beta \in \mathbf{Q}^{\models}(\alpha_1 \approx \beta_1, \ldots, \alpha_n \approx \beta_n),$$

then

$(\forall \underline{u})\ \Delta_0(\alpha, \beta, z, w, \underline{u}) \subseteq$

$\mathbf{Q}^{\models}((\forall \underline{u})\ \Delta_0(\alpha_1, \beta_1, z, w, \underline{u}) \cup \ldots \cup (\forall \underline{u})\ \Delta_0(\alpha_n, \beta_n, z, w, u)).$

The proof of this theorem is an easy modification of the proofs presented in Section 5.1. (1) implies that the equationally defined commutator for $\mathbf{Q}$ is additive in the lattice $\boldsymbol{Th}(\mathbf{Q}^{\models})$. Then (2) follows. In the proof of the implication (2) $\Rightarrow$ (1) one proves that (2).(i)–(2).(ii) imply that

$$[\mathbf{Q}^{\models}(X), \mathbf{Q}^{\models}(Y)] = \mathbf{Q}^{\models}(\bigcup\{(\forall \underline{u})\ \Delta_0(\alpha, \beta, \gamma, \delta, \underline{u}) : \alpha \approx \beta \in X, \gamma \approx \delta \in Y\})$$

for every pair X, Y of sets of equations.

(One argues as in the proof of the implication (2) $\Rightarrow$ (1) of Theorem 5.1.4.) We omit the details. Conditions (1) and (2) are therefore equivalent. □

Note 5.2.4. In condition (2).(ii) of the above theorem we may restrict ourselves to a set of standard rules forming an inferential base of $\mathbf{Q}^{\models}$, that is:

Let R be a set of standard rules such that $C^{eq}_R = \mathbf{Q}^{\models}$.

(2).(ii)$_R$ *Suppose that for every rule $\alpha_1 \approx \beta_1, \ldots, \alpha_n \approx \beta_n/\alpha \approx \beta$ of R and any variables z and w not occurring in the equations $\alpha_1 \approx \beta_1, \ldots, \alpha_n \approx \beta_n$ and $\alpha \approx \beta$ it is the case that*

$(\forall \underline{u})\ \Delta_0(\alpha, \beta, z, w, \underline{u}) \subseteq$

$\mathbf{Q}^{\models}((\forall \underline{u})\ \Delta_0(\alpha_1, \beta_1, z, w, \underline{u}) \cup \ldots \cup (\forall \underline{u})\ \Delta_0(\alpha_n, \beta_n, z, w, \underline{u})).$

Then (2).(i) *and* (2).(ii)$_R$ *imply that the equationally defined commutator for $\mathbf{Q}$ is additive. In fact,*

(∗) $[\mathbf{Q}^{\models}(X), \mathbf{Q}^{\models}(Y)] = \mathbf{Q}^{\models}(\bigcup\{(\forall \underline{u})\ \Delta_0(\alpha, \beta, \gamma, \delta, \underline{u}) : \alpha \approx \beta \in X, \gamma \approx \delta \in Y\}),$ *for any sets of equations X, Y.* □

Note that the equality (∗) implies that Δ_0 is a generating set for the commutator of $\mathbf{Q}$ because for $X = \{x \approx y\}$, $Y = \{z \approx w\}$ we get that

$$\mathbf{Q}^{\models}(x \approx y) \cap \mathbf{Q}^{\models}(z \approx w) = [\mathbf{Q}^{\models}(x \approx y), \mathbf{Q}^{\models}(z \approx w)] = \mathbf{Q}^{\models}((\forall \underline{u})\ \Delta_0(x, y, z, w, \underline{u})).$$

Conversely, if Δ_0 is a generating set and R is an inferential base for $\mathbf{Q}^{\models}$, then the additivity of the commutator and condition (2).(i) already imply (2).(ii)$_R$.

Remark. Some care is needed when one operates with standard rules. Each rule $r : \alpha_1 \approx \beta_1, \ldots, \alpha_n \approx \beta_n/\alpha \approx \beta$ is given in a schematic way—the pair $\langle\{\alpha_1 \approx \beta_1, \ldots, \alpha_n \approx \beta_n\}, \alpha \approx \beta\rangle$ is a scheme of the rule and all pairs belonging to this rule are of the form $\langle\{e\alpha_1 \approx e\beta_1, \ldots, e\alpha_n \approx e\beta_n\}, e\alpha \approx e\beta\rangle$, where e ranges over endomorphisms of the term algebra $\boldsymbol{Te}_\tau$. Any other scheme of r is of the form $\langle\{a\alpha_1 \approx a\beta_1, \ldots, a\alpha_n \approx a\beta_n\}, a\alpha \approx a\beta\rangle$, where a is an arbitrary automorphism of the term algebra. It then follows by structurality that condition (2).(ii)$_R$ is 'scheme-independent' in the following sense: if $\langle\{\alpha'_1 \approx \beta'_1, \ldots, \alpha'_n \approx \beta'_n\}, \alpha' \approx \beta'\rangle$ is an arbitrary scheme of r and z and w are any variables separated from the variables occurring in $\alpha'_1 \approx \beta'_1, \ldots, \alpha'_n \approx \beta'_n$, then (2).(ii)$_R$ also means that

$$(\forall \underline{u})\ \Delta_0(\alpha', \beta', z, w, \underline{u}) \subseteq$$
$$\mathbf{Q}^{\models}((\forall \underline{u})\ \Delta_0(\alpha'_1, \beta'_1, z, w, \underline{u}) \cup \ldots \cup (\forall \underline{u})\ \Delta_0(\alpha'_n, \beta'_n, z, w, \underline{u})). \qquad \square$$

Note 5.2.5. Taking as Δ_0 in Theorem 5.2.3 the (infinite) set Δ_c of quaternary commutator equations for $\mathbf{Q}$ supplied by the (relative) cube property and the centralization relation in the sense of the classical two-binary term condition, we conclude that if Δ_c satisfies (2).(ii), then it satisfies (∗), and hence the equationally defined commutator of $\mathbf{Q}$ is additive. □

Let $r : \alpha_1 \approx \beta_1, \ldots, \alpha_n \approx \beta_n / \alpha \approx \beta$ be a rule of $\mathbf{Q}^{\models}$ and let z and w be variables not occurring in the equations $\alpha_1 \approx \beta_1, \ldots, \alpha_n \approx \beta_n$ and $\alpha \approx \beta$. The set of rules

$$(\forall \underline{u})\ \Delta_0(\alpha_1, \beta_1, z, w, \underline{u}) \cup \ldots \cup (\forall \underline{u})\ \Delta_0(\alpha_n, \beta_n, z, w, \underline{u}) / \Delta_0(x, y, z, w, \underline{u})$$

is called the *Δ_0-transform of r*. This set of rules may be infinite (if the set $\Delta_0(x, y, z, w, \underline{u})$ is infinite). Moreover, if the string $\underline{u}$ of parameters is non-empty, the above rules may be infinitistic. If the Δ_0-transforms of r are rules of $\mathbf{Q}^{\models}$, then, taking into account the fact that the consequence $\mathbf{Q}^{\models}$ is finitary, a finitarization procedure can be applied to the Δ_0-transform of r so as to replace the above set by its finitary counterparts. More precisely, taking an arbitrary rule $(\forall \underline{u})\ \Delta_0(\alpha_1, \beta_1, z, w, \underline{u}) \cup \ldots \cup (\forall \underline{u})\ \Delta_0(\alpha_n, \beta_n, z, w, \underline{u}) / \gamma \approx \delta$ with $\gamma \approx \delta \in \Delta_0(x, y, z, w, \underline{u})$, there exists a finite subset

$$X_{\gamma\approx\delta} \subseteq (\forall \underline{u})\ \Delta_0(\alpha_1, \beta_1, z, w, \underline{u}) \cup \ldots \cup (\forall \underline{u})\ \Delta_0(\alpha_n, \beta_n, z, w, \underline{u})$$

such that $X_{\gamma\approx\delta}/\gamma \approx \delta$ is a standard rule of $\mathbf{Q}^{\models}$. The set of all such rules $X_{\gamma\approx\delta}/\gamma \approx \delta$ with $\gamma \approx \delta$ ranging over $\Delta_0(x, y, z, w, \underline{u})$ is called a *finitarization of* a given *Δ_0-transform of r*. This set is finite whenever $\Delta_0(x, y, z, w, \underline{u})$ is finite.

We underline the fact that the finitarization procedure makes sense if the Δ_0-transform of a given finitary rule r of $\mathbf{Q}^{\models}$ are (possibly infinitistic) rules of $\mathbf{Q}^{\models}$. In particular, this procedure can be applied for any rule of consequence $\mathbf{Q}^{\models}$ corresponding to any quasivariety $\mathbf{Q}$ whose equationally defined commutator is additive.

Theorem 5.2.6. *Let* $\mathbf{Q}$ *be a quasivariety (in a given signature* τ*) with the relative shifting property. Let* $\Delta_0(x, y, z, w, \underline{u})$ *be a generating set. Then for every rule* $r \in Birkhoff(\tau)$ *of Birkhoff's logic* B_τ*, the* Δ_0*-transform of* r *consist of rules of* $\mathbf{Q}^{\models}$ *as well.*

As $\mathbf{Q}^{\models}$ is stronger than the consequence $\boldsymbol{Va}(\mathbf{Q})^{\models}$, the latter being an axiomatic strengthening of Birkoff's logic B_τ, it follows that every Birkhoff's rule r is a rule of $\mathbf{Q}^{\models}$. (Generally, $\mathbf{Q}^{\models}$ may also have other rules not being rules of B_τ.) The theorem states that the Δ_0-transform of any Birkhoff's rule r yields a set of rules of $\mathbf{Q}^{\models}$.

Proof. As $\mathbf{Q}$ has the shifting property and $\Delta_0(x, y, z, w, \underline{u})$ is a generating set, Theorem 4.3.9 (see formula (1) in the proof of it) imply that $[X, Y] = \mathbf{Q}^{\models}(\bigcup\{(\forall \underline{u})\ \Delta_0(\alpha, \beta, \gamma, \delta, \underline{u}) : \alpha \approx \beta \in X, \gamma \approx \delta \in Y\})$ for every pair X, Y of theories of $\mathbf{Q}^{\models}$.

The Δ_0-transform of the axiomatic rule $/x \approx x$ consists of the equations $\Delta_0(x, x, z, w, \underline{u})$ which are obviously $\mathbf{Q}$-valid.

As $\mathbf{Q}^{\models}(x \approx y) = \mathbf{Q}^{\models}(y \approx x)$, we have that $\mathbf{Q}^{\models}(y \approx x) \cap \mathbf{Q}^{\models}(z \approx w) = \mathbf{Q}^{\models}(x \approx y) \cap \mathbf{Q}^{\models}(y \approx x)$. Consequently, $\mathbf{Q}^{\models}((\forall \underline{u})\ \Delta_0(y, x, z, w, \underline{u})) = \mathbf{Q}^{\models}((\forall \underline{u})\ \Delta_0(x, y, z, w, \underline{u}))$. The Δ_0-transforms of the rule $x \approx y/y \approx x$ consists of the rules $(\forall \underline{u})\ \ \Delta_0(x, y, z, w, \underline{u})/\ \Delta_0(y, x, z, w, \underline{u})$. Since $\Delta_0(y, x, z, w, \underline{u}) \subseteq \mathbf{Q}^{\models}(x \approx y) \cap \mathbf{Q}^{\models}(z \approx w) = \mathbf{Q}^{\models}((\forall \underline{u})\ \ \Delta_0(x, y, z, w, \underline{u}))$, the members of $(\forall \underline{u})\ \Delta_0(x, y, z, w, \underline{u})/\Delta_0(y, x, z, w, \underline{u})$ are $\mathbf{Q}^{\models}$-rules.

The case of other rules is less trivial. We shall prove the following two lemmas, interesting in their own right.

Lemma 5.2.7. *Let* $\mathbf{Q}$ *be a quasivariety possessing a Day implication system* $\Rightarrow_D$*. Let* x', y', z', z, w *be different variables. Then*

$$\mathbf{Q}^{\models}(x' \approx z') \cap \mathbf{Q}^{\models}(z \approx w) \subseteq \mathbf{Q}^{\models}(x' \approx y') \cap \mathbf{Q}^{\models}(z \approx w)$$
$$+_{\mathbf{Q}}\mathbf{Q}^{\models}(y' \approx z') \cap \mathbf{Q}^{\models}(z \approx w).$$

Proof (of the lemma). We write: $x \approx y \Rightarrow_D z \approx w = \{p_i(x, y, z, w) \approx q_i(x, y, z, w) : i \in I\}$. Let $\Delta_0 = \Delta_0(x, y, z, w, \underline{u})$ be a set of quaternary commutator equations for $\mathbf{Q}$ such that $\mathbf{Q}^{\models}(x \approx y) \cap \mathbf{Q}^{\models}(z \approx w) = \mathbf{Q}^{\models}(\Delta_0(x, y, z, w, \underline{u}))$. (The variables x, y are assumed to be different from x', y', z', z, w.) Then

$$\mathbf{Q}^{\models}(x' \approx z') \cap \mathbf{Q}^{\models}(z \approx w) = \mathbf{Q}^{\models}(\Delta_0(x', z', z, w, \underline{u})),$$

$$\mathbf{Q}^{\models}(x' \approx y') \cap \mathbf{Q}^{\models}(z \approx w) = \mathbf{Q}^{\models}(\Delta_0(x', y', z, w, \underline{u})),$$

$$\mathbf{Q}^{\models}(y' \approx z') \cap \mathbf{Q}^{\models}(z \approx w) = \mathbf{Q}^{\models}(\Delta_0(y', z', z, w, \underline{u})),$$

by the structurality of $\mathbf{Q}^{\models}$.

Thus the lemma equivalently states that the Δ_0-transform of the transitivity rule $x' \approx y', y' \approx z'/x' \approx z'$, viz.,

$$\Delta_0(x', y', z, w, \underline{u}) \cup \Delta_0(y', z', z, w, \underline{u})/\Delta_0(x', z', z, w, \underline{u}), \tag{1}$$

is a set of rules of $\mathbf{Q}^{\models}$ as well. The variables x', y', z', x, y, z, w are all different. For brevity, the consequence operation $\mathbf{Q}^{\models}$ is marked by C.

We define a set of equations as follows: in each equation $p_i(x, y, z, w) \approx q_i(x, y, z, w)$ belonging to $\Rightarrow_D$ the variables x_1 and y_1 are substituted for x, y, respectively, while for the variables z, w in the equation $p_i(x, y, z, w) \approx q_i(x, y, z, w)$ one substitutes the terms $p_j(x_2, y_2, x_3, y_3)$ and $q_j(x_2, y_2, x_3, y_3)$, respectively, for each j. The obtained equation in the developed form can be written down as

$$p_i(x_1, y_1, p_j(x_2, y_2, x_3, y_3), q_j(x_2, y_2, x_3, y_3)) \approx \\ q_i(x_1, y_1, p_j(x_2, y_2, x_3, y_3), q_j(x_2, y_2, x_3, y_3)). \tag{2}$$

The resulting set of equations, for all possible choices of i and j in I is marked here for brevity as

$$x_1 \approx y_1 \Rightarrow_D (x_2 \approx y_2 \Rightarrow_D x_3 \approx y_3), \tag{3}$$

(cf. the notation adopted in the proof of Theorem 3.6.2).

We then make further substitutions in the equations of (3). Let $\alpha(x, y, z, w, \underline{u}) \approx \beta(x, y, z, w, \underline{u})$ be an arbitrary but fixed equation belonging to $\Delta_0(x, y, z, w, \underline{u})$. We define the substitution

$$\begin{aligned} &x_1/\alpha(x', y', z, w, \underline{u}) \\ &y_1/\beta(x', y', z, w, \underline{u}) \\ &x_2/\alpha(y', z', z, w, \underline{u}) \\ &y_2/\beta(y', z', z, w, \underline{u}) \\ &x_3/\alpha(x', z', z, w, \underline{u}) \\ &y_3/\beta(x', z', z, w, \underline{u}). \end{aligned}$$

We apply this substitution to the equations belonging to (3). We then repeat the above substitution procedure for all equations $\alpha(x, y, z, w, \underline{u}) \approx \beta(x, y, z, w, \underline{u})$ in Δ_0. The resulting set of equations obtained from (3) is denoted by

$$\Delta_0(x', y', z, w, \underline{u}) \Rightarrow_D (\Delta_0(y', z', z, w, \underline{u}) \Rightarrow_D \Delta_0(x', z', z, w, \underline{u})). \tag{4}$$

The equations in the set (4) contain the variables x', y', z', z, w and possibly some other parameters. To simplify notation we shall mark the set (4) as

$$\Delta(x', y', z, w, \underline{v})$$

treating, e.g. z' together with the other variables different from x', y', z, w as parametric variables.

Claim 1. *$\Delta(x', y', z, w, \underline{v})$ is a set of quaternary commutator equations in the variables x', y' and z, w.*

Proof (of the claim). We must show that the equations $\Delta(x', x', z, w, \underline{v})$ and $\Delta(x', y', z, z, \underline{v})$ hold in $\mathbf{Q}$.

The set $\Delta(x', x', z, w, \underline{v})$ is equal to

$$\Delta_0(x', x', z, w, \underline{u}) \Rightarrow_D (\Delta_0(x', z', z, w, \underline{u}) \Rightarrow_D \Delta_0(x', z', z, w, \underline{u})). \quad (5)$$

The equations of $\Delta_0(x', x', z, w, \underline{u})$ are valid in $\mathbf{Q}$, because $\Delta_0(x', y', z, w, \underline{u})$ is a set of quaternary commutator equations in x', y' and z, w. In turn, the definition of (4) and condition (iD1) for $\Rightarrow_D$ implies that equations of $\Delta_0(x', z', z, w, \underline{u}) \Rightarrow_D \Delta_0(x', z', z, w, \underline{u})$ are valid in $\mathbf{Q}$. According to Note 1 following Theorem 3.5.1, the equations of $(p \approx q) \Rightarrow_D (r \approx s)$ are $\mathbf{Q}$-valid whenever the equations $p \approx q$ and $r \approx s$ are $\mathbf{Q}$-valid, Consequently, (5) is a set of $\mathbf{Q}$-valid equations.

The set $\Delta(x', y', z, z, \underline{v})$ is equal to

$$\Delta_0(x', y', z, z, \underline{u}) \Rightarrow_D (\Delta_0(y', z', z, z, \underline{u}) \Rightarrow_D \Delta_0(x', z', z, z, \underline{u})). \quad (6)$$

As the sets $\Delta_0(x', y', z, z, \underline{u})$, $\Delta_0(y', z', z, z, \underline{u})$ and $\Delta_0(x', z', z, z, \underline{u})$ consist of $\mathbf{Q}$-valid equations, it follows that (6) consists of $\mathbf{Q}$-valid equations only.

This proves the claim. □

It follows from the claim that $\Delta(x', y', z, w, \underline{v}) \subseteq C(x' \approx y') \cap C(z \approx w) = C(\Delta_0(x', y', z, w, \underline{u}))$. We immediately get from this inclusion, by enlarging the set on the right-hand side, that

Claim 2. $\Delta(x', y', z, w, \underline{v}) \subseteq C(\Delta_0(x', y', z, w, \underline{u}) \cup \Delta_0(y', z', z, w, \underline{u}))$. □

(1) then follows from the above claim. Indeed, according to Claim 2 we have that

$$\Delta_0(x', y', z, w, u) \Rightarrow_D (\Delta_0(y', z', z, w, \underline{u}) \Rightarrow_D \Delta_0(x', z', z, w, \underline{u})) \subseteq$$
$$C(\Delta_0(x', y', z, w, \underline{u}) \cup \Delta_0(y', z', z, w, \underline{u})). \quad (7)$$

Applying twice the detachment rule for $\Rightarrow_D$, it is easy to see that (7) implies

$$\Delta_0(x', z', z, w, \underline{u}) \subseteq C(\Delta_0(x', y', z, w, \underline{u}) \cup \Delta_0(y', z', z, w, \underline{u})).$$

So (1) holds. The lemma has been proved. □

In the following lemma $\Delta_0 = \Delta_0(x, y, z, w, \underline{u})$ is assumed to be a set of quaternary commutator equations for $\mathbf{Q}$ such that $\mathbf{Q}^{\models}(x \approx y) \cap \mathbf{Q}^{\models}(z \approx w) = \mathbf{Q}^{\models}(\Delta_0(x, y, z, w, \underline{u}))$.

Lemma 5.2.8. *Let $\mathbf{Q}$ be an arbitrary quasivariety. Let f be an m-ary operation symbol from the signature of $\mathbf{Q}$. Let $\underline{z} = z_1, \ldots, z_m$ be different variables disjoint from x, y, z, w. Then for any i $(i = 1, \ldots, m)$*

$$\Delta_0(f(\underline{z}_{i-1}x\underline{z}_{i+1}), f(\underline{z}_{i-1}y\underline{z}_{i+1}), z, w, \underline{u}) \subseteq \mathbf{Q}^{\models}(\Delta_0(x, y, z, w, \underline{u})) \qquad (8)_i$$

where $\underline{z}_{i-1} := z_1, \ldots, z_{i-1}$ *and* $\underline{z}_{i+1} := z_{i+1}, \ldots, z_m$.

Proof (of the lemma). Let $\Delta_0 = \Delta_0(x, y, z, w, \underline{u})$ be as above.

Let $\underline{z} = z_1, \ldots, z_m$ be different variables disjoint from x, y, z, w. Then for any i $(i = 1, \ldots, m)$ we trivially have:

$$\mathbf{Q}^{\models}(f(\underline{z}_{i-1}x\underline{z}_{i+1}) \approx f(\underline{z}_{i-1}y\underline{z}_{i+1})) \subseteq \mathbf{Q}^{\models}(x \approx y).$$

Hence, multiplying both sides by $\mathbf{Q}^{\models}(z \approx w)$ we get the inclusion:

$$\mathbf{Q}^{\models}(f(\underline{z}_{i-1}x\underline{z}_{i+1}) \approx f(\underline{z}_{i-1}y\underline{z}_{i+1})) \cap \mathbf{Q}^{\models}(z \approx w) \qquad (9)$$
$$\subseteq \mathbf{Q}^{\models}(x \approx y) \cap \mathbf{Q}^{\models}(z \approx w).$$

It follows that

$$\Delta_0(f(\underline{z}_{i-1}x\underline{z}_{i+1}), f(\underline{z}_{i-1}y\underline{z}_{i+1}), z, w, \underline{u}) \subseteq$$
$$\mathbf{Q}^{\models}(f(\underline{z}_{i-1}x\underline{z}_{i+1}) \approx f(z\underline{z}_{i-1}y\underline{z}_{i+1})) \cap \mathbf{Q}^{\models}(z \approx w) \subseteq$$
$$\mathbf{Q}^{\models}(x \approx y) \cap \mathbf{Q}^{\models}(z \approx w) = \mathbf{Q}^{\models}(\Delta_0(x, y, z, w, \underline{u})).$$

Hence $(8)_i$ follows. □

From the above remarks and the two lemmas the theorem follows. □

Corollary 5.2.9. *Let* $\mathbf{V}$ *be a congruence-modular variety. Then the equationally defined commutator for* $\mathbf{V}$ *is additive.*

It follows from Theorem 4.1.12 that the standard commutator coincides with the equationally defined commutator for any congruence-modular variety. Corollary 5.2.9 yields the classical result that the standard commutator in any CM variety $\mathbf{V}$ is additive (see, e.g., Freese and McKenzie 1987 , Proposition 4.3).

Proof. As $\mathbf{V}$ is congruence-modular, it has the relative shifting property. Let R be the set consisting of the rules of Birkhoff's logic B_τ and the axiomatic rules $Id(\mathbf{Q})$. (Each identity $\alpha \approx \beta$ valid in $\mathbf{Q}$ is identified with the axiomatic rule $/\alpha \approx \beta$.) R is thus an inferential base of the consequence operation $\mathbf{V}^{\models}$.

Let $\Delta_0(x, y, z, w, \underline{u})$ be a set of equations such that

$$\mathbf{V}^{\models}(x \approx y) \cap \mathbf{V}^{\models}(z \approx w) = \mathbf{V}^{\models}(\Delta_0(x, y, z, w, \underline{u})).$$

Let $\alpha \approx \beta$ be an equations which is valid in $\mathbf{V}$. Then, by the properties of commutator equations, the equations in $\Delta_0(\alpha, \beta, z, w, \underline{u})$ are $\mathbf{V}$-valid as well. This simply means that the Δ_0-transform of each axiomatic rule $\alpha \approx \beta$ in $Id(\mathbf{V})$, viz. the set $\Delta_0(\alpha, \beta, z, w, \underline{u})$, is a set of axiomatic rules of $\mathbf{V}^{\models}$.

In view of the above theorem, the Δ_0-transform of each Birkhoff's rule r consists of rules of $\mathbf{V}^{\models}$.

Thus, for every rule $r \in R$, the Δ_0-transform of r is a set of rules of $\mathbf{V}^{\models}$. By applying Note 5.2.4, the theorem follows. □

The above corollary also directly follows from Theorem 6.3.3 in the next paragraph and the fact that for varieties of algebras, the relative shifting property is equivalent to congruence modularity.

5.2.1 The Equationally-Defined Commutator on the Free Algebra $\boldsymbol{F}_{\mathbf{Q}}(\omega)$

This section is devoted to a syntactic characterization of additivity of the equationally defined commutator. We shall investigate special equational theories of quasivarieties, viz. theories generated by equalities of separated individual variables. We shall first provide some simple observations on the equational commutator on the free algebra $\boldsymbol{F}_{\mathbf{Q}}(\omega)$.

Let $\mathbf{Q}$ be a quasivariety and $C := \mathbf{Q}^{\models}$ the equational consequence associated with $\mathbf{Q}$. $C_0 := \boldsymbol{Va}(\mathbf{Q})^{\models}$ is the equational logic associated with the variety $\boldsymbol{Va}(\mathbf{Q})$.

If X is a set of equations, then $Var(X)$ is the set of individual variables occurring in the equations of X. Two sets of equations X and Y are said to have *separated variables* if $Var(X) \cap Var(Y) = \emptyset$.

C_0 is an axiomatic extension of Birkhoff's consequence B_τ in the signature τ of $\mathbf{Q}$ and $C_0(X) \subseteq C(X)$ for any set of equations X and $C_0(\emptyset) = C_0(\emptyset)$. Moreover, if $X = \{x_i \approx y_i : i \in I\}$ is a set of equations, where x_i, y_i $(i \in I)$ are individual variables, then $C_0(X) = C(X)$. The theories of the form $C(X)$ with $X = \{x_i \approx y_i : i \in I\}$ as above are determined by the equations of X in the following sense: suppose $X = \{x_i \approx y_i : i \in I\}$ and $Y = \{z_j \approx w_j : j \in J\}$, where the variables occurring in the equations of X are all pairwise different and the same holds for the variables occurring in the equations of Y. Then

$$C(X) = C(Y) \quad \text{if and only if} \quad X = Y,$$

provided that $\mathbf{Q}$ is non-trivial (see also Lemma 6.3.5.(2)).

The following simple lemma is crucial for the investigations into the additivity property of the equationally defined commutator.

Lemma 5.2.10. *Let $\mathbf{Q}$ be an arbitrary quasivariety. Let $X = \{x_i \approx y_i : i \in I\}$, where x_i, y_i, $i \in I$, are pairwise different individual variables. Let Y and Z be arbitrary sets of equations whose variables are separated from the variables of X, i.e., $Var(X) \cap (Var(Y) \cup Var(Z)) = \emptyset$. Then*

$$(\mathbf{Q}^{\models}(X) +_{\mathbf{Q}} \mathbf{Q}^{\models}(Y)) \cap (\mathbf{Q}^{\models}(X) +_{\mathbf{Q}} \mathbf{Q}^{\models}(Z)) = \mathbf{Q}^{\models}(X) +_{\mathbf{Q}} \mathbf{Q}^{\models}(Y) \cap \mathbf{Q}^{\models}(Z).$$

Proof (of the lemma). Let $C := \mathbf{Q}^{\models}$. The inclusion "$\supseteq$" is immediate. To prove the reverse inclusion, let us assume that $\alpha \approx \beta \in (C(X) +_{\mathbf{Q}} C(Y)) \cap (C(X) +_{\mathbf{Q}} C(Z))$, that is, $\alpha \approx \beta \in C(X \cup Y)$ and $\alpha \approx \beta \in C(X \cup Z)$. Let $e : \boldsymbol{Te}_\tau \to \boldsymbol{Te}_\tau$ be the endomorphism such that $ex_i = x_i$ and $ey_i = x_i$ for all $i \in I$, and e is the identity map on the remaining variables. In particular, e is the identity map on the variables occurring in Y and Z. Then, by the structurality of C and the fact that $eX \subseteq C(\emptyset)$, we get that $e\alpha \approx e\beta \in C(e(X \cup Y)) = C(eX \cup eY) = C(eY) = C(Y)$ and, likewise, $e\alpha \approx e\beta \in C(e(X \cup Z)) = C(eX \cup eZ) = C(eZ) = C(Z)$. Hence

$$e\alpha \approx e\beta \in C(Y) \cap C(Z). \tag{a}$$

To show that $\alpha \approx \beta \in C(X \cup (C(Y) \cap C(Z)))$, suppose $\boldsymbol{A} \in \mathbf{Q}$ and let $h : \boldsymbol{Te}_\tau \to \boldsymbol{A}$ be a homomorphism which validates the equations of $X \cup (C(Y) \cap C(Z))$. We claim that $h\alpha = h\beta$. We have that

$$hy_i = hx_i \quad \text{for all } i \in I, \tag{b}$$

because h validates X. As h also validates $C(Y) \cap C(Z)$, (a) gives that

$$he\alpha = he\beta. \tag{c}$$

But (b) and the definition of e yield that $het = ht$ for every term t. (Use induction on the complexity of terms.) In particular, $he\alpha = h\alpha$ and $he\beta = h\beta$. This and (c) give that $h\alpha = h\beta$. □

Corollary 5.2.11. *Let* $\mathbf{Q}$ *be an arbitrary quasivariety. Let* $X = \{x_i \approx y_i : i \in I\}$ *be a set of equations of pairwise different variables. Then for every finite non-empty family* $\{Y_1, \dots, Y_n\}$ *of arbitrary sets of equations of terms such that the variables occurring in the terms of* $Y_1 \cup \dots \cup Y_n$ *are separated from the variables of X, i.e.,* $Var(X) \cap (Var(Y_1) \cup \dots \cup Var(Y_n)) = \emptyset$, *it is the case that*

$$(*)_n \quad \mathbf{Q}^{\models}(X \cup Y_1) \cap \dots \cap \mathbf{Q}^{\models}(X \cup Y_n) = \mathbf{Q}^{\models}(X) +_{\mathbf{Q}} \mathbf{Q}^{\models}(Y_1) \cap \dots \cap \mathbf{Q}^{\models}(Y_n).$$

Proof. Let $C = \mathbf{Q}^{\models}$. The inclusion "$\supseteq$" is obvious. To prove the reverse inclusion, suppose that $\alpha \approx \beta \in C(X \cup Y_1) \cap \dots \cap C(X \cup Y_n)$. Then suitably modify the above proof of Lemma 5.2.10 to show that $\alpha \approx \beta \in C(X) +_{\mathbf{Q}} C(Y_1) \cap \dots \cap C(Y_n)$. □

In Section 5.3 we shall examine a condition which is dual of the statement of the above lemma. This dual property, however, does not universally hold for all quasivarieties. But, more importantly, it retains its validity for all RCM quasivarieties.

For finite sequences of equations and $C := \mathbf{Q}^{\models}$ we introduce the following abbreviations:

$$\underline{p}_m \approx \underline{q}_m := \langle p_1 \approx q_1, \dots, p_m \approx q_m \rangle,$$
$$C(\underline{p}_m \approx \underline{q}_m) := C(p_1 \approx q_1, \dots, p_m \approx q_m).$$

Thus if $\underline{r}_n \approx \underline{s}_n := \langle r_1 \approx s_1, \ldots, r_n \approx s_n \rangle$, then

$$C(\underline{p}_m \approx \underline{q}_m) \cap C(\underline{r}_n \approx \underline{s}_n) = C(p_1 \approx q_1, \ldots, p_m \approx q_m) \cap C(r_1 \approx s_1, \ldots, r_n \approx s_n).$$

We shall consider the following condition:

Let m and n be arbitrary positive integers. For any disjoint finite sequences $\underline{x}_m, \underline{y}_m, \underline{z}_n, \underline{w}_n$ of pairwise different individual variables, where $\underline{x}_m = x_1, \ldots, x_m$, $\underline{y}_m = y_1, \ldots, y_m, \underline{z}_n = z_1, \ldots, z_n, \underline{w}_n = w_1, \ldots, w_n$,

$$(\text{EqDistr})_{m,n} \quad C(\underline{x}_m \approx \underline{y}_m) \cap C(\underline{z}_n \approx \underline{w}_n) = C(\bigcup_{1 \leqslant i \leqslant m, 1 \leqslant j \leqslant n} C(x_i \approx y_i) \cap C(z_j \cap w_j)).$$

In particular, for $m = 2$ and $n = 1$, we obtain:

$$C(x_1 \approx y_1, x_2 \approx y_2) \cap C(z_1 \approx w_1,) =$$

$$(\text{EqDistr})_{2,1} \qquad C(C(x_1 \approx y_1) \cap C(z_1 \approx w_1) \cup C(x_2 \approx y_2) \cap C(z_1 \approx w_1)).$$

(EqDistr)$_{m,n}$ for $m \geqslant 1$ and $n \geqslant 1$ are certain restricted laws of distributivity tailored for the simplest atomic equations. Conditions (EqDistr)$_{m,n}$ do not continue to hold (with the exception of the trivial case $m = 1$ and $n = 1$) if the individual variables occurring in these laws are uniformly replaced by arbitrary terms. In particular, the laws (EqDistr)$_{m,n}$ should not be confused with congruence-distributivity. As we shall later see, the infinite number of equations (EqDistr)$_{m,n}$ *does not imply* the distributivity of the lattice of equational theories of **Q** (or, equivalently—the distributivity of the lattices of **Q**-congruences on the algebras of **Q**.) However, if **Q** is relatively congruence-modular, it validates (EqDistr)$_{m,n}$ for all $m \geqslant 1$ and all $n \geqslant 1$ (Theorems 5.2.16 and 6.3.2).

The above formulas continue to hold for infinite sets of equations (or pairs of free generators). Let $X = \{x_i \approx y_i : i \in I\}$, $Y = \{z_j \approx w_j : j \in J\}$ be possibly infinite sets of equations of variables, where the variables x_i, y_i ($i \in I$) are pairwise different and z_j, w_i ($j \in J$) are also pairwise different. We say that X and Y are *separated* if the equations in X and Y do not contain a common variable. This definition extends in an obvious way on sets X and Y of pairs of free generators of $\boldsymbol{F}$.

(EqDistr)$_\infty$ is the following condition:

(EqDistr)$_\infty$ *Let $X = \{x_i \approx y_i : i \in I\}$, $Y = \{z_j \approx w_j : j \in J\}$ be possibly infinite, separated set of equations of variables. Then*

$$C(X) \cap C(Y) = C(\bigcup_{i \in I, j \in J} C(x_i \approx y_i) \cap C(z_j \approx w_j)).$$

Lemma 5.2.12. *For any quasivariety* **Q**, *condition* (EqDistr)$_\infty$ *holds if and only if, for all positive m, n, condition* (EqDistr)$_{m,n}$ *holds.*

Proof. Use the fact that the consequence C is finitary. □

Notes.

(1). $(\mathrm{EqDistr})_\infty$ implies that for any separated sets X_1, X_2, Y of equations of variables:

$$(C(X_1) +_{\mathbf{Q}} C(X_2)) \cap C(Y) = C(X_1) \cap C(Y) +_{\mathbf{Q}} C(X_2) \cap C(Y), \qquad (*)$$

In fact, $(*)$ is equivalent to $(\mathrm{EqDistr})_\infty$ as one can easily check, because $(*)$ implies condition $(**)$ below:

$$(C(X_1) +_{\mathbf{Q}} C(X_2)) \cap (C(Y_1) +_{\mathbf{Q}} C(Y_2)) =$$
$$C(X_1) \cap C(Y_1) +_{\mathbf{Q}} C(X_1) \cap C(Y_2) +_{\mathbf{Q}} C(X_2) \cap C(Y_1) +_{\mathbf{Q}} C(X_2) \cap C(Y_2)). \qquad (**)$$

for any separated sets X_1, X_2, Y_1, Y_2 of equations of variables. $(*)$ yields $(\mathrm{EqDistr})_\infty$.
The dual equality to $(*)$, viz.

$$C(X_1) \cap C(X_2) +_{\mathbf{Q}} C(Y) = (C(X_1) +_{\mathbf{Q}} C(Y)) \cap (C(X_2) +_{\mathbf{Q}} C(Y)),$$

holds for *any* quasivariety $\mathbf{Q}$, because it follows from Lemma 5.2.10.

(2). As $C_0(X) = C(X)$ for any set $X = \{x_i \approx y_i : i \in I\}$, we may rewrite $(\mathrm{EqDistr})_{m,n}$ and $(\mathrm{EqDistr})_\infty$ in an equivalent form as

$$C_0(\underline{x}_m \approx \underline{y}_m) \cap C_0(\underline{z}_n \approx \underline{w}_n) = C(\bigcup_{1 \leqslant i \leqslant m, 1 \leqslant j \leqslant n} C_0(x_i \approx y_i) \cap C_0(z_j \approx w_j))$$

and

$$C_0(X) \cap C_0(Y) = C(\bigcup_{i \in I, j \in J} C_0(x_i \approx y_i) \cap C_0(z_j \approx w_j)),$$

respectively.

(3). We are concerned with congruences on the free algebra $\boldsymbol{F} := \boldsymbol{F}_{\mathbf{Q}}(\omega)$. If $X \subseteq F^2$, then $\Theta^{\boldsymbol{F}}(X)$ and $\Theta^{\boldsymbol{F}}_{\mathbf{Q}}(X)$ stand for the congruence of $\boldsymbol{F}$ generated by X and the $\mathbf{Q}$-congruence of $\boldsymbol{F}$ generated by X, respectively. If $X = \{\langle x_i, y_i\rangle : i \in I\}$ is a set of pairs of free generators of $\boldsymbol{F}$, then $\Theta^{\boldsymbol{F}}(X) = \Theta^{\boldsymbol{F}}_{\mathbf{Q}}(X)$ (Proposition 2.6).

Condition $(\mathrm{EqDistr})_{m,n}$ can be equivalently paraphrased as a property of the congruences generated by finite sets of pairs of free generators in the free algebra $\boldsymbol{F} := \boldsymbol{F}_{\mathbf{Q}}(\omega)$. In other words, the lattice of $\mathbf{Q}$-congruences of the free algebra $\boldsymbol{F}$ obeys the following form of distributivity:

$(\mathrm{FreeGenDistr})_{m,n}$ *Let m and n be arbitrary positive integers. For any disjoint finite sequences $\underline{x}, \underline{y}, \underline{z}, \underline{w}$ of pairwise different free generators* *of* $\boldsymbol{F}$, *where* $\underline{x} = x_1, \ldots, x_m$, $\underline{y} = y_1, \ldots, y_m$, $\underline{z} = z_1, \ldots, z_n$, $\underline{w} = w_1, \ldots, w_n$,

$$\Theta^{\boldsymbol{F}}(\langle x_1, y_1\rangle, \ldots, \langle x_m, y_m\rangle) \cap \Theta^{\boldsymbol{F}}(\langle z_1, w_1\rangle, \ldots, \langle z_n, w_n\rangle) = \Theta^{\boldsymbol{F}}_{\mathbf{Q}}(\bigcup_{1 \leqslant i \leqslant m, 1 \leqslant j \leqslant n} \Theta^{\boldsymbol{F}}(x_i, y_i) \cap \Theta^{\boldsymbol{F}}(z_j, w_j)).$$

As $\Theta^{\boldsymbol{F}}(\langle x_1, y_1\rangle, \ldots, \langle x_m, y_m\rangle)$ is a **Q**-congruence, therefore it is not necessary to put the subscript '$_{\mathbf{Q}}$' at '$\Theta^{\boldsymbol{F}}$'. But this subscript occurs on the right-hand side of $(\text{FreeGenDistr})_{m,n}$.

The law $(\text{FreeGenDistr})_{m,n}$ is equivalently expressed in terms of sets of pairs of free generators of the algebra $\boldsymbol{F}$ as follows.

$(\text{FreeGenDistr})_{\infty}$ *Let* $X = \{\langle x_i, y_i\rangle : i \in I\}$, $Y = \{\langle z_j, w_j\rangle : j \in J\}$ *be separated sets of pairs of free generators of* $\boldsymbol{F}$. *Then*

$$\Theta^{\boldsymbol{F}}(X) \cap \Theta^{\boldsymbol{F}}(Y) = \Theta^{\boldsymbol{F}}_{\mathbf{Q}}(\bigcup_{i \in I, j \in J} \Theta^{\boldsymbol{F}}(x_i, y_i) \cap \Theta^{\boldsymbol{F}}(z_j, w_j)).$$

(4) $(\text{FreeGenDistr})_{\infty}$ is equivalent, for any **Q**, to the following conditions:

Let X, Y, Z, W *be arbitrary pairwise separated sets of pairs of free generators. Then*

$$\begin{aligned} &\Theta^{\boldsymbol{F}}(X \cup Y) \cap \Theta^{\boldsymbol{F}}(Z \cup W) \\ &= \Theta^{\boldsymbol{F}}_{\mathbf{Q}}(\Theta^{\boldsymbol{F}}(X) \cap \Theta^{\boldsymbol{F}}(Z) \cup \Theta^{\boldsymbol{F}}(X) \cap \Theta^{\boldsymbol{F}}(W) \\ &\cup \Theta^{\boldsymbol{F}}(Y) \cap \Theta^{\boldsymbol{F}}(Z) \cup \Theta^{\boldsymbol{F}}(Y) \cap \Theta^{\boldsymbol{F}}(W)). \end{aligned}$$

All the conditions introduced in this note are therefore mutually equivalent, for any **Q**. □

Proposition 5.2.13. *Let* **Q** *be a quasivariety validating conditions* $(\text{EqDistr})_{m,n}$ *for all positive integers m and n. Then for every algebra* $\boldsymbol{A} \in \mathbf{Q}$ *and any* $\Phi, \Psi \in$ *Con*$_{\mathbf{Q}}(\boldsymbol{A})$,

$$[\Phi, \Psi]^{\boldsymbol{A}} = \sup\{[\Theta_{\mathbf{Q}}(a, b), \Theta_{\mathbf{Q}}(c, d)]^{\boldsymbol{A}} : a \equiv b\ (\Phi),\ c \equiv d\ (\Psi)\}.$$

(Here the supremum is taken in the lattice $\boldsymbol{Con}_{\mathbf{Q}}(\boldsymbol{A})$.)

Note. The statement of the above proposition, which is weaker than additivity, is the same as the thesis of Corollary 4.1.10. Thus, the equations $(\text{EqDistr})_{m,n}$, $m, n \geqslant 1$, yield the same result as the relative shifting property which was (implicitly) used in the proof of Corollary 4.1.10. However, it should be noted that conditions $(\text{EqDistr})_{m,n}$, $m, n \geqslant 1$, do not imply the relative shifting property. For let τ be the empty signature. Birkhoff's logic B_τ in this signature reduces to the pure identity theory and $B_\tau(\emptyset) = \{x \approx x : x \text{ is an individual variable}\}$. The corresponding quasivariety (which is actually a variety) is formed by the class of all non-empty sets. Congruences on any non-empty set A are equivalence relations of A. B_τ has trivial commutator equations. It follows that the equationally defined commutator $[\Phi, \Psi]^A$ of any two equivalence relations Φ, Ψ on any set A reduces to the diagonal relation of A. This trivial commutator is additive. But the relative shifting property fails for B_τ. □

Proof (of Proposition 5.2.13*).* Let

$$\underline{x} = x_1, x_2, \ldots$$

$$\underline{y} = y_1, y_2, \ldots$$

$$\underline{z} = z_1, z_2, \ldots$$

$$\underline{w} = w_1, w_2, \ldots$$

be four infinite sequences of pairwise different individual variables. As above, C stands for the consequence operation $\mathbf{Q}^{\models}$.

Let $\Delta_0(x, y, z, w, \underline{u})$ be a generating set of $C(x \approx y) \cap C(z \approx w)$, where x, y, z, w are different variables, i.e., $C((\forall \underline{u})\ \Delta_0(x, y, z, w, \underline{u})) = C(x \approx y) \cap C(z \approx w)$. Then $C((\forall \underline{u})\ \Delta_0(x_i, y_i, z_j, w_j, \underline{u})) = C(x_i \approx y_i) \cap C(z_j \approx w_j)$, for all i, j, by structurality.

Let Δ be $C(\underline{x} \approx \underline{y}) \cap C(\underline{z} \approx \underline{w})$. Conditions $(\text{EqDistr})_{m,n}$, $m, n \geqslant 1$, imply that

$$C(\Delta) = C(\bigcup_{m,n \geqslant 1} (\forall \underline{u})\ \Delta_0(x_m, y_m, z_n, w_n, \underline{u})). \tag{1}$$

Let X and Y be theories of C. In view of the comments following Definition 3.1.5 we have that

$$[X, Y] = C(\{\alpha(\underline{p}, \underline{q}, \underline{r}, \underline{s}, \underline{t}) \approx \beta(\underline{p}, \underline{q}, \underline{r}, \underline{s}, \underline{t}) : \alpha \approx \beta \in \Delta, \underline{p} \approx \underline{q} \in X, \underline{r} \approx \underline{s} \in Y, \underline{t} \in Te_\tau^{<\omega}\}). \tag{2}$$

But (1) and the structurality of C imply that

$$C(\{\alpha(\underline{p}, \underline{q}, \underline{r}, \underline{s}, \underline{t}) \approx \beta(\underline{p}, \underline{q}, \underline{r}, \underline{s}, \underline{t}) : \alpha \approx \beta \in \Delta, \underline{p} \approx \underline{q} \in X, \underline{r} \approx \underline{s} \in Y, \underline{t} \in Te_\tau^{<\omega}\}) = C(\bigcup\{\Delta_0(p, q, r, s, \underline{t}) : p \approx q \in X, r \approx s \subseteq Y, \underline{t} \in Te_\tau^{<\omega}\}). \tag{3}$$

(2) and (3) give that

$$[X,Y] = C(\bigcup\{\Delta_0(p,q,r,s,\underline{t}) : p \approx q \in X, r \approx s \subseteq Y, \underline{t} \in Te_\tau^{<\omega}\}) = \quad (4)$$
$$C(\bigcup\{(\forall \underline{u})\ \Delta_0(p,q,r,s,\underline{u}) : p \approx q \in X, r \approx s \in Y\}).$$

As $(\forall \underline{u})\ \Delta_0(p,q,r,s,\underline{u}) \subseteq [C(p \approx q), C(r \approx s)]$, for all terms p,q,r,s, (4) implies

$$[X,Y] \subseteq C(\bigcup\{[C(p \approx q), C(r \approx s)] : p \approx q \in X, r \approx s \in Y\}).$$

Since the opposite inclusion holds in the virtue of the monotonicity of the equationally defined commutator, we obtain that

$$[X,Y] = C(\bigcup\{[C(p \approx q), C(r \approx s)] : p \approx q \in X, r \approx s \in Y\}). \quad (5)$$

(5) carries over to **Q**-congruences of the free algebra $\boldsymbol{F} = \boldsymbol{F}_{\mathbf{Q}}(\omega)$, i.e.,

$$[\Phi, \Psi]^{\boldsymbol{F}} = \sup{}_{\mathbf{Q}}\{[\Theta_{\mathbf{Q}}(a,b), \Theta_{\mathbf{Q}}(c,d)]^{\boldsymbol{F}} : a \equiv b\ (\Phi),\ c \equiv d\ (\Psi)\},$$

for any $\Phi, \Psi \in Con_{\mathbf{Q}}(\boldsymbol{F})$. Then suitably modifying the argumentation presented in the proof of Lemma 5.2.2, one arrives at the thesis of the proposition. □

Problem. Show that the relative shifting property does not imply the condition $(\forall m,n \geqslant 1)(\mathrm{EqDistr})_{m,n}$. □

Note. Equations $(\mathrm{EqDistr})_{m,n}$ need not be preserved on passing to equational theories obtained by substituting arbitrary terms for the variables $x_1,\dots,x_m, y_1,\dots,y_m$, $z_1,\dots,z_n$, and $w_1,\dots,w_n$, because this would imply the distributivity of the lattice of theories of $\mathbf{Q}^{\models}$. In particular, $(\mathrm{EqDistr})_{m,n}$ does not hold if the strings $\underline{x}, \underline{y}, \underline{z}, \underline{w}$ do *not* form disjoint sets of variables. E.g., in view of the above proposition the equation

$$\mathbf{V}^{\models}(x_1 \approx y_1, x_2 \approx y_2) \cap \mathbf{V}^{\models}(z \approx w) =$$
$$\mathbf{V}^{\models}(\mathbf{V}^{\models}(x_1 \approx y_1) \cap \mathbf{V}^{\models}(z \approx w) \cup \mathbf{V}^{\models}(x_2 \approx y_2) \cap \mathbf{V}^{\models}(z \approx w))$$

holds for every CM variety **V** but the formula

$$\mathbf{V}^{\models}(x \approx z, x \approx w) \cap \mathbf{V}^{\models}(z \approx w) =$$
$$\mathbf{V}^{\models}(\mathbf{V}^{eq\models}(x \approx z) \cap \mathbf{V}^{\models}(z \approx w) \cup \mathbf{V}^{\models}(x \approx w) \cap \mathbf{V}^{\models}(z \approx w))$$

does not, because it implies congruence-distributivity of **V**. (To see this, note first that $\mathbf{V}^{\models}(x \approx z, x \approx w) \cap \mathbf{V}^{\models}(z \approx w) = \mathbf{V}^{\models}(z \approx w)$ and then repeat the proof of the well-known Jónsson's Theorem characterizing congruence-distributive varieties.)

□

The following corollary follows from the proof of Proposition 5.2.13. (cf. Theorem 5.1.4):

Corollary 5.2.14. *Let* $\mathbf{Q}$ *be a quasivariety validating conditions* $(\mathrm{EqDistr})_{m,n}$ *for all positive integers* m *and* n. *Let* x, y, z, w *be arbitrary different variables and* $\Delta_0(x, y, z, w, \underline{u})$ *a set of equations such that* $\mathbf{Q}^{\models}((\forall \underline{u})\ \Delta_0(x, y, z, w, \underline{u})) = \mathbf{Q}^{\models}(x \approx y) \cap \mathbf{Q}^{\models}(z \approx w)$. *Then for every algebra* $\mathbf{A} \in \mathbf{Q}$ *and any* $\Phi, \Psi \in Con_{\mathbf{Q}}(\mathbf{A})$,

$$[\Phi, \Psi]^{A} = \Theta^{A}_{\mathbf{Q}}(\bigcup\{(\forall \underline{e})\ \Delta_0(a, b, c, d, \underline{e}) : a \equiv b\ (\Phi),\ c \equiv d\ (\Psi)\}). \qquad \square$$

According to the definition of the equationally defined commutator (Definition 3.1.5), if $\mathbf{Q}$ is an *arbitrary* quasivariety $\mathbf{Q}$, $\mathbf{A} \in \mathbf{Q}$, and $\Phi, \Psi \in Con_{\mathbf{Q}}(\mathbf{A})$, then $[\Phi, \Psi]^{A}$ is the least $\mathbf{Q}$-congruence Ξ such that $Z_{\mathrm{com}}(\Phi, \Psi; \Xi)$, i.e., Φ centralizes Ψ relative to Ξ in the sense of *arbitrary* commutator equations. The above corollary is a stronger result: it states that if $\mathbf{Q}$ validates $(\mathrm{EqDistr})_{m,n}$ for all positive m and n, then $[\Phi, \Psi]^{A}$ is the least $\mathbf{Q}$-congruence Ξ such that $Z_{4,\mathrm{com}}(\Phi, \Psi; \Xi)$, i.e., Φ centralizes Ψ relative to Ξ in the sense of arbitrary *quaternary* commutator equations for $\mathbf{Q}$. In fact, the proof of Proposition 5.2.13 yields:

Corollary 5.2.15. *Let* $\mathbf{Q}$ *be a quasivariety validating conditions* $(\mathrm{EqDistr})_{m,n}$ *for all positive integers* m *and* n. *Then the centralization relations* Z_{com} *and* $Z_{4,\mathrm{com}}$ *coincide on* $Con_{\mathbf{Q}}(\mathbf{A})$ *in any algebra* $\mathbf{A} \in \mathbf{Q}$. $\square$

If one additionally assumes the relative shifting property for $\mathbf{Q}$, then the above equivalence extends to the remaining centralization relations examined in Chapter 4 (Theorem 4.1.2).

Yet another condition we shall introduce is a weakening of (C2). We shall restrict it to the free algebra $\boldsymbol{F}$; or, equivalently, to the term algebra $\boldsymbol{Te}_{\tau}$.

We recall that if $e : \boldsymbol{Te}_{\tau} \to \boldsymbol{Te}_{\tau}$ is an endomorphism, then $\ker_{\mathbf{Q}}(e) := e^{-1}(\mathbf{Q}^{\models}(\emptyset))$ (see the remarks preceding Proposition 2.13). Let m and n be positive integers. We define:

$(\mathrm{Epi})_{m,n}$ *Let* $\underline{x}_m, \underline{y}_m, \underline{z}_n, \underline{w}_n$ *be sequences of pairwise different individual variables, where* $\underline{x}_m = x_1, \ldots, x_m$, $\underline{y}_m = y_1, \ldots, y_m$, $\underline{z}_n = z_1, \ldots, z_n$, $\underline{w}_n = w_1, \ldots, w_n$. *Then for every epimorphism* $e : \boldsymbol{Te}_{\tau} \to \boldsymbol{Te}_{\tau}$,

$$\ker_{\mathbf{Q}}(e) +_{\mathbf{Q}} \mathbf{Q}^{\models}(\underline{x}_m \approx \underline{y}_m) \cap \mathbf{Q}^{\models}(\underline{z}_n \approx \underline{w}_n))$$
$$= e^{-1}([\mathbf{Q}^{\models}(e\underline{x}_m \approx e\underline{y}_m), \mathbf{Q}(e\underline{z}_n \approx e\underline{w}_n)]).$$

The universal closure $(\forall m, n \geqslant 1)(\mathrm{Epi})_{m,n}$ is equivalent to the following condition involving possibly infinite separated sets of equations of individual variables:

$(\mathrm{Epi})_{\infty}$ *Let* X *and* Y *be separated sets of equations of pairwise different individual variables.Then for any epimorphism* $e : \boldsymbol{Te}_{\tau} \to \boldsymbol{Te}_{\tau}$,

$$\ker_{\mathbf{Q}}(e) +_{\mathbf{Q}} \mathbf{Q}^{\models}(X) \cap \mathbf{Q}^{\models}(Y) = e^{-1}([\mathbf{Q}^{\models}(eX), \mathbf{Q}^{\models}(eY)]).$$

In view of Theorem 3.1.8, $(\mathrm{Epi})_\infty$ is equivalent to the condition:

Let X and Y be separated sets of equations of pairwise different individual variables. Then for any epimorphism $e : \boldsymbol{Te}_\tau \to \boldsymbol{Te}_\tau$,

$$\ker_{\mathbf{Q}}(e) +_{\mathbf{Q}} \mathbf{Q}^{\models}(X) \cap \mathbf{Q}^{\models}(Y) = \ker_{\mathbf{Q}}(e) +_{\mathbf{Q}} [\ker_{\mathbf{Q}}(e) +_{\mathbf{Q}} \mathbf{Q}^{\models}(X), \ker_{\mathbf{Q}}(e) +_{\mathbf{Q}} \mathbf{Q}^{\models}(X)].$$

Note. $(\mathrm{Epi})_{m,n}$ and $(\mathrm{Epi})_\infty$ can be equivalently paraphrased in terms of properties of congruences of the free algebra $\boldsymbol{F}$:

$(\mathrm{EpiFreeGen})_{m,n}$ *Let m and n be arbitrary positive integers. Let* $\underline{x}_m, \underline{y}_m, \underline{z}_n, \underline{w}_n$ *be sequences of pairwise different free generators of* $\boldsymbol{F}$, *where* $\underline{x}_m = x_1, \dots, x_m$, $\underline{y}_m = y_1, \dots, y_m$, $\underline{z}_n = z_1, \dots, z_n$, $\underline{w}_n = w_1, \dots, w_n$. *Then for every epimorphism* $h : \boldsymbol{F} \to \boldsymbol{F}$,

$$\ker(h) +_{\mathbf{Q}} \Theta^{\boldsymbol{F}}(\langle \underline{x}_m, \underline{y}_m \rangle) \cap \Theta^{\boldsymbol{F}}(\langle \underline{z}_n, \underline{w}_n \rangle) = h^{-1}([\Theta^{\boldsymbol{F}}_{\mathbf{Q}}(\langle h\underline{x}_m, h\underline{y}_m \rangle), \Theta^{\boldsymbol{F}}_{\mathbf{Q}}(\langle h\underline{z}_n, h\underline{w}_n \rangle)).$$

The universal closure $(\forall m, n \geqslant 1)(\mathrm{EpiFreeGen})_{m,n}$ is equivalent to the following condition formulated in terms of possibly infinite separated sets of pairs of free generators:

$(\mathrm{EpiFreeGen})_\infty$ *Let X and Y be separated sets of pairs of different free generators of* $\boldsymbol{F}$. *Then for every epimorphism* $h : \boldsymbol{F} \to \boldsymbol{F}$,

$$\ker(h) +_{\mathbf{Q}} \Theta^{\boldsymbol{F}}(X) \cap \Theta^{\boldsymbol{F}}(Y) = h^{-1}([\Theta^{\boldsymbol{F}}_{\mathbf{Q}}(hX), \Theta^{\boldsymbol{F}}_{\mathbf{Q}}(hY)]).$$

According to Theorem 3.1.8 $(\mathrm{EpiFreeGen})_\infty$ is equivalent to the condition:

Let X and Y be separated sets of pairs of different free generators of $\boldsymbol{F}$. *Then for every epimorphism* $h : \boldsymbol{F} \to \boldsymbol{F}$,

$$\ker(h) +_{\mathbf{Q}} \Theta^{\boldsymbol{F}}(X) \cap \Theta^{\boldsymbol{F}}(Y) = \ker(h) +_{\mathbf{Q}} [\ker(h) +_{\mathbf{Q}} \Theta^{\boldsymbol{F}}(X), \ker(h) +_{\mathbf{Q}} \Theta^{\boldsymbol{F}}(Y)].$$

Theorem 5.2.16. *Let* $\mathbf{Q}$ *be an arbitrary quasivariety. The following conditions are equivalent:*

(A) *The equationally defined commutator of* $\mathbf{Q}$ *is additive (on the algebras of* $\mathbf{Q}$);
(B) *The equationally defined commutator validates conditions* $(\mathrm{EqDistr})_{m,n}$ *and* $(\mathrm{Epi})_{m,n}$ *for all positive integers m and n.*
(C) *The equationally defined commutator validates conditions* $(\mathrm{EqDistr})_\infty$ *and* $(\mathrm{Epi})_\infty$.

Proof. We put $C := \mathbf{Q}^{\models}$. The equivalence of (B) and (C) has been already established.

(A) ⇒ (B). Assume (A). Fix m and let $\underline{x}, \underline{y}, \underline{z}, \underline{w}$ be as in (B). In view of Proposition 3.2.3 and additivity we have that

$$C(x_1 \approx y_1, \ldots, x_m \approx y_m) \cap C(z_1 \approx w_1, \ldots, z_n \approx w_n) =$$
$$[C(x_1 \approx y_1, \ldots, x_m \approx y_m), C(z_1 \ldots w_1, \ldots, z_n \approx w_n)] =$$
$$C(\bigcup_{1 \leqslant i \leqslant m, 1 \leqslant j \leqslant n} [C(x_i \approx y_i), C(z_j \approx w_j)]) =$$
$$C(\bigcup_{1 \leqslant i \leqslant m, 1 \leqslant j \leqslant n} C(x_i \approx y_i) \cap C(z_j \approx w_j)).$$

So $(\mathrm{EqDistr})_{m,n}$ holds.

Since that additivity of the equational commutator implies condition (C2), which in turn implies $(\mathrm{Epi})_{m,n}$, we see that (B) holds as well.

(C) ⇒ (A). The proof of this implication is longer. Throughout the proof of this implication $\mathbf{Q}$ is assumed to be a quasivariety satisfying (C).

Lemma 1. *Let $\Phi_1, \Phi_2, \Psi \in Con_{\mathbf{Q}}(\boldsymbol{F})$. Then $[\Phi_1 +_{\mathbf{Q}} \Phi_2, \Psi] = [\Phi_1, \Psi] +_{\mathbf{Q}} [\Phi_2, \Psi]$.*

Proof (of the lemma). It suffices to prove the lemma for finitely generated congruences $\Phi_1, \Phi_2, \Psi \in Con_{\mathbf{Q}}(\boldsymbol{F})$.

Select three finite sets X_1, X_2 and Y of mutually separated sets of pairs of free generators of $\boldsymbol{F}$ and a surjective endomorphism $h : \boldsymbol{F} \to \boldsymbol{F}$ such that $\Theta^{\boldsymbol{F}}_{\mathbf{Q}}(hX_1) = \Phi_1$, $\Theta^{\boldsymbol{F}}_{\mathbf{Q}}(hX_2) = \Phi_2$, and $\Theta^{\boldsymbol{F}}_{\mathbf{Q}}(hY) = \Psi$. Then:

$$h^{-1}([\Phi_1 +_{\mathbf{Q}} \Phi_2, \Psi]) =$$
$$h^{-1}([\Theta^{\boldsymbol{F}}_{\mathbf{Q}}(hX_1) +_{\mathbf{Q}} \Theta^{\boldsymbol{F}}_{\mathbf{Q}}(hX_2), \Theta^{\boldsymbol{F}}_{\mathbf{Q}}(hY)]) =$$
$$h^{-1}([\Theta^{\boldsymbol{F}}_{\mathbf{Q}}(h(X_1 \cup X_2), \Theta^{\boldsymbol{F}}_{\mathbf{Q}}(hY)]) = \quad \text{(by } (\mathrm{EpiFreeGen})_\infty\text{)}$$
$$\ker(h) +_{\mathbf{Q}} \Theta^{\boldsymbol{F}}(X_1 \cup X_2) \cap \Theta^{\boldsymbol{F}}(Y) = \quad \text{(by } (\mathrm{FreeGenDistr})_\infty\text{)}$$
$$\ker(h) +_{\mathbf{Q}} \Theta^{\boldsymbol{F}}(X_1) \cap \Theta^{\boldsymbol{F}}(Y) +_{\mathbf{Q}} \Theta^{\boldsymbol{F}}(X_2) \cap \Theta^{\boldsymbol{F}}(Y) =$$
$$\ker(h) +_{\mathbf{Q}} \Theta^{\boldsymbol{F}}(X_1) \cap \Theta^{\boldsymbol{F}}(Y) +_{\mathbf{Q}} \ker(h) +_{\mathbf{Q}} \Theta^{\boldsymbol{F}}(X_2) \cap \Theta^{\boldsymbol{F}}(Y) =$$
$$\text{(by } (\mathrm{EpiFreeGen})_\infty\text{)}$$
$$h^{-1}([\Phi_1, \Psi]) +_{\mathbf{Q}} h^{-1}([\Phi_2, \Psi]) =$$
$$h^{-1}([\Phi_1, \Psi] +_{\mathbf{Q}} [\Phi_2, \Psi]).$$

As h is surjective, it follows that $[\Phi_1 +_{\mathbf{Q}} \Phi_2, \Psi] = [\Phi_1, \Psi] +_{\mathbf{Q}} [\Phi_2, \Psi]$. □

Lemma 2. *Let $\boldsymbol{A}$ be a countably generated algebra in $\mathbf{Q}$ and $\Phi_1, \Phi_2, \Psi \in Con_{\mathbf{Q}}(\boldsymbol{A})$. Then $[\Phi_1 +_{\mathbf{Q}} \Phi_2, \Psi]^{\boldsymbol{A}} = [\Phi_1 +_{\mathbf{Q}} \Psi]^{\boldsymbol{A}} +_{\mathbf{Q}} [\Phi_2 +_{\mathbf{Q}} \Psi]^{\boldsymbol{A}}$.*

Proof. Let $h : \boldsymbol{F} \to \boldsymbol{A}$ be an arbitrary epimorphism. Then:

$$h^{-1}([\Phi_1 +_{\mathbf{Q}} \Phi_2, \Psi]^{A}) = \quad \text{(Corollary 3.1.9)}$$

$$\ker(h) +_{\mathbf{Q}} [h^{-1}(\Phi_1 +_{\mathbf{Q}} \Phi_2), h^{-1}(\Psi)]^{F} =$$

$$\ker(h) +_{\mathbf{Q}} [h^{-1}(\Phi_1) +_{\mathbf{Q}} h^{-1}(\Phi_2), h^{-1}(\Psi)]^{F} = \quad \text{(Lemma 1)}$$

$$\ker(h) +_{\mathbf{Q}} [h^{-1}(\Phi_1), h^{-1}(\Psi)]^{F} +_{\mathbf{Q}} [h^{-1}(\Phi_2), h^{-1}(\Psi)]^{F} =$$

$$\ker(h) +_{\mathbf{Q}} [h^{-1}(\Phi_1), h^{-1}(\Psi)]^{F} +_{\mathbf{Q}} \ker(h) +_{\mathbf{Q}} [h^{-1}(\Phi_2), h^{-1}(\Psi)]^{F} =$$

$$\text{(Corollary 3.1.9)}$$

$$h^{-1}([\Phi_1, \Psi]^{A}) +_{\mathbf{Q}} h^{-1}([\Phi_2, \Psi]^{A}) =$$

$$h^{-1}([\Phi_1, \Psi]^{A} +_{\mathbf{Q}} [\Phi_2, \Psi]^{A}).$$

Hence $[\Phi_1 +_{\mathbf{Q}} \Phi_2, \Psi]^{A} = [\Phi_1 +_{\mathbf{Q}} \Psi]^{A} +_{\mathbf{Q}} [\Phi_2 +_{\mathbf{Q}} \Psi]^{A}$. □

Lemma 3. *Let* $\mathbf{A}$ *be an arbitrary algebra in* $\mathbf{Q}$ *and* $\Phi_1, \Phi_2, \Psi \in Con_{\mathbf{Q}}(\mathbf{A})$. *Then* $[\Phi_1 +_{\mathbf{Q}} \Phi_2, \Psi]^{A} = [\Phi_1 +_{\mathbf{Q}} \Psi]^{A} +_{\mathbf{Q}} [\Phi_2 +_{\mathbf{Q}} \Psi]^{A}$.

Proof. Use Lemma 2 and Theorem 3.1.6.(viii). □

Lemma 3 concludes the proof of (A). The theorem has been proved. □

Note 5.2.17. Suppose $\mathbf{Q}$ is a quasivariety whose equationally defined commutator is additive. The additivity of the commutator entails yet another distributivity property of equational theories of $\mathbf{Q}^{\models}$. We shall discuss it briefly. Put $C := \mathbf{Q}^{\models}$. Let $\underline{x}, \underline{y}, \underline{z}, \underline{w}$ be finite disjoint sets of variables such that $|\underline{x}| = |\underline{y}| = m$, $|\underline{z}| = |\underline{w}|$ and let $\Sigma = \{p_i \approx q_i : i \in I\}$ be a non-empty set of equations such that

$$C(\underline{x} \approx \underline{y}) = C(\Sigma).$$

Then

$$C(\underline{x} \approx \underline{y}) \cap C(\underline{z} \approx \underline{w}) = C(\bigcup_{i \in I} C(p_i \approx q_i) \cap C(\underline{z} \approx \underline{w})). \qquad (\mathrm{Distr})_{\Sigma}$$

Proof. We have:

$$C(\underline{x} \approx \underline{y}) \cap C(\underline{z} \approx \underline{w}) = [C(\underline{x} \approx \underline{y}), C(\underline{z} \approx \underline{w})] =$$

$$[C(\Sigma), C(\underline{z} \approx \underline{w})] = \quad \text{(by additivity)}$$

$$C(\bigcup_{i \in I} [C(p_i \approx q_i), C(\underline{z} \approx \underline{w})]) \subseteq$$

$$C(\bigcup_{i \in I} C(p_i \approx q_i) \cap C(\underline{z} \approx \underline{w})) \subseteq$$

$$C(\Sigma) \cap C(\underline{z} \approx \underline{w}) = C(\underline{x} \approx \underline{y}) \cap C(\underline{z} \approx \underline{w}).$$

The above inclusions give that $C(\underline{x} \approx \underline{y}) \cap C(\underline{z} \approx \underline{w}) \subseteq C(\bigcup_{i \in I} C(p_i \approx q_i) \cap C(\underline{z} \approx \underline{w})) \subseteq C(\underline{x} \approx \underline{y}) \cap C(\underline{z} \approx \underline{w})$. Hence $(*)$ follows. □

It is not difficult to reformulate $(\mathrm{Distr})_\Sigma$ in terms of congruences of the free algebra $\boldsymbol{F}_{\mathbf{Q}}(\omega)$ generated by pairs of separated free generators.

5.3 Restricted Distributivity and Additivity of the Equationally-Defined Commutator

According to Lemma 5.2.10, if $X = \{x_i \approx y_i : i \in I\}$ is a set of equations of pairwise different variables and Y and Z are arbitrary sets of equations of terms whose variables are separated from the variables of X, that is, $Var(X) \cap Var(Y \cup Z) = \emptyset$, then for *any* quasivariety $\mathbf{Q}$ and for $C := \mathbf{Q}^{\models}$ it is the case that

$$(C(X) +_{\mathbf{Q}} C(Y)) \cap (C(X) +_{\mathbf{Q}} C(Z)) = C(X) +_{\mathbf{Q}} C(Y) \cap C(Z). \quad (*)$$

The dual form of this simple equality is of basic importance for the theory of the equationally defined commutator. We therefore define the following condition:

Let $\mathbf{Q}$ be a quasivariety and $C := \mathbf{Q}^{\models}$. Let $X = \{x_i \approx y_i : i \in I\}$, where x_i, y_i, $i \in I$, are pairwise different individual variables. Let Y and Z be arbitrary sets of equations of terms such that the variables occurring in the terms of $Y \cup Z$ are separated from the variables of X. Then:

$$C(X) \cap C(Y) +_{\mathbf{Q}} C(X) \cap C(Z) = C(X) \cap (C(Y) +_{\mathbf{Q}} C(Z)).$$

We call this condition the *restricted distributivity* of the lattice $\boldsymbol{Th}(\mathbf{Q}^{\models})$.

We shall discuss in this paragraph several aspects of this notion. The crucial fact we shall prove is Theorem 5.3.8 which states that the restricted distributivity of the lattice $\boldsymbol{Th}(\mathbf{Q}^{\models})$ implies the additivity of the equationally defined commutator for $\mathbf{Q}$. The proof of this theorem exhibits structural properties of the kernels of epimorphisms of the term algebra $\boldsymbol{Te}_\tau$.

The restricted distributivity can be suitably reformulated in terms of $\mathbf{Q}$-congruences of the free algebra $\boldsymbol{F} := \boldsymbol{F}_{\mathbf{Q}}(\omega)$ and separation of free generators. It follows that the restricted distributivity holds for the lattice $\boldsymbol{Th}(\mathbf{Q}^{\models})$ if and only if its counterpart formulated for $\mathbf{Q}$-congruences of $\boldsymbol{F}$ holds for the lattice $\boldsymbol{Con}_{\mathbf{Q}}(\boldsymbol{F})$.

Note. Restricted distributivity is a property that is essentially weaker than relative congruence-modularity. This issue is discussed in the next chapter. But here we want to elucidate another aspect of the former property. We return to the example from the note following the statement of Proposition 5.2.13.

Let τ be the empty signature. Birkhoff's logic B_τ in this signature is the pure identity theory. The corresponding quasivariety (which is actually a variety) is formed by the class of all non-empty sets and it is called the variety of sets. Each

B_τ-congruence on any non-empty set A is an equivalence relation on A. As B_τ has trivial commutator equations, the equationally defined commutator $[\Phi, \Psi]_A$ of any two equivalence relations Φ, Ψ on any set A, is equal to the diagonal relation of A. This trivial commutator is additive. The lattice $\boldsymbol{Th}(B_\tau)$ is the same as the lattice of equivalence relations on the countably infinite set of variables *Var*. Equations of variables are identified with ordered pairs of variables. From a more abstract perspective we consider the following situation. Let A be a non-empty set and let $X, Y, Z \subseteq A^2$ be sets of ordered pairs of elements of A. Suppose that X is separated from $Y \cup Z$ in the sense that the set of elements of A that occur in the pairs of X is disjoint from the set of elements of A that occur in the pairs of the union $Y \cup Z$. Let D be the operation of generating equivalence relations on A. Then $D(X) \cap D(Y)$, $D(X) \cap D(Z)$ and $D(X) \cap (D(Y) \cup D(Z))$ are all the diagonal relations of A. It follows that $D(X) \cap (D(Y) \cup D(Z)) = D(D(X) \cap D(Y) \cup D(X) \cap D(Z)) =$ the diagonal of A. This observation implies that

> *the lattice* $\boldsymbol{Th}(B_\tau)$ *is distributive in the restricted sense.*

As $\boldsymbol{Th}(B_\tau)$ is isomorphic with the set of all partitions of *Var*,

> $\boldsymbol{Th}(B_\tau)$ *does not satisfy any non-trivial lattice theoretic identity;*

in particular $\boldsymbol{Th}(B_\tau)$ is not modular.[1] Equivalently we may say that the ω-generated free algebra in the variety of sets validates restricted distributivity.

The above example shows that restricted distributivity is *not* expressible as a lattice-theoretic identity; it is a specific property of congruence lattices.

The above reasoning carries over to the equational logic B_τ whose signature τ contains only unary operation symbols. We therefore get:

> *If* τ *contains only unary operation symbols, then the lattice* $\boldsymbol{Th}(B_\tau)$ *is distributive in the restrictive sense.*

It is an open question how large the (hyper)class of quasivarieties $\mathbf{Q}$ is for which the lattice $\boldsymbol{Th}(\mathbf{Q}^{\models})$ validates the restricted distributivity. In other words, we ask for which $\mathbf{Q}$ Lemma 5.2.10 entails its dual form. In view of von Neumann's Theorem (Theorem 6.3.1) for modular lattices this question has a positive answer for any RCM quasivariety. We thus ask whether a lattice-theoretic property weaker than modularity also entails this form of duality. □

Theorem 5.3.1. *Let* $\mathbf{Q}$ *be a quasivariety such that* $\boldsymbol{Th}(\mathbf{Q}^{\models})$ *is distributive in the restricted sense. Let* $X = \{x_i \approx y_i : i \in I\}$ *be a set of equations of pairwise different variables and let, for* $n \geqslant 2$, $Y_1, \ldots, Y_n$ *be arbitrary sets of equations of terms such that the variables occurring in the terms of* $Y_1 \cup \ldots \cup Y_n$ *are separated from the variables of* X, *i.e.,* $Var(X) \cap (Var(Y_1) \cup \ldots \cup Var(Y_n)) = \emptyset$. *Then*

$$\mathbf{Q}^{\models}(X) \cap \mathbf{Q}^{\models}(Y_1) +_{\mathbf{Q}} \ldots +_{\mathbf{Q}} \mathbf{Q}^{\models}(X) \cap \mathbf{Q}^{\models}(Y_n) = \mathbf{Q}^{\models}(X) \cap \mathbf{Q}^{\models}(Y_1 \cup \ldots \cup Y_n). \quad (1)_n$$

[1] The lattice of partitions of any set satisfies, however, the condition of semimodularity. This condition is not expressible as a lattice-theoretic identity.

Proof. We put $C := \mathbf{Q}^{\models}$. The case $n = 2$ is covered by restricted distributivity. We assume the theorem holds for n. We prove it holds for $n+1$. Let $X, Y_1, \ldots, Y_n, Y_{n+1}$ satisfy the assumption of the theorem, i.e., $Var(X) \cap (Var(Y_1) \cup \ldots \cup Var(Y_n) \cup Var(Y_{n+1})) = \emptyset$. As $Var(X) \cap (Var(Y_1) \cup \ldots \cup Var(Y_n)) = \emptyset$, the induction hypothesis yields $(1)_n$. Moreover, by restricted distributivity we also have that

$$C(X) \cap C(Y_1 \cup \ldots \cup Y_n) +_{\mathbf{Q}} C(X) \cap C(Y_{n+1}) \\ = C(X) \cap C(Y_1 \cup \ldots \cup Y_n \cup Y_{n+1}). \tag{2}$$

It follows that

$$\begin{aligned} C(X) \cap C(Y_1) +_{\mathbf{Q}} \ldots +_{\mathbf{Q}} C(X) \cap C(Y_n) +_{\mathbf{Q}} C(X) \cap C(Y_{n+1}) = \quad &(\text{by } (1)_n) \\ C(X) \cap C(Y_1 \cup \ldots \cup Y_n) +_{\mathbf{Q}} C(X) \cap C(Y_{n+1}) = \quad &(\text{by } (2) \\ C(X) \cap C(Y_1 \cup \ldots \cup Y_n \cup Y_{n+1}). \end{aligned}$$

So $(1)_{n+1}$ holds. □

Corollary 5.3.2. *Let* $\mathbf{Q}$ *be a quasivariety. The following conditions are equivalent:*

(a) *The lattice* $\boldsymbol{Th}(\mathbf{Q}^{\models})$ *is distributive in the restricted sense.*

(b) *For any finite set of equations X, any non-empty finite sets of pairwise different variables of the same cardinality* $\underline{z} = z_1, \ldots, z_m$, $\underline{w} = w_1, \ldots, w_m$ *separated from Var(X) and for any equation* $r \approx s$, *if* $r \approx s \in \mathbf{Q}^{\models}(X)$, *then*

$$\mathbf{Q}^{\models}(r \approx s) \cap \mathbf{Q}^{\models}(\underline{z} \approx \underline{w}) \subseteq \mathbf{Q}^{\models}\Big(\bigcup_{p \approx q \in X} \mathbf{Q}^{\models}(p \approx q) \cap \mathbf{Q}^{\models}(\underline{z} \approx \underline{w})\Big).$$

(c) $\boldsymbol{Th}(\mathbf{Q}^{\models})$ *satisfies the conjunction of the following two conditions:*

(c_1) *For any finite set of equations X, for any two variables z and w separated from Var(X) and any equation* $r \approx s$, *if* $r \approx s \in \mathbf{Q}^{\models}(X)$, *then*

$$\mathbf{Q}^{\models}(r \approx s) \cap \mathbf{Q}^{\models}(z \approx w) \subseteq \mathbf{Q}^{\models}\Big(\bigcup_{p \approx q \in X} \mathbf{Q}^{\models}(p \approx q) \cap \mathbf{Q}^{\models}(z \approx w)\Big)$$

and

(c_2) *for any equation* $p \approx q$ *and for any non-empty finite sets of pairwise different variables of the same cardinality* $z_1, \ldots, z_m, w_1, \ldots, w_m$ *separated from* $Var(p \approx q)$,

$$\mathbf{Q}^{\models}(p \approx q) \cap \mathbf{Q}^{\models}(\underline{z} \approx \underline{w}) = \\ \mathbf{Q}^{\models}(p \approx q) \cap \mathbf{Q}^{\models}(z_1 \approx w_1) +_{\mathbf{Q}} \ldots +_{\mathbf{Q}} \mathbf{Q}^{\models}(p \approx q) \cap \mathbf{Q}^{\models}(z_m \approx w_m).$$

(Here $\underline{z} \approx \underline{w} := \{z_1 \approx w_1, \ldots, z_m \approx w_m\}$.)

Proof. We shall prove the following chains of implications:

$$(a) \Rightarrow (b) \Rightarrow (a), \quad (a) \Rightarrow (c) \Rightarrow (b).$$

We put $C := \mathbf{Q}^{\models}$.

(a) ⇒ (b). This is easy. Assume $r \approx s \in C(X)$, X finite. (a) gives that

$$C(r \approx s) \cap C(\underline{z} \approx \underline{w}) \subseteq C(X) \cap C(\underline{z} \approx \underline{w}) = C(\bigcup_{p \approx q \in X} C(p \approx q) \cap C(\underline{z} \approx \underline{w})).$$

So (b) holds.

(b) ⇒ (a). We assume (b). Due to the fact that the lattice $\boldsymbol{Th}(\mathbf{Q}^{\models})$ is algebraic, the restricted distributivity of $\boldsymbol{Th}(\mathbf{Q}^{\models})$ will be proved once we show that

$$C(p_1 \approx q_1, \dots, p_n \approx q_n) \cap C(\underline{z} \approx \underline{w}) \subseteq$$
$$C(p_1 \approx q_1) \cap C(\underline{z} \approx \underline{w}) +_{\mathbf{Q}} \dots +_{\mathbf{Q}} C(p_n \approx q_n) \cap C(\underline{z} \approx \underline{w}).$$

for any finite set $p_1 \approx q_1, \dots, p_n \approx q_n$ of equations and any finite disjoint sets $\underline{z} = z_1, \dots, z_m$, $\underline{w} = w_1, \dots, w_m$ of variables separated from the variables occurring in $p_1 \approx q_1, \dots, p_n \approx r_n$.

So fix $p_1 \approx q_1, \dots, p_n \approx q_n, \underline{z}, \underline{w}$ and suppose

$$r \approx s \in C(p_1 \approx q_1, \dots, p_n \approx q_n) \cap C(\underline{z} \approx \underline{w}).$$

Since $r \approx s \in C(p_1 \approx q_1, \dots, p_n \approx q_n)$, (b) gives that

$$C(r \approx s) \cap C(\underline{z} \approx \underline{w}) \subseteq$$
$$C(p_1 \approx q_1) \cap C(\underline{z} \approx \underline{w}) +_{\mathbf{Q}} \dots +_{\mathbf{Q}} C(p_n \approx q_n) \cap C(\underline{z} \approx \underline{w}). \quad (*)$$

As $r \approx s \in C(\underline{z} \approx \underline{w})$, we have that $r \approx s \in C(r \approx s) \cap C(\underline{z} \approx \underline{w})$. This and $(*)$ imply that

$$r \approx s \in C(p_1 \approx q_1) \cap C(\underline{z} \approx \underline{w}) +_{\mathbf{Q}} \dots +_{\mathbf{Q}} C(p_n \approx q_n) \cap C(\underline{z} \approx \underline{w}).$$

So (a) holds.

(a) ⇒ (c). This is also immediate.

(c) ⇒ (b). We assume (b) and consider the following formula $T(m)$ with m ranging over positive integers:

$T(m)$ For any finite set of equations X, and any pairwise different $2m$ variables $\underline{z}_m = z_1, \dots, z_m$ and $\underline{w}_m = w_1, \dots, w_m$ separated from $Var(X)$ and any equation $r \approx s$, if $r \approx s \in C(X)$, then

$$C(r \approx s) \cap C(\underline{z}_m \approx \underline{w}_m) \subseteq C(\bigcup_{p \approx q \in X} C(p \approx q) \cap C(\underline{z}_m \approx \underline{w}_m)).$$

Claim 1. *$T(m)$ implies that*

$$C(\bigcup_{p \approx q \in X} C(p \approx q) \cap C(\underline{z}_m \approx \underline{w}_m) = C(X) \cap C(\underline{z}_m \approx \underline{w}_m)$$

for all finite sets of equations X and any $2m$ variables $\underline{z}_m = z_1, \ldots, z_m$, $\underline{w}_m = w_1, \ldots, w_m$ separated from $Var(X)$.

Proof (of the claim). Assume $T(m)$. Suppose $r \approx s \in C(X) \cap C(\underline{z}_m \approx \underline{w}_m)$. As $r \approx s \in C(X)$, $T(m)$ gives that $C(r \approx s) \cap C(\underline{z}_m \approx \underline{w}_m) \subseteq C(\bigcup_{p \approx q \in X} C(p \approx q) \cap C(\underline{z}_m \approx \underline{w}_m))$. Since $r \approx s \in C(\underline{z}_m \approx \underline{w}_m)$, we have that $r \approx s \in C(r \approx s) \cap C(\underline{z}_m \approx \underline{w}_m)$. Hence $r \approx s \in C(\bigcup_{p \approx q \in X} C(p \approx q) \cap C(\underline{z}_m \approx \underline{w}_m))$. □

Claim 2. *$T(m)$ holds for all $m \geqslant 1$.*

Proof (of the claim).

Induction base. $T(1)$.

This holds in virtue of (c_1).

Inductive step. Fix $m \geqslant 1$. $T(m)$ implies $T(m+1)$.

We assume $T(m)$. To prove $T(m+1)$, let X be a finite set of equations X, let z, w and $\underline{z}_m = z_1, \ldots, z_m$, $\underline{w}_m = w_1, \ldots, w_m$ be pairwise different variables separated from $Var(X)$, and let $r \approx s$ be an equation such that $r \approx s \in C(X)$. We claim that

$$C(r \approx s) \cap C(\underline{z}_m \approx \underline{w}_m, z \approx w)$$
$$\subseteq C(\bigcup_{p \approx q \in X} C(p \approx q) \cap C(\underline{z}_m \approx \underline{w}_m, z \approx w)). \quad (1)$$

According to $T(m)$ and Claim 1 we have that

$$C(X) \cap C(\underline{z}_m \approx \underline{w}_m) = C(\bigcup_{p \approx q \in X} C(p \approx q) \cap C(\underline{z}_m \approx \underline{w}_m)). \quad (2)$$

To prove $T(m+1)$ we compute:

$$C(\bigcup_{p \approx q \in X} C(p \approx q) \cap C(\underline{z}_m \approx \underline{w}_m, z \approx w)) =$$

$$C(\bigcup_{p \approx q \in X} (C(p \approx q) \cap C(\underline{z}_m \approx \underline{w}_m) +_{\mathbf{Q}} C(z \approx w))) = \quad \text{(by } (c_2))$$

$$C(\bigcup_{p \approx q \in X} (C(p \approx q) \cap C(\underline{z}_m \approx \underline{w}_m) +_{\mathbf{Q}} C(p \approx q) \cap C(z \approx w))) =$$

$$C(\bigcup_{p\approx q\in X} (C(p \approx q) \cap C(\underline{z}_m \approx \underline{w}_m) +_{\mathbf{Q}} C((\bigcup_{p\approx q\in X} C(p \approx q) \cap C(z \approx w))) =$$

(by (2), $T(1)$, and Claim 1 for $m = 1$)

$$C(X) \cap C(\underline{z}_m \approx \underline{w}_m) +_{\mathbf{Q}} C(X) \cap C(z \approx w).$$

Thus

$$C(\bigcup_{p\approx q\in X} C(p \approx q) \cap C(\underline{z}_m \approx \underline{w}_m, z \approx w)) = \\ C(X) \cap C(\underline{z}_m \approx \underline{w}_m) +_{\mathbf{Q}} C(X) \cap C(z \approx w). \quad (3)$$

On the other hand, (c_2) also gives

$$C(r \approx s) \cap C(\underline{z}_m \approx \underline{w}_m, z \approx w) = \\ C(r \approx s) \cap C(\underline{z}_m \approx \underline{w}_m) +_{\mathbf{Q}} C(r \approx s) \cap C(z \approx w). \quad (4)$$

As $C(r \approx s) \cap C(\underline{z}_m \approx \underline{w}_m) \subseteq C(X) \cap C(\underline{z}_m \approx \underline{w}_m)$ and $C(r \approx s) \cap C(z \approx w) \subseteq C(X) \cap C(z \approx w)$, we see that (3) and (4) imply that

$$C(r \approx s) \cap C(\underline{z}_m \approx \underline{w}_m, z \approx w) \subseteq C(\bigcup_{p\approx q\in X} C(p \approx q) \cap C(\underline{z}_m \approx \underline{w}_m, z \approx w))$$

So (1) holds. This concludes the proof of Claim 2. □

From Claim 2 condition (b) follows. □

Note. In the above corollary, condition (c_1) cannot be replaced by the following statement:

For any finite set of equations X and any equations $a \approx b, c \approx d$, if $a \approx b \in \mathbf{Q}^{\models}(X)$, then

$$\mathbf{Q}^{\models}(a \approx b) \cap \mathbf{Q}^{\models}(c \approx d) \subseteq \mathbf{Q}^{\models}(\bigcup_{p\approx q\in X} \mathbf{Q}^{\models}(p \approx q) \cap \mathbf{Q}^{\models}(c \approx d))$$

because one then obtains a condition which is equivalent to relative congruence-distributivity of $\mathbf{Q}$ (see, e.g., Czelakowski 1985). □

Conjecture. *Let $\mathbf{Q}$ be a quasivariety such that the lattice $\boldsymbol{Th}(\mathbf{Q}^{\models})$ is distributive in the restricted sense. Let $X = \{x_i \approx y_i : i \in I\}$ be a non-empty set of equations of pairwise different variables and let Y be an arbitrary non-empty set of equations of terms such that the variables occurring in the terms of Y are separated from the variables of X, i.e., $Var(X) \cap (Var(Y) = \emptyset$. Then $\{\mathbf{Q}^{\models}(x_i \approx y_i) : i \in I\} \cup \{\mathbf{Q}^{\models}(p \approx q) : p \approx q \in Y\}$ generates a distributive sublattice of $\boldsymbol{Th}(\mathbf{Q}^{\models})$.*

A weaker version of the above conjecture holds for RCM quasivarieties—see Theorem 6.3.8. □

We recall condition $(\mathrm{Distr})_\Sigma$ from Note 5.2.17. Let $\mathbf{Q}$ be a quasivariety and $C := \mathbf{Q}^{eq\models}$. Let $\underline{x}, \underline{y}, \underline{z}, \underline{w}$ be finite disjoint sets of variables such that $|\underline{x}| = |\underline{y}| = m$, $|\underline{z}| = |\underline{w}|$ and let $\Sigma = \{p_i \approx q_i : i \in I\}$ be a non-empty set of equations such that

$$C(\underline{x} \approx \underline{y}) = C(\Sigma).$$

Then

$$C(\underline{x} \approx \underline{y}) \cap C(\underline{z} \approx \underline{w}) = C(\bigcup_{i \in I} C(p_i \approx q_i) \cap C(\underline{z} \approx \underline{w})). \qquad (\mathrm{Distr})_\Sigma$$

Corollary 5.3.3. *Let* $\mathbf{Q}$ *be an arbitrary quasivariety. Suppose the lattice* $\boldsymbol{Th}(\mathbf{Q}^{eq\models})$ *is distributive in the restricted sense. Then* $(\mathrm{Distr})_\Sigma$ *holds for any set of equations* Σ.

Note. $(\mathrm{Distr})_\Sigma$ does not directly follow from Theorem 5.3.1, because the variables occurring in the equations of Σ need not be separated from the variables of $\underline{z}$ and $\underline{w}$. □

Proof. It suffices to prove $(\mathrm{Distr})_\Sigma$ for Σ finite. So let

$$\Sigma = \{p_1 \approx q_1, \dots, p_k \approx q_k\}$$

be a finite set of equations such that $C(\Sigma) = C(\underline{x} \approx \underline{y})$ for some finite sets of pairwise different variables $\underline{x} = \{x_1, \dots, x_m\}$, $\underline{y} = \{y_1, \dots, y_m\}$.

Let $\underline{z} = \{z_1, \dots, z_n\}$ and $\underline{w} = \{w_1, \dots, w_n\}$ be sets of different variables. We claim that

$$C(\underline{x} \approx \underline{y}) \cap C(\underline{z} \approx \underline{w}) = C(p_1 \approx q_1) \cap C(\underline{z} \approx \underline{w}) \\ +_{\mathbf{Q}} \dots +_{\mathbf{Q}} C(p_k \approx q_k) \cap C(\underline{z} \approx \underline{w}). \qquad (1)$$

We write $\Sigma = \{p_1(\underline{x}, \underline{y}, \underline{u}) \approx q_1(\underline{x}, \underline{y}, \underline{u}), \dots, p_k(\underline{x}, \underline{y}, \underline{u}) \approx q_k(\underline{x}, \underline{y}, \underline{u})\}$, where $\underline{u}$ is the set of variables which occur in Σ and are different from those in $\underline{x}$ and $\underline{y}$. ($\underline{u}$ are parametric variables.)

We first select $2n$ pairwise different variables

$$\underline{z}' = \{z'_1, \dots, z'_n\}, \qquad \underline{w}' = \{w'_1, \dots, w'_n\},$$

separated from $\underline{x}, \underline{z}, \underline{u}, \underline{z}$ and $\underline{w}$. According to Theorem 5.3.1 we have that

$$C(\Sigma) \cap C(\underline{z}' \approx \underline{w}') = C(p_1 \approx q_1) \cap C(\underline{z}' \approx \underline{w}') +_{\mathbf{Q}} \dots +_{\mathbf{Q}} C(p_k \approx q_k) \cap C(\underline{z}' \approx \underline{w}').$$

As $C(\Sigma) = C(\underline{x} \approx \underline{y})$, we therefore obtain that

$$C(\underline{x} \approx \underline{y}) \cap C(\underline{z}' \approx \underline{w}') =$$
$$C(p_1 \approx q_1) \cap C(\underline{z}' \approx \underline{w}') +_{\mathbf{Q}} \ldots +_{\mathbf{Q}} C(p_k \approx q_k) \approx C(\underline{z}' \approx \underline{w}'). \quad (2)$$

Select a set of equations $\Delta_{mn}(\underline{x}, \underline{y}, \underline{z}', \underline{w}', \underline{v})$ such that

$$C(\Delta_{mn}(\underline{x}, \underline{y}, \underline{z}', \underline{w}', \underline{v})) = C(\underline{x} \approx \underline{y}) \cap C(\underline{z}' \approx \underline{w}'). \quad (3)$$

Let $e : \boldsymbol{Te}_\tau \to \boldsymbol{Te}_\tau$ be the endomorphism such that $ez_i' := z_i$, $ew_i' := w_i$ for $i = 1, \ldots, n$ and e is the identity map on the remaining variables. In particular, e is the identity map on the variables of $\underline{x}, \underline{y}, \underline{u}, \underline{z}, \underline{w}$ and $\underline{v}$. As

$$C(\Delta_{mn}(\underline{x}, \underline{y}, \underline{z}', \underline{w}', \underline{v})) =$$
$$C(p_1 \approx q_1) \cap C(\underline{z}' \approx \underline{w}') +_{\mathbf{Q}} \ldots +_{\mathbf{Q}} C(p_k \approx q_k) \cap C(\underline{z}' \approx \underline{w}'), \quad (4)$$

we see that by applying the above substitution to (4) we get

$$C(\Delta_{mn}(\underline{x}, \underline{y}, \underline{z}, \underline{w}, \underline{v})) \subseteq$$
$$C(p_1 \approx q_1) \cap C(\underline{z} \approx \underline{w}) +_{\mathbf{Q}} \ldots +_{\mathbf{Q}} C(p_k \approx q_k) \cap C(\underline{z} \approx \underline{w}),$$

by structurality. But $C(\Delta_{mn}(\underline{x}, \underline{y}, \underline{z}, \underline{w}, \underline{v})) = C(\underline{x} \approx \underline{y}) \cap C(\underline{z} \approx \underline{w})$. Hence

$$C(\underline{x} \approx \underline{y}) \cap C(\underline{z} \approx \underline{w}) \subseteq$$
$$C(p_1 \approx q_1) \cap C(\underline{z} \approx \underline{w}) +_{\mathbf{Q}} \ldots +_{\mathbf{Q}} C(p_k \approx q_k) \cap C(\underline{z} \approx \underline{w}). \quad (5)$$

Since $C(p_1 \approx q_1) \cap C(\underline{z} \approx \underline{w}) +_{\mathbf{Q}} \ldots +_{\mathbf{Q}} C(p_k \approx q_k) \cap C(\underline{z} \approx \underline{w}) \subseteq C(\Sigma) \cap C(\underline{z} \approx \underline{w})$, the corollary follows. □

According to Theorem 5.2.16, conditions $(\text{EqDistr})_{m,n}$ and $(\text{Epi})_{m,n}$, universally quantified over all positive integers m and n, jointly provide a necessary and sufficient condition for the equationally defined commutator to be additive. The crucial fact in the theory of the equationally defined commutator is that the restricted distributivity of the lattice $\boldsymbol{Th}(\mathbf{Q}^{\models})$ implies both $(\text{EqDistr})_{m,n}$ and $(\text{Epi})_{m,n}$, for all m, n. We shall prove these facts below.

Theorem 5.3.4. *Let* $\mathbf{Q}$ *be a quasivariety such that the lattice* $\boldsymbol{Th}(\mathbf{Q}^{\models})$ *is distributive in the restricted sense. Then the lattice* $\boldsymbol{Th}(\mathbf{Q}^{\models})$ *validates* $(\text{EqDistr})_{m,n}$ *for all positive* m, n.

Proof. We put $C := \mathbf{Q}^{eq\models}$. The restricted distributivity implies that

$$(C(X_1) +_{\mathbf{Q}} C(X_2)) \cap C(Y) = C(X_1) \cap C(Y) +_{\mathbf{Q}} C(X_2) \cap C(Y) \quad (*)$$

for any separated sets X_1, X_2, Y of equations of variables.

According to Notes following Lemma 5.2.12, $(*)$ entails the condition

$$(C(X_1) +_{\mathbf{Q}} C(X_2)) \cap (C(Y_1) +_{\mathbf{Q}} C(Y_2)) = \qquad (**)$$
$$C(X_1) \cap C(Y_1) +_{\mathbf{Q}} C(X_1) \cap C(Y_2) +_{\mathbf{Q}} C(X_2) \cap C(Y_1) +_{\mathbf{Q}} C(X_2) \cap C(Y_2)$$

for any separated sets X_1, X_2, Y_1, Y_2 of equations of variables. It is easy to see that $(**)$ yields $(\mathrm{EqDistr})_\infty$. The thesis follows. □

Note. Appendix B contains a (longer) proof of the above theorem for any RCM quasivariety. In this proof conditions $(\mathrm{EqDistr})_{m,n}$ are directly computed. The proof given there involves some purely syntactical techniques which are also useful in other contexts. □

The next step consists in showing that every quasivariety $\mathbf{Q}$ whose lattice $\boldsymbol{Th}(\mathbf{Q}^{\models})$ is distributive in the restricted sense validates $(\mathrm{Epi})_{m,n}$, for all positive m, n. The proof of this fact is more intricate.

Theorem 5.3.5. *Let $\mathbf{Q}$ be any quasivariety such that the lattice $\boldsymbol{Th}(\mathbf{Q}^{\models})$ is distributive in the restricted sense. Let X and Y be finite separated sets of equations of pairwise different variables of Var. Then for any epimorphism $e : \boldsymbol{Te}_\tau \to \boldsymbol{Te}_\tau$,*

$$[\mathbf{Q}^{\models}(eX), \mathbf{Q}^{\models}(eY)] = \mathbf{Q}^{\models}(e(\mathbf{Q}^{\models}(X) \cap \mathbf{Q}^{\models}(Y))). \qquad (1)$$

Notes. A. By taking the e-preimage of the theories on the both sides of (1) we obtain the equivalent equality

$$e^{-1}[\mathbf{Q}^{\models}(eX), \mathbf{Q}^{\models}(eY)] = \ker_{\mathbf{Q}}(e) +_{\mathbf{Q}} \mathbf{Q}^{\models}(X) \cap \mathbf{Q}^{\models}(Y).$$

Theorem 5.3.5 is therefore equivalent to $(\mathrm{Epi})_\infty$—see Corollary 5.3.6.

B. (1) cannot be replaced by the stronger identity:

$$\mathbf{Q}^{\models}(eX) \cap \mathbf{Q}^{\models}(eY) = \mathbf{Q}^{\models}(e(\mathbf{Q}^{\models}(X) \cap \mathbf{Q}^{\models}(Y))), \qquad (1)^*$$

holding for all finite sets of equations of separated variables X and Y, because $(1)^*$ implies, in the presence of $(\forall m, n \geqslant 1)$ $(\mathrm{EqDistr})_{m,n}$, the distributivity of the lattice $\boldsymbol{Th}(\mathbf{Q}^{\models})$. □

Proof (of the theorem). Write $C := \mathbf{Q}^{\models}$ and let us assume that $X = \underline{x} \approx \underline{y}$, $Y = \underline{z} \approx \underline{w}$, where $\underline{x} = \{x_1, \dots, x_m\}$, $\underline{y} = \{y_1, \dots, y_m\}$, $\underline{z} = \{z_1, \dots, z_n\}$ and $\underline{w} = \{w_1, \dots, w_n\}$ for some $m, n \geqslant 1$. We also write

$$e\underline{x} \approx e\underline{y} := \{ex_1 \approx ey_1, \dots, ex_m \approx ey_m\} \text{ and } e\underline{z} \approx e\underline{w} := \{ez_1 \approx ew_1, \dots, ez_n \approx ew_n\}.$$

We must show that

$$[C(e\underline{x} \approx e\underline{y}), C(e\underline{z} \approx e\underline{w})] = C(e(C(\underline{x} \approx \underline{y}) \cap C(\underline{z} \approx \underline{w}))). \qquad (2)$$

Passing to the e-preimages of the theories on both sides of (2), we get

Claim 1. (2) *is equivalent to*

$$\ker_{\mathbf{Q}}(e) +_{\mathbf{Q}} [\ker_{\mathbf{Q}}(e) +_{\mathbf{Q}} C(\underline{x} \approx \underline{y}), \ker_{\mathbf{Q}}(e) +_{\mathbf{Q}} C(\underline{z} \approx \underline{w})] =$$
$$\ker_{\mathbf{Q}}(e) +_{\mathbf{Q}} [C(\underline{x} \approx \underline{y}), C(\underline{z} \approx \underline{w})]. \quad (3)$$

(see the remarks following Corollary 5.2.15).

Proof (of the claim). We apply Corollary 3.2.5 and argue as follows.

(2) ⇒ (3). Assuming (2) we get:

$$\ker_{\mathbf{Q}}(e) +_{\mathbf{Q}} [\ker_{\mathbf{Q}}(e) +_{\mathbf{Q}} C(\underline{x} \approx \underline{y}), \ker_{\mathbf{Q}}(e) +_{\mathbf{Q}} C(\underline{z} \approx \underline{w})] =$$
$$\text{(by Corollary 3.2.5)}$$
$$e^{-1}([C(e\underline{x} \approx e\underline{y}), C(e\underline{z} \approx e\underline{w})]) = \text{(by (2))}$$
$$= e^{-1}C(e(C(\underline{x} \approx \underline{y}) \cap C(\underline{z} \approx \underline{w}))) =$$
$$\ker_{\mathbf{Q}}(e) +_{\mathbf{Q}} C(\underline{x} \approx \underline{y}) \cap C(\underline{z} \approx \underline{w}) = \ker_{\mathbf{Q}}(e) +_{\mathbf{Q}} [C(\underline{x} \approx \underline{y}), C(\underline{z} \approx \underline{w})].$$

So (3) holds and therefore (2) implies (3).

(3) ⇒ (2). Now assume (3). We then have that

$$e^{-1}([C(e\underline{x} \approx e\underline{y}), C(e\underline{z} \approx e\underline{w})]) = \text{(by Corollary 3.2.5)}$$
$$\ker_{\mathbf{Q}}(e) +_{\mathbf{Q}} [\ker_{\mathbf{Q}}(e) +_{\mathbf{Q}} C(\underline{x} \approx \underline{y}), \ker_{\mathbf{Q}}(e) +_{\mathbf{Q}} C(\underline{z} \approx \underline{w})] = \text{(by (3)}$$
$$\ker_{\mathbf{Q}}(e) +_{\mathbf{Q}} [C(\underline{x} \approx \underline{y}), C(\underline{z} \approx \underline{w})] =$$
$$\ker_{\mathbf{Q}}(e) +_{\mathbf{Q}} C(\underline{x} \approx \underline{y}) \cap C(\underline{z} \approx \underline{w}) =$$
$$e^{-1}(C(e(C(\underline{x} \approx \underline{y}) \cap C(\underline{z} \approx \underline{w})))).$$

It follows from the above equalities that $e^{-1}([C(e\underline{x} \approx e\underline{y}), C(e\underline{z} \approx e\underline{w})]) = e^{-1}(C(e(C(\underline{x} \approx \underline{y}) \cap C(\underline{z} \approx \underline{w}))))$. Hence $[C(e\underline{x} \approx e\underline{y}), C(e\underline{z} \approx e\underline{w})]) = C(e(C(\underline{x} \approx \underline{y}) \cap C(\underline{z} \approx \underline{w})))$ which means that (2) holds.

This proves the claim. □

Let $\Delta'(\underline{x}, \underline{y}, \underline{z}, \underline{w}, \underline{u})$ be any set of equations such that

$$C(\underline{x} \approx \underline{y}) \cap C(\underline{z} \approx \underline{w}) = C(\Delta'(\underline{x}, \underline{y}, \underline{z}, \underline{w}, \underline{u})). \quad (*)$$

Δ' may be infinite. We recall that if $\underline{u} \in Var^k$, where $k \leqslant \omega$, then

$$(\forall \underline{u})\Delta'(e\underline{x}, e\underline{y}, e\underline{z}, e\underline{w}, \underline{u}) := \bigcup\{\Delta'(\underline{x}/e\underline{x}, \underline{y}/e\underline{y}, \underline{z}/e\underline{z}, \underline{w}/e\underline{w}, \underline{u}/\underline{t})) : \underline{t} \in Te^k_\tau\}.$$

As $\Delta'(\underline{x}, \underline{y}, \underline{z}, \underline{w}, \underline{u}) \subseteq C(\underline{x} \approx \underline{y})$ and $\Delta'(\underline{x}, \underline{y}, \underline{z}, \underline{w}, \underline{u}) \subseteq C(\underline{z} \approx \underline{w})$, structurality gives that $(\forall \underline{u})\ \Delta'(\underline{x}, \underline{y}, \underline{z}, \underline{w}, \underline{u}) \subseteq C(\underline{x} \approx \underline{y}) \cap C(\underline{z} \approx \underline{w})$. Since $C(\Delta'(\underline{x}, \underline{y}, \underline{z}, \underline{w}, \underline{u})) \subseteq C((\forall \underline{u})\ \Delta'(\underline{x}, \underline{y}, \underline{z}, \underline{w}, \underline{u}))$, it follows that $C(\underline{x} \approx \underline{y}) \cap C(\underline{z} \approx \underline{w}) = C(\Delta'(\underline{x}, \underline{y}, \underline{z}, \underline{w}, \underline{u})) \subseteq C((\forall \underline{u}) \Delta'(\underline{x}, \underline{y}, \underline{z}, \underline{w}, \underline{u})) \subseteq C(\underline{x} \approx \underline{y}) \cap C(\underline{z} \approx \underline{w})$. Hence

$$C(\underline{x} \approx \underline{y}) \cap C(\underline{z} \approx \underline{w}) = C((\forall \underline{u}) \Delta'(\underline{x}, \underline{y}, \underline{z}, \underline{w}, \underline{u})). \qquad (**)$$

(2) (equivalently, (3)) continues to hold if e is replaced by *any other* epimorphism $e' : \boldsymbol{Te}_\tau \to \boldsymbol{Te}_\tau$ which agrees with e on the variables $\underline{x}, \underline{y}, \underline{z}, \underline{w}$ because both sides of (2) are unambiguously determined by the values of e on the variables $\underline{x}, \underline{y}, \underline{z}, \underline{w}$ displayed there. This remark is encapsulated in the following claim.

Claim 2. *Let $e' : \boldsymbol{Te}_\tau \to \boldsymbol{Te}_\tau$ be an epimorphism which coincides with e on the variables $\underline{x}, \underline{y}, \underline{z}, \underline{w}$. Then*

$$[C(e\underline{x} \approx e\underline{y}), C(e\underline{z} \approx e\underline{w})] = [C(e'\underline{x} \approx e'\underline{y}), C(e'\underline{z} \approx e'\underline{w})]$$

and

$$C(e(C(\underline{x} \approx \underline{y}) \cap C(\underline{z} \approx \underline{w}))) = C(e'(C(\underline{x} \approx \underline{y}) \cap C(\underline{z} \approx \underline{w}))).$$

Proof (of the claim). We obviously have $[C(e\underline{x} \approx e\underline{y}), C(e\underline{z} \approx e\underline{w})] = [C(e'\underline{x} \approx e'\underline{y}), C(e'\underline{z} \approx e'\underline{w})]$. To prove the other equality we define $\Delta'(\underline{x}, \underline{y}, \underline{z}, \underline{w}, \underline{u})$ as above and compute:

$$\begin{gathered}
C(e(C(\underline{x} \approx \underline{y}) \cap C(\underline{z} \approx \underline{w}))) = \text{ (by (**))}\\
C(e(C((\forall \underline{u})\ \Delta'(\underline{x}, \underline{y}, \underline{z}, \underline{w}, \underline{u})))) = \text{ (by structurality)}\\
C(e(\forall \underline{u})\ \Delta'(\underline{x}, \underline{y}, \underline{z}, \underline{w}, \underline{u})) =\\
C((\forall \underline{u})\ \Delta'(e\underline{x}, e\underline{y}, e\underline{z}, e\underline{w}, \underline{u})) =\\
C((\forall \underline{u})\ \Delta'(e'\underline{x}, e'\underline{y}, e'\underline{z}, e'\underline{w}, \underline{u})) =\\
C(e'(\forall \underline{u})\ \Delta'(\underline{x}, \underline{y}, \underline{z}, \underline{w}, \underline{u})) =\\
C(e'(C((\forall \underline{u})\ \Delta'(\underline{x}, \underline{y}, \underline{z}, \underline{w}, \underline{u})))) = \text{ (by (**))}\\
C(e'(C(\underline{x} \approx \underline{y}) \cap C(\underline{z} \approx \underline{w}))).
\end{gathered}$$

Consequently,

$$C(e(C(\underline{x} \approx \underline{y}) \cap C(\underline{z} \approx \underline{w}))) = C(e'(C(\underline{x} \approx \underline{y}) \cap C(\underline{z} \approx \underline{w}))).$$

This proves the claim. □

Thus to prove (2) (equivalently, (3)) it suffices to find *at least one* epimorphism $e' : \boldsymbol{Te}_\tau \to \boldsymbol{Te}_\tau$ which coincides with e on the variables $\underline{x}, \underline{y}, \underline{z}, \underline{w}$ and for which

$$[C(e'\underline{x} \approx e'\underline{y}), C(e'\underline{z} \approx e'\underline{w})] = C(e'(C(\underline{x} \approx \underline{y}) \cap C(\underline{z} \approx \underline{w}))).$$

We shall take a closer look at (3). We write

$$p_i(\underline{x}, \underline{y}, \underline{z}, \underline{w}, \underline{v}) := ex_i, \quad q_i(\underline{x}, \underline{y}, \underline{z}, \underline{w}, \underline{v}) := ey_i \quad \text{for } i = 1, \dots, m, \tag{4}$$

and

$$r_j(\underline{x}, \underline{y}, \underline{z}, \underline{w}, \underline{v}) := ez_j, \quad s_j(\underline{x}, \underline{y}, \underline{z}, \underline{w}, \underline{v}) := ew_j \quad \text{for } j = 1, \dots, n, \tag{5}$$

where $\underline{v}$ is the set of variables different from those in $\underline{x}, \underline{y}, \underline{z}$, and $\underline{w}$ that may occur in the above terms p_i, q_i, r_j, s_j $(i = 1, \dots, m, j = 1, \dots, n)$.

We apply Theorem 2.21 and the remarks following the formulation of this theorem. In view of Claim 2, we may assume that $Var \setminus \underline{x} \cup \underline{y} \cup \underline{z} \cup \underline{w} \subseteq V_e$. (Thus e assigns a variable to each variable belonging to $Var \setminus \underline{x} \cup \underline{y} \cup \underline{z} \cup \underline{w}$; but it may also happen that ev is a variable for some $v \in \underline{x} \cup \underline{y} \cup \underline{z} \cup \underline{w}$.) Moreover we may assume that e is the identity map on the set of parameters $\underline{v}$, and e bijectively maps V_e onto Var. We therefore have that $Var \setminus V_e \subseteq \underline{x} \cup \underline{y} \cup \underline{z} \cup \underline{w}$.

Let x'_i, y'_i, z'_j, w'_j be the (unique) variables such that $x_i = ex'_i$, $y_i = ey'_j$, $z_j = ez'_j$, $w_j = ew'_j$ for $i = 1, \dots, m$, $j = 1, \dots, n$. They are pairwise different. In view of Claim 2 we may also assume, without loss of generality, that all x'_i, y'_i, z'_j, w'_j are also different from $\underline{x} \cup \underline{y} \cup \underline{z} \cup \underline{w}$. We write:

$$\underline{x}' := \{x'_1, \dots, x'_m\}, \;\; \underline{y}' := \{y'_1, \dots, y'_m\}, \;\; \underline{z}' := \{z'_1, \dots, z'_m\}, \;\; \underline{w}' := \{w'_1, \dots, w'_m\}.$$

We also define:

$$p'_i := p_i(\underline{x}', \underline{y}', \underline{z}', \underline{w}', \underline{v}), \qquad q'_i := q_i(\underline{x}', \underline{y}', \underline{z}', \underline{w}', \underline{v}),$$
$$r'_j := r_j(\underline{x}', \underline{y}', \underline{z}', \underline{w}', \underline{v}), \qquad s'_j := s_j(\underline{x}', \underline{y}', \underline{z}', \underline{w}', \underline{v}),$$

for $i = 1, \dots, m$, $j = 1, \dots, n$. (The above terms are thus obtained by making appropriate substitutions in the terms p_i, q_i, r_j, s_j (for $i = 1, \dots, m$, $j = 1, \dots, n$), viz., replacing the variables $\underline{x}$ by $\underline{x}'$, $\underline{y}$ by $\underline{y}'$ etc.) Thus p'_i, q'_i, r'_j and s'_j are terms in the variables in V_e for which

$$ex_i = ep'_i, \quad ey_i = eq'_i, \quad ez_j = er'_j, \quad ew_j = es'_j$$

for $i = 1, \dots, m$, $j = 1, \dots, n$. (Here "=" means the identity of terms.)

We also put:

$$\underline{p} := \langle p_1(\underline{x}, \underline{y}, \underline{z}, \underline{w}, \underline{v}), \dots, p_m(\underline{x}, \underline{y}, \underline{z}, \underline{w}, \underline{v})\rangle,$$

$$\underline{q} := \langle q_1(\underline{x}, \underline{y}, \underline{z}, \underline{w}, \underline{v}), \ldots, q_m(\underline{x}, \underline{y}, \underline{z}, \underline{w}, \underline{v})\rangle,$$
$$\underline{r} := \langle r_1(\underline{x}, \underline{y}, \underline{z}, \underline{w}, \underline{v}), \ldots, r_n(\underline{x}, \underline{y}, \underline{z}, \underline{w}, \underline{v})\rangle,$$
$$\underline{s} := \langle s_1(\underline{x}, \underline{y}, \underline{z}, \underline{w}, \underline{v}), \ldots, s_n(\underline{x}, \underline{y}, \underline{z}, \underline{w}, \underline{v})\rangle,$$

$$\underline{p}' := \langle p_1(\underline{x}', \underline{y}', \underline{z}', \underline{w}', \underline{v}'), \ldots, p_m(\underline{x}', \underline{y}', \underline{z}', \underline{w}', \underline{v}')\rangle,$$
$$\underline{q}' := \langle q_1(\underline{x}', \underline{y}', \underline{z}', \underline{w}', \underline{v}'), \ldots, q_m(\underline{x}', \underline{y}', \underline{z}', \underline{w}', \underline{v}')\rangle,$$
$$\underline{r}' := \langle r_1(\underline{x}', \underline{y}', \underline{z}', \underline{w}', \underline{v}'), \ldots, r_n(\underline{x}', \underline{y}', \underline{z}', \underline{w}', \underline{v}')\rangle,$$
$$\underline{s}' := \langle s_1(\underline{x}', \underline{y}', \underline{z}', \underline{w}', \underline{v}'), \ldots, s_n(\underline{x}', \underline{y}', \underline{z}', \underline{w}', \underline{v}')\rangle.$$

It may happen that some terms occurring in the concatenation of the sequence $\underline{p}, \underline{q}, \underline{r}, \underline{s}$ repeat. It therefore follows from Theorem 2.21 that $\ker_{\mathbf{Q}}(e)$ is generated by

$$\begin{aligned} A_0 = \{&x_i \approx p_i(\underline{x}', \underline{y}', \underline{z}', \underline{w}', \underline{v}),\ y_i \approx q_i(\underline{x}', \underline{y}', \underline{z}', \underline{w}', \underline{v}) : i = 1, \ldots, m\} \cup \\ &\{z_j \approx r_j(\underline{x}', \underline{y}', \underline{z}', \underline{w}', \underline{v}),\ w_j \approx s_j(\underline{x}', \underline{y}', \underline{z}', \underline{w}', \underline{v}) : j = 1, \ldots, n\} \cup \\ &\{x'' \approx y'' : x'', y'' \in \underline{x} \cup \underline{y} \cup \underline{z} \cup \underline{w} \text{ and the terms } ex'' \text{ and } ey'' \\ &\qquad\text{(among those in } \underline{p}, \underline{q}, \underline{r}, \underline{s}\text{) are identical}\}. \end{aligned} \tag{6}$$

i.e.,

$$\ker_{\mathbf{Q}}(e) = C(A_0), \tag{7}$$

(The set $\{x'' \approx y'' : x'', y'' \in \underline{x} \cup \underline{y} \cup \underline{z} \cup \underline{w}$ and the terms ex'' and ey'' are identical$\}$ may contain only trivial equations.) Thus

$$\ker_{\mathbf{Q}}(e) = P +_{\mathbf{Q}} R, \tag{8}$$

where

$$\begin{aligned} P := C(\{&x_i \approx p_i(\underline{x}', \underline{y}', \underline{z}', \underline{w}', \underline{v}), y_i \approx q_i(\underline{x}', \underline{y}', \underline{z}', \underline{w}', \underline{v}) : i = 1, \ldots, m\} \cup \\ &\{z_j \approx r_j(\underline{x}', \underline{y}', \underline{z}', \underline{w}', \underline{v}), w_j \approx s_j(\underline{x}', \underline{y}', \underline{z}', \underline{w}', \underline{v}) : j = 1, \ldots, n\}) \text{ and} \end{aligned} \tag{9}$$
$$\begin{aligned} R := C(\{&x'' \approx y'' : x'', y'' \in \underline{x} \cup \underline{y} \cup \underline{z} \cup \underline{w} \text{ and the terms } ex'' \text{ and } ey'' \\ &\text{among those in } \underline{p}, \underline{q}, \underline{r}, \underline{s} \text{ are identical}\}). \end{aligned}$$

In virtue of (8) we see that to prove (3) it suffices to show that

$$\ker_{\mathbf{Q}}(e) +_{\mathbf{Q}} [P +_{\mathbf{Q}} C(\underline{x} \approx \underline{y}) +_{\mathbf{Q}} R, P +_{\mathbf{Q}} C(\underline{z} \approx \underline{w}) +_{\mathbf{Q}} R] = P +_{\mathbf{Q}} C(\underline{x} \approx \underline{y}) \cap C(\underline{z} \approx \underline{w}) +_{\mathbf{Q}} R. \quad (10)$$

But

$$P +_{\mathbf{Q}} C(\underline{x} \approx \underline{y}) = P +_{\mathbf{Q}} C(\underline{p}' \approx \underline{q}') \quad \text{and} \quad P +_{\mathbf{Q}} C(\underline{z} \approx \underline{w}) = P +_{\mathbf{Q}} C(\underline{r}' \approx \underline{s}').$$

We may therefore rewrite (10) as

$$\ker_{\mathbf{Q}}(e) +_{\mathbf{Q}} [P +_{\mathbf{Q}} C(\underline{p}' \approx \underline{q}') +_{\mathbf{Q}} R, P +_{\mathbf{Q}} C(\underline{r}' \approx \underline{s}') +_{\mathbf{Q}} R] = P +_{\mathbf{Q}} C(\underline{x} \approx \underline{y}) \cap C(\underline{z} \approx \underline{w}) +_{\mathbf{Q}} R,$$

that is,

$$\ker_{\mathbf{Q}}(e) +_{\mathbf{Q}} [\ker_{\mathbf{Q}}(e) +_{\mathbf{Q}} C(\underline{p}' \approx \underline{q}'), \ker_{\mathbf{Q}}(e) +_{\mathbf{Q}} C(\underline{r}' \approx \underline{s}')] = \ker_{\mathbf{Q}}(e) +_{\mathbf{Q}} C(\underline{x} \approx \underline{y}) \cap C(\underline{z} \approx \underline{w}). \quad (11)$$

We recall that $\Delta'(\underline{x}, \underline{y}, \underline{z}, \underline{w}, \underline{u})$ is the set of equations defined in $(*)$ above. Thus $(**)$ holds. We then proceed as follows. e is the identity map on the set of parameters $\underline{v}$. As the assignment $x_i \to x_i', y_i \to y_j', z_j \to z_j', w_j \to w_j'$ for $i = 1, \ldots, m, j = 1, \ldots, n$, is well-defined and one-to-one, we extend it to a choice function g from Var to the family $\{u' \in Var : eu' = u\} : u \in Var\}$ so that $gu \in \{u' \in Var : eu' = u\}$, for all $u \in Var$ and g is the identity map on $\underline{v}$. Let $V := g[Var]$ and let $\boldsymbol{T}$ be the subalgebra of $\boldsymbol{Te}_\tau$ generated by the variables of V (see Section 3.3). Since $\underline{x} \cup \underline{y} \cup \underline{z} \cup \underline{w} \cup \underline{u}'$ is a subset of V, the terms of $\underline{p}', \underline{q}', \underline{r}', \underline{s}'$ belong to the term algebra $\boldsymbol{T}$ generated by V. Theorem 3.3.5 applies to this situation and it implies that the LHS of (11) is equal to $\ker_{\mathbf{Q}}(e) +_{\mathbf{Q}} [C(\underline{p}' \approx \underline{q}'), C(\underline{r}' \approx \underline{s}')]$. Thus (11) is equivalent to

$$\ker_{\mathbf{Q}}(e) +_{\mathbf{Q}} [C(\underline{p}' \approx \underline{q}'), C(\underline{r}' \approx \underline{s}')] = \ker_{\mathbf{Q}}(e) +_{\mathbf{Q}} C(\underline{x} \approx \underline{y}) \cap C(\underline{z} \approx \underline{w}). \quad (12)$$

But the inclusion "$\supseteq$" in (12) is immediate because:

$$\ker_{\mathbf{Q}}(e) +_{\mathbf{Q}} [C(\underline{p}' \approx \underline{q}'), C(\underline{r}' \approx \underline{s}')] =$$
$$P +_{\mathbf{Q}} R +_{\mathbf{Q}} [C(\underline{p}' \approx \underline{q}'), C(\underline{r}' \approx \underline{s}')] \supseteq$$
$$P +_{\mathbf{Q}} R +_{\mathbf{Q}} C((\forall \underline{u})\ \Delta'(\underline{p}', \underline{q}', \underline{r}', \underline{s}', \underline{u})) \supseteq$$
$$P +_{\mathbf{Q}} C((\forall \underline{u})\ \Delta'(\underline{p}', \underline{q}', \underline{r}', \underline{s}', \underline{u})) \supseteq$$
$$C((\forall \underline{u})\ \Delta'(\underline{x}, \underline{y}, \underline{z}, \underline{w})) = C(\underline{x} \approx \underline{y}) \cap C(\underline{z} \approx \underline{w}),$$

by the definition of P. We thus see that (12) is equivalent to the inclusion:

$$[C(\underline{p}' \approx \underline{q}'), C(\underline{r}' \approx \underline{s}')] \subseteq \ker_{\mathbf{Q}}(e) +_{\mathbf{Q}} C(\underline{x} \approx \underline{y}) \cap C(\underline{z} \approx \underline{w}),$$

i.e.,

$$[C(\underline{p}' \approx \underline{q}'), C(\underline{r}' \approx \underline{s}')] \subseteq P +_{\mathbf{Q}} C(\underline{x} \approx \underline{y}) \cap C(\underline{z} \approx \underline{w}) +_{\mathbf{Q}} R. \tag{13}$$

Once we show (13), our theorem will be proved. To show (13), it suffices to check that $C(\underline{p}' \approx \underline{q}')$ centralizes $C(\underline{r}' \approx \underline{s}')$ relative the theory $T := P +_{\mathbf{Q}} C(\underline{x} \approx \underline{y}) \cap C(\underline{z} \approx \underline{w})$ in the sense of the centralization relation $Z_{4,\mathrm{com}}$, defined as in Chapter 2 (see Corollary 5.2.14). $Z_{4,\mathrm{com}}$ is the centralization relation in the sense of quaternary commutator equations of $\mathbf{Q}$ (see Chapter 1). (13) then follows, because, by definition, the equationally defined commutator $[C(\underline{p}' \approx \underline{q}'), C(\underline{r}' \approx \underline{s}')]$ is the *least* equational theory Σ of C such that $C(\underline{p}' \approx \underline{q}')$ centralizes $C(\underline{r}' \approx \underline{s}')$ relative to Σ in the sense Z_{com} and hence in the sense of $Z_{4,\mathrm{com}}$, by Corollary 5.2.15. As the theory T is included in $\ker_{\mathbf{Q}}(e) +_{\mathbf{Q}} C(\underline{x} \approx \underline{y}) \cap C(\underline{z} \approx \underline{w})$, (13) will follow. The following lemma elaborates this idea.

Lemma. *Let*

$$A := C(\underline{p}' \approx \underline{q}'), \quad B := C(\underline{r}' \approx \underline{s}').$$

$$T := C(\underline{x} \approx \underline{p}', \underline{y} \approx \underline{q}', \underline{z} \approx \underline{r}', \underline{w} \approx \underline{s}') +_{\mathbf{Q}} C(\underline{x} \approx \underline{y}) \cap C(\underline{z} \approx \underline{w}).$$

Then $Z_{4,\mathrm{com}}(A, B; T)$.

Proof (of the lemma).

Claim A. *Suppose that* $\underline{a}, \underline{b}, \underline{c}, \underline{d}$ *are arbitrary sequences of terms such that*

$$\underline{a} \approx \underline{b} \in C(\underline{p}' \approx \underline{q}') \quad \text{and} \quad \underline{c} \approx \underline{d} \in C(\underline{r}' \approx \underline{s}'), \tag{14}$$

where $\underline{a} \approx \underline{b} = \langle a_1 \approx b_1, \dots, a_m \approx b_m \rangle$ *and* $\underline{c} \approx \underline{d} = \langle c_1 \approx d_1, \dots, c_n \approx d_n \rangle$. *Then*

$$C((\forall \underline{u})\ \Delta'(\underline{a}, \underline{b}, \underline{c}, \underline{d}, \underline{u})) \subseteq T. \tag{15}$$

Proof (of the claim). Let

$$Y := \{\underline{x} \approx \underline{p}'\} \cup \{\underline{y} \approx \underline{q}'\} \quad \text{and} \quad Z := \{\underline{z} \approx \underline{r}'\} \cup \{\underline{w} \approx \underline{s}'\}.$$

The first conjunct of (14) gives that

$$\underline{a} \approx \underline{b} \in C(\underline{p}' \approx \underline{q}') \subseteq C(Y) +_{\mathbf{Q}} C(\underline{x} \approx \underline{y}).$$

Hence

$$C(\underline{a} \approx \underline{b}) \subseteq C(Y) +_{\mathbf{Q}} C(\underline{x} \approx \underline{y}). \tag{16}$$

As the variables $\underline{z}$ and $\underline{w}$ are separated from the variables occurring in $Y \cup \underline{x} \approx \underline{y}$, we apply Theorem 5.3.1 to the theory $C(Y) +_{\mathbf{Q}} C(\underline{x} \approx \underline{y})$. (16) then yields that

$$\begin{aligned} C((\forall \underline{u})\ \Delta'(\underline{a}, \underline{b}, \underline{z}, \underline{w}, \underline{u})) \subseteq C(\underline{a} \approx \underline{b}) \cap C(\underline{z} \approx \underline{w}) \subseteq \\ (C(Y) +_{\mathbf{Q}} C(\underline{x} \approx \underline{y})) \cap C(\underline{z} \approx \underline{w}) = \quad \text{(by Theorem 5.3.1)} \\ C(Y) \cap C(\underline{z} \approx \underline{w}) +_{\mathbf{Q}} C(\underline{x} \approx \underline{y}) \cap C(\underline{z} \approx \underline{w}) = \\ C(Y) \cap C(\underline{z} \approx \underline{w}) +_{\mathbf{Q}} C((\forall \underline{u})\ \Delta'(\underline{x}, \underline{y}, \underline{z}, \underline{w}, \underline{u})) \subseteq \\ C(Y) +_{\mathbf{Q}} C((\forall \underline{u})\ \Delta'(\underline{x}, \underline{y}, \underline{z}, \underline{w}, \underline{u})). \end{aligned}$$

Hence

$$C((\forall \underline{u})\ \Delta'(\underline{a}, \underline{b}, \underline{z}, \underline{w}, \underline{u})) \subseteq C(Y) +_{\mathbf{Q}} C((\forall \underline{u})\ \Delta'(\underline{x}, \underline{y}, \underline{z}, \underline{w}, \underline{u})). \tag{17}$$

By applying the substitution $\underline{z}/\underline{c}$, $\underline{w}/\underline{d}$ to (17), we get

$$C((\forall \underline{u})\ \Delta'(\underline{a}, \underline{b}, \underline{c}, \underline{d}, \underline{u})) \subseteq C(Y) +_{\mathbf{Q}} C((\forall \underline{u})\ \Delta'(\underline{x}, \underline{y}, \underline{c}, \underline{d}, \underline{u})), \tag{18}$$

by structurality.

In turn, the second conjunct of (15) gives that

$$\underline{c} \approx \underline{d} \in C(\underline{r}' \approx \underline{s}') \subseteq C(Z) +_{\mathbf{Q}} C(\underline{z} \approx \underline{w}).$$

Hence

$$C(\underline{c} \approx \underline{d}) \subseteq C(Z) +_{\mathbf{Q}} C(\underline{z} \approx \underline{w}). \tag{19}$$

As $\underline{x}$ and $\underline{y}$ are separated from the variables occurring in $Z \cup \underline{z} \approx \underline{w}$, we again apply Theorem 5.3.1 but this time to the theory $C(Z) +_{\mathbf{Q}} C(\underline{z} \approx \underline{w})$. (19) then yields that:

$$\begin{aligned} C((\forall \underline{u})\ \Delta'(\underline{x}, \underline{y}, \underline{c}, \underline{d}, \underline{u})) \subseteq C(\underline{c} \approx \underline{d}) \cap C(\underline{x} \approx \underline{y}) \subseteq \\ (C(Z) +_{\mathbf{Q}} C(\underline{z} \approx \underline{w})) \cap C(\underline{x} \approx \underline{y}) = \quad \text{(by Theorem 5.3.1)} \\ C(Z) \cap C(\underline{x} \approx \underline{y}) +_{\mathbf{Q}} C(\underline{z} \approx \underline{w}) \cap C(\underline{x} \approx \underline{y}) = \\ C(Z) \cap C(\underline{x} \approx \underline{y}) +_{\mathbf{Q}} C((\forall \underline{u})\ \Delta'(\underline{x}, \underline{y}, \underline{z}, \underline{w}, \underline{u})) \subseteq \\ C(Z) +_{\mathbf{Q}} C((\forall \underline{u})\ \Delta'(\underline{x}, \underline{y}, \underline{z}, \underline{w}, \underline{u})). \end{aligned}$$

Hence

$$C((\forall \underline{u})\ \Delta'(\underline{x}, \underline{y}, \underline{c}, \underline{d}, \underline{u})) \subseteq C(Z) +_{\mathbf{Q}} C((\forall \underline{u})\ \Delta'(\underline{x}, \underline{y}, \underline{z}, \underline{w}, \underline{u})). \tag{20}$$

Combining (18) and (20) we get that

$$C((\forall \underline{u})\ \Delta'(\underline{a}, \underline{b}, \underline{c}, \underline{d}, \underline{u})) \subseteq C(Y) +_{\mathbf{Q}} C((\forall \underline{u})\ \Delta'(\underline{x}, \underline{y}, \underline{c}, \underline{d}, \underline{u})) \subseteq$$
$$C(Y) +_{\mathbf{Q}} C(Z) +_{\mathbf{Q}} C((\forall \underline{u})\ \Delta'(\underline{x}, \underline{y}, \underline{z}, \underline{w}, \underline{u})) =$$
$$C(Y) +_{\mathbf{Q}} C(Z) +_{\mathbf{Q}} C(\underline{x} \approx \underline{y}) \cap C(\underline{z} \approx \underline{w}) = T.$$

Thus $C((\forall \underline{u})\ \Delta'(\underline{a}, \underline{b}, \underline{c}, \underline{d}, \underline{u})) \subseteq T$. As $T \subseteq \ker_{\mathbf{Q}}(e) +_{\mathbf{Q}} C(\underline{x} \approx \underline{y}) \cap C(\underline{z} \approx \underline{w})$, the claim follows. □

Claim B. $Z_{4,\mathrm{com}}(A, B; T)$.

Proof. Let x, y, z, w be four distinct variables and let $\Delta_0(x, y, z, w, \underline{u}_0)$ be a set of equations such that $C(x \approx y) \cap C(z \approx w) = C((\forall \underline{u}_0)\ \Delta_0(x, y, z, w, \underline{u}_0))$.

To prove the claim, we apply Corollary 5.2.14. We then have:

$$[A, B] =$$
$$[C(\underline{p}' \approx \underline{q}'), C(\underline{r}' \approx \underline{s}')] = C(\bigcup\{(\forall \underline{u}_0)\ \Delta_0(a, b, c, d, \underline{u}_0) :$$
$$a \approx b \in C(\underline{p}' \approx \underline{q}') \text{ and } c \approx d \in C(\underline{r}' \approx \underline{s}')\}).$$

As $C(x \approx y) \cap C(z \approx w) = C((\forall \underline{u}_0)\ \Delta_0(x, y, z, w, \underline{u}_0))$, the fact $Z_{4,\mathrm{com}}(A, B; T)$ will be proved once we show that

$$(\forall \underline{u}_0)\ \Delta_0(a, b, c, d, \underline{u}_0) \subseteq T \quad \text{whenever}$$
$$a \approx b \in C(\underline{p}' \approx \underline{q}') \ \text{ and } \ c \approx d \in C(\underline{r}' \approx \underline{s}'). \tag{21}$$

As $\mathbf{Q}$ validates $(\mathrm{EqDistr})_{m,n}$ (by Theorem 5.3.5), we also have that

$$C((\forall \underline{u})\ \Delta'(\underline{x}, \underline{y}, \underline{z}, \underline{w}, \underline{u})) = C(\bigcup_{1 \leqslant i \leqslant m, 1 \leqslant j \leqslant n} (\forall \underline{u}_0)\ \Delta_0(x_i, y_i, z_j, w_j, \underline{u}_0)). \tag{22}$$

To show (21), assume $a \approx b \in C(\underline{p}' \approx \underline{q}')$ and $c \approx d \in C(\underline{r}' \approx \underline{s}')$. Let

$$\underline{a} := \langle a, \ldots, a \rangle, \quad \underline{b} := \langle b, \ldots, b \rangle, \quad (a \text{ and } b \text{ repeated } m \text{ times})$$
$$\underline{c} := \langle c, \ldots, c \rangle, \quad \underline{d} := \langle d, \ldots, d \rangle, \quad (c \text{ and } d \text{ repeated } n \text{ times}).$$

As $\underline{a} \approx \underline{b} \in C(\underline{p}' \approx \underline{q}')$ and $\underline{c} \approx \underline{d} \in C(\underline{r}' \approx \underline{s}')$, Claim A gives that

$$C((\forall \underline{u})\ \Delta'(\underline{a}, \underline{b}, \underline{c}, \underline{d}, \underline{u})) \subseteq T.$$

But (22) and the structurality of C give that

$$C((\forall \underline{u}_0)\ \Delta_0(a, b, c, d, \underline{u}_0)) = C((\forall \underline{u})\ \Delta'(\underline{a}, \underline{b}, \underline{c}, \underline{d}, \underline{u})).$$

It then follows that $C((\forall \underline{u}_0)\ \Delta_0(a, b, c, d, \underline{u}_0)) \subseteq T$. So Claim B holds. □

The lemma has been proved. This concludes the proof of the theorem. □

Corollary 5.3.6. *Let* $\mathbf{Q}$ *be a quasivariety such that* $\boldsymbol{Th}(\mathbf{Q}^{\models})$ *is distributive in the restricted sense. Then* $\boldsymbol{Th}(\mathbf{Q}^{\models})$ *validates* $(\mathrm{Epi})_{m,n}$, *for all positive* m, n.

Proof. Let X and Y be separated finite sets of equations of individual variables and $e : \boldsymbol{Te}_\tau \to \boldsymbol{Te}_\tau$ an epimorphism. Put: $C := \mathbf{Q}^{\models}$.

According to the above theorem, we have that

$$[C(eX), C(eY)] = C(e(C(X) \cap C(Y))).$$

Taking the e-preimages of both sides we get that

$$e^{-1}([C(eX), C(eY)]) = e^{-1}(C(e(C(X) \cap C(Y)))) = \ker_{\mathbf{Q}}(e) +_{\mathbf{Q}} C(X) \cap C(Y).$$

Hence

$$\ker_{\mathbf{Q}}(e) +_{\mathbf{Q}} C(X) \cap C(Y) = e^{-1}([C(eX), C(eY)])$$

So $(\mathrm{Epi})_{m,n}$ holds for all positive m, n. □

We thus arrive at the crucial result of this book:

Theorem 5.3.7. *Let* $\mathbf{Q}$ *be a quasivariety such that the lattice* $\boldsymbol{Th}(\mathbf{Q}^{\models})$ *is distributive in the restricted sense. The equationally defined commutator for* $\mathbf{Q}$ *is additive.*

Proof. The thesis directly follows from Theorem 5.3.4, Corollary 5.3.6 and Theorem 5.2.16. □

The question as to whether the additivity of the equationally defined commutator for $\mathbf{Q}$ is essentially weaker than the restricted distributivity of $\boldsymbol{Th}(\mathbf{Q}^{\models})$ appears to be open. In other words, we ask if there is an example of a quasivariety $\mathbf{Q}$ whose equationally defined commutator is additive but where the lattice $\boldsymbol{Th}(\mathbf{Q}^{\models})$ is not distributive in the restricted sense.

In the next chapter we shall show that for every relatively congruence-modular quasivariety $\mathbf{Q}$, the lattice $\boldsymbol{Th}(\mathbf{Q}^{\models})$ is distributive in the restricted sense. From this fact and Theorem 5.3.7 we shall immediately get that for every relatively congruence-modular quasivariety the equationally defined commutator is additive.

We close this section with a comparison of the properties of the commutator $[\Phi, \Psi]_{edc(\mathbf{Va}(\mathbf{Q}))}$ with those of the commutator $[\Phi, \Psi]_{edc(\mathbf{Q})}$ for $\Phi, \Psi \in Con_{\mathbf{Q}}(\mathbf{A})$ in any algebra $\mathbf{A}$—see Section 3.1, Definition 3.1.10. We first pose the following:

Problem. Suppose that the equationally defined commutator for $\mathbf{Q}$ is additive. Let Φ and Ψ be $\mathbf{Q}$-congruences of τ-algebra $\mathbf{A}$. Is it true that $[\Phi, \Psi]_{edc(\mathbf{Va}(\mathbf{Q}))}$ is a $\mathbf{Q}$-congruence?

Leaving aside the problem, we shall just prove the Extension Principle for the Equationally defined Commutator here.

Theorem 5.3.8. *Let* $\mathbf{Q}$ *be a quasivariety whose equationally defined commutator is additive. Then for any algebra* $\mathbf{A} \in \mathbf{Va}(\mathbf{Q})$ *and any congruences* $\Phi, \Psi \in Con(\mathbf{A})$,

$$\Theta^{\mathbf{A}}_{\mathbf{Q}}([\Phi, \Psi]_{edc(\mathbf{Va}(\mathbf{Q}))}) = [\Theta^{\mathbf{A}}_{\mathbf{Q}}(\Phi), \Theta^{\mathbf{A}}_{\mathbf{Q}}(\Psi)]_{edc(\mathbf{Q})}.$$

Proof. We shall first prove the above theorem for principal congruences:

Lemma 5.3.9. *Let* $\mathbf{A} \in \mathbf{Va}(\mathbf{Q})$ *and* $a, b, c, d \in A$. *Then*

$$[\Theta^{\mathbf{A}}_{\mathbf{Q}}(a, b), \Theta^{\mathbf{A}}_{\mathbf{Q}}(c, d)]_{edc(\mathbf{Q})} = \Theta^{\mathbf{A}}_{\mathbf{Q}}([\Theta(a, b), \Theta(c, d)]_{edc(\mathbf{Va}(\mathbf{Q}))}).$$

Proof. The inclusion "$\supseteq$" is immediate because $\Theta^{\mathbf{A}}_{\mathbf{Q}}(a, b) \supseteq \Theta(a, b)$, $\Theta^{\mathbf{A}}_{\mathbf{Q}}(c, d) \supseteq \Theta(c, d)$ and hence

$$[\Theta^{\mathbf{A}}_{\mathbf{Q}}(a, b), \Theta^{\mathbf{A}}_{\mathbf{Q}}(c, d)]_{edc(\mathbf{Q})} = \Theta^{\mathbf{A}}_{\mathbf{Q}}([\Theta^{\mathbf{A}}_{\mathbf{Q}}(a, b), \Theta^{\mathbf{A}}_{\mathbf{Q}}(c, d)]_{edc(\mathbf{Va}(\mathbf{Q}))}) \supseteq \Theta^{\mathbf{A}}_{\mathbf{Q}}([\Theta(a, b), \Theta(c, d)]_{edc(\mathbf{Va}(\mathbf{Q}))}),$$

by the monotonicity of $[\,\cdot\,]_{edc(\mathbf{Va}(\mathbf{Q}))}$.

To prove the reverse inclusion, suppose that $\Delta(x, y, z, w, \underline{u})$ is a generating set for the equationally defined commutator of $\mathbf{Q}$. As $[\cdot]_{edc(\mathbf{Q})}$ is additive on $\mathbf{Q}$-congruences of the algebras of $\mathbf{Va}(\mathbf{Q})$, we have that

$$[\Theta^{\mathbf{A}}_{\mathbf{Q}}(a, b), \Theta^{\mathbf{A}}_{\mathbf{Q}}(c, d)]_{edc(\mathbf{Q})} = \Theta^{\mathbf{A}}_{\mathbf{Q}}(\bigcup\{(\forall \underline{e})\ \Delta(a, b, c, d, \underline{e}) : \underline{e} \in A^k\})$$

in the lattice $\mathbf{Con}_{\mathbf{Q}}(\mathbf{A})$. We must therefore show that

$$\Theta^{\mathbf{A}}_{\mathbf{Q}}(\bigcup\{(\forall \underline{e})\ \Delta(a, b, c, d, \underline{e}) : \underline{e} \in A^k\}) \subseteq \Theta^{\mathbf{A}}_{\mathbf{Q}}([\Theta(a, b), \Theta(c, d)]_{edc(\mathbf{Va}(\mathbf{Q}))}).$$

To prove the above inclusion it suffices to show that

$$\langle p(a, b, c, d, \underline{e}), q(a, b, c, d, \underline{e})\rangle \in [\Theta(a, b), \Theta(c, d)]_{edc(\mathbf{Va}(\mathbf{Q}))}, \qquad (*)$$

for all $p \approx q \in \Delta(x, y, z, w, \underline{u})$ and all sequences $\underline{e}$ of elements of A.

Fix $p \approx q$ and $\underline{e}$. We have:

$$p(a, a, c, d, \underline{e}) \equiv p(a, b, c, d, \underline{e}) \ (\text{mod } \Theta(a, b)) \tag{1}$$

$$q(a, b, c, c, \underline{e}) \equiv q(a, b, c, d, \underline{e}) \ (\text{mod } \Theta(c, d)) \tag{2}$$

Since $p \approx q$ is a commutator equation for $\boldsymbol{Va}(\mathbf{Q})$ and $\boldsymbol{A} \in \boldsymbol{Va}(\mathbf{Q})$, we also have that

$$p(a, a, c, d, \underline{e}) = q(a, a, c, d, \underline{e}) \tag{3}$$

and

$$p(a, b, c, c, \underline{e}) = q(a, b, c, c, \underline{e}). \tag{4}$$

(1) and (3) imply that

$$p(a, b, c, d, \underline{e}) \equiv q(a, a, c, d, \underline{e}) \ (\text{mod } \Theta(a, b)). \tag{5}$$

In turn, (2) and (4) give that

$$q(a, b, c, d, \underline{e}) \equiv p(a, b, c, c, \underline{e}) \ (\text{mod } \Theta(c, d)). \tag{6}$$

But trivially,

$$q(a, a, c, d, \underline{e}) \equiv q(a, b, c, d, \underline{e}) \ (\text{mod } \Theta(a, b)) \tag{7}$$

and

$$p(a, b, c, d, \underline{e}) \equiv p(a, b, c, c, \underline{e}) \ (\text{mod } \Theta(c, d)). \tag{8}$$

(5) and (7) imply that

$$p(a, b, c, d, \underline{e}) \equiv q(a, b, c, d, \underline{e}) \ (\text{mod } \Theta(a, b)),$$

while (6) and (8) give that

$$p(a, b, c, d, \underline{e}) \equiv q(a, b, c, d, \underline{e}) \ (\text{mod } \Theta(c, d)).$$

Hence

$$p(a, b, c, d, \underline{e}) \equiv q(a, b, c, d, \underline{e}) \ (\text{mod } [\Theta(a, b), \Theta(c, d)]_{edc(\boldsymbol{Va}(\mathbf{Q}))}),$$

by the definition of $[\Theta(a, b), \Theta(c, d)]_{edc(\boldsymbol{Va}(\mathbf{Q}))}$. So $(*)$ holds.
This proves the lemma. □

We pass to the proof of the theorem. The inclusion

$$\Theta^A_{\mathbf{Q}}([\Phi, \Psi]_{edc(\mathbf{Va}(\mathbf{Q}))}) \subseteq [\Theta^A_{\mathbf{Q}}(\Phi), \Theta^A_{\mathbf{Q}}(\Psi)]_{edc(\mathbf{Q})}$$

is immediate. To prove the opposite inclusion, we argue as follows.

$$[\Theta^A_{\mathbf{Q}}(\Phi), \Theta^A_{\mathbf{Q}}(\Psi)]_{edc(\mathbf{Q})} = \quad \text{(by the additivity of } [\,\cdot\,]_{edc(\mathbf{Q})})$$

$$\sup{}_{\mathbf{Q}}(\bigcup\{[\Theta^A_{\mathbf{Q}}(a,b), \Theta^A_{\mathbf{Q}}(c,d)]_{edc(\mathbf{Q})} : \langle a,b\rangle \in \Phi, \langle c,d\rangle \in \Psi\}) =$$

$$\text{(by the lemma)}$$

$$\Theta^A_{\mathbf{Q}}(\bigcup\{\Theta^A_{\mathbf{Q}}([\Theta(a,b), \Theta(c,d)]_{edc(\mathbf{Va}(\mathbf{Q}))}) : \langle a,b\rangle \in \Phi, \langle c,d\rangle \in \Psi\}) =$$

$$\Theta^A_{\mathbf{Q}}(\bigcup\{[\Theta(a,b), \Theta(c,d)]_{edc(\mathbf{Va}(\mathbf{Q}))}) : \langle a,b\rangle \in \Phi, \langle c,d\rangle \in \Psi\}) =$$

$$\Theta^A_{\mathbf{Q}}(\Theta^A(\bigcup\{[\Theta(a,b), \Theta(c,d)]_{edc(\mathbf{Va}(\mathbf{Q}))}) : \langle a,b\rangle \in \Phi, \langle c,d\rangle \in \Psi\}) \subseteq$$

$$\Theta^A_{\mathbf{Q}}([\Phi, \Psi]_{edc(\mathbf{Va}(\mathbf{Q}))}).$$

This concludes the proof of the theorem. □

5.4 Semilattices and Restricted Distributivity

Let **S** be the variety of semilattices. Each semilattice is endowed with a single binary operation $\cdot$ that is idempotent, commutative, and associative. **S** is minimal and it is not congruence-modular.

Theorem 5.4.1. *Let* **S** *be the variety of semilattices. The lattice* $\mathbf{Th}(\mathbf{S}^{\models})$ *obeys the law of distributivity in the restricted sense.*

Proof. We shall sketch the idea of the proof. Hereafter the term 'term' refers to any term in the signature of **S**. We shall use simplified notation and for any terms p, q we write pq instead $(p \cdot q)$ omitting the dot $\cdot$ and parentheses as much as possible. In particular we write x_1x_2 in place of $(x_1 \cdot x_2)$. By a *block* of variables we shall mean any term of the form $(\ldots(x_1x_2)\ldots x_n)$, where $x_1, x_2, \ldots, x_n$ is a sequence (without repetitions) of pairwise different variables. Thus, if σ is a permutation of the set $\{1, \ldots, n\}$, then $(\ldots(x_{\sigma(1)}x_{\sigma(2)})\ldots x_{\sigma(n)})$ is also a block. It is clear that for any term $t = t(x_1, x_2, \ldots, x_n)$, where all variables of t are displayed, the equality $t(x_1, x_2, \ldots, x_n) \approx (\ldots(x_1x_2)\ldots x_n)$ is valid in **S**, that is, every term is 'equivalent' to a block.

To simplify matters we assume that **2** is the two-element meet-semilattice. (This assumption is not restrictive—we could work with **2** treated as a join-semilattice as well. But then in the proofs we present below one should replace 0 by 1 and 1 by 0.)

Lemma 1. *Let s_1 and s_2 be arbitrary terms. Then $s_1 \approx s_2$ is valid in* **S** *if and only if $Var(s_1) = Var(s_2)$.*

Proof. Immediate – use **2**. □

We put: $C := \mathbf{S}^{\models}$.

Lemma 2. *Let x and y be different variables and $p \approx q \in C(x \approx y)$. If $p \approx q$ is not in $C(\emptyset)$, then there is (a possibly empty) block of variables $\underline{a}$ not containing x and y such that*

$$\textit{either}\quad C(p \approx q) = C(x\underline{a} \approx y\underline{a}) \quad \textit{or} \quad C(p \approx q) = C(x\underline{a} \approx yx\underline{a})$$
$$\textit{or} \quad C(p \approx q) = C(xy\underline{a} \approx y\underline{a}).$$

Proof (of the claim). We first note that for an arbitrary block of variables $\underline{a}$ not containing x and y we have that

$$x\underline{a} \approx y\underline{a} \in C(x \approx y), \quad x\underline{a} \approx yx\underline{a} \in C(x \approx y), \quad xy\underline{a} \approx y\underline{a} \in C(x \approx y).$$

The claim shows that up to deductive equivalence with respect to C these are the only possible cases.

As $p \approx q \notin C(\emptyset)$, we have that $Var(p) \neq Var(q)$. We first show

$$\{x, y\} \not\subseteq Var(p) \cap Var(q). \tag{a}$$

Suppose that $\{x, y\} \subseteq Var(p) \cap Var(q)$. Hence we may write that p and q are of the form $xy\underline{a}$ and $xy\underline{b}$, where $\underline{a}$ and $\underline{b}$ are blocks of variables not containing x and y. As $xy\underline{a} \approx xy\underline{b} \in C(x \approx y)$, we get that $xx\underline{a} \approx xx\underline{b}$ is valid in **S**. Hence $x\underline{a} \approx x\underline{b}$ is also valid in **S**, by idempotency. It follows that $\underline{a}$ and $\underline{b}$ are blocks of the same variables (i.e., $\underline{a}$ and $\underline{b}$ are equal up to a permutation of variables). Consequently, $C(p \approx q) = C(xy\underline{a} \approx xy\underline{a})$. Hence $p \approx q \in C(\emptyset)$. A contradiction.

$$\{x, y\} \cap (Var(p) \cup Var(q)) \neq \emptyset. \tag{b}$$

Suppose that $\{x, y\} \cap (Var(p) \cup Var(q)) = \emptyset$. Hence neither x nor y occurs in p and q. As $p \approx q \notin C(\emptyset)$, there is an assignment h in a semilattice **A** such that $hp \neq hq$. Since the sets of variables $\{x, y\}$ and $Var(p) \cup Var(q)$ are disjoint, we may at the same time assume that $hx = hy$. It follows that $p \approx q \notin C(x \approx y)$. A contradiction.

$\{x, y\} \subseteq Var(p)$, $Var(q)$ contains exactly one variable from $\{x, y\}$ *or* (c)

$\{x, y\} \subseteq Var(q)$, $Var(p)$ contains exactly one variable from $\{x, y\}$ *or*

both $Var(p)$, $Var(q)$ contain exactly one (but not the same) variable from $\{x, y\}$.

In view of (a) and (b), to prove (c) it suffices to *exclude* the following four cases:

$$\{x, y\} \subseteq Var(p), \quad \{x, y\} \cap Var(q) = \emptyset, \tag{d1}$$

$$\{x, y\} \subseteq Var(q), \quad \{x, y\} \cap Var(p) = \emptyset, \tag{d2}$$

$$Var(p) \cap \{x, y\} \text{ is a singleton}, \quad \{x, y\} \cap Var(q) = \emptyset, \tag{d3}$$

$$Var(q) \cap \{x, y\} \text{ is a singleton}, \quad \{x, y\} \cap Var(p) = \emptyset. \tag{d4}$$

As to (d1), suppose $\{x, y\} \subseteq Var(p)$, $\{x, y\} \cap Var(q) = \emptyset$. Hence there are blocks of variables $\underline{a}$ and $\underline{b}$ not containing x and y such that $C(p \approx q) = C(xy\underline{a} \approx \underline{b})$. As $p \approx q \in C(x \approx y)$, we therefore get that $xy\underline{a} \approx \underline{b} \in C(x \approx y)$. Let us take an assignment h in the two-element semilattice **2** such that $h\underline{a} = h\underline{b} = 1$ and $hx = hy = 0$. Then $hxy\underline{a} = 0$. As $h\underline{b} = 1$, it follows that $xy\underline{a} \approx \underline{b} \notin C(x \approx y)$. A contradiction.

(d2) is proved similarly.

As to (d3), suppose first that $x \in Var(p)$, $y \notin Var(p)$ and $\{x, y\} \cap Var(q) = \emptyset$. Hence there are blocks of variables $\underline{a}$ and $\underline{b}$ not containing x and y such that $C(p \approx q) = C(x\underline{a} \approx \underline{b})$. As $p \approx q \in C(x \approx y)$, we get that $x\underline{a} \approx \underline{b} \in C(x \approx y)$. Let h be an assignment in the two-element semilattice **2** such that $h\underline{a} = h\underline{b} = 1$ and $hx = hy = 0$. Then $hx\underline{a} = 0$. As $h\underline{b} = 1$, it follows that $x\underline{a} \approx \underline{b} \notin C(x \approx y)$. A contradiction.

The case when $x \notin Var(p)$, $y \in Var(p)$ and $\{x, y\} \cap Var(q) = \emptyset$ is similarly handled.

(d4) is left as an exercise.

From (a), (b), (c) the claim follows. Indeed, assume first that $\{x, y\} \subseteq Var(p)$ and $Var(q)$ contains exactly one variable from $\{x, y\}$, say x. Then $C(p \approx q) = C(xy\underline{a} \approx x\underline{b})$ for some blocks of variables $\underline{a}$ and $\underline{b}$ not containing x and y. We then have that $xy\underline{a} \approx x\underline{b} \in C(x \approx y)$. Substituting x for y we therefore get that $xx\underline{a} \approx x\underline{b} \in C(\emptyset)$. Hence $x\underline{a} \approx x\underline{b} \in C(\emptyset)$. It follows that the blocs $\underline{a}$ and $\underline{b}$ contain the same variables. So without loss of generality we may assume that they are identical. We therefore have that $C(p \approx q) = C(xy\underline{a} \approx x\underline{a})$.

The other cases are handled in a similar way. They all yield one of the cases of the statement of the claim. For instance, let us also check the last case when both $Var(p)$ and $Var(q)$ contain exactly one (but not the same) variable from $\{x, y\}$. In this situation $C(p \approx q) = C(x\underline{a} \approx y\underline{b})$ for some blocks $\underline{a}$ and $\underline{b}$ not containing x and y. As $x\underline{a} \approx y\underline{b} \in C(x \approx y)$, we get that $x\underline{a} \approx x\underline{b} \in C(\emptyset)$. From this it follows that $\underline{a}$ and $\underline{b}$ contain the same variables. Hence $\underline{a} \approx \underline{b} \in C(\emptyset)$. Consequently, $C(p \approx q) = C(x\underline{a} \approx y\underline{a})$. □

Lemma 3. *Let X be any finite set of equations of terms and let Y be any finite set of equations of variables separated from X, that is, $Var(X) \cap Var(Y) = \emptyset$. Then $C(X) \cap C(Y) = C(\emptyset)$.*

Proof. The lemma is reformulated as the statement:

Let m and n be arbitrary positive integers. For any sequences of terms $\underline{p}_m = p_1, \ldots, p_m$, $\underline{q}_m = q_1, \ldots, q_m$ and any sequences $\underline{x}_n = x_1, \ldots, x_n$,

$\underline{y}_n = y_1, \ldots, y_n$ *of pairwise different individual variables separated from the variables occurring in the terms of* $\underline{p}_m$ *and* $\underline{q}_m$,

$$(T)_{m,n} \qquad C(\underline{p}_m \approx \underline{q}_m) \cap C(\underline{x}_n \approx \underline{y}_n) = C(\emptyset).$$

Proof. We prove $(T)_{m,n}$ by double induction on m and n.

Claim 1. *$(T)_{m,1}$ holds for all m.*

Proof (of the claim). Fix $m \geqslant 1$ and let $\underline{p}_m = p_1, \ldots, p_m$, $\underline{q}_m = q_1, \ldots, q_m$ be sequences of terms. Let x and y be two variables separated from the variables of $\underline{p}_m$ and $\underline{q}_m$. To prove $(T)_{m,1}$ suppose by way of contradiction that $r \approx s \in C(\underline{p}_m \approx \underline{q}_m) \cap C(x \approx y)$ but $r \approx s \notin C(\emptyset)$. As $r \approx s \in C(x \approx y)$, Lemma 2 applies. We therefore consider three cases.

Case 1. $C(p \approx q) = C(x\underline{a} \approx y\underline{a})$.

So we have $x\underline{a} \approx y\underline{a} \in C(\underline{p}_m \approx \underline{q}_m)$. As x and y do not occur in $\underline{a}$, $\underline{p}_m$ and $\underline{q}_m$, we take an assignment h in $\mathbf{2}$ such that $hx = 1$, $hy = 0$ and $h\underline{a} = hp_1 = \ldots = hp_m = hq_1 = \ldots = hq_m = 1$. As $hx\underline{a} = 1$ and $hy\underline{a} = 0$, we see that $x\underline{a} \approx y\underline{a} \notin C(\underline{p}_m \approx \underline{q}_m)$. A contradiction.

Case 2. $C(p \approx q) = C(x\underline{a} \approx yx\underline{a})$.

Then we have that $x\underline{a} \approx yx\underline{a} \in C(\underline{p}_m \approx \underline{q}_m)$. Taking an assignment h in $\mathbf{2}$ such that $hx = 1$, $hy = 1$ and $h\underline{a} = hp_1 = \ldots = hp_m = hq_1 = \ldots = hq_m = 1$, we get that $hx\underline{a} = 1$ and $hyx\underline{a} = 0$. Hence $x\underline{a} \approx yx\underline{a} \notin C(\underline{p}_m \approx \underline{q}_m)$. A contradiction.

Case 3. $C(p \approx q) = C(xy\underline{a} \approx y\underline{a})$.

Then $xy\underline{a} \approx y\underline{a} \in C(\underline{p}_m \approx \underline{q}_m)$. Then taking an assignment h in $\mathbf{2}$ such that $hx = 0$, $hy = 1$ and $h\underline{a} = hp_1 = \ldots = hp_m = hq_1 = \ldots = hq_m = 1$, we get that $hxy\underline{a} = 0$ and $hy\underline{a} = 1$. It follows that $x\underline{a} \approx yx\underline{a} \notin C(\underline{p}_m \approx \underline{q}_m)$. A contradiction.
This proves the claim. □

Claim 2. *Fix m. Then $(T)_{m,n}$ holds for all n.*

Proof. This is proved by induction on n. The case $n = 1$ is established by Claim 1. Fix $n \geqslant 1$ and assume $(T)_{m,n}$. We shall prove $(T)_{m,n+1}$. Fix terms $\underline{p}_m = p_1, \ldots, p_m$, $\underline{q}_m = q_1, \ldots, q_m$ and variables $\underline{x}_n = x_1, \ldots, x_n$, $\underline{y}_n = y_1, \ldots, y_n$ and let x and y be two new variables separated from the variables of $\underline{p}_m$ and $\underline{q}_m$. We claim that $C(\underline{p}_m \approx \underline{q}_m) \cap C(\underline{x}_n \approx \underline{y}_n, x \approx y) = C(\emptyset)$. Assume $r \approx s \in C(\underline{p}_m \approx \underline{q}_m) \cap C(\underline{x}_n \approx \underline{y}_n, x \approx y)$ and suppose $r \approx s$ is not in $C(\emptyset)$. Let e be a substitution such that $ex = x$, $ey = x$ and e is the identity map on the remaining variables. Then $er \approx es \in C(e\underline{p}_m \approx e\underline{q}_m) \cap C(\underline{x}_n \approx \underline{y}_n) = C(\underline{p}_m \approx \underline{q}_m) \cap C(\underline{x}_n \approx \underline{y}_n) = C(\emptyset)$, because e is the identity map on the variables of $\underline{p}_m \approx \underline{q}_m$ and by $(T)_{m,n}$. Hence $er \approx es \in C(\emptyset)$, i.e., $r \approx s \in \ker_{\mathbf{S}}(e)$. But from the definition of e and Theorem 2.21 it follows that $\ker_{\mathbf{S}}(e) = C(x \approx y)$. Thus $r \approx s \in C(x \approx y)$. As $r \approx s \in C(\underline{p}_m \approx \underline{q}_m)$, we therefore get that $r \approx s \in C(\underline{p}_m \approx \underline{q}_m) \cap C(x \approx y)$. But as the variables x

and y are separated from the variables of $\underline{p}_m$ and $\underline{q}_m$, $(T)_{m,1}$ yields that $C(\underline{p}_m \approx \underline{q}_m) \cap C(x \cap y) = C(\emptyset)$. So $r \approx s \in C(\emptyset)$. A contradiction. □

From Claims 1–2 the lemma follows. □

The above lemma implies the theorem. For let $X = \{x_i \approx y_i : i \in I\}$, where x_i, y_i, $i \in I$, are pairwise different individual variables. Let Y and Z be arbitrary sets of equations of terms such that the variables occurring in the terms of $Y \cup Z$ are separated from the variables of X. Then Lemma 3 gives that $C(X) \cap (C(Y) +_{\mathbf{Q}} C(Z)) = C(\emptyset)$, $C(X) \cap C(Y) = C(\emptyset)$ and $C(X) \cap C(Z) = C(\emptyset)$. Hence

$$C(X) \cap C(Y) +_{\mathbf{Q}} C(X) \cap C(Z) = C(X) \cap (C(Y) +_{\mathbf{Q}} C(Z)),$$

which means that $\boldsymbol{Th}(\mathbf{S}^{\models})$ validates the restricted distributivity. □

The above Lemma 3 shows that the law of restricted distributivity of $\boldsymbol{Th}(\mathbf{S}^{\models})$ trivializes because its constituents are equal to the theory $C(\emptyset)$. This fact also implies:

Corollary 5.4.2. *The equationally defined commutator for the variety of semilattices* $\mathbf{S}$ *is the zero commutator: for every semilattice* $\mathbf{A}$ *and any two congruences* $\Phi, \Psi \in Con(\mathbf{A})$, $[\Phi, \Psi]_{edc(\mathbf{S})} = \mathbf{0_A}$.

Proof. This follows from the fact that for any positive integers m and n and any sequences of pairwise distinct individual variables $\underline{x} = x_1, \dots, x_m$, $\underline{y} = y_1, \dots, y_m$ and $\underline{z} = z_1, \dots, z_n$, $\underline{w} = w_1, \dots, w_n$ it is the case that

$$\mathbf{S}^{\models}(\underline{x} \approx \underline{y}) \cap \mathbf{S}^{\models}(\underline{z} \approx \underline{w}) = \mathbf{S}^{\models}(\emptyset).$$

This is a particular case of Lemma 3. Thus the set of commutator equations for $\mathbf{S}$ reduces to the identities of $\mathbf{S}$. □

5.5 A Characterization of the Equationally-Defined Commutator

The following theorem states that if the equationally defined commutator is additive, then it is the largest one among all additive operations defined in the lattices of relative congruences and satisfying (C2):

Theorem 5.5.1. *Let* $\mathbf{Q}$ *be a quasivariety of algebras whose equationally defined commutator is additive. Suppose that* $*$ *is a binary operation defined in the lattices* $\boldsymbol{Con}_{\mathbf{Q}}(\mathbf{A})$, $\mathbf{A} \in \mathbf{Q}$, *such that for any algebra* $\mathbf{A} \in \mathbf{Q}$, $*$ *satisfies*

(C1)* $\Phi *^A \Psi \subseteq \Phi \cap \Psi$,
(C2)* $\Phi *^A \sup_{\mathbf{Q}}(X) = \sup_{\mathbf{Q}}\{\Phi *^A \Xi : \Xi \in X\}$,

for any $\Phi, \Psi \in Con_{\mathbf{Q}}(A)$ *and any set* $X \subseteq Con_{\mathbf{Q}}(\mathbf{A})$.

If, furthermore, $*$ *satisfies*

(C3)* *If* $h : \mathbf{A} \to \mathbf{B}$ *is a surjective homomorphism between* $\mathbf{Q}$*-algebras and* $\Phi, \Psi \in Con_{\mathbf{Q}}(\mathbf{A})$, *then*

$$\ker(h) +_{\mathbf{Q}} (\Phi \ast^{A} \Psi) = h^{-1}(h\Phi \ast^{A} h\Psi),$$

then the operation $\ast$ *is included in the equationally defined commutator, i.e.,* $\Phi \ast \Psi \subseteq [\Phi, \Psi]^{A}$, *for any* $\mathbf{A} \in \mathbf{Q}$ *and any* $\Phi, \Psi \in Con_{\mathbf{Q}}(\mathbf{A})$.

(In (C1)* the supremum is taken in the lattice $\boldsymbol{Con}_{\mathbf{Q}}(\mathbf{A})$. (C1)* and (C2)* are the $\ast$-counterparts of (C1) and (C2) for the equationally defined commutator.)

To prove the theorem, one repeats the reasoning carried out for the equationally defined commutator in Section 3.1. One first selects four distinct variables x, y, z, w and defines $\Delta_0(x, y, z, w, \underline{u})$ to be any set of equations such that $\mathbf{Q}^{\models}(\Delta_0) = \mathbf{Q}^{\models}(x \approx y) \ast \mathbf{Q}^{\models}(z \approx w)$. (The definition of $\ast$ makes sense in the lattice of equational theories of $\mathbf{Q}^{\models}$ as well. But Δ_0 need not be a generating set for the equationally defined commutator of $\mathbf{Q}$.)

(C0)* implies that $\mathbf{Q}^{\models}(\Delta_0) \subseteq \mathbf{Q}^{\models}(x \approx y) \cap \mathbf{Q}^{\models}(z \approx w)$, i.e., Δ_0 is a set of quaternary commutator equations of $\mathbf{Q}$. In turn, (C1)* and (C2)* imply that

For every algebra $\mathbf{A} \in \mathbf{Q}$ *and for every pair of sets* $X, Y \subseteq A^2$,

$$\Theta^{A}_{\mathbf{Q}}(X) \ast^{A} \Theta^{A}_{\mathbf{Q}}(Y) = \Theta^{A}_{\mathbf{Q}}(\bigcup\{(\forall \underline{e})\ \Delta^{A}_{0}(a, b, c, d, \underline{e}) : \langle a, b\rangle \in X, \langle c, d\rangle \in Y\}). \quad \text{(a)}$$

(Argue as in the (1) $\Rightarrow$ (2)-part of the proof of Theorem 5.1.2.)

(a) and the fact that the equationally defined commutator of $\mathbf{Q}$ is additive imply that

For every algebra $\mathbf{A} \in \mathbf{Q}$ *and for every pair of sets* $X, Y \in A^2$,

$$\Theta^{A}_{\mathbf{Q}}(X) \ast^{A} \Theta^{A}_{\mathbf{Q}}(Y) \subseteq [\Theta^{A}_{\mathbf{Q}}(X), \Theta^{A}_{\mathbf{Q}}(Y)]^{A}, \quad \text{(b)}$$

because Δ_0 is a set of quaternary commutator equations.

(b) readily implies that $\Phi \ast^{A} \Psi \subseteq [\Phi, \Psi]^{A}$, for any $\mathbf{A} \in \mathbf{Q}$ and any $\Phi, \Psi \in Con_{\mathbf{Q}}(\mathbf{A})$. □

Chapter 6
Modularity and Related Topics

6.1 Relative Congruence-Modularity and the Equationally-Defined Commutator

Definition 6.1.1. A lattice $\boldsymbol{L} = (L, \wedge, \vee)$ is *modular* if it satisfies the following *modularity* identity:

$$(x \wedge z) \vee (y \wedge z) \approx z \wedge (x \vee (y \wedge z)) \tag{M}$$

or, equivalently, it satisfies the dual identity

$$(x \vee z) \wedge (y \vee z) \approx z \vee (x \wedge (y \vee z)). \tag{M^d}$$

It is easy to see that $\boldsymbol{L}$ is modular if and only if the following implication is true in $\boldsymbol{L}$:

$$z \leqslant x \rightarrow z \vee (x \wedge y) \approx x \wedge (y \vee z). \quad \square \tag{M$'$}$$

A quasivariety $\mathbf{Q}$ of algebras is called *relatively congruence-modular* (RCM, for short) if, for any algebra $\boldsymbol{A} \in \mathbf{Q}$, the lattice $\boldsymbol{Con}_{\mathbf{Q}}(\boldsymbol{A})$ is modular.

One can prove that $\mathbf{Q}$ is RCM if and only if the lattice of $\mathbf{Q}$-congruences on the free algebra in $\mathbf{Q}$ with ω free generators $\boldsymbol{F}_{\mathbf{Q}}(\omega)$ is modular.

A lattice $\boldsymbol{L}$ is *distributive* it satisfies the following distributivity identity

$$(x \wedge z) \vee (y \wedge z) \approx (x \vee y) \wedge z \tag{D}$$

or, equivalently, it satisfies the dual identity

$$(x \vee z) \wedge (y \vee z) \approx (x \wedge y) \vee z. \tag{D^d}$$

J. Czelakowski, *The Equationally-Defined Commutator*,
DOI 10.1007/978-3-319-21200-5_6

A quasivariety $\mathbf{Q}$ of algebras is called *relatively congruence-distributive* (abbreviated RCD) if, for any algebra $\boldsymbol{A} \in \mathbf{Q}$, the lattice $\boldsymbol{Con}_{\mathbf{Q}}(\boldsymbol{A})$ is distributive.

$\mathbf{Q}$ is RCD if and only if the lattice of $\mathbf{Q}$-congruences on the free algebra $\boldsymbol{F}_{\mathbf{Q}}(\omega)$ is distributive.

Examples of congruence-modular *varieties*: any variety of groups, any variety of rings, any variety of equivalence algebras (see Section 6.4). More generally, any point-regular variety is congruence-modular. Any congruence-permutable variety is also congruence-modular.

General fact due to Kearnes and McKenzie (1992): *If* $\mathbf{Q}$ *is a locally finite, RCM subquasivariety of a semisimple congruence-modular variety, then* $\mathbf{Q}$ *is a variety.*

Examples of relatively congruence-modular *quasivarieties*. According to Kearnes and McKenzie (1992), every RCM quasivariety of semigroups satisfies the cancellation laws: $xy \approx xz \rightarrow y \approx z$ and a non-trivial equation. Conversely, any quasivariety of semigroups axiomatized by the cancellation laws and a non-trivial equation is RCM. □

The following theorem, already mentioned in this book, is a non-trivial result in the theory of the commutator.

Theorem 6.1.2 (Kearnes and McKenzie (1992)). *Let* $\mathbf{Q}$ *be an RCM quasivariety of algebras. Then, for every algebra* $\boldsymbol{A} \in \mathbf{Q}$*, the commutator in the sense of Kearnes and McKenzie satisfies conditions* (C1) *and* (C2) *on the lattice* $\boldsymbol{Con}_{\mathbf{Q}}(\boldsymbol{A})$. □

We recall that conditions (C1) and (C2) have been defined in Section 5.1.

But according to Theorem 4.1.12, for any RCM quasivariety the Kearnes-McKenzie and equationally defined commutators coincide. We thus have:

Corollary 6.1.3. *Let* $\mathbf{Q}$ *be an RCM quasivariety of algebras. For any algebra* $\boldsymbol{A} \in \mathbf{Q}$*, the equationally defined commutator is additive on the lattice* $\boldsymbol{Con}_{\mathbf{Q}}(\boldsymbol{A})$. □

Open Problem Suppose that a quasivariety $\mathbf{Q}$ has the relative shifting property and the equationally defined commutator for $\mathbf{Q}$ satisfies (C1). Is $\mathbf{Q}$ an RCM quasivariety?

It is easy to see that (C1) alone does not imply relative congruence-modularity. For, let τ be the empty signature (no operation symbols). Let us take Birkhoff's logic B_τ (see remarks placed in Section 5.3). Its models coincide with "bare" non-empty sets (no operations on them) with equivalence relations as congruences. B_τ has the zero equational commutator—the equationally defined commutator $[\Phi, \Psi]^A$ of any two equivalence relations is the identity relation in A. Trivially, the commutator satisfies (C1). But the class of lattices of equivalence relations on all sets does not satisfy a non-trivial lattice-theoretic identity. It follows that B_τ is not congruence-modular.

The above variety fails to have the shifting property because otherwise it would be congruence-modular, which is excluded.

Corollary 6.1.3 is a consequence of the fact that for any RCM quasivariety $\mathbf{Q}$, the equationally defined commutator and the commutator in the sense of Kearnes-McKenzie coincide (Theorem 4.1.12). Hence the former is additive because the latter has this property. But the proof of Theorem 4.1.12 is indirect because it makes heavy use of the results from the theory of centralization relations presented in Chapter 4.

The rest of this section is devoted to a *direct* proof of the fact that, for any RCM quasivariety, the equationally defined commutator is additive, without resort to the theory of Kearnes and McKenzie.

We begin with the following simple observations concerning modular lattices.

Let $x, x_1, \ldots, x_n, \ldots$ be an infinite sequence of pairwise distinct individual variables of the language of lattice theory. We inductively define the following infinite sequence of terms

$$x * (x_1, \ldots, x_n) \quad \text{for } n = 1, 2, \ldots.$$

(see Erné 1988):

$$x * (x_1) := x \wedge x_1$$
$$x * (x_1, \ldots, x_n, x_{n+1}) := x \wedge (x_{n+1} \vee (x * (x_1, \ldots, x_n))),$$

for all $n \geqslant 1$.

Thus

$$x * (x_1, x_2) \quad \text{is} \quad x \wedge (x_2 \vee (x \wedge x_1)),$$
$$x * (x_1, x_2, x_3) \quad \text{is} \quad x \wedge (x_3 \vee (x \wedge (x_2 \vee (x \wedge x_1)))) \quad \text{etc.}$$

Every lattice $\boldsymbol{L}$ validates the inequalities

$$(x \wedge x_1) \vee \ldots \vee (x \wedge x_n) \leqslant x * (x_1, \ldots, x_n) \leqslant x \wedge (x_1 \vee \ldots \vee x_n)$$

for all $n \geqslant 1$. (The proof is by induction on n.)

Theorem 6.1.4. *A lattice* $\boldsymbol{L}$ *is modular if and only if* $\boldsymbol{L}$ *universally validates the equations*

$$x * (x_1, \ldots, x_n) \approx (x \wedge x_1) \vee \ldots \vee (x \wedge x_n)$$

for all $n \geqslant 1$.

Notice that if $\boldsymbol{L}$ is modular and $(a, b_1, \ldots, b_n) \in L^{n+1}$ then, in view of the above theorem, $a * (b_1, \ldots, b_n) = a * (b_{\sigma(1)}, \ldots, b_{\sigma(n)})$ for any permutation σ of the set $\{1, 2, \ldots, n\}$, because $(a \wedge b_1) \vee \ldots \vee (a \wedge b_n)$ is invariant under permutations of $\{1, 2, \ldots, n\}$. (The interpretation of $*$ in $\boldsymbol{L}$ is denoted by the same symbol.)

Proof. ($\Rightarrow$). Assume $\boldsymbol{L}$ is modular and $a \in L$. We prove by induction:

$$T(n){:}\quad a * (b_1, \ldots, b_n) = (a \wedge b_1) \vee \ldots \vee (a \wedge b_n), \qquad \text{for all } (b_1, \ldots, b_n) \in L^n.$$

$T(1)$ is trivial. To show $T(2)$, assume $(b_1, b_2) \in L^2$. Let $c := a \wedge b_1$. As $c \leqslant a$, modularity gives that $a \wedge (b_2 \vee (a \wedge b_1)) = a \wedge (b_2 \vee c) = (a \wedge b_2) \vee c = (a \wedge b_2) \vee (a \wedge b_1)$. So $T(2)$ holds.

Assume $T(n)$ holds. Let $(b_1, \ldots, b_n, b_{n+1}) \in L^{n+1}$. We compute:

$$a * (b_1, \ldots, b_n, b_{n+1}) := a \wedge (b_{n+1} \vee (a * (b_1, \ldots, b_n))) = \quad \text{(IH)} \qquad \text{(a)}$$
$$a \wedge (b_{n+1} \vee ((a \wedge b_1) \vee \ldots \vee (a \wedge b_n))).$$

Putting $c := (a \wedge b_1) \vee \ldots \vee (a \wedge b_n)$, we have that $c \leqslant a$. Hence, by modularity,

$$a \wedge (b_{n+1} \vee ((a \wedge b_1) \vee \ldots \vee (a \wedge b_n))) = a \wedge (b_{n+1} \vee c) = \qquad \text{(b)}$$
$$a \wedge b_{n+1}) \vee c = (a \wedge b_{n+1}) \vee (a \wedge b1) \vee \ldots \vee (a \wedge b_n) =$$
$$(a \wedge b_1) \vee \ldots \vee (a \wedge b_n) \vee (a \wedge b_{n+1}).$$

Combining together (a) and (b) we see that $T(n + 1)$ holds.

($\Rightarrow$). As $T(2)$ holds, we have that

$$a \wedge (b_2 \vee (a \wedge b_1)) = (a \wedge b_2) \vee (a \wedge b_1),$$

for all $a, b_1, b_2 \in L$. Hence

$$a \wedge (b_2 \vee (b_1 \wedge a)) = (b_2 \wedge a) \vee (b_1 \wedge a).$$

It follows that $\boldsymbol{L}$ validates modularity law

$$(x \wedge z) \vee (y \wedge z) \approx z \wedge (x \vee (y \wedge z)). \qquad \square$$

Note. On the basis of the axioms of lattice theory, modularity is equivalent to the identity $x * (x_1, x_2) \approx (x \wedge x_1) \vee (x \wedge x_2)$. The remaining identities $x * (x_1, \ldots, x_n) \approx (x \wedge x_1) \vee \ldots \vee (x \wedge x_n)$, $n \geqslant 3$, are consequences of it. $\square$

Let $*_d$ be the term operation dual to $*$. Thus

$$x *_d (x_1) := x \vee x_1$$
$$x *_d (x_1, \ldots, x_n, x_{n+1}) := x \vee (x_{n+1} \wedge (x *_d (x_1, \ldots, x_n))),$$

for all $n \geqslant 1$ and $(x_1, \ldots, x_n, x_{n+1}) \in L^{n+1}$.

In particular,

$$x *_d (x_1, x_2) \quad \text{is} \quad x \vee (x_2 \wedge (x \vee x_1)),$$
$$x *_d (x_1, x_2, x_3) = x \vee (x_3 \wedge (x \vee (x_2 \wedge (x \vee x_1)))) \quad \text{etc.}$$

Every lattice $\boldsymbol{L}$ validates the inequalities

$$x \vee (x_1 \wedge \ldots \wedge x_n) \leqslant x *_d (x_1, \ldots, x_n) \leqslant (x \vee x_1) \wedge \ldots \wedge (x \vee x_n)$$

for all $n \geqslant 1$.

The reverse of the second inequality is equivalent to modularity:

Corollary 6.1.5. *A lattice* $\mathbf{L}$ *is modular if and only if* $\mathbf{L}$ *universally validates the equations*

$$x *_d (x_1, \ldots, x_n) \approx (x \vee x_1) \wedge \ldots \wedge (x \vee x_n),$$

for all $n \geqslant 1$.

Proof. This can directly verified by emulating the proof of Theorem 6.1.4 for its dual counterpart or directly by using Theorem 6.1.4 and the duality principle for modular lattices in the following version (see Appendix A): a lattice equation $s \approx t$ is universally valid in the class of modular lattices if and only if its dual $s^d \approx t^d$ is universally valid in this class. □

Corollary 6.1.6. *Let* $\mathbf{Q}$ *be an RCM quasivariety,* $\boldsymbol{A} \in \mathbf{Q}$*,* $\Phi \in Con_{\mathbf{Q}}(\boldsymbol{A})$*. Let* $\boldsymbol{B} := \boldsymbol{A}/\Phi$ *and* $h : \boldsymbol{A} \to \boldsymbol{B}$ *be the canonical homomorphism. Then for any* $n \geqslant 1$ *and* $a_1, \ldots, a_n, b_1, \ldots, b_n \in A$,

$$\Phi *_d (\Theta^{\boldsymbol{A}}_{\mathbf{Q}}(a_1, b_1), \Theta^{\boldsymbol{A}}_{\mathbf{Q}}(a_2, b_2), \ldots, \Theta^{\boldsymbol{A}}_{\mathbf{Q}}(a_n, b_n)) = h^{-1}(\bigcap_{1 \leqslant i \leqslant n} \Theta^{\boldsymbol{B}}_{\mathbf{Q}}(a_i/\Phi, b_i/\Phi)).$$

Proof.

$$\Phi *_d (\Theta^{\boldsymbol{A}}_{\mathbf{Q}}(a_1, b_1), \Theta^{\boldsymbol{A}}_{\mathbf{Q}}(a_2, b_2), \ldots, \Theta^{\boldsymbol{A}}_{\mathbf{Q}}(a_n, b_n)) = \quad \text{(by Corollary 6.1.5)}$$

$$\bigcap_{1 \leqslant i \leqslant n} (\Phi +_{\mathbf{Q}} \Theta^{\boldsymbol{A}}_{\mathbf{Q}}(a_i, b_i)) = \bigcap_{1 \leqslant i \leqslant n} (h^{-1}(\Theta^{\boldsymbol{B}}_{\mathbf{Q}}(a_i/\Phi, b_i/\Phi)) =$$

$$h^{-1}(\bigcap_{1 \leqslant i \leqslant n} \Theta^{\boldsymbol{B}}_{\mathbf{Q}}(a_i/\Phi, b_i/\Phi)). \quad \square$$

We also note the following fact:

Proposition 6.1.7. *Let* $\mathbf{Q}$ *be an RCM quasivariety. Then* $\mathbf{Q}$ *has the relative shifting property.*

Proof. In view of Lemma 3.5.2, it suffices to show that:

For any (equivalently, for some) different variables x, y, z, w,

$$z \approx w \in \mathbf{Q}^{\models}(\{x \approx y\} \cup \mathbf{Q}^{\models}(x \approx y, z \approx w) \cap \mathbf{Q}^{\models}(x \approx z, y \approx w)).$$

Fix x, y, z, w and put:

$$C := \mathbf{Q}^{\models}(\{x \approx y\}), \quad A := \mathbf{Q}^{\models}(x \approx y, z \approx w), \quad B := \mathbf{Q}^{\models}(x \approx z, y \approx w).$$

Then $z \approx w \in A$, $z \approx w \in \mathbf{Q}^{\models}(B \cup C)$ and $C \subseteq A$. By modularity, $A \cap \mathbf{Q}^{\models}(B \cup C) = (A \cap B) \vee C$. Hence $z \approx w \in \mathbf{Q}^{\models}(\{x \approx y\} \cup \mathbf{Q}^{\models}(x \approx y, z \approx w) \cap \mathbf{Q}^{\models}(x \approx z, y \approx w))$. □

The relative shifting property is essentially a weaker property than relative congruence-modularity. There exists a congruence-modular variety **V** with a subquasivariety **Q** that fails to be relatively congruence-modular (Kearnes and McKenzie 1992, Example 3.3.) Since the relative shifting property is inherited by subquasivarieties (Corollary 3.5.3), we see that **Q** has the relative shifting property and **Q** is not an RCM quasivariety.

6.2 The Extension Principle

We recall that a quasivariety **Q** satisfies the *Extension Principle* if for every algebra $\mathbf{A} \in \mathbf{Q}$, the operator $\Theta_{\mathbf{Q}}(\,\cdot\,)$ is a homomorphism from the lattice $\boldsymbol{Con}(\mathbf{A})$ to the lattice $\boldsymbol{Con}_{\mathbf{Q}}(\mathbf{A})$. Equivalently, **Q** satisfies the Extension Principle if for every algebra $\mathbf{A} \in \mathbf{Q}$ and for every pair of congruences $\Phi, \Psi \in Con(\mathbf{A})$, it is the case that $\Theta_{\mathbf{Q}}(\Phi \cap \Psi) = \Theta_{\mathbf{Q}}(\Phi) \cap \Theta_{\mathbf{Q}}(\Psi)$.

Kearnes and McKenzie (1992) have proved a surprising result:

> *A quasivariety* **Q** *is RCM if and only if it has the relative shifting property and satisfies the Extension Principle.*

We shall first broaden the scope of the Extension Principle a little bit as compared with the original proof given in Kearnes and McKenzie (1992).

Let τ be a fixed algebraic signature, let **Q** be a quasivariety of τ-algebras, and $\mathbf{A}$ a τ-algebra, not necessarily in **Q**. If $\Phi \in Con(\mathbf{A})$, then Φ' denotes the least **Q**-congruence of $\mathbf{A}$ that includes Φ. (This congruence is usually denoted by $\Theta^{\mathbf{A}}_{\mathbf{Q}}(\Phi)$ but we shall use the more compact notation Φ' because **Q** is clear from context.). Thus Φ' is the least congruence Ψ of $\mathbf{A}$ such that $\Phi \subseteq \Psi$ and $\mathbf{A}/\Psi \in \mathbf{Q}$. Clearly, $\mathbf{0}'_{\mathbf{A}}$ is the least **Q**-congruence of $\mathbf{A}$. If $\mathbf{A} \in \mathbf{Q}$, then $\mathbf{0}'_{\mathbf{A}} = \mathbf{0}_{\mathbf{A}}$.

Theorem 6.2.1. (The Extension Principle). *Let* **Q** *be an RCM quasivariety. Then for every algebra* $\mathbf{A} \in \mathbf{Va}(\mathbf{Q})$ *and any congruences* $\Phi, \Psi \in Con(\mathbf{A})$, $(\Phi \cap \Psi)' = \Phi' \cap \Psi'$.

If $(\Phi \cap \Psi)' = \Phi' \cap \Psi'$ holds for any $\Phi, \Psi \in Con(\mathbf{A})$, we say that $\mathbf{A}$ *satisfies the Extension Principle with respect to* **Q**.

Before proving the theorem we shall present some general remarks on quasivarieties.

Proposition 6.2.2. *Let* $\mathbf{Q}$ *be an arbitrary quasivariety of a signature* τ. *Let* $h : \mathbf{A} \to \mathbf{B}$ *be a surjective homomorphism between* τ*-algebras. Then* $(h^{-1}(\Xi))' = h^{-1}(\Xi')$, *for all* $\Xi \in Con(\mathbf{B})$.

(Here, for any $a, b \in A$, $\langle a, b\rangle \in h^{-1}(\Xi)$ if and only if $\langle ha, hb\rangle \in \Xi$.)

Proof (of the proposition). Assume $\Xi \in Con(\mathbf{B})$. As $\mathbf{B}/\Xi' \in \mathbf{Q}$ and $\mathbf{A}/h^{-1}(\Xi') \cong \mathbf{B}/\Xi'$, we have that $h^{-1}(\Xi')$ is a $\mathbf{Q}$-congruence of $\mathbf{A}$. As $h^{-1}(\Xi) \subseteq h^{-1}(\Xi')$, it follows that $(h^{-1}(\Xi))' \subseteq h^{-1}(\Xi')$, because $(h^{-1}(\Xi))'$ is the least $\mathbf{Q}$-congruence containing $h^{-1}(\Xi)$.

It remains to show that $h^{-1}(\Xi') \subseteq (h^{-1}(\Xi))'$.

We have that for every $X \subseteq A^2$

$$h(\Theta^{\mathbf{A}}_{\mathbf{Q}}(X) +_{\mathbf{Q}} \ker_{\mathbf{Q}}(h)) = \Theta^{\mathbf{B}}_{\mathbf{Q}}(hX).$$

Hence, putting $X := h^{-1}(\Xi)$, we get that

$$h((h^{-1}(\Xi))' +_{\mathbf{Q}} \ker_{\mathbf{Q}}(h)) = (hh^{-1}(\Xi))'. \tag{a}$$

But $\ker(h) \subseteq h^{-1}(\Xi)$. Indeed, suppose $\langle a, b\rangle \in \ker(h)$, i.e., $ha = hb$. Then trivially, $\langle ha, hb\rangle \in \Xi$, which gives that $\langle a, b\rangle \in h^{-1}(\Xi)$. It follows that $\ker_{\mathbf{Q}}(h) \subseteq (h^{-1}(\Xi))'$. This and (a) give that

$$h((h^{-1}(\Xi))') = (hh^{-1}(\Xi))'.$$

Since h is surjective, $hh^{-1}(\Xi) = \Xi$ and hence $(hh^{-1}(\Xi))' = \Xi'$. Thus

$$h((h^{-1}(\Xi))') = \Xi'. \tag{b}$$

To show the inclusion $h^{-1}(\Xi') \subseteq (h^{-1}(\Xi))'$, assume $\langle a, b\rangle \in h^{-1}(\Xi')$, i.e., $\langle ha, hb\rangle \in \Xi'$. Hence, by (b), we get that $\langle ha, hb\rangle \in h((h^{-1}(\Xi))')$. Consequently, there exist $x, y \in A$ such that $ha = hx$, $hb = hy$ and $\langle x, y\rangle \in (h^{-1}(\Xi))'$. But $\langle a, x\rangle \in \ker(h) \subseteq h^{-1}(\Xi) \subseteq (h^{-1}(\Xi))'$ and $\langle b, y\rangle \in \ker(h) \subseteq h^{-1}(\Xi) \subseteq (h^{-1}(\Xi))'$. Hence

$$\langle a, x\rangle \in (h^{-1}(\Xi))', \quad \langle b, y\rangle \in (h^{-1}(\Xi))', \quad \langle x, y\rangle \in (h^{-1}(\Xi))'. \tag{c}$$

It follows from (c) that $\langle a, b\rangle \in (h^{-1}(\Xi))'$.

This proves the lemma. □

We prove the following fact providing a necessary and sufficient condition for the Extension Principle to hold in a variety.

Theorem 6.2.3. *Let* $\mathbf{Q}$ *be an arbitrary quasivariety. Then every algebra in the variety* $\mathbf{Va}(\mathbf{Q})$ *satisfies the Extension Principle with respect to* $\mathbf{Q}$ *if and only if the free algebra* $\mathbf{F} := \mathbf{F}_{\mathbf{Q}}(\omega)$ *satisfies the Extension Principle relative to* $\mathbf{Q}$.

The "⇒"-part of the proof is obvious. The proof of the reverse implication "⇐" is based on two simple lemmas. Throughout the proof **Q** is an arbitrary but fixed quasivariety of a signature τ.

Lemma 1. *Let $h : \boldsymbol{A} \to \boldsymbol{B}$ be an epimorphism between algebras, where $\boldsymbol{A} \in \boldsymbol{Va}(\mathbf{Q})$. If $\boldsymbol{A}$ satisfies the Extension Principle with respect to* **Q***, then so does $\boldsymbol{B}$.*

Proof (of the lemma). Let $\Phi, \Psi \in Con(\boldsymbol{B})$. Then:

$$h^{-1}((\Phi \cap \Psi)') = \text{(by Proposition 6.2.2) } (h^{-1}(\Phi \cap \Psi))' =$$
$$(h^{-1}(\Phi) \cap h^{-1}(\Psi))' = \text{(by the assumption)}$$
$$(h^{-1}(\Phi))' \cap (h^{-1}(\Psi))' = \text{(by Proposition 6.2.2)}$$
$$h^{-1}(\Phi') \cap h^{-1}(\Psi') = h^{-1}(\Phi' \cap \Psi').$$

As h is "onto" it follows that $(\Phi \cap \Psi)' = \Phi' \cap \Psi'$. □

Lemma 2. *Suppose that every countably generated algebra in $\boldsymbol{Va}(\mathbf{Q})$ satisfies the Extension Principle. Then every algebra in $\boldsymbol{Va}(\mathbf{Q})$ satisfies this principle with respect to* **Q**.

Proof (of the lemma). Let $\boldsymbol{A}$ be an algebra in $\boldsymbol{Va}(\mathbf{Q})$ and $\Phi, \Psi \in Con(\boldsymbol{A})$. Suppose $a \equiv b\ (\Phi' \cap \Psi')$. We shall show that $a \equiv b\ ((\Phi \cap \Psi)')$.

As $a \equiv b\ (\Phi')$ and $a \equiv b\ (\Psi')$, there exist finite sequences

$$\langle a'_1, b'_1 \rangle, \ldots, \langle a'_m, b'_m \rangle \in \Phi$$

and

$$\langle a''_1, b''_1 \rangle, \ldots, \langle a''_n, b''_n \rangle \in \Psi$$

such that $a \equiv b\ (\Theta^{\boldsymbol{A}}_{\mathbf{Q}}(\langle a'_1, b'_1 \rangle, \ldots, \langle a'_m, b'_m \rangle))$ and $a \equiv b\ (\Theta^{\boldsymbol{A}}_{\mathbf{Q}}(\langle a''_1, b''_1 \rangle, \ldots, \langle a''_n, b''_n \rangle))$. Then there exists a countably generated subalgebra $\boldsymbol{B}$ of $\boldsymbol{A}$ which includes the pairs $\langle a'_1, b'_1 \rangle, \ldots, \langle a'_m, b'_m \rangle, \langle a''_1, b''_1 \rangle, \ldots, \langle a''_n, b''_n \rangle$ such that $a \equiv b(\Theta^{\boldsymbol{B}}_{\mathbf{Q}}(\langle a'_1, b'_1 \rangle, \ldots, \langle a'_m, b'_m \rangle))$ and $a \equiv b\ (\Theta^{\boldsymbol{B}}_{\mathbf{Q}}(\langle a''_1, b''_1 \rangle, \ldots, \langle a''_n, b''_n \rangle))$.

Let $\Phi_0 := \Theta^{\boldsymbol{B}}_{\mathbf{Q}}(\langle a'_1, b'_1 \rangle, \ldots, \langle a'_m, b'_m \rangle)$ and $\Psi_0 := \Theta^{\boldsymbol{B}}_{\mathbf{Q}}(\langle a''_1, b''_1 \rangle, \ldots, \langle a''_n, b''_n \rangle)$. Thus $a \equiv b\ (\Theta^{\boldsymbol{B}}_{\mathbf{Q}}(\Phi_0))$ and $a \equiv b\ (\Theta^{\boldsymbol{B}}_{\mathbf{Q}}(\Psi_0))$. As $\boldsymbol{B}$ satisfies the Extension Principle, $a \equiv b\ (\Theta^{\boldsymbol{B}}_{\mathbf{Q}}(\Phi_0 \cap \Psi_0))$. But $\Theta^{\boldsymbol{B}}_{\mathbf{Q}}(\Phi_0 \cap \Psi_0) \subseteq B^2 \cap \Theta^{\boldsymbol{A}}_{\mathbf{Q}}(\Phi_0 \cap \Psi_0) \subseteq \Theta^{\boldsymbol{A}}_{\mathbf{Q}}(\Phi \cap \Psi)$. Hence $a \equiv b\ ((\Phi \cap \Psi)')$. □

To conclude the proof of Theorem 6.2.3, let us assume that the free algebra $\boldsymbol{F} = \boldsymbol{F}_{\mathbf{Q}}(\omega)\ (= \boldsymbol{F}_{\boldsymbol{Va}(\mathbf{Q})}(\omega))$ satisfies the Extension Principle relative to **Q**. It follows from Lemma 1 that every countably generated algebra in $\boldsymbol{Va}(\mathbf{Q})$, being a homomorphic image of $\boldsymbol{F}$, satisfies the Extension Principle. Then Lemma 2 yields that every algebra in $\boldsymbol{Va}(\mathbf{Q})$ satisfies this principle. □

We shall pass to the proof of Theorem 6.2.1. To prove the theorem it suffices to show that the free algebra $\boldsymbol{F} := \boldsymbol{F}_{\mathbf{Q}}(\omega)$ satisfies this principle relative to **Q**.

We are concerned with congruences on the free algebra $\boldsymbol{F} := \boldsymbol{F}_{\mathbf{Q}}(\omega)$. If $X \subseteq F^2$, then $\Theta^{\boldsymbol{F}}(X)$ and $\Theta^{\boldsymbol{F}}_{\mathbf{Q}}(X)$ stand for the congruence of $\boldsymbol{F}$ generated by X and the $\mathbf{Q}$-congruence of $\boldsymbol{F}$ generated by X, respectively. If $X = \{\langle x_i, y_i\rangle : i \in I\}$ is a set of pairs of free generators of $\boldsymbol{F}$, then $\Theta^{\boldsymbol{F}}(X) = \Theta^{\boldsymbol{F}}_{\mathbf{Q}}(X)$, by Proposition 2.6.

Lemma 3. *Suppose $\Phi \in Con(\boldsymbol{F})$. Let X be a set of pairs of free generators of $\boldsymbol{F}$ and $h : \boldsymbol{F} \to \boldsymbol{F}$ an epimorphism such that $\Phi = \Theta^{\boldsymbol{F}}(hX)$. Then*

$$h^{-1}(\Phi) = \Theta^{\boldsymbol{F}}(X) + \ker(h) \quad \textit{and} \quad h^{-1}(\Phi') = \Theta^{\boldsymbol{F}}(X) +_{\mathbf{Q}} \ker(h).$$

Proof (of the lemma). It is clear that the required epimorphism h and X always exist. (One may simply take X to be an infinite set of pairs of free generators, map first X onto Φ and then, assuming that there are infinitely many free generators not involved in X, map surjectively the remaining free generators onto F.) Thus $\Phi = h\Theta^{\boldsymbol{F}}(X)$. But the claim below shows that then $\Phi = \Theta^{\boldsymbol{F}}(hX)$.

Claim. *If $\Phi = h(\Theta^{\boldsymbol{F}}(X))$, then $\Phi = \Theta^{\boldsymbol{F}}(hX)$.*

Proof (of the claim). We have: $\Phi = h(\Theta^{\boldsymbol{F}}(X))$ and $hX \subseteq h(\Theta^{\boldsymbol{F}}(X))$. Proposition 2.9 gives: $\Phi = h(\Theta^{\boldsymbol{F}}(X)) \subseteq \Theta^{\boldsymbol{F}}(hX) \subseteq \Theta^{\boldsymbol{F}}(h(\Theta^{\boldsymbol{F}}(X)) = \Theta^{\boldsymbol{F}}(\Phi) = \Phi$. □

It should be also noted that for any epimorphism $h : \boldsymbol{F} \to \boldsymbol{F}$, $\ker(h)$ is a $\mathbf{Q}$-congruence because the quotient algebra $\boldsymbol{F}/\ker(h)$ is isomorphic with $\boldsymbol{F}$ and $\boldsymbol{F} \in \mathbf{Q}$.

We have:

$$\Phi = \Theta^{\boldsymbol{F}}(hX) = h(\Theta^{\boldsymbol{F}}(X) + \ker(h)),$$

by the claim and Proposition 2.10 applied to $\boldsymbol{Va}(\mathbf{Q})$.

Passing to the h-preimages, we get: $h^{-1}(\Phi) = h^{-1}(h(\Theta^{\boldsymbol{F}}(X) + \ker(h))) = \Theta^{\boldsymbol{F}}(X) + \ker(h)$, because h is surjective.

The proof of the other equality is similar:

$$\Phi' = (\Theta^{\boldsymbol{F}}(hX))' = \Theta^{\boldsymbol{F}}_{\mathbf{Q}}(hX) = h(\Theta^{\boldsymbol{F}}(X) +_{\mathbf{Q}} \ker(h)),$$

by Proposition 2.10 applied to $\mathbf{Q}$ (see Note following Proposition 2.10). Taking the h-preimages, we get

$$h^{-1}(\Phi') = h^{-1}(h(\Theta^{\boldsymbol{F}}(X) +_{\mathbf{Q}} \ker(h))) = \Theta^{\boldsymbol{F}}(X) +_{\mathbf{Q}} \ker(h). \quad \square$$

We shall pass to the proof of the Extension Principle for RCM quasivarieties. We shall make use of Theorem 6.2.3. The following lemma is crucial:

Lemma 4. *Let $\mathbf{Q}$ be an RCM quasivariety. Then the free algebra $\boldsymbol{F} = \boldsymbol{F}_{\mathbf{Q}}(\omega)$ satisfies the Extension Principle.*

Proof (of the lemma). The lemma is proved in two steps.

Claim 1. *Let $\Phi, \Psi \in Con(\boldsymbol{F})$. Then $(\Phi \cap \Psi)' = \Phi' \cap \Psi'$.*

Proof (of the claim). It is clear that $(\Phi \cap \Psi') \subseteq \Phi' \cap \Psi'$. To prove the reverse inclusion, let X and Y be separated sets of pairs of free generators of $\boldsymbol{F}$ and let $h : \boldsymbol{F} \to \boldsymbol{F}$ be an epimorphism such that $\Phi = \Theta^{\boldsymbol{F}}(hX)$ and $\Psi = \Theta^{\boldsymbol{F}}(hY)$. Then

$$h^{-1}((\Phi \cap \Psi')') = \text{(by Proposition 6.2.2)}$$
$$(h^{-1}(\Phi \cap \Psi'))' =$$
$$(h^{-1}(\Phi) \cap h^{-1}(\Psi'))' = \text{(by Lemma 3)}$$
$$((\ker(h) + \Theta^{\boldsymbol{F}}(X)) \cap (\ker(h) +_{\mathbf{Q}} \Theta^{\boldsymbol{F}}(Y)))'.$$

On the other hand,

$$h^{-1}(\Phi' \cap \Psi') = h^{-1}(\Phi') \cap h^{-1}(\Psi') = \text{(by Proposition 6.2.2)}$$
$$(h^{-1}(\Phi))' \cap (h^{-1}(\Psi))' = \text{(by Lemma 3)}$$
$$(\ker(h) +_{\mathbf{Q}} \Theta^{\boldsymbol{F}}(X)) \cap (\ker(h) +_{\mathbf{Q}} \Theta^{\boldsymbol{F}}(Y)).$$

It follows that the inclusion $\Phi' \cap \Psi' \subseteq (\Phi \cap \Psi')'$ is equivalent to

$$(\ker(h) +_{\mathbf{Q}} \Theta^{\boldsymbol{F}}(X)) \cap (\ker(h) +_{\mathbf{Q}} \Theta^{\boldsymbol{F}}(Y)) \subseteq$$
$$\Theta^{\boldsymbol{F}}_{\mathbf{Q}}((\ker(h) + \Theta^{\boldsymbol{F}}(X)) \cap (\ker(h) +_{\mathbf{Q}} \Theta^{\boldsymbol{F}}(Y))). \quad (1)$$

Since $\mathbf{Q}$ is RCM, the congruence on the left-hand side of (1) equals $\ker(h) +_{\mathbf{Q}} (\Theta^{\boldsymbol{F}}(X) \cap (ker(h) +_{\mathbf{Q}} \Theta^{\boldsymbol{F}}(Y)))$ (see Corollary 6.1.5). Consequently, (1) is equivalent to the inclusion

$$\ker(h) +_{\mathbf{Q}} (\Theta^{\boldsymbol{F}}(X) \cap (\ker(h) +_{\mathbf{Q}} \Theta^{\boldsymbol{F}}(Y))) \subseteq$$
$$\Theta^{\boldsymbol{F}}_{\mathbf{Q}}((\ker(h) + \Theta^{\boldsymbol{F}}(X)) \cap (\ker(h) +_{\mathbf{Q}} \Theta^{\boldsymbol{F}}(Y))). \quad (2)$$

As $\ker(h)$ is included in the congruence on the right-hand side of (2), to prove (2), it suffices to show that

$$\Theta^{\boldsymbol{F}}(X) \cap (\ker(h) +_{\mathbf{Q}} \Theta^{\boldsymbol{F}}(Y)) \subseteq \Theta^{\boldsymbol{F}}_{\mathbf{Q}}((\ker(h) + \Theta^{\boldsymbol{F}}(X)) \cap (\ker(h) +_{\mathbf{Q}} \Theta^{\boldsymbol{F}}(Y))).$$

But this inclusion trivially holds. □

Claim 2. *Let $\Phi, \Psi \in Con(\boldsymbol{F})$. Then $(\Phi \cap \Psi)' = \Phi' \cap \Psi'$.*

Proof (of the claim). The inclusion $(\Phi \cap \Psi') \subseteq \Phi' \cap \Psi'$ is obvious. To prove the reverse inclusion, suppose X and Y are separated sets of pairs of free generators of

$\boldsymbol{F}$ and let $h : \boldsymbol{F} \to \boldsymbol{F}$ be an epimorphism such that $\Phi = \Theta^{\boldsymbol{F}}(hX)$ and $\Psi = \Theta^{\boldsymbol{F}}(hY)$. Then, arguing as in Claim 1, we get that

$$h^{-1}((\Phi \cap \Psi)') = ((\ker(h) + \Theta^{\boldsymbol{F}}(X)) \cap (\ker(h) + \Theta^{\boldsymbol{F}}(Y)))'$$

and

$$h^{-1}(\Phi' \cap \Psi') = (\ker(h) +_{\mathbf{Q}} \Theta^{\boldsymbol{F}}(X)) \cap (\ker(h) +_{\mathbf{Q}} \Theta^{\boldsymbol{F}}(Y)).$$

It follows that the inclusion $\Phi' \cap \Psi' \subseteq (\Phi \cap \Psi)'$ is equivalent to

$$(\ker(h) +_{\mathbf{Q}} \Theta^{\boldsymbol{F}}(X)) \cap (\ker(h) + \Theta^{\boldsymbol{F}}(Y)) \subseteq \Theta^{\boldsymbol{F}}_{\mathbf{Q}}((\ker(h) + \Theta^{\boldsymbol{F}}(X)) \cap (\ker(h) + \Theta^{\boldsymbol{F}}(Y))). \quad (3)$$

Corollary 6.1.5 gives that

$$(\ker(h) +_{\mathbf{Q}} \Theta^{\boldsymbol{F}}(X)) \cap (\ker(h) +_{\mathbf{Q}} \Theta^{\boldsymbol{F}}(Y)) = \ker(h) *_d (\Theta^{\boldsymbol{F}}(Y), \Theta^{\boldsymbol{F}}(X)) = \ker(h) +_{\mathbf{Q}} (\Theta^{\boldsymbol{F}}(X) \cap (\ker(h) +_{\mathbf{Q}} \Theta^{\boldsymbol{F}}(Y))).$$

Consequently, (3) is equivalent to the inclusion

$$\ker(h) +_{\mathbf{Q}} (\Theta^{\boldsymbol{F}}(X) \cap (\ker(h) +_{\mathbf{Q}} \Theta^{\boldsymbol{F}}(Y))) \subseteq \Theta^{\boldsymbol{F}}_{\mathbf{Q}}((\ker(h) + \Theta^{\boldsymbol{F}}(X)) \cap (\ker(h) + \Theta^{\boldsymbol{F}}(Y))). \quad (4)$$

To prove (4), it suffices to show that

$$(\Theta^{\boldsymbol{F}}(X) \cap (\ker(h) +_{\mathbf{Q}} \Theta^{\boldsymbol{F}}(Y))) \subseteq \Theta^{\boldsymbol{F}}_{\mathbf{Q}}((\ker(h) + \Theta^{\boldsymbol{F}}(X)) \cap (\ker(h) + \Theta^{\boldsymbol{F}}(Y))). \quad (5)$$

We shall apply Claim 1 to the congruences $\Phi_0 := \ker(h) + \Theta^{\boldsymbol{F}}(Y)$ and $\Psi_0 := \Theta^{\boldsymbol{F}}(X)$. As Ψ_0 is a **Q**-congruence, Claim 1 gives that $\Phi_0' \cap \Psi_0 = \Phi_0' \cap \Psi_0' = (\Phi_0 \cap \Psi_0)' = (\Phi_0 \cap \Psi_0)'$, i.e., $\Phi_0' \cap \Psi_0 = (\Phi_0 \cap \Psi_0)'$. Hence

$$\Theta^{\boldsymbol{F}}(X) \cap (\ker(h) +_{\mathbf{Q}} \Theta^{\boldsymbol{F}}(Y)) = \Theta^{\boldsymbol{F}}_{\mathbf{Q}}(\Theta^{\boldsymbol{F}}(X) \cap (\ker(h) + \Theta^{\boldsymbol{F}}(Y))) \subseteq \Theta^{\boldsymbol{F}}_{\mathbf{Q}}((\ker(h) + \Theta^{\boldsymbol{F}}(X)) \cap (\ker(h) + \Theta^{\boldsymbol{F}}(Y))).$$

So (5) holds. This proves the claim and concludes the proof of the lemma. □

Theorem 6.2.1 follows from Lemma 4 and Theorem 6.2.3. □

6.3 Distributivity in the Lattice of Equational Theories of RCM Quasivarieties

The theorems presented in this section exhibit various interrelations holding between RCM quasivarieties $\mathbf{Q}$ and some restricted forms of distributivity law for $\mathbf{Q}^{\models}$-theories generated by equations of separated variables (equivalently, for the congruences of the free algebra $\boldsymbol{F}_{\mathbf{Q}}(\omega)$ generated by separated pairs of free generators).

The following observation is due to von Neumann (1936–1937).

Theorem 6.3.1. *Let $\boldsymbol{L}$ be a modular lattice and $a, b, c \in L$. The following conditions are equivalent:*

(1) *The sublattice of $\boldsymbol{L}$ generated by $\{a, b, c\}$ is distributive.*
(2) $a \wedge (b \vee c) = (a \wedge b) \vee (a \wedge c)$.
(3) $a \vee (b \wedge c) = (a \vee b) \wedge (a \vee c)$.

Proof. The proof of the equivalence of (1) and (2) can be found in Grätzer (1978), Theorem 12, Chapter IV.

(1) $\Rightarrow$ (3). This is trivial.

(3) $\Rightarrow$ (1). Let $\boldsymbol{L}^d$ be the lattice dual to $\boldsymbol{L}$. Since $\boldsymbol{L}$ is modular, $\boldsymbol{L}^d$ is modular as well. The sublattice $\boldsymbol{L}_0$ of $\boldsymbol{L}$ generated $\{a, b, c\}$ has the same carrier as the sublattice $\boldsymbol{L}_0^d$ of $\boldsymbol{L}^d$ dual to $\boldsymbol{L}_0$, that is, $\boldsymbol{L}_0^d$ is generated by the same set $\{a, b, c\}$. Moreover, by duality, $\boldsymbol{L}_0$ is distributive if and only if $\boldsymbol{L}_0^d$ is distributive. The equivalence of (1) and (2) when applied to $\boldsymbol{L}^d$ implies that the sublattice $\boldsymbol{L}_0^d$ is distributive if and only if $a \vee (b \wedge c) = (a \vee b) \wedge (a \vee c)$. Hence, assuming (3), we obtain that $\boldsymbol{L}_0^d$ is distributive. It follows that $\boldsymbol{L}_0$ is distributive as well. So (1) holds. □

The following theorem is a basic tool in the commutator theory for RCM quasivarieties:

Theorem 6.3.2. *Let $\mathbf{Q}$ be an RCM quasivariety. The lattice $\boldsymbol{Th}(\mathbf{Q}^{\models})$ is distributive in the restricted sense.*

Proof. Put $C := \mathbf{Q}^{\models}$. Let $X = \{x_i \approx y_i : i \in I\}$ be a set of equations of pairwise different variables and let Y and Z be arbitrary sets of equations of terms whose variables are separated from the variables of X, that is, $Var(X) \cap Var(Y \cup Z) = \emptyset$. According to Lemma 5.2.10,

$$(C(X) +_{\mathbf{Q}} C(Y)) \cap (C(X) +_{\mathbf{Q}} C(Z)) = C(X) +_{\mathbf{Q}} C(Y) \cap C(Z). \qquad (*)$$

Theorem 6.3.1 and $(*)$ imply that the sublattice of $\boldsymbol{Th}(\mathbf{Q}^{\models})$ generated by $\{C(X), C(Y), C(Z)\}$ is distributive. Hence

$$C(X) \cap C(Y) +_{\mathbf{Q}} C(X) \cap C(Z) = C(X) \cap (C(Y) +_{\mathbf{Q}} C(Z)).$$

Thus $\boldsymbol{Th}(\mathbf{Q}^{\models})$ validates the restrictive distributivity. □

It follows from the above observation that Theorem 5.3.1 applies to any RCM quasivariety. We shall make use of it several times.

Theorem 6.3.3. *Let* $\mathbf{Q}$ *be an RCM quasivariety. The equationally defined commutator for* $\mathbf{Q}$ *is additive.*

Proof. In view of the above theorem the lattice $\boldsymbol{Th}(\mathbf{Q}^{\models})$ is distributive in the restricted sense. Then apply Theorem 5.3.7. □

Notes 6.3.4.

(1). According to Theorem 4.1.12, the equationally defined commutator for any RCM quasivariety coincides with the commutator in the sense of Kearnes and McKenzie. (The latter commutator is simply referred to as *the* commutator in the literature.) Theorem 6.3.3 thus provides a new proof of the additivity of the commutator for any RCM quasivariety.

(2). Although there are non-RCM quasivarieties for which the thesis of Theorem 6.3.3 holds, the above proof of this theorem cannot be extended onto non-RCM quasivarieties. The above proof makes use of Theorems 6.3.1–6.3.2 and is based on the fact that any modular lattice $\boldsymbol{L}$ validates the following equivalence:

$$\text{for any } a, b, c \in L,\ a \wedge (b \vee c) = (a \wedge b) \vee (a \wedge c) \ \text{ if and only if } \quad (*)$$
$$a \vee (b \wedge c) = (a \vee b) \wedge (a \vee c).$$

The equivalence $(*)$ fails to hold in any non-modular lattice (take the pentagon). In fact, $(*)$ characterizes modularity: a lattice $\boldsymbol{L}$ validates $(*)$ if and only if $\boldsymbol{L}$ is modular. It follows that Theorem 6.3.1 does not hold for non-modular lattices. Consequently, if $\mathbf{Q}$ is not an RCM quasivariety, Theorem 6.3.1 cannot be used to show that the lattice $\boldsymbol{Th}(\mathbf{Q}^{\models})$ is distributive in the restricted sense. It is an open question as to whether there is a *lattice-theoretic* condition weaker than modularity and implying the restricted distributivity of the lattice $\boldsymbol{Th}(\mathbf{Q}^{\models})$.

□

In the second part of this paragraph we shall isolate some distributive sublattices of the lattice $\boldsymbol{Th}(\mathbf{Q}^{\models})$ for any RCM quasivariety $\mathbf{Q}$. We begin with some preliminary observations.

Lemma 6.3.5. *Let* $\mathbf{Q}$ *be an arbitrary non-trivial quasivariety. Let* $\{x_i \approx y_i : i \in I\}$ *be a non-empty set consisting of equations of pairwise different variables. For each non-empty set* $J \subseteq I$ *define*

$$\boldsymbol{\Sigma}(J) := \mathbf{Q}^{\models}(\{x_i \approx y_i : i \in J\}).$$

Then for all non-empty subsets $A, B, C \subseteq I$*:*

(1) $A \subseteq B$ *if and only if* $\boldsymbol{\Sigma}(A) \subseteq \boldsymbol{\Sigma}(B)$:
(2) $A = B$ *if and only if* $\boldsymbol{\Sigma}(A) = \boldsymbol{\Sigma}(B)$;
(3) $\boldsymbol{\Sigma}(A \cup B) = \boldsymbol{\Sigma}(A) +_{\mathbf{Q}} \boldsymbol{\Sigma}(B)$;
(4) $\boldsymbol{\Sigma}(A \cap B) \subseteq \boldsymbol{\Sigma}(A) \cap \boldsymbol{\Sigma}(B)$.
(5) *if* $A \cap (B \cup C) = \emptyset$, *then*
$$(\boldsymbol{\Sigma}(A) +_{\mathbf{Q}} \boldsymbol{\Sigma}(B)) \cap (\boldsymbol{\Sigma}(A) +_{\mathbf{Q}} \boldsymbol{\Sigma}(C)) = \boldsymbol{\Sigma}(A) +_{\mathbf{Q}} \boldsymbol{\Sigma}(B) \cap \boldsymbol{\Sigma}(C).$$

Proof. As the set of individual variables is countably infinite, the set I is countable too.

(1). The implication '($\Rightarrow$)' is immediate. To prove the reverse, assume $\boldsymbol{\Sigma}(A) \subseteq \boldsymbol{\Sigma}(B)$ and suppose A is not included in B. There is an equation $x_i \approx y_i$ such that $x_i \approx y_i \in A$ and $x_i \approx y_i \notin B$. As $\mathbf{Q}$ is non-trivial and $x_i \approx y_i$ is separated from B, there is an algebra $\boldsymbol{A} \in \mathbf{Q}$ and a valuation h of variables in $\boldsymbol{A}$ which validates B and for which $hx_i \neq hy_i$. Hence $x_i \approx y_i \notin \boldsymbol{\Sigma}(B)$. Consequently, $x_i \approx y_i \notin \boldsymbol{\Sigma}(A)$, a contradiction.

(2), (3), and (4) are immediate. (5) follows from Lemma 5.2.10. □

Note. We shall later show that the assumption that $A \cap (B \cup C)$ is non-empty in (5) can be dropped if $\mathbf{Q}$ is RCM. Consequently, in any RCM quasivariety, the family $\{\boldsymbol{\Sigma}(J) : \emptyset \neq J \subseteq I\}$ generates a bounded and closed distributive sublattice of $\boldsymbol{Th}(\mathbf{Q}^{\models})$ (Theorem 6.3.8 below). Moreover, (4) can be replaced by the equality:

$$\boldsymbol{\Sigma}(A) \cap \boldsymbol{\Sigma}(B) = \mathbf{Q}^{\models}(\bigcup\{\mathbf{Q}^{\models}(x_i \approx y_i) \cap \mathbf{Q}^{\models}(x_j \approx y_j) : i, j \in A \cap B\}). \quad \square$$

Lemma 6.3.6. *Let $\mathbf{Q}$ be an arbitrary quasivariety. Let $\{x_i \approx y_i : i \in I\}$ be a non-empty set consisting of equations of pairwise different variables. If I is infinite, then*

$$\bigcap_{i \in I} \mathbf{Q}^{\models}(x_i \approx y_i) = \mathbf{Q}^{\models}(\emptyset).$$

Proof. As the set of individual variables is countably infinite, the lemma is equivalently reformulated as follows:

Lemma 4.3.6*. *Let $\{x_n : n \in \omega\}$ be an infinite set of individual variables. Then*

$$\bigcap_{n \in \omega} \mathbf{Q}^{\models}(x_{2n} \approx x_{2n+1}) = \mathbf{Q}^{\models}(\emptyset).$$

The inclusion '$\supseteq$' is obvious. To prove the reverse inclusion, assume $p \approx q \in \bigcap_{n \in \omega} \mathbf{Q}^{\models}(x_{2n} \approx x_{2n+1})$. We write $p = p(x_0, \ldots, x_{2n+1}, \underline{u})$, $q = q(x_0, \ldots, x_{2n+1}, \underline{u})$, where $x_0, \ldots, x_{2n+1}$ contains all variables among $\{x_n : n \in \omega\}$ which at most occur in p and q and $\underline{u}$ is the set of the remaining variables outside $\{x_n : n \in \omega\}$ occurring in p and q. (The set $\{x_n : n \in \omega\}$ need not contain all individual variables.)

The assumption implies in particular that

$$p(x_0, \ldots, x_{2n+1}, \underline{u}) \approx q(x_0, \ldots, x_{2n+1}, \underline{u}) \in \mathbf{Q}^{\models}(x_{2n+2} \approx x_{2n+3}).$$

Let e be an endomorphism of $\boldsymbol{Te}_\tau$ which is the identity map on the variables $x_0, \ldots, x_{2n+2}, \underline{u}$, and $ex_{2n+3} = x_{2n+2}$. e is well defined because the variables of $x_0, \ldots, x_{2n+1}, x_{2n+2}, x_{2n+3}$ and $\underline{u}$ are all pairwise different. Hence, by structurality,

$$p(x_0, \ldots, x_{2n+1}, \underline{u}) \approx q(x_0, \ldots, x_{2n+1}, \underline{u}) \in \mathbf{Q}^{\models}(ex_{2n+2} \approx ex_{2n+3}) = \mathbf{Q}^{\models}(\emptyset). \quad \square$$

Lemma 6.3.7. *Let* $\{x_0, x_1, \ldots, x_{2n+1}\}$ *be a finite set of individual variables. Then* $\mathbf{Q}^{\models}(x_0 \approx x_1) \cap \mathbf{Q}^{\models}(x_2 \approx x_3) \cap \ldots \cap \mathbf{Q}^{\models}(x_{2n} \approx x_{2n+1}) \neq \mathbf{Q}^{\models}(\emptyset)$ *if and only if there exist terms* $p = p(x_0, \ldots, x_{2n+1}, \underline{u})$, $q = q(x_0, \ldots, x_{2n+1}, \underline{u})$ *containing the variables* $x_0, \ldots, x_{2n+1}$ *and possibly some other variables* $\underline{u}$ *such that* $p \approx q \notin \mathbf{Q}^{\models}(\emptyset)$ *but* $p(x_{2i+1}/x_{2i}) \approx q(x_{2i+1}/x_{2i}) \in \mathbf{Q}^{\models}(\emptyset)$ *for* $i = 0, \ldots, n$.

Proof. ($\Rightarrow$). If $\mathbf{Q}^{\models}(x_0 \approx x_1) \cap \ldots \cap \mathbf{Q}^{\models}(x_{2n} \approx x_{2n+1}) \neq \mathbf{Q}^{\models}(\emptyset)$, we select an equation $p \approx q \in \mathbf{Q}^{\models}(x_0 \approx x_1) \cap \ldots \cap \mathbf{Q}^{\models}(x_{2n} \approx x_{2n+1}) \setminus \mathbf{Q}^{\models}(\emptyset)$. Let $\underline{x}$ be the set of variables among $x_0, x_1, \ldots, x_{2n+1}$ that occur in $p \approx q$ and let $\underline{u}$ be the set of the remaining variables occurring in this equation.

Suppose that the list $\underline{x}$ is shorter than $x_0, x_1, \ldots, x_{2n+1}$. Hence for some k ($0 \leqslant k \leqslant 2n+1$) the variable x_k is not in $\underline{x}$. If k is even, $k = 2l$, then $p \approx q \in \mathbf{Q}^{\models}(x_{2l} \approx x_{2l+1})$. We then define a substitution e which is the identity map on the variables of $\underline{x}$ and $\underline{u}$ such $ex_{2l} = x_{2l+1}$ and $ex_{2l+1} = x_{2l+1}$. Hence, by structurality, $p \approx q \in \mathbf{Q}^{\models}(ex_{2l} \approx ex_{2l+1}) = \mathbf{Q}^{\models}(\emptyset)$. A contradiction. A similar argument is applied when k is odd, $k = 2l+1$. Then work with the variables x_{2l} and x_{2l+1}. In this case we also arrive at a contradiction.

It is clear that $p(x_{2i+1}/x_{2i}) \approx q(x_{2i+1}/x_{2i}) \in \mathbf{Q}^{\models}(\emptyset)$ for $i = 0, \ldots, n$.

($\Leftarrow$). Assume $p = p(x_0, \ldots, x_{2n+1}, \underline{u})$ and $q = q(x_0, \ldots, x_{2n+1}, \underline{u})$ are terms satisfying the RHS and suppose *a contrario* that $\mathbf{Q}^{\models}(x_0 \approx x_1) \cap \ldots \cap \mathbf{Q}^{\models}(x_{2n} \approx x_{2n+1}) = \mathbf{Q}^{\models}(\emptyset)$. As $p \approx q \notin \mathbf{Q}^{\models}(\emptyset)$, it follows that $p \approx q \notin \mathbf{Q}^{\models}(x_{2i} \approx x_{2i+1})$ for *some* i $0 \leqslant i \leqslant n$). Consequently, $p(x_{2i+1}/x_{2i}) \approx q(x_{2i+1}/x_{2i}) \notin \mathbf{Q}^{\models}(\emptyset)$. A contradiction. □

Before reading the proof of the next theorem the reader is advised to have a look at Appendix A.

Let $\boldsymbol{M}$ and $\boldsymbol{L}$ be complete lattices such that M is a subset of L. $\boldsymbol{M}$ is a *closed sublattice* of $\boldsymbol{L}$ if for every *non-empty* subset X of M, the elements $\sup(X)$ and $\inf(X)$, as computed in $\boldsymbol{L}$, are actually in M. If $\boldsymbol{M}$ is a closed sublattice of $\boldsymbol{L}$, then $\boldsymbol{M}$ is a sublattice of $\boldsymbol{L}$ but $\boldsymbol{M}$ may not be a bounded sublattice of $\boldsymbol{L}$—the zero and the unit elements of $\boldsymbol{M}$ need not coincide with their counterparts in $\boldsymbol{L}$. If moreover, the zero and the unit elements of $\boldsymbol{L}$ belong to $\boldsymbol{M}$, the sublattice $\boldsymbol{M}$ is called a *complete sublattice* of $\boldsymbol{L}$. Thus a sublattice $\boldsymbol{M}$ of $\boldsymbol{L}$ is a complete sublattice of $\boldsymbol{L}$ if for every subset X of M, the elements $\sup(X)$ and $\inf(X)$, when computed in $\boldsymbol{L}$, belong to M.

If $\boldsymbol{L}$ is a complete lattice and X a non-empty subset of L, then there exist the least closed sublattice of $\boldsymbol{L}$ that includes X and the least complete sublattice of $\boldsymbol{L}$ that contains X.

Theorem 6.3.8. *Let* $\mathbf{Q}$ *be an RCM quasivariety. Let* $\{x_i \approx y_i : i \in I\}$ *be any non-empty set of equations of pairwise different variables. The least closed and bounded sublattice* $\boldsymbol{L}$ *of* $\boldsymbol{Th}(\mathbf{Q}^{\models})$ *that contains the family* $X = \{\mathbf{Q}^{\models}(x_i \approx y_i) : i \in I\}$ *is algebraic and distributive.*

Notes.

1. The theorem trivially holds if $\mathbf{Q}$ is an RCD quasivariety.
2. Under the assumption that $\mathbf{Q}$ is RCM, the statement of the theorem is equivalently paraphrased in terms of free generators of the free algebra $\boldsymbol{F}_{\mathbf{Q}}(\omega)$ as follows:

Let $\{\langle x_i, y_i\rangle : i \in I\}$ be any non-empty set of pairs of pairwise distinct free generators of $\mathbf{F_Q}(\omega)$. The closed sublattice of $\mathbf{Con_Q}(\mathbf{F_Q}(\omega))$ with zero and unity containing the family of $\mathbf{Q}$-congruences $\{\Theta(x_i, y_i) : i \in I\}$ is algebraic and distributive.

3. The assumption that the free generators occurring in the pairs $\{\langle x_i, y_i\rangle : i \in I\}$ are pairwise different cannot be dropped in the above theorem. If $\mathbf{Q}$ is a variety, the fact that for *any* non-empty set $\{\langle x_i, y_i\rangle : i \in I\}$ of pairs of free generators $\{\Theta(x_i, y_i) : i \in I\}$ generates a distributive sublattice of the lattice $\boldsymbol{Con}_{\mathbf{Q}}(\boldsymbol{F}_{\mathbf{Q}}(\omega))$ implies that $\mathbf{Q}$ is a congruence-distributive—see the proof of the well-known Jónsson's characterization of CD varieties. □

Proof (of the theorem). Some additional facts concerning modular lattices are needed. We first prove:

Lemma 1. *The family $\boldsymbol{X} = \{\mathbf{Q}^{\models}(x_i \approx y_i) : i \in I\}$ generates a distributive sublattice of $\boldsymbol{Th}(\mathbf{Q}^{\models})$.*

Proof (of the lemma). We shall apply the classical result due to Jónsson (1955) (with simplifications made by Balbes 1969) which characterizes distributive sublattices of modular lattices (see also Tamura 1971):

A non-empty subset X of a modular lattice generates a distributive sublattice if and only if

$$(a_1 \vee \ldots \vee a_m) \wedge (b_1 \wedge \ldots \wedge b_n) = a_1 \wedge (b_1 \wedge \ldots \wedge b_n) \vee \ldots \vee a_m \wedge (b_1 \wedge \ldots \wedge b_n) \tag{1}$$

whenever $a_1, \ldots, a_m, b_1, \ldots, b_n \in X$, for all positive integers m, n.

Accordingly, we show that (1) holds for any $m, n \geqslant 1$ and any finite non-empty subsets $\{a_1, \ldots, a_m\}$, $\{b_1, \ldots, b_n\}$ of $\boldsymbol{X} = \{\mathbf{Q}^{\models}(x_i \approx y_i) : i \in I\}$. The family $\boldsymbol{X}$ is countable. Note that if the intersection $\{a_1, \ldots, a_m\} \cap \{b_1, \ldots, b_n\}$ is non-empty, then the theory on the left-hand side of (1) equals $b_1 \cap \ldots \cap b_n$, which in turn is equal to the equational theory on the right-hand side of (1), so in this case (1) holds. It therefore suffices to consider the case when the sets $\{a_1, \ldots, a_m\}$, $\{b_1, \ldots, b_n\}$ are disjoint.

We prove the following statement $(T)_{m,n}$ by double induction on m and n:

For any disjoint subsets $\{a_1, \ldots, a_m\}, \{b_1, \ldots, b_n\}$ of X,

$$(a_1 +_{\mathbf{Q}} \cdots +_{\mathbf{Q}} a_m) \cap (b_1 \cap \ldots \cap b_n) =$$

$$a_1 \cap (b_1 \cap \ldots \cap b_n) +_{\mathbf{Q}} \cdots +_{\mathbf{Q}} a_m \cap (b_1 \cap \ldots \cap b_n). \tag*{$(T)_{m,n}$}$$

$(T)_{1,1}$ is immediate. $(T)_{m,1}$ follows from Theorems 6.3.2 and 5.3.1. $(T)_{1,n}$ is also immediate for all n.

Claim 1. $(T)_{m,2}$ *holds for all $m \geqslant 1$.*

Proof (of the claim). $(T)_{1,2}$ is obvious. We prove $(T)_{2,2}$, that is, we show that

$(T)_{2,2}$ For any disjoint subsets $\{a_1, a_2\}$, $\{b_1, b_2\}$ of X,

$$(a_1 +_{\mathbf{Q}} a_2) \cap (b_1 \cap b_2) = a_1 \cap (b_1 \cap b_2) +_{\mathbf{Q}} a_2 \cap (b_1 \cap b_2).$$

We put: $C := \mathbf{Q}^{\models}$. Let $a_i := C(x_i \approx y_i)$ for $i = 1, 2$ and $b_j := C(z_j \approx w_j)$ for $j = 1, 2$. We claim that

$$C(x_1 \approx y_1, x_2 \approx y_2) \cap C(z_1 \approx w_1) \cap C(z_2 \approx w_2) =$$
$$C(x_1 \approx y_1) \cap C(z_1 \approx w_1) \cap C(z_2 \approx y_w) +_{\mathbf{Q}} C(x_2 \approx y_2) \cap C(z_1 \approx w_1) \cap C(z_2 \approx w_2).$$

We apply Theorem 5.3.1. Let

$$X := C(z_1 \approx w_1) \cap C(z_2 \approx w_2), \quad Y := C(x_1 \approx y_1), \quad Z := C(x_2 \approx y_2).$$

We have:

$$C(x_1 \approx y_1, x_2 \approx y_2) \cap C(z_1 \approx w_1) \cap C(z_2 \approx w_2) =$$
$$(Y +_{\mathbf{Q}} Z) \cap X = \quad \text{(by Theorem 5.3.1)}$$
$$Y \cap X +_{\mathbf{Q}} Z \cap X =$$
$$C(x_1 \approx y_1) \cap C(z_1 \approx w_1) \cap C(z_2 \approx y_2) +_{\mathbf{Q}} C(x_2 \approx y_2) \cap C(z_1 \approx w_1) \cap C(z_2 \approx w_2).$$

So $(T)_{2,2}$ holds.

Let $m \geqslant 2$. We prove that $(T)_{m,2}$ implies $(T)_{m+1,2}$. Assume $(T)_{m,2}$. Let $\{a_1, \ldots, a_m, a_{m+1}\}$, $\{b_1, b_2\}$ be disjoint subsets of X. We put:

$$X := b_1 \cap b_2, \quad Y := a_1 +_{\mathbf{Q}} \ldots +_{\mathbf{Q}} a_m, \quad Z := a_{m+1}.$$

We compute:

$$(a_1 +_{\mathbf{Q}} \ldots +_{\mathbf{Q}} a_m +_{\mathbf{Q}} a_m) \cap (b_1 \cap b_2) = (Y +_{\mathbf{Q}} Z) \cap X = \text{(by Theorem 5.3.1)}$$
$$Y \cap X +_{\mathbf{Q}} Z \cap X = (a_1 +_{\mathbf{Q}} \ldots +_{\mathbf{Q}} a_m) \cap (b_1 \cap b_2) +_{\mathbf{Q}} a_{m+1} \cap (b_1 \cap b_2) = \text{(by } (T)_{m,2})$$
$$a_1 \cap (b_1 \cap b_2) +_{\mathbf{Q}} \ldots +_{\mathbf{Q}} a_m \cap (b_1 \cap b_2) +_{\mathbf{Q}} a_{m+1} \cap (b_1 \cap b_2).$$

So $(T)_{m+1,2}$ holds. □

It remains to prove:

Claim 2. *$(T)_{m,n}$ holds for all $m, n \geqslant 2$.*

Proof (of the claim). Fix $m \geqslant 2$ and assume that $(T)_{m,n}$ holds, where $n \geqslant 2$. We prove $(T)_{m,n+1}$. Let $\{a_1, \ldots, a_m\}$, $\{b_1, \ldots, b_n, b_{n+1}\}$ be disjoint subsets of X. We put:

$$X_i := a_i \cap (b_1 \cap \ldots \cap b_n) \quad \text{for } i = 1, \ldots, m, \qquad Y := b_{n+1}.$$

We compute:

$$(a_1 +_{\mathbf{Q}} \ldots +_{\mathbf{Q}} a_m) \cap (b_1 \cap b_2 \cap \ldots \cap b_{n+1}) =$$
$$(a_1 +_{\mathbf{Q}} \ldots +_{\mathbf{Q}} a_m) \cap (b_1 \cap \ldots \cap b_n) \cap b_{n+1} = \quad \text{(by } (T)_{m,n})$$
$$(a_1 \cap (b_1 \cap \ldots \cap b_n) +_{\mathbf{Q}} \ldots +_{\mathbf{Q}} a_m \cap (b_1 \cap \ldots \cap b_n)) \cap b_{n+1} =$$
$$(X_1 +_{\mathbf{Q}} \ldots +_{\mathbf{Q}} X_m) \cap Y = \quad \text{(by Theorem 5.3.1)}$$
$$(X_1 \cap Y) +_{\mathbf{Q}} \ldots +_{\mathbf{Q}} (X_m \cap Y) =$$
$$a_1 \cap (b_1 \cap \ldots \cap b_n \cap b_{n+1}) +_{\mathbf{Q}} \ldots +_{\mathbf{Q}} a_m \cap (b_1 \cap \ldots \cap b_n \cap b_{n+1}).$$

So $(T)_{m,n+1}$ holds.

This completes the proof of the lemma. □

Let $\boldsymbol{L}_0$ be the (distributive) sublattice of $\boldsymbol{Th}(\mathbf{Q}^{\models})$ generated by the family $X = \{\mathbf{Q}^{\models}(x_i \approx y_i) : i \in I\}$.

Note. For each finite and non-empty set $J \subseteq I$ define $\boldsymbol{L}_0(J)$ to be the (finite) sublattice of $\boldsymbol{L}_0$ generated by the set $\{\mathbf{Q}^{\models}(x_i \approx y_i) : i \in J\}$. $\boldsymbol{L}_0$ is thus the union of the directed family of lattices $\{\boldsymbol{L}_0(J) : \emptyset \neq J \subseteq I,\ J \text{ finite}\}$. □

If the set I is finite, the lattice $\boldsymbol{L}_0$ is finite bounded (and hence algebraic) with the zero $\mathbf{0} = \bigcap\{\mathbf{Q}^{\models}(x_i \approx y_i) : i \in I\}$ and the unit $\mathbf{1} = \mathbf{Q}^{\models}(\{x_i \approx y_i : i \in I\})$ and the theorem follows. Note that $\mathbf{1}$ need not be the unit element of the lattice $\boldsymbol{Th}(\mathbf{Q}^{\models})$ which is $Eq(\tau)$. Likewise, $\mathbf{0}$ need not coincide with $\mathbf{Q}^{eq\models}(\emptyset)$ being the zero of $\boldsymbol{Th}(\mathbf{Q}^{\models})$. But nevertheless $\boldsymbol{L}_0$ is a closed sublattice of $\boldsymbol{Th}(\mathbf{Q}^{\models})$.

If I is infinite, some other constructions are needed. Suppose that I is infinite. Then, by Lemma 6.3.6,

$$\bigcap\{\mathbf{Q}^{\models}(x_i \approx y_i) : i \in I\} = \mathbf{Q}^{\models}(\emptyset).$$

It is easy to see that due to distributivity, the lattice $\boldsymbol{L}_0$ consists of all elements of the form

$$a_1 +_{\mathbf{Q}} \ldots +_{\mathbf{Q}} a_m, \tag{1}$$

where $m \geqslant 1$ and

$$a_k := \bigcap\{\mathbf{Q}^{\models}(x_i \approx y_i) : i \in A_k\} \tag{2}$$

for some finite non-empty sets $A_k \subseteq I$ for $k = 1, \ldots, m$.

The lattice $\boldsymbol{L}_0$ is augmented with the zero element $\mathbf{0} := \mathbf{Q}^{\models}(\emptyset)$ and the unit element $\mathbf{1} := \mathbf{Q}^{\models}(\{x_i \approx y_i : i \in I\})$. This extended lattice is still distributive. (To simplify matters we shall assume that $\mathbf{0}$ and $\mathbf{1}$ are already in $\boldsymbol{L}_0$.) Of course, $\mathbf{0}$ is also the zero element of $\boldsymbol{Th}(\mathbf{Q}^{\models})$ but $\mathbf{1}$ need not coincide with the top element $Eq(\tau)$ of $\boldsymbol{Th}(\mathbf{Q}^{\models})$.

Let $(Ideal(\boldsymbol{L}_0), \subseteq)$ be the extension of $\boldsymbol{L}_0$ by ideals. For each ideal $\boldsymbol{I}$ of $\boldsymbol{L}_0$ define:

$$h(\boldsymbol{I}) := \sup(\boldsymbol{I}),$$

where $\sup(\boldsymbol{I})$ is taken in the lattice $\boldsymbol{Th}(\mathbf{Q}^{\models})$. Thus $\sup(\boldsymbol{I}) = \mathbf{Q}^{\models}(\bigcup \boldsymbol{I})$. Note that $h(\{\mathbf{0}\}) = \mathbf{0}$ and $h(\boldsymbol{L}_0) = \mathbf{Q}^{\models}(\bigcup_{i \in I} \mathbf{Q}^{\models}(x_i \approx y_i)) = \mathbf{Q}^{\models}(\{x_i \approx y_i : i \in I\})$. The last theory need not coincide with the top element of $\boldsymbol{Th}(\mathbf{Q}^{\models})$.

Lemma 2. *h is a closed embedding of the algebraic lattice* $(Ideal(\boldsymbol{L}_0), \subseteq)$ *into* $\boldsymbol{Th}(\mathbf{Q}^{\models})$. *Moreover h preserves the zero elements.*

Proof. h is well defined.

Claim 1. $\boldsymbol{I} \subseteq \boldsymbol{J}$ *if and only if* $h(\boldsymbol{I}) \leqslant h(\boldsymbol{J})$, *for any* $\boldsymbol{I}, \boldsymbol{J} \in Ideal(\boldsymbol{L}_0)$.

Proof (of the claim). The implication ($\Rightarrow$) is obvious.

($\Leftarrow$). Assume $h(\boldsymbol{I}) \leqslant h(\boldsymbol{J})$. Let $a \in \boldsymbol{I}$. Trivially $a \leqslant \sup(\boldsymbol{I})$. Hence $a \leqslant \sup(\boldsymbol{J})$. As $\boldsymbol{Th}(\mathbf{Q}^{\models})$ is algebraic, there is a finite subset $\boldsymbol{J}_f \subseteq \boldsymbol{J}$ such that $a \leqslant \sup(\boldsymbol{J}_f)$. Since $\boldsymbol{J}_f \subseteq \boldsymbol{J}$, $\sup(\boldsymbol{J}_f)$ belongs to $\boldsymbol{J}$. Hence $a \in \boldsymbol{J}$. This proves the inclusion $\boldsymbol{I} \subseteq \boldsymbol{J}$. □

Claim 2. *h is injective.*

Proof (of the claim). This directly follows from Claim 1. □

Claim 3. *Let* $\{\boldsymbol{I}_t : t \in T\}$ *be a non-empty family of ideals of* $\boldsymbol{L}_0$. *Then*

$$h(\sup(\{\boldsymbol{I}_t : t \in T\}) = \sup(\{h(\boldsymbol{I}_t) : t \in T\}) \tag{3}$$

and

$$h(\inf(\{\boldsymbol{I}_t : t \in T\}) = \inf(\{h(\boldsymbol{I}_t) : t \in T\}) \tag{4}$$

Proof (of the claim). (3) is obvious.

(4). Claim 1 implies that $h(\bigcap_{t \in T} \boldsymbol{I}_t) \leqslant \bigcap_{t \in T} h(\boldsymbol{I}_t)$. We have: $h(\inf(\{\boldsymbol{I}_t : t \in T\}) = h(\bigcap_{t \in T} \boldsymbol{I}_t)$ and $\inf(\{h(\boldsymbol{I}_t) : t \in T\}) = \inf(\{h(\boldsymbol{I}_t) : t \in T\})$ in the lattice $\boldsymbol{Th}(\mathbf{Q}^{\models})$.

Let a be a compact element of the lattice $\boldsymbol{Th}(\mathbf{Q}^{\models})$ and suppose $a \leqslant \bigcap_{t \in T} h(\boldsymbol{I}_t)$. This means that $a \leqslant \sup(\boldsymbol{I}_t)$ for all $t \in T$. Hence, for every $t \in T$, there is a finite subset $\boldsymbol{I}_{f_t} \subseteq \boldsymbol{I}_t$ such that $a \leqslant \sup(\boldsymbol{I}_{f_t})$ in the algebraic lattice $\boldsymbol{Th}(\mathbf{Q}^{\models})$. But evidently, $\sup(\boldsymbol{I}_{f_t}) \in \boldsymbol{I}_t$ for all $t \in T$. It follows that $a \in \boldsymbol{I}_t$ for all $t \in T$, and consequently, $a \in \bigcap_{t \in T} \boldsymbol{I}_t$. This implies that $a \leqslant \sup\{\bigcap_{t \in T} \boldsymbol{I}_t\}) = h(\bigcap_{t \in T} \boldsymbol{I}_t)$. Thus $\bigcap_{t \in T} h(\boldsymbol{I}_t) \leqslant h(\bigcap_{t \in T} \boldsymbol{I}_t)$. □

The above three claims prove Lemma 2. □

Note that Claim 3.(4) does *not* hold if $\{\boldsymbol{I}_t : t \in T\}$ is empty.

Let $\boldsymbol{L}$ be the h-image of the lattice $(Ideal(\boldsymbol{L}_0), \subseteq)$. In view of Lemma 2, $\boldsymbol{L}$ is an algebraic, distributive lattice, and a closed sublattice of $\boldsymbol{Th}(\mathbf{Q}^{\models})$ with the same zero element as $\boldsymbol{Th}(\mathbf{Q}^{\models})$. The zero in the lattice $\boldsymbol{L}$ is $\bigcap\{\mathbf{Q}^{\models}(x_i \approx y_i) : i \in I\} = \mathbf{Q}^{\models}(\emptyset)$. But one can say more about the structure of $\boldsymbol{L}$.

Claim 4. *The set of non-zero elements of* $\mathbf{L}$ *consists of all elements of the form*

$$\sup\{a_t : t \in T\}, \tag{5}$$

where T *ranges over non-empty sets of positive integers and*

$$a_t := \bigcap\{\mathbf{Q}^{\models}(x_i \approx y_i) : i \in A_t\}$$

for some finite non-empty set $A_t \subseteq I$*, for all* $t \in T$.

(If A_t is infinite, we get $a_t = \mathbf{0} = \mathbf{Q}^{\models}(\emptyset)$, by Lemma 6.3.6, so we may disregard this case. The supremum is taken in $\boldsymbol{Th}(\mathbf{Q}^{\models})$.)

Proof (of the claim). The set of compact elements in $\boldsymbol{L}$ is equal to $\boldsymbol{L}_0$, that is, compact elements in $\boldsymbol{L}$ are the zero $\mathbf{0}$ together with all elements of the form (5) with T finite (see (1) and (2)). It follows that for any ideal $\boldsymbol{I}$ of $\boldsymbol{L}_0$, the element $h(\boldsymbol{I}) = \sup(\boldsymbol{I})$ is of the form (5) for an appropriate T (T may be now infinite.) Hence every non-zero element of $\boldsymbol{L}$ is of the form (5). □

It directly follows from the above claim that $\boldsymbol{L}$ is a closed sublattice of every complete sublattice $\boldsymbol{L}'$ of $\boldsymbol{Th}(\mathbf{Q}^{\models})$ such that $\boldsymbol{L}'$ includes the family $\boldsymbol{X} = \{\mathbf{Q}^{\models}(x_i \approx y_i) : i \in I\}$. Thus $\boldsymbol{L}$ is the least closed sublattice of $\boldsymbol{Th}(\mathbf{Q}^{\models})$ (with the same zero as $\boldsymbol{Th}(\mathbf{Q}^{\models})$) that includes $\boldsymbol{X}$. As $\boldsymbol{L}$ is algebraic and distributive, the theorem follows. □

Though $\boldsymbol{L}$ is a (distributive) sublattice of $\boldsymbol{Th}(\mathbf{Q}^{\models})$ if $\mathbf{Q}$ is RCM, it is unclear when $\boldsymbol{L}$ is closed with respect to the commutator operation. (If $\mathbf{Q}$ is RCD, the answer is positive, because the commutator operation reduces to the meet of $\mathbf{Q}$-theories.)

6.4 Generating Sets of Commutator Equations in Varieties

The main focus on the book is on quasivarieties but readers may also be interested to learn that the results of this book lead to novel conclusions in the simpler context of *varieties*. This section is devoted to the exposition of some results in this area.

The theorems placed in Chapter 5 show a significant role of generating sets in the theory of the equationally defined commutator.

Trivially, for any quasivariety $\mathbf{Q}$, the infinite set $\mathbf{Q}^{\models}(x \approx y) \cap \mathbf{Q}^{\models}(z \approx w)$ is a generating set for the equationally defined commutator of $\mathbf{Q}$. If $\mathbf{Q}$ has the relative shifting property, yet another generating set was defined in Section 4.3. According to Theorem 4.1.11 and Proposition 4.3.10, the set

$$\Delta_0 := \bigcup\{(\forall \underline{u})\ \Delta_c(p, q, r, s, \underline{u}) : p \approx q \in \mathbf{Q}^{\models}(x \approx y),\ r \approx s \in \mathbf{Q}^{\models}(z \approx w)\}$$

is also a generating set for the equationally defined commutator of $\mathbf{Q}$. Here Δ_c is the set of quaternary commutator equations for $\mathbf{Q}$ supplied by the relative cube property in the sense of a classical two-binary term condition.

In this section further examples of generating sets are presented for congruence modular *varieties*.

6.4.1 Gumm Terms

The following theorem is due to H. P. Gumm (1983):

Theorem 6.4.1. *A variety* $\mathbf{V}$ *is congruence-modular iff there exists a natural number* n *and ternary terms* $q_0(x,y,z),\ldots,q_n(x,y,z),p(x,y,z)$ *in the language of* $\mathbf{V}$ *such that the following equations hold in* $\mathbf{V}$*:*

(1) $q_0(x,y,z) \approx x$,
(2) $q_i(x,y,x) \approx x$ *for* $0 \leqslant i \leqslant n$,
(3) $q_i(x,y,y) \approx q_{i+1}(x,y,y)$ *for* i *even*,
(4) $q_i(x,x,y) \approx q_{i+1}(x,x,y)$ *for* i *odd*,
(5) $q_n(x,y,y) \approx p(x,y,y)$,
(6) $p(x,x,y) \approx y$. □

Note. In the above theorem n may be assumed to be an even natural number because if it is odd then equations (3) and (5) imply that $q_{n-1}(x,y,y) \approx p(x,y,y)$ and thus the term q_n could be omitted (see Freese and McKenzie 1987, p. 60). □

Let $\mathbf{V}$ be a congruence-modular variety with Gumm terms $q_0(x,y,z),\ldots,$ $q_n(x,y,z)$, $p(x,y,z)$. Let $t = t(x,y,u_1,\ldots,u_m)$ be an $(m+2)$-ary term (for any m) in the language of $\mathbf{V}$. We write $\underline{u}$ for the m-tuple $u_1,\ldots,u_m$. We then define the following set Δ_t of $2n+3$ equations in the variables $x,y,z,w,\underline{u}$:

$$p(t(x,z,\underline{u}),t(y,z,\underline{u}),t(y,w,\underline{u})) \approx p(t(x,w,\underline{u}),t(y,w,\underline{u}),t(y,w,\underline{u})), \tag{a}$$

$$q_i(t(x,w,\underline{u}),t(x,z,\underline{u}),t(y,w,\underline{u})) \approx q_i(t(x,w,\underline{u}),t(x,w,\underline{u}),t(y,w,\underline{u})), \tag{b}$$

$$q_i(t(x,w,\underline{u}),t(y,z,\underline{u}),t(y,w,\underline{u})) \approx q_i(t(x,w,\underline{u}),t(y,w,\underline{u}),t(y,w,\underline{u})), \tag{c}$$

for $i = 0,\ldots,n$.

Let Δ_0 be the union of the sets Δ_t with t ranging over $(m+2)$-ary terms (for all m). The set Δ_0 is infinite.

Theorem 6.4.2. *Let* $\mathbf{V}$ *be a congruence-modular variety. The set* Δ_0*, defined as above, is a generating set of quaternary commutator equations for the commutator of* $\mathbf{V}$.

Proof. The theorem is basically due to McKenzie (1987), Corollary 2.10.(2). He proves that for any algebra $\boldsymbol{A} \in \mathbf{V}$ and any congruences Φ and Ψ on $\boldsymbol{A}$,

$$[\Phi,\Psi] = \Theta^{\boldsymbol{A}}(\bigcup\{(\forall \underline{e})\ \Delta_0(a,b,c,d,\underline{e}) : a \equiv b\ (\Phi),\ c \equiv d\ (\Psi)\}) \tag{d}$$

and, moreover, for any $a,b,c,d \in A$,

$$[\Theta^{\boldsymbol{A}}(a,b),\Theta^{\boldsymbol{A}}(c,d)] = \Theta^{\boldsymbol{A}}((\forall \underline{e})\ \Delta_0(a,b,c,d,\underline{e})). \tag{e}$$

In particular, taking the free algebra $\boldsymbol{F} = \boldsymbol{F}_{\mathbf{V}}(\omega)$ we get from (ii) that

$$\Theta^{F}([x],[y]) \cap \Theta^{F}([z],[w])] =$$
$$[\Theta^{F}([x],[y]), \Theta^{F}([z],[w])] = \Theta^{F}((\forall \underline{e})\ \Delta_0([x],[y],[z],[w],\underline{e})).$$

In view of Proposition 4.3.5 the above equations imply that

$$\mathbf{V}^{\models}((\forall \underline{u})\ \Delta_0(x,y,z,w,\underline{u})) = \mathbf{V}^{\models}(x \approx y) \cap \mathbf{V}^{\models}(z \approx w). \tag{f}$$

This proves that $\Delta_0(x,y,z,w,\underline{u})$ is a generating set for the commutator of **V**.

Due to congruence-modularity of **V** and the fact that the equational commutator for **V** is additive, (e) also gives that for any algebra $\boldsymbol{A}$ in **V** and all sets $X, Y \subseteq A^2$,

$$[\Theta^{A}(X), \Theta^{A}(Y)] = \Theta^{A}(\bigcup\{(\forall \underline{e})\ \Delta_0(a,b,c,d,\underline{e}) : \langle a,b\rangle \in X, \langle c,d\rangle \in Y\}).$$

(To prove the last equality, one may simply repeat the (1) ⇒ (2)-part of the proof of Theorem 5.1.2. (e) is the same as Claim 1 there. Then automatically repeat the proof of Claims 2–4.) □

Notes. 1. As $\Delta_0(x,y,z,w,\underline{s}) \subseteq \Delta_0(x,y,z,w,\underline{u})$ for any string $\underline{s}$ of terms, we obtain that $\Delta_0(x,y,z,w,\underline{u}) = (\forall \underline{u})\ \Delta_0(x,y,z,w,\underline{u})$ in the algebra $\boldsymbol{Te}_\tau$. Consequently, we get a stronger equality than (f), viz.

$$\mathbf{V}^{\models}(\Delta_0(x,y,z,w,\underline{u})) = \mathbf{V}^{\models}(x \approx y) \cap \mathbf{V}^{\models}(z \approx w).$$

2. In view of Theorem 4.3.9, (e) implies (d) because **V** has the shifting property. Indeed, as shown above, (e) implies that $\Delta_0(x,y,z,w,\underline{u})$ is a generating set for the commutator of **V** (cf. Corollary 5.1.3). Hence, applying Theorem 4.3.9 we get (d).

The fact that $\Delta_0(x,y,z,w,\underline{u})$ are commutator equations for **V** can be also directly checked. We shall do that below.

Given a term $t(x,y,u_1,\ldots,u_m)$, we first show that (a)–(c) are (quaternary) commutator equations for **V**. As to (a), we must show that after the identification of x and y in (a), the resulting equation

$$p(t(x,z,\underline{u}), t(x,z,\underline{u}), t(x,w,\underline{u})) \approx p(t(x,w,\underline{u}), t(x,w,\underline{u}), t(x,w,\underline{u})) \tag{i}$$

is valid in **V**.

It follows from (6) that the equations

$$p(t(x,z,\underline{u}), t(x,z,\underline{u}), t(x,w,\underline{u})) \approx t(x,w,\underline{u})$$

and

$$p(t(x,w,\underline{u}), t(x,w,\underline{u}), t(x,w,\underline{u})) \approx t(x,w,\underline{u})$$

hold in **V**. Taking into account left sides of the above two equations, we see that (i) holds.

We must also show that after identification of z and w in (a), the equation

$$p(t(x,z,\underline{u}),t(y,z,\underline{u}),t(y,w,\underline{u})) \approx p(t(x,z,\underline{u}),t(y,z,\underline{u}),t(y,w,\underline{u})) \qquad \text{(ii)}$$

is valid in $\mathbf{V}$. But (ii) is a tautology of any equational logic.

As to (b), we must show that after the identification of x and y in (b), the resulting equation

$$q_i(t(x,w,\underline{u}),t(x,z,\underline{u}),t(x,w,\underline{u})) \approx q_i(t(x,w,\underline{u}),t(x,w,\underline{u}),t(x,w,\underline{u})) \qquad \text{(iii)}$$

is valid in $\mathbf{V}$ for $i = 0,\dots,n$.

Fix i. It follows from (2) that the equations

$$q_i(t(x,w,\underline{u}),t(x,z,\underline{u}),t(x,w,\underline{u})) \approx t(x,w,\underline{u})$$

and

$$q_i(t(x,w,\underline{u}),t(x,w,\underline{u}),t(x,w,\underline{u})) \approx t(x,w,\underline{u})$$

are valid in $\mathbf{V}$. Taking into account left sides of the above two equations, we see that (iii) holds in $\mathbf{V}$.

We must also show that after identification of z and w in (b), the resulting equation

$$q_i(t(x,z,\underline{u}),t(x,z,\underline{u}),t(y,z,\underline{u})) \approx q_i(t(x,z,\underline{u}),t(x,z,\underline{u}),t(y,z,\underline{u})) \qquad \text{(iv)}$$

is $\mathbf{V}$-valid for $i = 0,\dots,n$. But (iv) is a tautology of any equational logic.

As to (c), we must show that after the identification of x and y in (c), the resulting equation

$$q_i(t(x,w,\underline{u}),t(x,z,\underline{u}),t(x,w,\underline{u})) \approx q_i(t(x,w,\underline{u}),t(x,w,\underline{u}),t(x,w,\underline{u})) \qquad \text{(v)}$$

is valid in $\mathbf{V}$ for $i = 0,\dots,n$. But (v) is exactly the equation (iii), which, as shown above, holds in $\mathbf{V}$.

We must also show that after identification of z and w in (c), the resulting equation

$$q_i(t(x,z,\underline{u}),t(y,z,\underline{u}),t(y,z,\underline{u})) \approx q_i(t(x,z,\underline{u}),t(y,z,\underline{u}),t(y,z,\underline{u})) \qquad \text{(vi)}$$

is $\mathbf{V}$-valid for $i = 0,\dots,n$. But (vi) is a tautology of any equational logic.

This proves that Δ_0 is a set of commutator equations for $\mathbf{V}$. □

6.4.2 Varieties with Congruence-Permutable Congruences

Let $\mathbf{V}$ be a variety with permutable congruences. $\mathbf{V}$ is congruence-modular and, in view of the well-known Mal'cev's Theorem, there exists a ternary term $p(x,y,z)$ such that $\mathbf{V}$ validates the identities $p(x,y,y) \approx x$ and $p(x,x,y) \approx y$. $p(x,y,z)$ is called a *Mal'cev term* for $\mathbf{V}$.

Any equational class with permutable congruences is also called a *Mal'cev variety*. Commutator theory for Mal'cev varieties is presented in Bergman (2011).

It is easy to see that the single equation

$$z \approx p(w, p(y, x, z), z)$$

forms a Day implication system for $\mathbf{V}^{\models}$.

Let $q_0(x, y, z)$ be the variable x. The set $\{q_0(x, y, z), p(x, y, z)\}$ forms a system of Gumm terms for $\mathbf{V}$ with $n = 0$. (Conditions (3)–(4) of Theorem 6.4.1 are vacuously satisfied.)

Theorem 6.4.2 implies that the infinite set $\Delta_0(x, y, z, w, \underline{u})$ consisting of all equations of the form

$$p(t(x, z, \underline{u}), t(y, z, \underline{u}), t(y, w, \underline{u})) \approx p(t(x, w, \underline{u}), t(y, w, \underline{u}), t(y, w, \underline{u})), \qquad (*)$$

where $t(x, y, \underline{u})$ ranges over arbitrary terms, is a generating set for the commutator of $\mathbf{V}$.

Since $\mathbf{V}$ validates $p(x, y, y) \approx x$, the term on the right side of (*) can be replaced by the equivalent term $t(x, w, \underline{u})$ (on the basis of $\mathbf{V}$). We therefore get:

Corollary 6.4.3. *Let* $\mathbf{V}$ *be a congruence-permutable variety with a Mal'cev term* $p(x, y, z)$. *The infinite set* $\Delta_0(x, y, z, w, \underline{u})$ *consisting of all equations of the form*

$$p(t(x, z, \underline{u}), t(y, z, \underline{u}), t(y, w, \underline{u})) \approx t(x, w, \underline{u}),$$

where $t(x, y, \underline{u})$ *ranges over arbitrary terms, is a generating set for the commutator of* $\mathbf{V}$. □

6.4.3 Groups

For each group $\mathbf{A} = (A, \cdot, ^{-1}, 1)$ (in the multiplicative notation) and $a, b \in A$ we define: $a \leftrightarrow b := a \cdot b^{-1}$ (the dot will be omitted). For any normal subgroup $F \subseteq A$, we also define:

$$\boldsymbol{\Omega}_{\mathbf{A}}(F) := \{\langle a, b\rangle : a \leftrightarrow b \in F\}.$$

$\boldsymbol{\Omega}_{\mathbf{A}}(F)$ is a congruence of $\mathbf{A}$ and it is the largest congruence of $\mathbf{A}$ compatible with F. (The notation adopted in abstract algebraic logic is applied here. A congruence $\Phi \in Con(\mathbf{A})$ is compatible with F if for all $a, b \in A$, $a \equiv b \pmod{\Phi}$ implies that $a \in F$ if and only if $b \in F$.)

The operator $\boldsymbol{\Omega}_{\mathbf{A}}$ establishes an isomorphism between the lattice of normal subgroups of $\mathbf{A}$ and the lattice of congruences of $\mathbf{A}$. The mapping $\boldsymbol{H}_{\mathbf{A}}$ given by

$$\boldsymbol{H}_{\mathbf{A}}(\Phi) := \{a \in A : a \equiv 1 \pmod{\Phi}\}, \qquad \Phi \in Con(\mathbf{A}),$$

is the inverse of $\boldsymbol{\Omega}_{\mathbf{A}}$.

In group theory, one investigates the formally simpler commutator defined for normal subgroups rather than for congruences. The "standard" commutator of normal subgroups F and G of $\mathbf{A}$ is defined as follows:

$$[F, G] := \text{ the subgroup of } \mathbf{A} \text{ generated by the set } \{ab \leftrightarrow ba : a \in F, b \in G\}. \quad (1)$$

(As $yx \leftrightarrow xy \approx (xy \leftrightarrow yx)^{-1}$ holds in all groups, it follows that $[F, G]$ includes the set $\{ba \leftrightarrow ab : a \in F, b \in G\}$.)

Since the variety $\mathbf{G}$ of groups is congruence-modular, it is endowed with the additive commutator operation for congruences. This commutator coincides with the equationally defined commutator for $\mathbf{G}$.

Apart from the property of preserving lattice operations, the operator $\boldsymbol{\Omega}_{\mathbf{A}}$ preserves the commutator operations in the following sense: for any normal subgroups F and G of $\mathbf{A}$,

$$\boldsymbol{\Omega}_{\mathbf{A}}([F, G]) = [\boldsymbol{\Omega}_{\mathbf{A}}(F), \boldsymbol{\Omega}_{\mathbf{A}}(G)], \quad (2)$$

where on the right side, the commutator is defined for congruences of $\mathbf{A}$. We also have

$$\boldsymbol{H}_{\mathbf{A}}([\Phi, \Psi]) = [\boldsymbol{H}_{\mathbf{A}}(\Phi), \boldsymbol{H}_{\mathbf{A}}(\Psi)],$$

for all $\Phi, \Psi \in Con(\mathbf{A})$.

Since groups have permutable congruences with $p(x, y, z) := x(y^{-1}z)$ as a Mal'cev term, Corollary 6.4.3 defines an infinite generating set of commutator equations for $\mathbf{G}$, viz. the infinite set $\Delta_0(x, y, z, w, \underline{u})$ consisting of all equations of the form

$$t(x, z, \underline{u}) \cdot (t(y, z, \underline{u})^{-1} \cdot t(y, w, \underline{u})) \approx t(x, w, \underline{u}) \quad (3)$$

where $t(x, y, \underline{u})$ ranges over arbitrary terms, is a generating set for the commutator of $\mathbf{G}$.

After making appropriate simplifications in (3), we get the following sequence of equivalent equations (modulo the axioms of groups), viz.

$$t(x, z, \underline{u}) \cdot (t(y, z, \underline{u})^{-1} \cdot t(y, w, \underline{u})) \approx t(x, w, \underline{u})$$

$$(t(x, z, \underline{u}) \cdot t(y, z, \underline{u})^{-1}) \cdot t(y, w, \underline{u}) \approx t(x, w, \underline{u})$$

$$t(x, z, \underline{u}) \cdot t(y, z, \underline{u})^{-1} \approx t(x, w, \underline{u}) \cdot t(y, w, \underline{u})^{-1}$$

It follows that the equations

$$t(x, z, \underline{u}) \cdot t(y, z, \underline{u})^{-1} \approx t(x, w, \underline{u}) \cdot t(y, w, \underline{u})^{-1}$$

with t ranging over arbitrary terms, form a generating set for the (congruence) commutator of **G**. This set is infinite. But it can be shown that in fact the commutator for **G** has a one-element generating set, viz. the set consisting of the following equation (with one parameter u; parentheses omitted):

$$z^{-1}wu^{-1}x^{-1}yu \approx u^{-1}x^{-1}yuz^{-1}w \tag{4}$$

(see, e.g., the remarks following Corollary 2.10 in McKenzie 1987). □

6.4.4 Equivalence Algebras

An *intuitionistic equivalence algebra* is an algebra $\boldsymbol{A} = (A, \leftrightarrow)$ (endowed with one binary operation) satisfying the following equations:

$$(x \leftrightarrow x) \leftrightarrow y \approx y \tag{e1}$$

$$((x \leftrightarrow y) \leftrightarrow z) \leftrightarrow z \approx (x \leftrightarrow y) \leftrightarrow (y \leftrightarrow z), \tag{e2}$$

$$((x \leftrightarrow y) \leftrightarrow ((x \leftrightarrow z) \leftrightarrow z)) \leftrightarrow ((x \leftrightarrow z) \leftrightarrow z) \approx x \leftrightarrow y \tag{e3}$$

(see, e.g., Idziak et al. 2011). **EA** is the variety of intuitionistic equivalence algebras. **EA** is pointed ($1 := x \leftrightarrow x$ is a distinguished constant of **EA**) and point-regular (relative to 1). **EA** is the algebraic counterpart of the purely equivalential fragment of the intuitionistic propositional calculus. **EA** is locally finite. Moreover, every quasivariety contained in **EA** is a variety—a result proved by Słomczyńska (1996).

In the theory of **EA** one usually adopts the conventions of ignoring the operation sign $\leftrightarrow$ in the terms of Te_τ and also left bracketing which means that the terms lacking parentheses are to be associated with the left. For example, axioms (e1)–(e3) are written down as

(e1) $xxy \approx y$, (e2) $xyzz \approx xy(yz)$, (e3) $xy(xzz)(xzz) \approx xy$.

A *filter* F of an equivalence algebra $\boldsymbol{A}$ is a subset $F \subseteq A$ such that, for all $a, b \in A$,

(f1) $1 \in F$, (f2) if $a, ab \in F$, then $b \in F$ (f3) if $a \in F$, then $abb \in F$.

((f3) is referred to as the *Tax property*.)

For each equivalence algebra $\boldsymbol{A}$ and a filter $F \subseteq A$, one defines:

$$\boldsymbol{\Omega}_A(F) := \{\langle a, b\rangle : ab \in F\}.$$

$\boldsymbol{\Omega}_A(F)$ is a congruence of $\boldsymbol{A}$. The operator $\boldsymbol{\Omega}_A$ establishes an isomorphism between the lattice of filters of $\boldsymbol{A}$ and the lattice of congruences of $\boldsymbol{A}$. The mapping $\boldsymbol{H}_A$ given by

$$H_A(\Phi) := \{a \in A : a \equiv 1 \ (\mathrm{mod}\ \Phi)\}, \qquad \Phi \in Con(A),$$

is the inverse of $\boldsymbol{\Omega}_A$.

Since the variety **EA** is point-regular, it is congruence-modular. In fact, **EA** has permutable congruences with $p(x, y, z) := xyz(xzzx)$ being a Mal'cev term (Idziak 1991). **EA** is therefore endowed with the additive commutator operation. This commutator coincides with the equationally defined commutator for **EA**.

In the theory of **EA** one investigates the formally simpler commutator defined for filters rather than for congruences. This is due to the fact that the operator $\boldsymbol{\Omega}_A$ preserves the commutator operations in the following sense: for any filters F and G of $\boldsymbol{A}$,

$$\boldsymbol{\Omega}_A([F, G]) = [\boldsymbol{\Omega}_A(F), \boldsymbol{\Omega}_A(G)], \tag{1}$$

where on the right side, the commutator is defined for congruences of $\boldsymbol{A}$. We also have

$$\boldsymbol{H}_A([\Phi, \Psi]) = [\boldsymbol{H}_A(\Phi), \boldsymbol{H}_A(\Phi)],$$

for all $\Phi, \Psi \in Con(\boldsymbol{A})$.

Applying (1), it can be proved that the "standard" commutator of filters F and G of $\boldsymbol{A}$ is defined as follows:

$$[F, G] = \text{the filter of } \boldsymbol{A} \text{ generated by the set} \{abba, baab : a \in F, b \in G\}. \tag{2}$$

(Idziak et al. 2011).

(2) *cannot* be replaced by the stronger formula:

$$[Fi(X), Fi(Y)] = \text{the filter of } \boldsymbol{A} \text{ generated by the set} \tag{3}$$
$$\{abba, baab : a \in X, b \in Y\}, \text{ for any sets } X, Y \subseteq A.$$

(Idziak, Słomczyńska and Wroński, personal correspondence.) $Fi(Z)$ stands for the filter of $\boldsymbol{A}$ generated by Z.

Taking into account the above Mal'cev term $p(x, y, z)$ for **EA** and Corollary 6.4.3, we see that the infinite set $\Delta_0(x, y, z, w, \underline{u})$ consisting of all equations of the form

$$t(x, z, \underline{u})\ t(y, z, \underline{u})\ t(y, w, \underline{u})\ (t(x, z, \underline{u})\ t(y, w, \underline{u})\ t(y, w, \underline{u})\ t(x, z, \underline{u})) \approx t(x, w, \underline{u}).$$

where $t(x, y, \underline{u})$ ranges over arbitrary terms, is a generating set for the (congruence) commutator of **EA**. □

Open problem. Let **Q** be a quasivariety of algebras with the relative congruence extension property. Then the equationally defined commutator has a generating set of equations $\Delta_0(x, y, z, w)$ in four variables only (without parameters). □

6.5 Various Commutators in Varieties

In the literature one finds a definition of the "Term Condition" (TC) commutator, the one called $C(\Phi,\Psi)$ in Freese and McKenzie (1987) (and $[\Phi,\Psi)]$ in the second edition). $C(\Phi,\Psi)$ is the commutator arising from the centralization relation Z_1 of the present book (but in the context of *all* algebras of a given signature). To avoid confusion with the equationally defined commutator, we shall mark this commutator as $[\cdot]_1$. Thus, for any algebra $\mathbf{A}$ and any congruences Φ,Ψ of A,

$$[\Phi,\Psi]_1 := \text{ the least congruence } \Xi \text{ of } \mathbf{A} \text{ such that } Z_1(\Phi,\Psi;\Xi) \text{ holds.}$$

$Z_1(\Phi,\Psi;\Xi)$ means here that for any positive integers m,n, for any term $t(\underline{x},\underline{y},\underline{u})$ with $\underline{x}=x_1,\ldots,x_m$, $\underline{y}=y_1,\ldots,y_n$, for any two m-tuples $\underline{a},\underline{b}$ of elements of A and any two n-tuples $\underline{c},\underline{d}$ of elements of A such that $\underline{a}\equiv\underline{b} \pmod{\Phi}$ and $\underline{c}\equiv d \pmod{\Psi}$,

$$t(\underline{a},\underline{c},\underline{e})\equiv t(\underline{a},\underline{d},\underline{e}) \pmod{\Xi} \quad \text{implies} \quad t(\underline{b},\underline{c},\underline{e})\equiv t(\underline{b},\underline{d},\underline{e}) \pmod{\Xi},$$

for any sequence $\underline{e}$ of elements of A of the length of $\underline{u}$.

The definition of $Z_1(\Phi,\Psi;\Xi)$ is often rendered in a matrix form as follows. Let Φ, Ψ, and Ξ be in $Con(\mathbf{A})$.

$M(\Phi,\Psi)$ is the set of all square 2×2 matrices

$$\begin{pmatrix} t(\underline{a},\underline{c},\underline{e}) & t(\underline{a},\underline{d},\underline{e}) \\ t(\underline{b},\underline{c},\underline{e}) & t(\underline{b},\underline{d},\underline{e}) \end{pmatrix}$$

where $t(\underline{x},\underline{y},\underline{u})$ with $\underline{x}=x_1,\ldots,x_m$, $\underline{y}=y_1,\ldots,y_n$ is an arbitrary term of Te_τ with positive m and n, $\underline{a},\underline{b}$ are arbitrary m-tuples of elements of A such that $\underline{a}\equiv\underline{b} \pmod{\Phi}$, $\underline{c},\underline{d}$ are arbitrary n-tuples of elements of A such that $\underline{c}\equiv\underline{d} \pmod{\Psi}$, and $\underline{e}$ is an arbitrary sequence of elements of A of the length of $\underline{u}$.

It is then easy to check that

$$Z_1(\Phi,\Psi;\Xi) \quad \Leftrightarrow \quad \text{for every matrix } \begin{pmatrix} u_{11} & u_{12} \\ u_{21} & u_{22} \end{pmatrix} \text{ in } M(\Phi,\Psi),$$

$$u_{11}\equiv u_{12} \pmod{\Xi} \quad \text{implies} \quad u_{21}\equiv u_{22} \pmod{\Xi}.$$

Let $M(\Phi,\Psi)^T$ be the set of *transposed* matrices of $M(\Phi,\Psi)$.

Lemma 6.5.1. *$M(\Phi,\Psi)^T = M(\Psi,\Phi)$ for all Φ,Ψ in $Con(\mathbf{A})$.*

Proof. Let

$$\begin{pmatrix} t(\underline{a},\underline{c},\underline{e}) & t(\underline{a},\underline{d},\underline{e}) \\ t(\underline{b},\underline{c},\underline{e}) & t(\underline{b},\underline{d},\underline{e}) \end{pmatrix}$$

be a matrix in $M(\Phi, \Psi)$. We show that its transposition, viz.,

$$\begin{pmatrix} t(\underline{a}, \underline{c}, \underline{e}) & t(\underline{b}, \underline{c}, \underline{e}) \\ t(\underline{a}, \underline{d}, \underline{e}) & t(\underline{b}, \underline{d}, \underline{e}) \end{pmatrix}$$

belongs to $M(\Psi, \Phi)$.

We may write $t = t(\underline{x}, \underline{y}, \underline{u})$ with $\underline{x} = x_1, \ldots, x_m$, $\underline{y} = y_1, \ldots, y_n$. We select sequences of new different variables $\underline{x}'$ and $\underline{y}'$ such that $|\underline{x}'| = |\underline{y}|$ and $|\underline{y}'| = |\underline{x}|$ and put:

$$t' = t'(\underline{x}', \underline{y}', \underline{u}) := t(\underline{x}/\underline{y}', \underline{y}/\underline{x}', \underline{u}).$$

We have:

$$\begin{aligned} t'(\underline{c}, \underline{a}, \underline{e}) &= t'(\underline{x}'/\underline{c}, \underline{y}'/\underline{a}, \underline{u}/\underline{e}) = t(\underline{x}/\underline{a}, \underline{y}/\underline{c}, \underline{u}/\underline{e}) = t(\underline{a}, \underline{c}, \underline{e}) \\ t'(\underline{d}, \underline{a}, \underline{e}) &= t'(\underline{x}'/\underline{d}, \underline{y}'/\underline{a}, \underline{u}/\underline{e}) = t(\underline{x}/\underline{a}, \underline{y}/\underline{d}, \underline{u}/\underline{e}) = t(\underline{a}, \underline{d}, \underline{e}) \\ t'(\underline{c}, \underline{b}, \underline{e}) &= t'(\underline{x}'/\underline{c}, \underline{y}'/\underline{b}, \underline{u}/\underline{e}) = t(\underline{x}/\underline{b}, \underline{y}/\underline{c}, \underline{u}/\underline{e}) = t(\underline{b}, \underline{c}, \underline{e}) \\ t'(\underline{d}, \underline{b}, \underline{e}) &= t'(\underline{x}'/\underline{d}, \underline{y}'/\underline{b}, \underline{u}/\underline{e}) = t(\underline{x}/\underline{b}, \underline{y}/\underline{d}, \underline{u}/\underline{e}) = t(\underline{b}, \underline{d}, \underline{e}). \end{aligned}$$

Hence the matrix

$$\begin{pmatrix} t(\underline{a}, \underline{c}, \underline{e}) & t(\underline{a}, \underline{d}, \underline{e}) \\ t(\underline{b}, \underline{c}, \underline{e}) & t(\underline{b}, \underline{d}, \underline{e}) \end{pmatrix} \quad \text{is identical with} \quad \begin{pmatrix} t'(\underline{c}, \underline{a}, \underline{e}) & t'(\underline{d}, \underline{a}, \underline{e}) \\ t'(\underline{c}, \underline{b}, \underline{e}) & t'(\underline{d}, \underline{b}, \underline{e}) \end{pmatrix}.$$

After transposition we get that

$$\begin{pmatrix} t(\underline{a}, \underline{c}, \underline{e}) & t(\underline{b}, \underline{c}, \underline{e}) \\ t(\underline{a}, \underline{d}, \underline{e}) & t(\underline{b}, \underline{d}, \underline{e}) \end{pmatrix} \quad \text{is identical with} \quad \begin{pmatrix} t'(\underline{c}, \underline{a}, \underline{e}) & t'(\underline{c}, \underline{b}, \underline{e}) \\ t'(\underline{d}, \underline{a}, \underline{e}) & t'(\underline{d}, \underline{b}, \underline{e}) \end{pmatrix}.$$

But the second matrix belongs to $M(\Psi, \Phi)$. This proves that $M(\Phi, \Psi)^T \subseteq M(\Psi, \Phi)$.

By a symmetric argument we get that $M(\Psi, \Phi)^T \subseteq M(\Phi, \Psi)$. Hence $M(\Psi, \Phi) = M(\Psi, \Phi)^{TT} \subseteq M(\Phi, \Psi)^T$. This shows that $M(\Phi, \Psi)^T = M(\Psi, \Phi)$. □

The definition of $[\Phi, \Psi]_1$ is *absolute*; the commutator $[\Phi, \Psi]_1$ does not depend on a variety to which $\boldsymbol{A}$ belongs. The commutator $[\cdot]_1$, defined by the centralization Z_1, is studied in the literature by its own right. Generally, $[\cdot]_1$ need not be a symmetric function!

We take a step farther and define yet another commutator, viz., the "Two-Term Condition" (2TC) commutator $[\cdot]_2$ determined by the centralization Z_2 in the sense of the two-term condition in *arbitrary* algebras. Thus, for any algebra $\boldsymbol{A}$ and any congruences Φ, Ψ of $\boldsymbol{A}$,

$$[\Phi, \Psi]_2 := \text{ the least congruence } \Xi \text{ of } \boldsymbol{A} \text{ such that } Z_2(\Phi, \Psi; \Xi) \text{ holds.}$$

($Z_2(\Phi, \Psi; \Xi)$ means here that for any $m, n \geqslant 1$, any two m-tuples $\underline{a}, \underline{b}$, any two n-tuples $\underline{c}, \underline{d}$ of elements of A such that $\underline{a} \equiv \underline{b} \pmod{\Phi}$ and $\underline{c} \equiv \underline{d} \pmod{\Psi}$, any terms $f(\underline{x}, \underline{y}, \underline{v})$, $g(\underline{x}, \underline{y}, \underline{v})$, where $|\underline{x}| = m$, $|\underline{y}| = n$, and any sequence $\underline{e}$ of elements of A of the length of $\underline{v}$,

$$f(\underline{a}, \underline{c}, \underline{e}) \equiv g(\underline{a}, \underline{c}, \underline{e}) \pmod{\Xi}$$
$$f(\underline{a}, \underline{d}, \underline{e}) \equiv g(\underline{a}, \underline{d}, \underline{e}) \pmod{\Xi}$$
$$f(\underline{b}, \underline{c}, \underline{e}) \equiv g(\underline{b}, \underline{c}, \underline{e}) \pmod{\Xi}$$

imply

$$f(\underline{b}, \underline{d}, \underline{e}) \equiv g(\underline{b}, \underline{d}, \underline{e}) \pmod{\Xi}.)$$

By the same token, the definition of $[\Phi, \Psi]_2$ is *absolute*; the commutator $[\Phi, \Psi]_2$ does not depend on a variety to which $\mathbf{A}$ belongs. $[\,\cdot\,]_2$ is, however, a symmetric function (see Note 2 following Proposition 4.1.1).

For the sake of the present comments, if $\mathbf{V}$ is a variety (not necessarily CM), $\mathbf{A} \in \mathbf{V}$, and $\Phi, \Psi \in Con(\mathbf{A})$, then $[\Phi, \Psi]_{4,edc(\mathbf{V})}$ is the commutator defined by means of *quaternary* commutator equations in the sense of $\mathbf{V}$. Thus

$$[\Phi, \Psi]_{4,edc(\mathbf{V})} := \text{ the least congruence } \Xi \text{ of } \mathbf{A} \text{ such that } Z_{4,\mathrm{com}}(\Phi, \Psi; \Xi) \text{ holds.}$$

Here $Z_{4,\mathrm{com}}(\Phi, \Psi; \Xi)$ means that for any quadruple a, b, c, d of elements of A, the conditions

$$a \equiv b \pmod{\Phi} \text{ and } c \equiv d \pmod{\Psi} \text{ imply } p(a, b, c, d, \underline{e}) \equiv q(a, b, c, d, \underline{e}) \pmod{\Xi}$$

for any *quaternary* commutator equation $p(x, y, z, w, \underline{u}) \approx q(x, y, z, w, \underline{u})$ in the sense of $\mathbf{V}$ and any sequence $\underline{e}$ of elements of A of the length of $\underline{u}$.

We recall that $[\Phi, \Psi]_{edc(\mathbf{V})}$ is the commutator defined by means of *arbitrary* commutator equations in the sense of $\mathbf{V}$. Thus, if $\mathbf{A} \in \mathbf{V}$, and $\Phi, \Psi \in Con(\mathbf{A})$, then

$$[\Phi, \Psi]_{edc(\mathbf{V})} := \text{ the least congruence } \Xi \text{ of } \mathbf{A} \text{ such that } Z_{\mathrm{com}}(\Phi, \Psi; \Xi) \text{ holds}$$

(see Section 4.1).

Theorem 4.3.2 implies:

Theorem 6.5.2. *Let $\mathbf{V}$ be a congruence-modular variety. Then for any algebra $\mathbf{A} \in \mathbf{V}$ and any congruences Φ, Ψ of $\mathbf{A}$,*

$$[\Phi, \Psi]_1 = [\Phi, \Psi]_2 = [\Phi, \Psi]_{edc(\mathbf{V})} = [\Phi, \Psi]_{4,edc(\mathbf{V})}. \qquad \square$$

A part of the above result was proved by Kiss (1992)—he showed that $[\cdot]_1 = [\cdot]_2$ holds in any CM variety.

Since the above definitions of congruences $[\Phi, \Psi]_1$ and $[\Phi, \Psi]_2$ are variety independent, it therefore follows from the above theorem that if $\mathbf{V}_1$ and $\mathbf{V}_2$ are CM varieties, then for any algebra $\boldsymbol{A} \in \mathbf{V}_1 \cap \mathbf{V}_2$ and any $\Phi, \Psi \in Con(\boldsymbol{A})$, the equationally defined commutators of Φ, Ψ in the sense of $\mathbf{V}_1$ and $\mathbf{V}_2$, respectively, coincide, that is, $[\Phi, \Psi]_{edc(\mathbf{V}_1)} = [\Phi, \Psi]_{edc(\mathbf{V}_2)}$ despite the fact that the sets of commutator equations of $\mathbf{V}_1$ and $\mathbf{V}_2$, respectively, are generally different. Since the definition of congruences $[\Phi, \Psi]_1$ and $[\Phi, \Psi]_2$ do not depend on $\mathbf{V}$ but merely on the structural properties of the algebra $\boldsymbol{A}$, we see that the equationally defined commutator of Φ, Ψ is *stable* in the sense that it is the *same* congruence whenever it is computed in any CM variety to which $\boldsymbol{A}$ belongs.

Let $\mathbf{V}$ be a variety, not necessarily congruence-modular. Then $[\Phi, \Psi]_{4,edc(\mathbf{V})} \subseteq [\Phi, \Psi]_{edc(\mathbf{V})}$ for any algebra $\boldsymbol{A} \in \mathbf{V}$ and any $\Phi, \Psi \in Con(\boldsymbol{A})$. In turn, Lemma 4.1.3 gives that $[\Phi, \Psi]_{edc(\mathbf{V})} \subseteq [\Phi, \Psi]_2$. We may therefore abbreviate the above inclusions as

$$[\cdot]_{4,edc(\mathbf{V})} \leqslant [\cdot]_{edc(\mathbf{V})} \leqslant [\cdot]_2$$

for any variety $\mathbf{V}$.

As far as *non*-CM varieties $\mathbf{V}$ are concerned, it is not clear to what extent the commutators $[\cdot]_1$, $[\cdot]_2$, $[\cdot]_{4,edc(\mathbf{V})}$ and $[\cdot]_{edc(\mathbf{V})}$ are different. If $\mathbf{V}$ is the "empty" variety consisting of all non-empty sets (no operations), then $[\cdot]_{4,edc(\mathbf{V})}$ and $[\cdot]_{edc(\mathbf{V})}$ coincide. $[\cdot]_{4,edc(\mathbf{V})} = [\cdot]_{edc(\mathbf{V})}$ also holds in "free" varieties satisfying no non-trivial identity, because in this case the equationally defined commutator is a zero commutator. These are Birkhoff's logics in all possible signatures. (More, generally, if $\mathbf{V}$ is a variety whose free algebra $\boldsymbol{F}_{\mathbf{V}}(\omega)$ satisfies conditions (FreeGenDistr)$_{m,n}$ for all positive integers m and n, as defined in Section 5.2, then $[\cdot]_{4,edc(\mathbf{V})} = [\cdot]_{edc(\mathbf{V})}$. This follows from Corollary 5.2.15.) We do not know how the above commutators behave in monadic varieties (with only unary fundamental operations).

The *symmetric commutator* of Φ and Ψ, written $[\Phi, \Psi]_{sym}$, is the least Ξ such that $Z_1(\Phi, \Psi; \Xi)$ and $Z_1(\Psi, \Phi; \Xi)$ hold. Let $[\cdot]_{sym}$ be the symmetrization of $[\cdot]_1$. We have $[\Phi, \Psi]_1 \subseteq [\Phi, \Psi]_{sym} = [\Psi, \Phi]_{sym}$. Moreover, both operations $[\cdot]_1$ and $[\cdot]_{sym}$ are monotone in each variable.

According to Freese and McKenzie (1987), Proposition 4.2, the equation $[\Phi, \Psi]_1 = [\Phi, \Psi]_{sym}$ holds in every congruence modular variety. (This also directly follows from Theorem 4.2.2.) This equation holds for an algebra exactly when the operation $[\cdot]_{sym}$ is symmetric. It is shown in Kearnes (1995) that the identity $[\cdot]_1 = [\cdot]_{sym}$ holds in any variety with a difference term, but it fails in some varieties with a weak difference term, such as the variety of inverse semigroups.

If $\mathbf{V}$ is an idempotent variety satisfying a non-trivial Mal'cev condition, then $[\cdot]_2 = [\cdot]_{sym}$ throughout $\mathbf{V}$ (see Kearnes and Szendrei 1998). But as $[\cdot]_{edc(\mathbf{V})} \leqslant [\cdot]_2$, we have that $[\cdot]_{edc(\mathbf{V})} \leqslant [\cdot]_{sym}$ in such varieties $\mathbf{V}$. In particular, if $[\cdot]_1$ is symmetric in $\mathbf{V}$, then $[\cdot]_{edc(\mathbf{V})} \leqslant [\cdot]_1$ in the algebras of $\mathbf{V}$.

The variety **S** of semilattices is minimal and it is not congruence-modular. Kearnes (2000) has proved that, up to term equivalence, the only minimal idempotent varieties that are not congruence-modular are the variety of sets and the variety of semilattices. (The variety of sets has the empty signature—this is the variety with no fundamental operations—see also Section 5.3.) $[\cdot]_{edc(\mathbf{S})}$ is the zero commutator, the fact established by Corollary 5.4.2.

The following corollary is an immediate consequence of the above result of Kearnes and of Theorems 5.4.1 and 6.3.2:

Theorem 6.5.3. *For every minimal idempotent variety* **V***, the lattice* $\mathbf{Con}(\mathbf{F}_{\mathbf{V}}(\omega))$ *validates the law of restricted distributivity. Consequently, the equationally defined commutator for* **V** *is additive.* □

The following theorem is due to Keith Kearnes (unpublished):

Theorem 6.5.4. *Suppose that* **V** *is a locally finite variety and that* $[\cdot]_{edc(\mathbf{V})} = [\cdot]_1$ *throughout* **V** *(or just only* $[\cdot]_{edc(\mathbf{V})} = [\cdot]_{sym}$ *throughout* **V***). Then:*

(a) **V** *omits type* **5** *in the sense of Hobby and McKenzie (1988).*
(b) *If* **V** *satisfies some non-trivial idempotent Mal'cev condition (that is,* **V** *omits type* **1**)*, then congruences of algebras in* **V** *satisfy some non-trivial lattice identity (that is,* **V** *omits types* **1** *and* **5** *(see also Kearnes 2001, in particular Theorem* 2.4).
(c) *If* **V** *is congruence meet semidistributive and satisfies some non-trivial idempotent Mal'cev condition, then* **V** *is congruence join semidistributive (Kearnes 2001, Theorem* 2.6). □

Since (b) and (c) do not mention tame congruence theory, it is conceivable that they hold for *arbitrary* varieties, not just locally finite varieties. This is an open problem and perhaps a solution can be obtained in a relatively simple way using the methods of Kearnes and Kiss (2013) .

Yet another direction of investigation is to see how the commutators behave in varieties **V** which are not congruence modular but whose congruence lattices still satisfy a non-trivial lattice theoretic identity. Interesting test cases are 4-permutable varieties, e.g., Polin's variety (see Day and Freese 1980) or the variant of Polin's variety with Abelian groups as "internal algebras" (see Example A_3 in Exercise 6.23 in Hobby and McKenzie 1988). The crucial issue is that of the additivity of $[\cdot]_{edc(\mathbf{V})}$ (see Chapter 5).

Chapter 7
Additivity of the Equationally-Defined Commutator and Relatively Congruence-Distributive Subquasivarieties

7.1 Relatively Finitely Subdirectly Irreducible Algebras

Let $\mathbf{Q}$ be a quasivariety. The notation $\boldsymbol{A} \cong \boldsymbol{A}' \subseteq_{\mathrm{SD}} \Pi_{i \in I} \boldsymbol{B}_i$ means that the algebra $\boldsymbol{A}$ is isomorphic to a subdirect product of a system $\boldsymbol{B}_i$, $i \in I$, of algebras in $\mathbf{Q}$.

A non-trivial algebra $\boldsymbol{A} \in \mathbf{Q}$ is *(finitely) subdirectly irreducible relative* to $\mathbf{Q}$ if it is not isomorphic to a subdirect product of a (finite) system $\boldsymbol{B}_i$, $i \in I$, of algebras in $\mathbf{Q}$, unless at least one of the algebras $\boldsymbol{B}_i$ is isomorphic with $\boldsymbol{A}$. When $\mathbf{Q}$ is clear from context, $\boldsymbol{A}$ is also called *relatively (finitely) subdirectly irreducible*.

$\boldsymbol{A} \in \mathbf{Q}$ is relatively (finitely) subdirectly irreducible iff the identity congruence $\mathbf{0}_A$ is (finitely) meet-irreducible in the lattice $\boldsymbol{Con}_{\mathbf{Q}}(A)$. Since $\boldsymbol{Con}_{\mathbf{Q}}(A)$ is algebraic, every $\mathbf{Q}$-congruence on A is the meet of a family of finitely meet-irreducible $\mathbf{Q}$-congruences.

Theorem 7.1.1. *Let $\mathbf{Q}$ be a quasivariety. Any algebra in $\mathbf{Q}$ is isomorphic with a subdirect product of relatively finitely subdirectly irreducible members of $\mathbf{Q}$.* □

The class of relatively (finitely) subdirectly irreducible algebras of $\mathbf{Q}$ is denoted by $\mathbf{Q}_{\mathrm{RSI}}$ ($\mathbf{Q}_{\mathrm{RFSI}}$, respectively). It is clear that $\mathbf{Q}_{\mathrm{RSI}} \subseteq \mathbf{Q}_{\mathrm{RFSI}}$. Moreover, if $\mathbf{Q}$ is a subquasivariety of a quasivariety $\mathbf{Q}'$, then $\mathbf{Q} \cap \mathbf{Q}'_{\mathrm{RSI}} \subseteq \mathbf{Q}_{\mathrm{RSI}}$ and $\mathbf{Q} \cap \mathbf{Q}'_{\mathrm{RFSI}} \subseteq \mathbf{Q}_{\mathrm{RFSI}}$.

Theorem 7.1.2. *Let $\mathbf{K}$ be any class of algebras. Then $\boldsymbol{Qv}(\mathbf{K})_{\mathrm{RFSI}} \subseteq \boldsymbol{SP}_u(\mathbf{K})$.*

Proof. See, e.g., Czelakowski and Dziobiak (1990). □

The following observation is due to Pigozzi (1988), Lemma 2.1; see also the proof of Proposition 7.2.2 below.

Proposition 7.1.3. *Let $\mathbf{Q}$ be a quasivariety and let $\Delta(x, y, z, w, \underline{u})$ be a possibly infinite set of quaternary equations (with parameters). The following conditions are equivalent:*

J. Czelakowski, *The Equationally-Defined Commutator*,
DOI 10.1007/978-3-319-21200-5_7

(1) *For all* $\mathbf{A} \in \mathbf{Q}$ *and for all* $a, b, c, d \in A$,

$$\Theta^{A}_{\mathbf{Q}}(a,b) \cap \Theta^{A}_{\mathbf{Q}}(c,d) = \Theta^{A}_{\mathbf{Q}}((\forall \underline{e})\ \Delta^{A}(a,b,c,d,\underline{u}));$$

(2) *For all* $\mathbf{A} \in \mathbf{Q}$ *and for all* $a, b, c, d \in A$,

$$\Theta^{A}_{\mathbf{Q}}(a,b) \cap \Theta^{A}_{\mathbf{Q}}(c,d) = \mathbf{0}_A \quad \Leftrightarrow \quad \mathbf{A} \models (\forall \underline{u}) \bigwedge \Delta(x,y,z,w,\underline{u})[a,b,c,d];$$

(3) $\mathbf{Q}_{\mathrm{RFSI}} \models (\forall xyzw)((\forall \underline{u})\ \Delta(x,y,z,w,\underline{u}) \leftrightarrow x \approx y \vee z \approx w)$.

Proof. The proof is omitted. □

A quasivariety **Q** has *(universally) parameterized equationally definable relative principal meets* (*parameterized* EDPM for short) if any (and hence all) of the conditions Proposition 7.1.3 hold for some set $\Delta(x,y,z,w,\underline{u})$ of parameterized quaternary equations.

A class of algebras is *elementary* if it is definable by a (possibly infinite) set of sentences of a first-order language.

The following theorem was independently proved by Pigozzi (1988) and Czelakowski and Dziobiak (1990):

Theorem 7.1.4. *A quasivariety* **Q** *has parameterized EDPM if and only if* **Q** *is relatively congruence-distributive.*

Q *has parameterized EDPM with respect to a finite set* $\Delta(x,y,z,w,\underline{u})$ *if and only if* **Q** *is relatively congruence-distributive and* $\mathbf{Q}_{\mathrm{RFSI}}$ *is an elementary class.* □

If **Q** is relatively congruence-distributive, then each member of $\mathbf{Q}_{\mathrm{RFSI}}$ is finitely subdirectly irreducible in the absolute sense, which means that $\mathbf{Q}_{\mathrm{RFSI}} \subseteq \mathbf{Va}(\mathbf{Q})_{\mathrm{FSI}}$. (In fact, this inclusion holds for any relatively congruence-modular quasivariety **Q**—see Kearnes and McKenzie 1992.)

The following theorem provides a purely syntactic characterization of quasivarieties with EDPM.

Theorem 7.1.5. *Let* **Q** *be a quasivariety. Let* Γ *be a set of proper quasi-equations such that* $\mathbf{Q} = \mathbf{Mod}(Id(\mathbf{Q}) \cup \Gamma)$. *Let* $\Delta(x,y,z,w,\underline{u})$ *be a parameterized set of quaternary equations. The following assertions are equivalent:*

(1) **Q** *has EDPM with respect to* $\Delta(x,y,z,w,\underline{u})$.
(2) *The following are sets of rules of* $\mathbf{Q}^{\models}$:

(2)(a)	$x \approx y / \Delta(x,y,z,w,\underline{u})$	(absorption rules)
(2)(b)	$(\forall \underline{u})\ \Delta(x,y,z,w,\underline{u}) / \Delta(z,w,x,y,\underline{u})$	(commutativity rules)
(2)(c)	$(\forall \underline{u})\ \Delta(x,y,x,y,\underline{u}) / x \approx y$	(idempotency rule).

Moreover, for every proper quasi-equation $\rho : \alpha_1 \approx \beta_1, \ldots, \alpha_n \approx \beta_n \to \alpha \approx \beta$ *belonging to* $\Gamma \cup Birkhoff(\tau)$, *its* Δ*-transform*

(2)(d)$_\rho$ $\quad (\forall \underline{u})\ \Delta(\alpha_1, \beta_1, z_1, w_1, \underline{u}) \cup \ldots \cup (\forall \underline{u})\ \Delta(\alpha_n, \beta_n, z_1, w_1, \underline{u}) / \Delta(\alpha, \beta, z_1, w_1, \underline{u})$

is a set of rules of $\mathbf{Q}^{\models}$. *(Here* z_1 *and* w_1 *are arbitrary but fixed variables not occurring in* ρ.*)*

Note. The rules of (2)(b), the rule (2)(c), and the rules of (2)(d)$_\rho$ are infinite. Since the equational consequence operation $\mathbf{Q}^{\models}$ determined by $\mathbf{Q}$ is finitary, the above rules are subject to the finitarization procedure and can be replaced by sets of finitary rules valid in $\mathbf{Q}$.

If the set $\Delta(x, y, z, w, \underline{u})$ is finite, there are only finitely many rules in (2)(a) and (2)(b). Moreover, for every rule ρ, the set of rules (2)(d)$_\rho$ is finite. □

Open problem. Let $\Delta(x, y, z, w, \underline{u})$ be a set of equations. Give a set of inference rules characterizing the consequence operation $\mathbf{Q}^{\models}$ of the largest quasivariety $\mathbf{Q}$ whose equationally defined commutator is additive and $\Delta(x, y, z, w, \underline{u})$ is a generating set. It is convenient to reformulate syntactic conditions (2)(a), (2)(b) and, for each rule ρ, (2)(d)$_\rho$ uniformly in terms of equational inference rules. The rule (2)(c) is a problem. □

Proof. (1) $\Rightarrow$ (2). Assume (1). We use the facts and notation presented in the proof of the implication (B) $\Rightarrow$ (C) of Theorem 3.5.1. Taking the free algebra $\boldsymbol{F} := \boldsymbol{F}_{\mathbf{Q}}(\omega)$ we see that (1) gives that $\Theta^{\boldsymbol{F}}_{\mathbf{Q}}([x], [y]) \cap \Theta^{\boldsymbol{F}}_{\mathbf{Q}}([z], [w]) = \Theta^{\boldsymbol{F}}_{\mathbf{Q}}((\forall \underline{e})\ \Delta^{\boldsymbol{F}}([x], [y], [z], [w], \underline{e}))$. This readily implies that $\mathbf{Q}^{\models}(\Delta(x, y, z, w, \underline{u})) \subseteq \mathbf{Q}^{\models}(x \approx y)$. Hence (2)(a) holds.

EDPM implies that $\Theta^{\boldsymbol{F}}_{\mathbf{Q}}((\forall \underline{e}) \Delta^{\boldsymbol{F}}([x], [y], [z], [w], \underline{e})) = \Theta^{\boldsymbol{F}}_{\mathbf{Q}}([x], [y]) \cap \Theta^{\boldsymbol{F}}_{\mathbf{Q}}([z], [w]) = \Theta^{\boldsymbol{F}}_{\mathbf{Q}}([z], [w]) \cap \Theta^{\boldsymbol{F}}_{\mathbf{Q}}([x], [y]) = \Theta^{\boldsymbol{F}}_{\mathbf{Q}}((\forall \underline{e})\ \Delta^{\boldsymbol{F}}([z], [w], [x], [y], \underline{e}))$. Hence

$$\mathbf{Q}^{\models}((\forall \underline{u})\ \Delta(x, y, z, w, \underline{u})) = \mathbf{Q}^{\models}((\forall \underline{u})\ \Delta(z, w, x, y, \underline{u})),$$

by Proposition 2.5. From the above equality 2(b) follows.

In view of Theorem 7.1.4, the quasivariety $\mathbf{Q}$ is relatively congruence-distributive. Suppose $\rho : \alpha_1 \approx \beta_1 \wedge \ldots \wedge \alpha_n \approx \beta_n \rightarrow \alpha \approx \beta$ is a quasi-equation of $\mathbf{Q}$. Hence $\alpha \approx \beta \in \mathbf{Q}^{\models}(\alpha_1 \approx \beta_1, \ldots, \alpha_n \approx \beta_n)$. This fact implies (2)(d)$_\rho$. Indeed, we have

$$\Theta^{\boldsymbol{F}}_{\mathbf{Q}}([\alpha], [\beta]) \subseteq \Theta^{\boldsymbol{F}}_{\mathbf{Q}}(\langle[\alpha_1], [\beta_1]\rangle, \ldots, \langle[\alpha_n], [\beta_n]\rangle).$$

It follows that

$$\Theta^{\boldsymbol{F}}_{\mathbf{Q}}([\alpha], [\beta]) \cap \Theta^{\boldsymbol{F}}_{\mathbf{Q}}([\gamma], [\delta]) \subseteq$$
$$\Theta^{\boldsymbol{F}}_{\mathbf{Q}}([\alpha_1], [\beta_1]) \cap \Theta^{\boldsymbol{F}}_{\mathbf{Q}}([\gamma], [\delta]) +_{\mathbf{Q}} \ldots +_{\mathbf{Q}} \Theta^{\boldsymbol{F}}_{\mathbf{Q}}([\alpha_n], [\beta_n]) \cap \Theta^{\boldsymbol{F}}_{\mathbf{Q}}([\gamma], [\delta]),$$

by the distributivity of the lattice $\boldsymbol{Con}_{\mathbf{Q}}(\boldsymbol{F})$. EDPM then gives that

$$\Theta^{\boldsymbol{F}}_{\mathbf{Q}}((\forall \underline{e})\ \Delta^{\boldsymbol{F}}([\alpha],[\beta],[\gamma],[\delta],\underline{e})) \subseteq$$
$$\Theta^{\boldsymbol{F}}_{\mathbf{Q}}((\forall \underline{e})\ \Delta^{\boldsymbol{F}}([\alpha_1],[\beta_1],[\gamma],[\delta],\underline{e})) +_{\mathbf{Q}} \ldots +_{\mathbf{Q}} \Theta^{\boldsymbol{F}}_{\mathbf{Q}}((\forall \underline{e})\ \Delta^{\boldsymbol{F}}([\alpha_n],[\beta_n],[\gamma],[\delta],\underline{e})) =$$
$$\Theta^{\boldsymbol{F}}_{\mathbf{Q}}((\forall \underline{e})\ \Delta^{\boldsymbol{F}}([\alpha_1],[\beta_1],[\gamma],[\delta],\underline{e}) \cup \ldots \cup (\forall \underline{e})\ \Delta^{\boldsymbol{F}}([\alpha_n],[\beta_n],[\gamma],[\delta],\underline{e})).$$

Applying Proposition 2.5 to the above inclusion we get that

$$(\forall \underline{u})\ \Delta(\alpha,\beta,\gamma,\delta,\underline{u}) \subseteq$$
$$\mathbf{Q}^{\models}((\forall \underline{u})\ \Delta(\alpha_1,\beta_1,\gamma,\delta,\underline{u}) \cup \ldots \cup (\forall \underline{u})\ \Delta(\alpha_n,\beta_n,\gamma,\delta,\underline{u}))$$

for every equation $\gamma \approx \delta$. So (2)(d)$_\rho$ holds.

EDPM also implies that

$$\Theta^{\boldsymbol{F}}_{\mathbf{Q}}((\forall \underline{e})\ \Delta^{\boldsymbol{F}}([x],[y],[x],[y],\underline{e})) = \Theta^{\boldsymbol{F}}_{\mathbf{Q}}([x],[y]).$$

Hence

$$\mathbf{Q}^{\models}(x \approx y) = \mathbf{Q}^{\models}((\forall \underline{u})\ \Delta(x,y,x,y,\underline{u})),$$

again by Proposition 2.5. So (2)(c) holds.

(2) ⇒ (1). Assume (2).

Claim 1. (2)(e) $\quad z \approx w/\Delta(x,y,z,w,\underline{u})$

is a set of rules of $\mathbf{Q}^{\models}$.

Proof (of the claim). By (2)(b) and structurality,

$$(\forall \underline{u})\ \Delta(z,w,x,y,\underline{u})/\Delta(x,y,z,w,\underline{u}) \tag{*}$$

are rules of $\mathbf{Q}^{\models}$. In turn, (2)(a) and structurality give that

$$z \approx w/\Delta(z,w,x,y,\underline{u})$$

are rules of $\mathbf{Q}^{\models}$. It follows that

$$z \approx w/(\forall \underline{u})\ \Delta(z,w,x,y,\underline{u}) \tag{**}$$

is a set of rules of $\mathbf{Q}^{\models}$. (*) and (**) imply that

$$z \approx w/\Delta(x,y,z,w,\underline{u})$$

is a set of rules of $\mathbf{Q}^{\models}$ as well. □

Note that (2)(a) and (2)(e) jointly state that $\Delta(x,y,z,w,\underline{u})$ is a set of commutator equations for **Q**.

Claim 2. (2)(e) $(\forall\underline{u})\ \Delta(z,w,x,y,\underline{u})/\Delta(x,y,z,w,\underline{u})$

is a set of rules of $\mathbf{Q}^{\models}$.

Proof (of the claim). This follows from (2)(b) and structurality. □

(2)(b), (2)(e) and structurality imply that Δ satisfies the following equality:

$$\mathbf{Q}^{\models}((\forall\underline{u})\ \Delta(x,y,z,w,\underline{u})) = \mathbf{Q}^{\models}((\forall\underline{u})\ \Delta(z,w,x,y,\underline{u})) \qquad \text{(commutativity)}.$$

Consequently, for any algebra $\boldsymbol{A} \in \mathbf{Q}$ and all $a,b,c,d \in A$,

$$\Theta^{\boldsymbol{A}}_{\mathbf{Q}}((\forall\underline{e})\ \Delta(a,b,c,d,\underline{e})) = \Theta^{\boldsymbol{A}}_{\mathbf{Q}}((\forall\underline{e})\ \Delta(c,d,a,b,\underline{e})).$$

We now pass to the proof that **Q** has EDPM with respect to $\Delta(x,y,z,w,\underline{u})$. Let $\boldsymbol{A}$ be an algebra in **Q** and $a,b,c,d \in A$. The inclusion

$$\Theta^{\boldsymbol{A}}_{\mathbf{Q}}((\forall\underline{e})\ \Delta(a,b,c,d,e)) \subseteq \Theta^{\boldsymbol{A}}_{\mathbf{Q}}(a,b) \cap \Theta^{\boldsymbol{A}}_{\mathbf{Q}}(c,d)$$

holds because (2)(a) and (2)(f) are sets of rules of $\mathbf{Q}^{\models}$.

The proof of the reverse inclusion is harder. Suppose that $\langle e,f\rangle \in \Theta^{\boldsymbol{A}}_{\mathbf{Q}}(a,b) \cap \Theta^{\boldsymbol{A}}_{\mathbf{Q}}(c,d)$. According to Theorem 2.1, there exist finite sequences of pairs of elements of A

$$\langle a_1,b_1\rangle,\ldots,\langle a_m,b_m\rangle \tag{1}$$

and

$$\langle c_1,d_1\rangle,\ldots,\langle c_n,d_n\rangle \tag{2}$$

being **Q**-proofs of $\langle e,f\rangle$ from $\langle a,b\rangle$ and $\langle c,d\rangle$, respectively.

Taking (1) into account, we first inductively prove that

$$\Theta^{\boldsymbol{A}}_{\mathbf{Q}}((\forall\underline{e})\ \Delta(a_i,b_i,c,d,\underline{e})) \subseteq \Theta^{\boldsymbol{A}}_{\mathbf{Q}}((\forall\underline{e})\ \Delta(a,b,c,d,\underline{e})) \tag{3}$$

for $i = 1,2,\ldots,m$.

For $i = 1$, we have that $\langle a_1,b_1\rangle = \langle a,b\rangle$ or $a_1 = b_1$. If $\langle a_1,b_1\rangle = \langle a,b\rangle$, then evidently $(\forall\underline{e})\ \Delta(a_1,b_1,c,d,\underline{e}) \subseteq (\forall\underline{e})\ \Delta(a,b,c,d,\underline{e})$. If $a_1 = b_1$, then $(\forall\underline{e})\ \Delta(a_1,b_1,\ c,d,\underline{e}) = \mathbf{0}_A \subseteq (\forall\underline{e})\ \Delta(a,b,c,d,\underline{e})$, because in view of (2)(a), $x \approx y/\Delta(x,y,z,w,\underline{u})$ is a set of rules of **Q**. Now fix $i > 1$ and suppose that

$$\Theta^{\boldsymbol{A}}_{\mathbf{Q}}((\forall\underline{e})\ \Delta(a_j,b_j,c,d,\underline{e})) \subseteq \Theta^{\boldsymbol{A}}_{\mathbf{Q}}((\forall\underline{e})\ \Delta(a,b,c,d,\underline{e})) \quad \text{for all } j < i. \tag{4}$$

It remains to consider the case when there exist a set $J \subseteq \{1, \ldots, i-1\}$, a quasi-equation $\rho : r_1(\underline{x}) \approx s_1(\underline{x}) \wedge \ldots \wedge r_k(\underline{x}) \approx s_k(\underline{x}) \rightarrow r(\underline{x}) \approx s(\underline{x})$ in $\Gamma \cup Birkhoff(\tau)$, where $\underline{x} = x_1, \ldots, x_p$, and a sequence $\underline{c} = c_1, \ldots, c_p$ such that $\{\langle r_t(\underline{c}), s_t(\underline{c})\rangle : 1 \leqslant t \leqslant k\} = \{\langle a_j, b_j\rangle : j \in J\}$ and $\langle r(\underline{c}), s(\underline{c})\rangle = \langle a_i, b_i\rangle$. These relations together with (4) imply that

$$\Theta^{A}_{\mathbf{Q}}((\forall \underline{e})\ \Delta(r_t(\underline{c}), s_t(\underline{c}), c, d, \underline{e})) \subseteq \Theta^{A}_{\mathbf{Q}}((\forall \underline{e})\ \Delta(a, b, c, d, \underline{e})) \quad \text{for all } t,\ 1 \leqslant t \leqslant k. \tag{5}$$

Since (2)(d)$_\rho$ is a set of rules of $\mathbf{Q}^{\models}$, we get that

$$\Theta^{A}_{\mathbf{Q}}((\forall \underline{e})\ \Delta(r(\underline{c}), s(\underline{c}), c, d, \underline{e})) \subseteq \Theta^{A}_{\mathbf{Q}}(\bigcup\{(\forall \underline{e})\ \Delta(r_t(\underline{c}), s_t(\underline{c}), c, d, \underline{e}) : 1 \leqslant t \leqslant k\}). \tag{6}$$

(5) and (6) imply that

$$\Theta^{A}_{\mathbf{Q}}((\forall \underline{e})\ \Delta(r(\underline{c}), s(\underline{c}), c, d, \underline{e})) \subseteq \Theta^{A}_{\mathbf{Q}}((\forall \underline{e})\ \Delta(a, b, c, d, \underline{e})),$$

i.e.,

$$\Theta^{A}_{\mathbf{Q}}((\forall \underline{e})\ \Delta(a_i, b_i, c, d, \underline{e})) \subseteq \Theta^{A}_{\mathbf{Q}}((\forall \underline{e})\ \Delta(a, b, c, d, \underline{e})),$$

So (3) holds. In particular, for $i = m$, we obtain

$$\Theta^{A}_{\mathbf{Q}}((\forall \underline{e})\ \Delta(e, f, c, d, \underline{e})) \subseteq \Theta^{A}_{\mathbf{Q}}((\forall \underline{e})\ \Delta(a, b, c, d, \underline{e})). \tag{7}$$

Having established (7), we inductively prove that

$$\Theta^{A}_{\mathbf{Q}}((\forall \underline{e})\ \Delta(e, f, c_j, d_j, \underline{e})) \subseteq \Theta^{A}_{\mathbf{Q}}((\forall \underline{e})\ \Delta(e, f, c, d, \underline{e})) \tag{8}$$

for all j, $(1 \leqslant j \leqslant n)$.

For $j = 1$ we have that $\langle c_1, d_1\rangle = \langle c, d\rangle$ or $c_1 = d_1$. If $\langle c_1, d_1\rangle = \langle c, d\rangle$, then evidently $(\forall \underline{e})\ \Delta(e, f, c_1, d_1, \underline{e})\ \Delta(\forall \underline{e})\ \Delta(e, f, c, d, \underline{e})$. If $c_1 = d_1$, then $(\forall \underline{e})\ \Delta(e, f, c_1, d_1, \underline{e}) = \mathbf{0}_A \subseteq (\forall \underline{e})\ \Delta(e, f, c, d, \underline{e})$, because by (2)(e) (Claim 2), $z \approx w \vdash \Delta(x, y, z, w, \underline{u})$ is a set of rules of $\mathbf{Q}^{\models}$. Now fix $j > 1$ and suppose that

$$\Theta^{A}_{\mathbf{Q}}((\forall \underline{e})\ \Delta(e, f, c_i, d_i, \underline{e})) \subseteq \Theta^{A}_{\mathbf{Q}}((\forall \underline{e})\ \Delta(e, f, c, d, \underline{e})) \quad \text{for all } i < j. \tag{9}$$

It remains to consider the case where there exists a set $I \subseteq \{1, \ldots, j-1\}$, a quasi-equation $\rho : r_1(\underline{x}) \approx s_1(\underline{x}) \wedge \ldots \wedge r_k(\underline{x}) \approx s_k(\underline{x}) \rightarrow r(\underline{x}) \approx s(\underline{x})$ in $\Gamma \cup \mathit{Birkhoff}(\tau)$, where $\underline{x} = x_1, \ldots, x_p$, and a sequence $\underline{c} = c_1, \ldots, c_p$ such that $\{\langle r_t(\underline{c}), s_t(\underline{c})\rangle : 1 \leqslant t \leqslant k\} = \{\langle a_i, b_i\rangle : i \in I\}$ and $\langle r(\underline{c}), s(\underline{c})\rangle = \langle a_j, b_j\rangle$. These relations together with (9) imply that

$$\Theta^{A}_{\mathbf{Q}}((\forall \underline{e})\ \Delta(e, f, r_t(\underline{c}), s_t(\underline{c}), \underline{e})) \subseteq \Theta^{A}_{\mathbf{Q}}((\forall \underline{e})\ \Delta(e, f, c, d, \underline{e})) \quad \text{for all } t,\ 1 \leqslant t \leqslant k. \tag{10}$$

But according to (2)(d)$_\rho$

$$(\forall \underline{u})\ \Delta(r_1, s_1, z_1, w_1, \underline{u}) \cup \ldots \cup (\forall \underline{u})\ \Delta(r_k, s_k, z_1, w_1, \underline{u}) \vdash \Delta(r, s, z_1, w_1, \underline{u}) \tag{$*$}$$

is a set of rules of $\mathbf{Q}^{\models}$, where z_1 and w_1 are different variables not occurring in $\underline{x}$. Since

$$\mathbf{Q}^{\models}((\forall \underline{u})\ \Delta(x, y, z, w, \underline{u})) = \mathbf{Q}^{\models}((\forall \underline{u})\ \Delta(z, w, x, y, \underline{u}))$$

(see the remarks following Claim 2), and hence

$$(\forall \underline{u})\ \Delta(x, y, z, w, \underline{u}) / \Delta(z, w, x, y, \underline{u}) \quad \text{and} \quad (\forall \underline{u})\ \Delta(z, w, x, y, \underline{u}) / \Delta(x, y, z, w, \underline{u})$$

are sets of rules of $\mathbf{Q}^{\models}$, it follows from $(*)$ that

$$(\forall \underline{u})\ \Delta(z, w, r_1, s_1, \underline{u}) \cup \ldots \cup (\forall \underline{u})\ \Delta(z, w, r_k, s_k, \underline{u}) / \Delta(z, w, r, s, \underline{u})$$

is a set of rules of $\mathbf{Q}^{\models}$ as well. Consequently,

$$\Theta^{A}_{\mathbf{Q}}((\forall \underline{e})\ \Delta(e, f, r(\underline{c}), s(\underline{c}), \underline{e})) \subseteq \Theta^{A}_{\mathbf{Q}}\Big(\bigcup \{(\forall \underline{e})\ \Delta(e, f, r_t(\underline{c}), s_t(\underline{c}), \underline{e}) : 1 \leqslant t \leqslant k\}\Big). \tag{11}$$

(10) and (11) imply that

$$\Theta^{A}_{\mathbf{Q}}((\forall \underline{e})\ \Delta(e, f, r(\underline{c}), s(\underline{c}), \underline{e})) \subseteq \Theta^{A}_{\mathbf{Q}}((\forall \underline{e})\ \Delta(e, f, c, d, \underline{e}))$$

i.e.,

$$\Theta^{A}_{\mathbf{Q}}((\forall \underline{e})\ \Delta(e, f, c_j, d_j, \underline{e})) \subseteq \Theta^{A}_{\mathbf{Q}}((\forall \underline{e})\ \Delta(e, f, c, d, \underline{e})).$$

So (8) holds. In particular, for $j = n$, we obtain

$$\Theta^A_{\mathbf{Q}}((\forall \underline{e})\ \Delta(e,f,e,f,\underline{e})) \subseteq \Theta^A_{\mathbf{Q}}((\forall \underline{e})\ \Delta(e,f,c,d,\underline{e})).$$

Hence, taking (7) into account, we get that

$$\Theta^A_{\mathbf{Q}}((\forall \underline{e})\ \Delta(e,f,e,f,\underline{e})) \subseteq \Theta^A_{\mathbf{Q}}((\forall \underline{e})\ \Delta(a,b,c,d,\underline{e})). \tag{12}$$

Since (2)(c) is a rule of $\mathbf{Q}^{\models}$, we have that

$$\langle e,f\rangle \in \Theta^A_{\mathbf{Q}}((\forall \underline{e})\ \Delta(e,f,e,f,\underline{e})). \tag{13}$$

(12) and (13) imply that $\langle e,f\rangle \in \Theta^A_{\mathbf{Q}}((\forall \underline{e})\ \Delta(a,b,c,d,\underline{e}))$.

This proves the inclusion $\Theta^A_{\mathbf{Q}}((\forall \underline{e})\ \Delta(a,b,c,d,\underline{e})) \supseteq \Theta^A_{\mathbf{Q}}(a,b) \cap \Theta^A_{\mathbf{Q}}(c,d)$.

It follows that $\Theta^A_{\mathbf{Q}}(a,b) \cap \Theta^A_{\mathbf{Q}}(c,d) = \Theta^A_{\mathbf{Q}}((\forall \underline{e})\ \Delta(a,b,c,d,\underline{e}))$. Since $\mathbf{A}$ is an arbitrary algebra in $\mathbf{Q}$, we see that $\mathbf{Q}$ has EDPM with respect to the set of equations $\Delta(x,y,z,w,\underline{u})$.

The proof of the theorem is completed. □

The following observation is a simple corollary to Theorem 7.1.4:

Theorem 7.1.6. *Let* $\mathbf{Q}$ *be an RCD quasivariety. Then the equational commutator of any two relative congruences coincides with their meet, i.e., for any algebra* $\mathbf{A} \in \mathbf{Q}$ *and for all* $\Phi, \Psi \in Con_{\mathbf{Q}}(\mathbf{A})$, $[\Phi, \Psi]^A = \Phi \cap \Psi$.

Proof. As $\mathbf{Q}$ has EDPM with respect to some set $\Delta(x,y,z,w,\underline{u})$, it follows that $\Delta(x,y,z,w,\underline{u})$ is a set of quaternary commutator equations for $\mathbf{Q}$. Hence, for all $\mathbf{A} \in \mathbf{Q}$ and for all $a,b,c,d \in A$,

$$\Theta^A_{\mathbf{Q}}(a,b) \cap \Theta^A_{\mathbf{Q}}(c,d) = \Theta^A_{\mathbf{Q}}((\forall \underline{e})\ \Delta_A(a,b,c,d,\underline{e})) \subseteq$$
$$[\Theta^A_{\mathbf{Q}}(a,b), \Theta^A_{\mathbf{Q}}(c,d)]^A \subseteq \Theta^A_{\mathbf{Q}}(a,b) \cap \Theta^A_{\mathbf{Q}}(c,d).$$

It follows that

$$[\Theta^A_{\mathbf{Q}}(a,b), \Theta^A_{\mathbf{Q}}(c,d)]^A = \Theta^A_{\mathbf{Q}}(a,b) \cap \Theta^A_{\mathbf{Q}}(c,d).$$

Then, making use of the fact that the lattice $\boldsymbol{Con}_{\mathbf{Q}}(\mathbf{A})$ is algebraic and distributive, we get that $[\Phi, \Psi]^A = \Phi \cap \Psi$, for all $\Phi, \Psi \in Con_{\mathbf{Q}}(\mathbf{A})$. □

Corollary 7.1.7. *Let* $\mathbf{Q}$ *be a quasivariety whose equationally defined commutator is additive.* $\mathbf{Q}$ *is RCD if and only if* $[\Phi, \Psi]^A = \Phi \cap \Psi$ *for any algebra* $\mathbf{A} \in \mathbf{Q}$ *and for all* $\Phi, \Psi \in Con_{\mathbf{Q}}(\mathbf{A})$. □

7.2 Prime Algebras

Let $\mathbf{Q}$ be a quasivariety and $\boldsymbol{A}$ an algebra in $\mathbf{Q}$. Let Φ be a $\mathbf{Q}$-congruence on $\boldsymbol{A}$. Φ is said to be *prime* (in the lattice $\boldsymbol{Con}_{\mathbf{Q}}(\boldsymbol{A})$) if, for any congruences $\Phi_1, \Phi_2 \in Con_{\mathbf{Q}}(\boldsymbol{A})$, $[\Phi_1, \Phi_2]^A \subseteq \Phi$ implies that $\Phi_1 \subseteq \Phi$ or $\Phi_2 \subseteq \Phi$. (Here $[\Phi_1, \Phi_2]^A$ denotes the equationally defined commutator of the congruences Φ_1, Φ_2 in $\boldsymbol{A}$ in the sense of $\mathbf{Q}$.)

$\boldsymbol{A} \in \mathbf{Q}$ is said to be *prime* (in $\mathbf{Q}$) if the identity congruence $\mathbf{0}_A$ is prime in $\boldsymbol{Con}_{\mathbf{Q}}(\boldsymbol{A})$. Thus $\boldsymbol{A}$ is prime in $\mathbf{Q}$ if and only if $[\Phi_1, \Phi_2] = \mathbf{0}_A$ holds for no pair of nonzero congruences $\Phi_1, \Phi_2 \in Con_{\mathbf{Q}}(\boldsymbol{A})$.

$\mathbf{Q}_{\mathrm{PRIME}}$ denotes the class of all prime algebras in $\mathbf{Q}$.

Proposition 7.2.1. *For every quasivariety* $\mathbf{Q}$*,* $\mathbf{Q}_{\mathrm{PRIME}} \subseteq \mathbf{Q}_{\mathrm{RFSI}}$.

Proof. Suppose $\boldsymbol{A} \in \mathbf{Q}_{\mathrm{PRIME}}$ and let $\Theta^A_{\mathbf{Q}}(a, b) \cap \Theta^A_{\mathbf{Q}}(c, d) = \mathbf{0}_A$ for some $a, b, c, d \in A$. Then $[\Theta^A_{\mathbf{Q}}(a, b), \Theta^A_{\mathbf{Q}}(c, d)] = \mathbf{0}_A$. It follows that $a = b$ or $c = d$. So $\boldsymbol{A} \in \mathbf{Q}_{\mathrm{RFSI}}$. □

If $\mathbf{Q}$ is an RCD quasivariety, Theorem 7.1.6 implies that

$$\mathbf{Q}_{\mathrm{PRIME}} = \mathbf{Q}_{\mathrm{RFSI}}.$$

Proposition 7.2.2. *Let* $\mathbf{Q}$ *be quasivariety whose equationally defined commutator is additive. Let* $\Delta(x, y, z, w, \underline{u})$ *be a generating set. Let* $\boldsymbol{A}$ *be an algebra in* $\mathbf{Q}$ *and* $\Phi \in Con_{\mathbf{Q}}(\boldsymbol{A})$*. Then the following conditions are equivalent:*

(i) *Φ is prime in $\boldsymbol{Con}_{\mathbf{Q}}(\boldsymbol{A})$;*
(ii) *For all $a, b, c, d \in A$,*

$$[\Theta^A_{\mathbf{Q}}(a, b), \Theta^A_{\mathbf{Q}}(c, d)] \subseteq \Phi \quad \textit{implies} \quad \langle a, b\rangle \in \Phi \ \textit{or} \ \langle c, d\rangle \in \Phi;$$

(iii) *For all $a, b, c, d \in A$,*

$$\Theta^A_{\mathbf{Q}}((\forall \underline{e})\ \Delta(a, b, c, d, e)) \subseteq \Phi \quad \textit{implies} \quad \langle a, b\rangle \in \Phi \ \textit{or}\ \langle c, d\rangle \in \Phi.$$

Proof. Obviously (i) implies (ii).

(ii) $\Rightarrow$ (iii). Assume (ii). $\Delta(x, y, z, w, \underline{u})$ is a generating set for $\mathbf{Q}$, i.e., $\mathbf{Q}^{\models}((\forall \underline{u})\ \Delta_0(x, y, z, w, \underline{u})) = \mathbf{Q}^{\models}(x \approx y) \cap \mathbf{Q}^{\models}(z \approx w)$. But this implies that $[\Theta^A_{\mathbf{Q}}(a, b), \Theta^A_{\mathbf{Q}}(c, d)]^A = \Theta^A_{\mathbf{Q}}((\forall \underline{e})\ \Delta(a, b, c, d, \underline{e}))$ for all $a, b, c, d \in A$, by Corollary 5.1.3. From (ii) and this observation condition (iii) follows.

(iii) $\Rightarrow$ (i). Assume (iii) holds for Φ and suppose $\Phi_1, \Phi_2 \in Con_{\mathbf{Q}}(\boldsymbol{A})$ so that $[\Phi_1, \Phi_2] \subseteq \Phi$. Let us assume that $\Phi_1 = \Theta^A_{\mathbf{Q}}(X)$ and $\Phi_2 = \Theta^A_{\mathbf{Q}}(Y)$ for some sets X and Y. Then, by additivity, $[\Phi_1, \Phi_2] = \Theta^A_{\mathbf{Q}}(\bigcup\{(\forall \underline{e})\ \Delta^A_0(a, b, c, d, \underline{e}) : \langle a, b\rangle \in X, \langle c, d\rangle \in Y\})$ (see Theorem 5.1.2). It follows that for any $\langle a, b\rangle \in X$, $\langle c, d\rangle \in Y$, it is the case that $\langle a, b\rangle \in \Phi$ or $\langle c, d\rangle \in \Phi$. Hence, $X \subseteq \Phi$ or $Y \subseteq \Phi$. This gives that $\Phi_1 \subseteq \Phi$ or $\Phi_2 \subseteq \Phi$. □

Corollary 7.2.3. *Let* $\mathbf{Q}$ *and* $\Delta(x, y, z, w, \underline{u})$ *be as above. Then for any algebra* $\mathbf{A} \in \mathbf{Q}$ *and any* $\Phi \in Con_{\mathbf{Q}}(\mathbf{A})$, Φ *is prime in* $\mathbf{Con}_{\mathbf{Q}}(\mathbf{A})$ *if and only if the algebra* $\mathbf{A}/\Phi$ *is prime in* $\mathbf{Q}$*. Moreover, the following conditions are equivalent:*

(i) $\mathbf{A}$ *is prime in* $\mathbf{Q}$*;*
(ii) $\mathbf{A} \models (\forall xyzw)((\forall \underline{u}) \bigwedge \Delta(x, y, z, w, \underline{u}) \leftrightarrow x \approx y \vee z \approx w)$. □

The aim of this chapter is to prove the following theorem:

Theorem 7.2.4. *Let* $\mathbf{Q}$ *be a quasivariety whose equationally defined commutator is additive and has a finite generating set. The class* $\boldsymbol{SP}(\mathbf{Q}_{\mathrm{PRIME}})$ *is the largest RCD quasivariety included in* $\mathbf{Q}$*.*

Proof. We have: $\mathbf{Q}_{\mathrm{PRIME}} \subseteq \boldsymbol{SP}(\mathbf{Q}_{\mathrm{PRIME}}) \subseteq \mathbf{Q}$.

Let $\Delta(x, y, z, w, \underline{u})$ be a finite generating set of quaternary commutator equations for $\mathbf{Q}$. As Δ is finite, the sentence $(\forall xyzw)((\forall \underline{u}) \bigwedge \Delta(x, y, x, y, \underline{u}) \leftrightarrow x \approx y)$ is elementary. It then follows from Corollary 7.2.3 that the class $\mathbf{Q}_{\mathrm{PRIME}}$ is closed under the formation of ultraproducts. Hence

$$\mathbf{Q}^* := \boldsymbol{SP}(\mathbf{Q}_{\mathrm{PRIME}})$$

is a quasivariety included in $\mathbf{Q}$.

Lemma 7.2.5. $\mathbf{Q}^*$ *has EDPM with respect to* $\Delta(x, y, z, w, \underline{u})$ *and hence it is an RCD quasivariety.*

Proof (of the lemma). We shall apply Theorem 7.1.5 to $\mathbf{Q}^*$.

Claim 1. (2)(a) *is a set of rules of* $\mathbf{Q}^{*\models}$.

Proof (of the claim). Since $\Delta(x, y, z, w, \underline{u})$ is a set of quaternary commutator equations for $\mathbf{Q}$, it follows that $x \approx y/\Delta(x, y, z, w, \underline{u})$ is a set of rules of $\mathbf{Q}^{\models}$. Consequently, $x \approx y/\Delta(x, y, z, w, \underline{u})$ is a set of rules of $\mathbf{Q}^{*\models}$, because $\mathbf{Q}^* \subseteq \mathbf{Q}$. □

Claim 2. (2)(b) *is a set of rules of* $\mathbf{Q}^*$.

Proof (of the claim). Since $\Delta(x, y, z, w, \underline{u})$ is a generating set of equations for the equationally defined commutator of $\mathbf{Q}$, it follows that $\mathbf{Q}^{\models}((\forall \underline{u})\, \Delta(x, y, z, w, \underline{u})) = \mathbf{Q}^{\models}((\forall \underline{u})\, \Delta(z, w, x, y, \underline{u}))$—see Corollary 5.1.3. Consequently, $(\forall \underline{u})\, \Delta(x, y, z, w, \underline{u})/\Delta(z, w, x, y, \underline{u})$ is a set of rules of $\mathbf{Q}^{\models}$. Hence it is a set of rules of $\mathbf{Q}^{*\models}$. □

Claim 3. (2)(c) *is a rule of* $\mathbf{Q}^{*\models}$.

Proof (of the claim). According to Corollary 7.2.3.(iii), the class $\mathbf{Q}_{\mathrm{PRIME}}$ validates the universal-existential first-order sentence

$$(\forall xy)((\forall \underline{u}) \bigwedge \Delta(x, y, x, y, \underline{u}) \rightarrow x \approx y). \qquad (*)$$

As $\mathbf{Q}_{\mathrm{PRIME}}$ is closed under ultraproducts, a straightforward ultraproduct argument shows that there exists a finite set of finite sequences of terms $\underline{t}_1, \ldots, \underline{t}_n$ (all of the same length) such that the class $\mathbf{Q}_{\mathrm{PRIME}}$ validates the quasi-identity

$$\sigma: \quad (\forall xy)(\forall \underline{u})(\Delta(x,y,x,y,\underline{t}_1) \wedge \ldots \wedge \Delta(x,y,x,y,\underline{t}_n) \rightarrow x \approx y).$$

Hence σ is valid in the quasivariety $\mathbf{Q}^* = \boldsymbol{SP}(\mathbf{Q}_{\mathrm{PRIME}})$. Consequently,

$$\Delta(x,y,x,y,\underline{t}_1) \cup \ldots \cup \Delta(x,y,x,y,\underline{t}_n)/x \approx y.$$

is a rule of $\mathbf{Q}^{*\models}$ and hence

$$(\forall \underline{u})\ \Delta(x,y,x,y,\underline{u})/x \approx y$$

is a rule of $\mathbf{Q}^{*\models}$ as well.

(Yet another justification of the claim runs as follows. Since the sentence $(*)$ is valid in $\mathbf{Q}_{\mathrm{PRIME}}$, $(\forall \underline{u})(\Delta(x,y,x,y,\underline{u})/x \approx y$ is a rule of the consequence operation $\mathbf{Q}^{\models}_{\mathrm{PRIME}}$. As $\mathbf{Q}_{\mathrm{PRIME}}$ is closed with respect to $\boldsymbol{P_u}$, the consequence operation $\mathbf{Q}^{\models}_{\mathrm{PRIME}}$ is finitary. Hence there exists a finite subset $\Sigma_f \subset (\forall \underline{u})\ \Delta(x,y,x,y,\underline{u})$ such that $\Sigma_f/x \approx y$ is a rule of $\mathbf{Q}^{\models}_{\mathrm{PRIME}}$. It follows that $\Sigma_f/x \approx y$ is also a rule of $\boldsymbol{SP}(\mathbf{Q}_{\mathrm{PRIME}})^{\models} = \mathbf{Q}^{*\models}$. Consequently, $(\forall \underline{u})\ \Delta(x,y,x,y,\underline{u})/x \approx y$ is a rule of $\mathbf{Q}^{*\models}$.)

□

Claim 4. *For every quasi-equation ρ of $\mathbf{Q}^*$, $(2)(\mathrm{d})_\rho$ is a set of rules of $\mathbf{Q}^{*\models}$.*

Proof (of the claim). Suppose that a quasi-equation $\rho : r_1 \approx s_1 \wedge \ldots \wedge r_k \approx s_k \rightarrow r \approx s$ is valid in $\mathbf{Q}^{*\models}$. We must show that

$$(\forall \underline{u})\ \Delta(r_1,s_1,z,w,\underline{u}) \cup \ldots \cup (\forall \underline{u})\ \Delta(r_k,s_k,z,w,\underline{u})/\Delta(r,s,z,w,\underline{u}) \qquad (2)(\mathrm{d})_\rho$$

is a set of rules of $\mathbf{Q}^{*\models}$, where z, w are different variables not occurring in ρ. It suffices to show that the rules of $(2)(\mathrm{d})_\rho$ are validated in the algebras of $\mathbf{Q}_{\mathrm{PRIME}}$, because, having established this fact, arguing as in the proof of Claim 3 one shows that $(2)(\mathrm{d})_\rho$ is validated in $\mathbf{Q}^*$.

Let $\boldsymbol{A} \in \mathbf{Q}_{\mathrm{PRIME}}$ and suppose $\underline{x} = x_1, \ldots, x_p$ is a sequence of variables occurring in the equations of r. Let $\underline{c} = c_1, \ldots, c_p$ be a sequence of elements of A and let $c, d \in A$ so that

$$\boldsymbol{A} \models (\forall \underline{u})\ \Delta(r_j,s_j,z,w,\underline{u})[\underline{c},c,d] \qquad \text{for } j = 1,\ldots,k.$$

This means that

$(\forall \underline{e})\ \Delta^{\boldsymbol{A}}(r_1(\underline{c}),s_1(\underline{c}),c,d,\underline{e}) \cup \ldots \cup (\forall \underline{e})\ \Delta(r_k(\underline{c}),s_k(\underline{c}),c,d,\underline{e})$ is a subset of $\mathbf{0}_A$.

But Corollary 7.2.3.(iii) implies that for every j $(1 \leqslant j \leqslant k)$, either $r_j(\underline{c}) = s_j(\underline{c})$ or $c = d$. Hence $c = d$ or $r_j(\underline{c}) = s_j(\underline{c})$ for all j $(1 \leqslant j \leqslant k)$. If $c = d$, then taking into account the fact that $z \approx w/\Delta(x,y,z,w,\underline{u})$ is a set of rules of $\mathbf{Q}^{*\models}$, we have that $(\forall \underline{e})\ \Delta(r(\underline{c}),s(\underline{c}),c,d,\underline{e})$ is a subset of $\mathbf{0}_A$. If $r_j(\underline{c}) = s_j(\underline{c})$ for all j $(1 \leqslant j \leqslant k)$, then $r(\underline{c}) = s(\underline{c})$, because ρ is a rule of $\mathbf{Q}^{*\models}$. It follows that $(\forall \underline{e})\ \Delta(r(\underline{c}),s(\underline{c}),c,d,\underline{e})$

is also a subset of $\mathbf{0}_A$, because $x \approx y/\Delta(x, y, z, w, \underline{u})$ is a set of rules of $\mathbf{Q}^{*\models}$. This proves the claim. □

The above claims and Theorem 7.1.5 imply that $\mathbf{Q}^*$ has EDPM with respect to $\Delta(x, y, x, y, \underline{u})$ and hence $\mathbf{Q}^*$ is an RCD quasivariety. This proves the lemma. □

Lemma 7.2.6. $\mathbf{Q}^*_{\mathrm{RFSI}} = \mathbf{Q}_{\mathrm{PRIME}}$.

Proof (of the lemma). (⊆). As $\mathbf{Q}^*$ has EDPM with respect to $\Delta(x, y, x, y, \underline{u})$, Proposition 7.1.3 implies that for any algebra $\boldsymbol{A} \in \mathbf{Q}^*$,

$$\boldsymbol{A} \in \mathbf{Q}^*_{\mathrm{RFSI}} \quad \Leftrightarrow$$

$$\boldsymbol{A} \models (\forall xyzw)((\forall \underline{u}) \bigwedge \Delta(x, y, z, w, \underline{u}) \leftrightarrow x \approx y \vee z \approx w). \qquad (*)$$

Hence, if $\boldsymbol{A} \in \mathbf{Q}^*_{\mathrm{RFSI}}$, then $\boldsymbol{A} \in \mathbf{Q}$ and $(*)$ holds. So $\boldsymbol{A} \in \mathbf{Q}_{\mathrm{PRIME}}$, by Corollary 7.2.3.
(⊇). Suppose $\boldsymbol{A} \in \mathbf{Q}_{\mathrm{PRIME}}$. As $\mathbf{Q}_{\mathrm{PRIME}} \subseteq \mathbf{Q}^*$, we have that $\boldsymbol{A} \in \mathbf{Q}^*$. But

$$\boldsymbol{A} \models (\forall xyzw)((\forall \underline{u}) \bigwedge \Delta(x, y, z, w, \underline{u}) \leftrightarrow x \approx y \vee z \approx w).$$

As $\mathbf{Q}^*$ has EDPM with respect to $\Delta(x, y, z, w, \underline{u})$, we get that $\boldsymbol{A} \in \mathbf{Q}^*_{\mathrm{RFSI}}$, by Proposition 7.1.3. □

Lemma 7.2.7. *The class* $\boldsymbol{SP}(\mathbf{Q}_{\mathrm{PRIME}})$ *is a unique RCD quasivariety* $\mathbf{Q}'$ *with the following properties:* $\mathbf{Q}_{\mathrm{PRIME}} \subseteq \mathbf{Q}' \subseteq \mathbf{Q}$ *and* $\mathbf{Q}'$ *has EDPM with respect to* $\Delta(x, y, z, w, \underline{u})$. *Consequently,* $\mathbf{Q}'_{\mathrm{RFSI}} = \mathbf{Q}_{\mathrm{PRIME}}$.

Proof (of the lemma). Suppose $\mathbf{Q}'$ is a quasivariety such that $\mathbf{Q}_{\mathrm{PRIME}} \subseteq \mathbf{Q}' \subseteq \mathbf{Q}$ and $\mathbf{Q}'$ has EDPM with respect to $\Delta(x, y, z, w, \underline{u})$. Since $\mathbf{Q}'$ is an RCD quasivariety, we have that $\mathbf{Q}'_{\mathrm{RFSI}} = \mathbf{Q}'_{\mathrm{PRIME}}$.

Claim. $\mathbf{Q}'_{\mathrm{PRIME}} \subseteq \mathbf{Q}_{\mathrm{PRIME}}$.

Proof (of the claim). Suppose $\boldsymbol{A} \in \mathbf{Q}'_{\mathrm{PRIME}}$. So $\boldsymbol{A} \in \mathbf{Q}'_{\mathrm{RFSI}}$. Hence $\boldsymbol{A} \in \mathbf{Q}' \subseteq \mathbf{Q}$ and $\boldsymbol{A}$ validates the sentence $(\forall xyzw)((\forall \underline{u}) \bigwedge \Delta(x, y, z, w, \underline{u}) \leftrightarrow x \approx y \vee z \approx w)$. Hence $\boldsymbol{A} \in \mathbf{Q}_{\mathrm{PRIME}}$. □

We then have: $\mathbf{Q}' = \boldsymbol{SP}(\mathbf{Q}'_{\mathrm{RFSI}}) = \boldsymbol{SP}(\mathbf{Q}'_{\mathrm{PRIME}}) \subseteq \boldsymbol{SP}(\mathbf{Q}_{\mathrm{PRIME}})$. On the other hand, as $\mathbf{Q}_{\mathrm{PRIME}} \subseteq \mathbf{Q}'$, we also have that $\boldsymbol{SP}(\mathbf{Q}_{\mathrm{PRIME}}) \subseteq \boldsymbol{SP}(\mathbf{Q}') = \mathbf{Q}'$. So

$$\mathbf{Q}' = \boldsymbol{SP}(\mathbf{Q}_{\mathrm{PRIME}}).$$

As $\mathbf{Q}'$ coincides with $\boldsymbol{SP}(\mathbf{Q}_{\mathrm{PRIME}}) = \mathbf{Q}^*$, we get that $\mathbf{Q}'_{\mathrm{RFSI}} = \mathbf{Q}^*_{\mathrm{RFSI}} = \mathbf{Q}_{\mathrm{PRIME}}$, by Lemma 7.2.6. □

The class $\mathbf{Q}_{\mathrm{PRIME}}$ does not determine the quasivariety $\mathbf{Q}$ in an unambiguous way. In other words, there exist *different* quasivarieties whose equationally defined commutators are additive (and are determined by the same set of quaternary equations with parameters) such that the corresponding classes of prime algebras coincide. For let $\mathbf{Q}$ be any RCM quasivariety which is not relatively congruence-distributive and whose equationally defined commutator is determined by some finite set $\Delta(x, y, z, w, \underline{u})$ of quaternary commutator equations (see Theorem 8.1.1). It follows from Lemma 7.2.7 that the (distinct) quasivarieties $\mathbf{Q}$ and $\boldsymbol{SP}(\mathbf{Q}_{\mathrm{PRIME}})$ have the same classes of prime algebras.

The following lemma is crucial:

Lemma 7.2.8. *Let* $\mathbf{Q}$ *be a quasivariety whose equationally defined commutator additive and has a finite generating set* $\Delta(x, y, z, w, \underline{u})$. *Let* $\mathbf{Q}'$ *be a quasivariety included in* $\mathbf{Q}$. *Then the following conditions are equivalent:*

(1) $\mathbf{Q}'$ *is RCD.*
(2) $\mathbf{Q}'$ *has EDPM with respect to* $\Delta(x, y, z, w, \underline{u})$.
(3) $\mathbf{Q}'_{\mathrm{RFSI}} \subseteq \mathbf{Q}_{\mathrm{PRIME}}$.

Proof (of the lemma). (3) $\Rightarrow$ (1). Suppose $\mathbf{Q}'_{\mathrm{RFSI}} \subseteq \mathbf{Q}_{\mathrm{PRIME}}$. Then evidently $\mathbf{Q}' = \boldsymbol{SP}(\mathbf{Q}'_{\mathrm{RFSI}}) \subseteq \boldsymbol{SP}(\mathbf{Q}_{\mathrm{PRIME}})$. But $\mathbf{Q}_{\mathrm{PRIME}}$ and hence also the class $\mathbf{Q}'_{\mathrm{RFSI}}$ satisfy the universal-existential sentence $(\forall xyzw)((\forall \underline{u}) \bigwedge \Delta(x, y, z, w, \underline{u}) \leftrightarrow x \approx y \vee z \approx w)$. Proposition 7.1.3 then implies that $\mathbf{Q}'$ has EDPM with respect to $\Delta(x, y, x, y, \underline{u})$. Hence $\mathbf{Q}'$ is RCD.

(1) $\Rightarrow$ (2). Suppose $\mathbf{Q}'$ is RCD. As $\mathbf{Q}'$ satisfies the Extension Principle (see Note 3.4.2.(4) and Section 6.2) and $\mathbf{Q}' \subseteq \mathbf{Q}$, we have that for any algebra $\boldsymbol{A} \in \mathbf{Q}'$ and all $a, b, c, d \in A$,

$$\begin{gathered}
\Theta^{\boldsymbol{A}}_{\mathbf{Q}'}(a, b) \cap \Theta^{\boldsymbol{A}}_{\mathbf{Q}'}(c, d) = \\
\Theta^{\boldsymbol{A}}_{\mathbf{Q}'}(\Theta^{\boldsymbol{A}}_{\mathbf{Q}}(a, b)) \cap \Theta^{\boldsymbol{A}}_{\mathbf{Q}'}(\Theta^{\boldsymbol{A}}_{\mathbf{Q}}(c, d)) = \\
\Theta^{\boldsymbol{A}}_{\mathbf{Q}'}(\Theta^{\boldsymbol{A}}_{\mathbf{Q}}(a, b) \cap \Theta^{\boldsymbol{A}}_{\mathbf{Q}}(c, d)) = \\
\Theta^{\boldsymbol{A}}_{\mathbf{Q}'}(\Theta^{\boldsymbol{A}}_{\mathbf{Q}}((\forall \underline{e})\ \Delta(a, b, c, d, \underline{e}))) = \\
\Theta^{\boldsymbol{A}}_{\mathbf{Q}'}((\forall \underline{e})\ \Delta(a, b, c, d, \underline{e})).
\end{gathered}$$

This shows that $\mathbf{Q}'$ has EDPM with respect to $\Delta(x, y, x, y, \underline{u})$.

(2) $\Rightarrow$ (3). Assume (2). Applying Proposition 7.1.3 to $\mathbf{Q}'$, we obtain that $\mathbf{Q}'_{\mathrm{RFSI}}$ validates the sentence $(\forall xyzw)((\forall \underline{u}) \bigwedge \Delta(x, y, z, w, \underline{u}) \leftrightarrow x \approx y \vee z \approx w)$. As $\mathbf{Q}'_{\mathrm{RFSI}} \subseteq \mathbf{Q}$, it follows that $\mathbf{Q}'_{\mathrm{RFSI}} \subseteq \mathbf{Q}_{\mathrm{PRIME}}$. □

We can now conclude the proof of Theorem 7.2.4.

In view of Lemma 7.2.5, $\boldsymbol{SP}(\mathbf{Q}_{\mathrm{PRIME}})$ is an RCD quasivariety included in $\mathbf{Q}$.

Now suppose $\mathbf{Q}'$ is an RCD quasivariety such that $\mathbf{Q}' \subseteq \mathbf{Q}$. Lemma 7.2.8 implies that $\mathbf{Q}'_{\mathrm{RFSI}} \subseteq \mathbf{Q}_{\mathrm{PRIME}}$. It follows that $\mathbf{Q}' = \boldsymbol{SP}(\mathbf{Q}'_{\mathrm{RFSI}}) \subseteq \boldsymbol{SP}(\mathbf{Q}_{\mathrm{PRIME}})$.

The theorem has been proved. □

Theorem 7.1.5 provides a syntactic characterization of the equational logic $\mathbf{Q}^{\models}$ associated with any quasivariety $\mathbf{Q}$ with EDPM. Thus for $\mathbf{Q}$ to have EDPM with respect to $\Delta(x, y, z, w, \underline{u})$, it is necessary that for any standard rule $r : \alpha_1 \approx \beta_1, \ldots, \alpha_n \approx \beta_n / \alpha \approx \beta$ of $\mathbf{Q}^{\models}$, the Δ-transform

$$(\forall \underline{u})\ \Delta(\alpha_1, \beta_1, z_1, w_1, \underline{u}) \cup \ldots \cup (\forall \underline{u})\ \Delta(\alpha_n, \beta_n, z_1, w_1, \underline{u}) / \Delta(x, y, z_1, w_1, \underline{u}) \qquad (*)$$

be a set of rules of $\mathbf{Q}^{\models}$ as well. (Here z_1, w_1 are arbitrary but fixed variables not occurring in the above scheme of r.) The rules of (*) are infinitistic when Δ is infinite or the set of parametric variables $\underline{u}$ is non-empty. But after applying the finitarization procedure with respect to $\mathbf{Q}^{\models}$, we see that (*) yields a (possibly infinite) set of standard rules of $\mathbf{Q}^{\models}$. Let us denote this set by $\Delta(r)$. But again, the Δ-transform of each rule of $\Delta(r)$ defines a new set of rules of $\mathbf{Q}^{\models}$. The finitarization procedure, when applied to the Δ-transforms of the rules $\Delta(r)$, produces a family of sets of standard rules of $\mathbf{Q}^{\models}$ which we shall denote by $\Delta^2(r)$. Continuing this pattern, that is, applying the Δ-transforms and then the finitarization procedure, we define an infinite sequence $\Delta^n(r)$ of families of sets of standard rules of $\mathbf{Q}^{\models}$. The question arises: Is it necessary to iterate the above procedure infinitely many times? Does the above procedure terminate in finitely many steps in the sense that the rules obtained at some level n suffice to generate the rules of the level $n + 1$? The answer is: Yes—even one step suffices. This is due to the fact that the system $\mathbf{Q}^{\models}$ associated with $\mathbf{Q}$ with EDPC validates certain rules determined by Δ which are called associativity rules for $\mathbf{Q}^{\models}$. We shall briefly elucidate this issue (but without going into details), because it is one of crucial points in the proof of the finite basis theorem for finitely generated RCD quasivarieties due to Pigozzi (1988) (see also Czelakowski and Dziobiak 1990).

What are associativity rules? For didactic reasons, we shall first present them on the level of propositional logic. Let $\boldsymbol{L}$ be a propositional language endowed with one binary connective $\vee$ (and possibly some other connectives). The *associativity rule* for $\vee$ is the two-sided one-premiss rule of inference $(p \vee q) \vee r // p \vee (q \vee r)$ in $\boldsymbol{L}$ (applied in both directions), where p, q, r are propositional variables. This rule is obviously valid in classical and intuitionistic propositional logics for $\vee$ being the disjunction connective.

Let $r\ :\ \alpha_1, \ldots, \alpha_n/\alpha$ be any (proper) rule of inference in $\boldsymbol{L}$. The (right) $\vee$-*transform* of r is the rule $r\vee : \alpha_1 \vee p, \ldots, \alpha_n \vee p/\alpha \vee p$, where p is a variable not occurring in the formulas $\alpha_1, \ldots, \alpha_n, \alpha$. In turn, the $\vee$-transform of $r\vee$ is the rule $r\vee\vee : (\alpha_1 \vee p) \vee q, \ldots, (\alpha_n \vee p) \vee q/(\alpha \vee p) \vee q$, where q is a fresh variable not occurring in the formulas $\alpha_1 \vee p, \ldots, \alpha_n \vee p, \alpha \vee p$. Continuing, we define $\vee$-transforms of r of higher ranks. But in the presence of the associativity rule for $\vee$, the rule $r\vee\vee$ (as well as the other successive $\vee$-transforms of r) is already derivable from r and $r\vee$. For we have: from the formulas $(\alpha_1 \vee p) \vee q, \ldots, (\alpha_n \vee p) \vee q$ we derive the formulas $\alpha_1 \vee (p \vee q), \ldots, \alpha_n \vee (p \vee q)$, by associativity. Then, using the transform $r\vee$ and making a suitable substitution, from the last set of formulas we derive the single formula $\alpha \vee (p \vee q)$. Applying again the associativity rule (but in the reverse order) to $\alpha \vee (p \vee q)$, we derive $(\alpha \vee p) \vee q$. Thus, in logical terms, $r\vee\vee$ is a secondary rule of the propositional logical system based on r, $r\vee$ and the associativity rule for $\vee$.

The above remarks can be appropriately lifted to the level of equational logics and applied to the system $\mathbf{Q}^{\models}$ associated with any quasivariety $\mathbf{Q}$ that has EDPC.

But the role of $\vee$ is played there by any set of equations $\Delta(x, y, z, w, \underline{u})$ which defines EDPC for **Q**. A detailed elaboration of this idea is presented in Czelakowski and Dziobiak (1990) (in case when Δ does not involve parameters) and in the seminal Pigozzi's paper [1988] in the general case. To simplify our narration, we shall consider only finite sets $\Delta(x, y, z, w, \underline{u})$.

Let $\Delta(x, y, z, w, \underline{u})$ be a finite set of quaternary equations in x, y, z, w and in some parametric variables $\underline{u}$. Let **Q** be a quasivariety with EDPM with respect to $\Delta(x, y, z, w, \underline{u})$. By the *associativity rules* determined by $\Delta(x, y, z, w, \underline{u})$ we shall mean the following two sets of rules:

$$\bigcup\{(\forall\underline{u})\ \Delta(\alpha, \beta, z_2, w_2, \underline{u}) : \alpha \approx \beta \in (\forall\underline{u})\ \Delta(x, y, z_1, w_1, \underline{u})\}/$$
$$\bigcup\{\Delta(x, y, \alpha, \beta, \underline{u}_1) : \alpha \approx \beta \in \Delta(z_1, w_1, z_2, w_2, \underline{u}_2)\} \quad (1)$$

and

$$\bigcup\{(\forall\underline{u})\ \Delta(x, y, \alpha, \beta, \underline{u}) : \alpha \approx \beta \in (\forall\underline{u})\ \Delta(z_1, w_1, z_2, w_2, \underline{u})\}/$$
$$\bigcup\{\Delta(\alpha, \beta, z_2, w_2, \underline{u}_1) : \alpha \approx \beta \in \Delta(x, y, z_1, w_1, \underline{u}_2)\}, \quad (2)$$

where $\underline{u}_1$ and $\underline{u}_1$ are separated k-element sets of variables different form x, y, z, w, and k is the cardinality of $\underline{u}$. Here $(\forall\underline{u})\Delta(\alpha, \beta, z_2, w_2, \underline{u})$ abbreviates $\bigcup\{\Delta(x/\alpha, y/\beta, z/z_2, w/w_2, \underline{u}/\underline{t}) : \underline{t}$ is a sequence of terms whose length is $k\}$. (Note that the sets (1) and (2) are finite because Δ is finite.)

The crucial fact is that if a quasivariety **Q** has EDPM with respect to Δ, then (1) and (2) are rules of $\mathbf{Q}^{\models}$ (see Pigozzi 1988). This directly follows from Proposition 7.2.2 and the fact that finitely meet-irreducible **Q**-congruences in the algebras of **Q** coincide with prime congruences.

The above rules are infinitistic if Δ involves parameters $\underline{u}$, but they are replaced by their finitistic variants for $\mathbf{Q}^{\models}$ by applying the above finitarization procedures. The so modified rules are called *finitary associativity rules* for $\mathbf{Q}^{\models}$ determined by Δ. They are of course standard rules of $\mathbf{Q}^{\models}$. The above standard rules can be equivalently replaced by two sets of quasi-equations.

The following observation supplements Theorem 7.2.4. Let **Q** be a quasivariety whose equationally defined commutator is additive. Suppose that **Q** has a finite generating set $\Delta = \Delta(x, y, z, w, \underline{u})$ of quaternary commutator equations of **Q** with k being the length of $\underline{u}$. (Thus $\mathbf{Q}^{\models}((\forall\underline{u})\ \Delta(x, y, x, y, \underline{u})) = \mathbf{Q}^{\models}(x \approx y) \cap \mathbf{Q}^{\models}(z \approx w)$.) The quasivariety $\boldsymbol{SP}(\mathbf{Q}_{\mathrm{PRIME}})$ is RCD and it has EDPM with respect to Δ (Lemma 7.2.5).

Let R_0 be an inferential base for the equational logic $\mathbf{Q}^{\models}$, i.e., $\mathbf{Q}^{\models} = C^{eq}_{R_0}$. The absorption rules

$$x \approx y/\Delta(x, y, z, w, \underline{u})$$

and the following finitarizations (with respect to $\mathbf{Q}^{\models}$) of the commutativity rules

$$\Delta(x, y, z, w, \underline{T})/\Delta(z, w, x, y, \underline{u}),$$

where $\underline{T}$ is a finite set of k-tuples of terms, are all secondary rules in the system $C^{eq}_{R_0}$, i.e., they all are provable by means of R_0. The same remark applies to the Birkhoff's rules *Birkhoff*(τ) and the axiomatic rules $Id(\mathbf{Q})$. Moreover, Theorem 5.2.3.(2).(ii) implies that for every proper rule

$$\rho : \alpha_1 \approx \beta_1, \ldots, \alpha_n \approx \beta_n \rightarrow \alpha \approx \beta$$

of the system $C^{eq}_{R_0}$,

$$\Delta(\alpha_1, \beta_1, z, w, \underline{T}) \cup \ldots \cup \Delta(\alpha_n, \beta_n, z, w, \underline{T})/\Delta(\alpha, \beta, z, w, \underline{u})$$

is a set of standard rules of $C^{eq}_{R_0}$ as well, for some set $\underline{T}$ (depending on the rule ρ). (Here z and w are arbitrary variables not occurring in $\alpha_1, \beta_1, \ldots, \alpha_n, \beta_n, \alpha, \beta$.) All these rules are also secondary in $C^{eq}_{R_0}$.

Summing up, as the equationally defined commutator for $\mathbf{Q}$ is additive and Δ is a finite generating set, it follows from Theorem 5.2.3.(2).(ii) that the finitarizations of the Δ-transform of the rules of $\mathbf{Q}^{\models}$ (with respect to $\mathbf{Q}^{\models}$) are already rules of $\mathbf{Q}^{\models}$ and hence they are secondary rules with respect to R_0 (because R_0 is a base of $\mathbf{Q}^{\models}$). Hence all these rules are also rules of the stronger logic $\boldsymbol{SP}(\mathbf{Q}_{\mathrm{PRIME}})^{\models}$.

In view of Corollary 7.2.3, the infinite idempotency rule $(\forall \underline{u})\Delta(x, y, x, y, \underline{u})/ x \approx y$ is a rule of $\boldsymbol{SP}(\mathbf{Q}_{\mathrm{PRIME}})^{\models}$. Since the equational logic $\boldsymbol{SP}(\mathbf{Q}_{\mathrm{PRIME}})^{\models}$ is finitary, there exists a finite set $\underline{T}$ of k-tuples of terms such that $\Delta(x, y, x, y, \underline{T})/x \approx y$ is a standard rule of $\boldsymbol{SP}(\mathbf{Q}_{\mathrm{PRIME}})^{\models}$. (Formally, $\Delta(x, y, x, y, \underline{T})/x \approx y$ is the rule $\bigcup\{\Delta(x, y, x, y, \underline{t}) : \underline{t} \in \underline{T}\}/x \approx y$.) Let us denote the last rule by r_{idem}. We call r_{idem} the *idempotency rule*. Thus r_{idem} is a finitarization of $(\forall \underline{u})\ \Delta(x, y, x, y, \underline{u})/x \approx y$ with respect to $\boldsymbol{SP}(\mathbf{Q}_{\mathrm{PRIME}})^{\models}$. Equivalently, the quasi-identity σ_{idem} corresponding to r_{idem} is valid in the quasivariety $\boldsymbol{SP}(\mathbf{Q}_{\mathrm{PRIME}})$.

According to Theorem 7.1.5, the Δ-transform of r_{idem} produces rules of $\boldsymbol{SP}(\mathbf{Q}_{\mathrm{PRIME}})^{\models}$. We take the set of rules $\Delta_{fin}(r_{idem})$ being finitarizations (with respect to $\boldsymbol{SP}(\mathbf{Q}_{\mathrm{PRIME}})^{\models}$) of the rules of $\Delta(r_{idem})$. We also take the finitary associativity rules for $\boldsymbol{SP}(\mathbf{Q}_{\mathrm{PRIME}})^{\models}$ determined by Δ together with finitarizations of their Δ-transforms with respect to $\boldsymbol{SP}(\mathbf{Q}_{\mathrm{PRIME}})^{\models}$. The resulting set of standard rules is denoted by R_1. (Note that if Δ is finite, R_1 is finite as well.)

Let $R := R_0 \cup R_1$. Let $\mathbf{Q}'$ be the quasivariety axiomatized by the quasi-equations corresponding to the rules of R. The rules of R form a base of the consequence $\mathbf{Q}'^{\models}$, i.e., $\mathbf{Q}'^{\models} = C^{eq}_R$. As R is a base for $\mathbf{Q}'^{\models}$, the identities belonging to $Id(\mathbf{Q}')$ are all provable in the system C^{eq}_R.

Due to the presence of the above finitary associativity rules in R_1, the iterations of Δ-transform of *arbitrary* rank of the rules of R produce rules of C^{eq}_R. Thus, in view of the above remarks, condition (2) of Theorem 7.1.5 is satisfied for the system C^{eq}_R.

It follows from this theorem that $\mathbf{Q}'$ has EDPM with respect to $\Delta(x, y, z, w, \underline{u})$. Thus $\mathbf{Q}'$ is a relatively congruence-distributive quasivariety included in $\mathbf{Q}$.

We thus arrive at the following theorem:

Theorem 7.2.9. *Let* $\mathbf{Q}$ *be a quasivariety whose equationally defined commutator is additive. Suppose that there exists a finite generating set* $\Delta = \Delta(x, y, z, w, \underline{u})$ *of quaternary commutator equations of* $\mathbf{Q}$. *The quasivariety* $\boldsymbol{SP}(\mathbf{Q}_{\mathrm{PRIME}})$ *is axiomatized relative to* $\mathbf{Q}$ *by the quasi-identities corresponding to the set of standard rules formed by finitarizations of the idempotence rule and the associativity rules with respect to* Δ *together with finitarizations of the* Δ*-transform of these rules.*

Moreover $\boldsymbol{SP}(\mathbf{Q}_{\mathrm{PRIME}})$ *is finitely based relative to* $\mathbf{Q}$.

Proof. The rules of R are so defined that they are also rules of $\boldsymbol{SP}(\mathbf{Q}_{\mathrm{PRIME}})^{\models}$. Hence the quasi-identities corresponding to the rules of R are all valid in $\boldsymbol{SP}(\mathbf{Q}_{\mathrm{PRIME}})$. As the quasi-equations corresponding to R form a base of $\mathbf{Q}'$, we obtain that $\boldsymbol{SP}(\mathbf{Q}_{\mathrm{PRIME}}) \subseteq \mathbf{Q}'$.

In virtue of the above inclusion, Lemma 7.2.7 and the fact that $\mathbf{Q}'$ is RCD, we get that $\mathbf{Q}' = \boldsymbol{SP}(\mathbf{Q}_{\mathrm{PRIME}})$. This concludes the proof of the theorem. □

Note. RCD subquasivarieties of various quasivarieties are investigated, e.g., in Czelakowski and Dziobiak (1990). Kearnes (1990) provides characterizations of RCD subquasivarieties of congruence-modular *varieties*. □

Open Problem If $\mathbf{K}$ is a class of algebras closed under the formation of subdirect products, then for any algebra $\boldsymbol{A}$ of the type of $\mathbf{K}$, the family $Con_{\mathbf{K}}(\boldsymbol{A})$ of $\mathbf{K}$-congruences on $\boldsymbol{A}$ forms a complete lattice (but not necessarily algebraic). $\boldsymbol{P}_{\mathrm{S}}(\mathbf{K})$ is the class of subdirect products of families of members of $\mathbf{K}$.

Suppose that $\mathbf{Q}$ is a quasivariety whose equationally defined commutator is additive. Let $\Delta(x, y, x, y, \underline{u})$ be a generating set. (It is not assumed that Δ is finite.)

Is it true that $\boldsymbol{P}_{\mathrm{S}}(\mathbf{Q}_{\mathrm{PRIME}})$ is the largest subclass $\mathbf{K} \subseteq \mathbf{Q}$ closed with respect to the operation $\boldsymbol{P}_{\mathrm{S}}$ such that for all $\boldsymbol{A} \in \mathbf{K}$ and all $a, b, c, d \in A$,

$$\Theta^{\boldsymbol{A}}_{\mathbf{K}}(a, b) \cap \Theta^{\boldsymbol{A}}_{\mathbf{K}}(c, d) = \Theta^{\boldsymbol{A}}_{\mathbf{K}}((\forall \underline{e})\ \Delta^{\boldsymbol{A}}(a, b, c, d, \underline{e}))?$$

Conclude that then the lattices of $\boldsymbol{P}_{\mathrm{S}}(\mathbf{Q}_{\mathrm{PRIME}})$-congruences are distributive. Decide the analogous problem for the class $\boldsymbol{SP}(\mathbf{Q}_{\mathrm{PRIME}})$. □

7.3 Semiprime Algebras

Let $\mathbf{Q}$ be a quasivariety and $\boldsymbol{A}$ an algebra in $\mathbf{Q}$. Let Φ be a $\mathbf{Q}$-congruence on $\boldsymbol{A}$. Φ is called *semiprime* if for every congruence $\Psi \in Con_{\mathbf{Q}}(\boldsymbol{A})$, $[\Psi, \Psi] = \Phi$ implies that $\Psi = \Phi$.

$\boldsymbol{A} \in \mathbf{Q}$ is *semiprime* (in $\mathbf{Q}$) if the identity congruence $\mathbf{0}_A$ is semiprime in $\boldsymbol{Con}_{\mathbf{Q}}(\boldsymbol{A})$. (Equivalently, $[\Phi, \Phi] = \mathbf{0}_A$ holds for no nonzero $\Phi \in Con_{\mathbf{Q}}(\boldsymbol{A})$.)

$\mathbf{Q}_{\mathrm{SEMIPRIME}}$ denotes the class of semiprime members of $\mathbf{Q}$. Evidently, $\mathbf{Q}_{\mathrm{PRIME}} \subseteq \mathbf{Q}_{\mathrm{SEMIPRIME}}$.

Proposition 7.3.1. *For every quasivariety* $\mathbf{Q}$, $\mathbf{Q}_{\mathrm{PRIME}} = \mathbf{Q}_{\mathrm{RFSI}} \cap \mathbf{Q}_{\mathrm{SEMIPRIME}}$.

Proof. $\mathbf{Q}_{\mathrm{PRIME}} \subseteq \mathbf{Q}_{\mathrm{RFSI}}$, by Proposition 7.2.1. Let $\boldsymbol{A} \in \mathbf{Q}_{\mathrm{RFSI}} \cap \mathbf{Q}_{\mathrm{SEMIPRIME}}$. To show that $\boldsymbol{A}$ is prime suppose that $[\Phi_1, \Phi_2] = \mathbf{0}_A$ for some $\Phi_1, \Phi_2 \in Con_{\mathbf{Q}}(\boldsymbol{A})$. Then $[\Phi_1 \cap \Phi_2, \Phi_1 \cap \Phi_2] = \mathbf{0}_A$. It follows that $\Phi_1 \cap \Phi_2 = \mathbf{0}_A$, because $\boldsymbol{A}$ is semiprime. But then $\Phi_1 = \mathbf{0}_A$ or $\Phi_2 = \mathbf{0}_A$, because $\boldsymbol{A} \in \mathbf{Q}_{\mathrm{RFSI}}$. Thus $\boldsymbol{A} \in \mathbf{Q}_{\mathrm{PRIME}}$. □

Proposition 7.3.2. *Let* $\mathbf{Q}$ *be quasivariety with the additive equationally defined commutator and generating set* $\Delta(x, y, z, w, \underline{u})$. *Let* $\boldsymbol{A}$ *be an algebra in* $\mathbf{Q}$ *and* $\Phi \in Con_{\mathbf{Q}}(\boldsymbol{A})$. *Then the following conditions are equivalent:*

(i) Φ *is semiprime in* $\boldsymbol{Con}_{\mathbf{Q}}(\boldsymbol{A})$;
(ii) *For all* $a, b \in A$, $[\Theta^{A}_{\mathbf{Q}}(a, b), \Theta^{A}_{\mathbf{Q}}(a, b)]^{A} \subseteq \Phi$ *implies* $\langle a, b\rangle \in \Phi$;
(iii) *For all* $a, b \in A$, $\Theta^{A}_{\mathbf{Q}}((\forall \underline{e})\ \Delta(a, b, a, b, \underline{e})) \subseteq \Phi$ *implies* $\langle a, b\rangle \in \Phi$.

Proof. Immediate. □

Corollary 7.3.3. *Let* $\mathbf{Q}$ *be as above. Then for any algebra* $\boldsymbol{A} \in \mathbf{Q}$ *and any* $\Phi \in Con_{\mathbf{Q}}(\boldsymbol{A})$, Φ *is semiprime in* $\boldsymbol{Con}_{\mathbf{Q}}(\boldsymbol{A})$ *if and only if the algebra* $\boldsymbol{A}/\Phi$ *is semiprime in* $\mathbf{Q}$. *Moreover, the following are equivalent:*

(iv) $\boldsymbol{A}$ *is semiprime in* $\mathbf{Q}$;
(v) *The congruence* $\mathbf{0}_A$ *is semiprime*;
(vi) $\boldsymbol{A} \models (\forall xy)((\forall \underline{u}) \bigwedge \Delta(x, y, x, y, \underline{u}) \leftrightarrow x \approx y)$. □

If $\mathbf{Q}$ is an RCD quasivariety, then Corollaries 7.3.3 and 7.1.7 imply that

$$\mathbf{Q}_{\mathrm{SEMIPRIME}} = \mathbf{Q}.$$

The following observation supplements Theorem 7.2.4.

Theorem 7.3.4. *Let* $\mathbf{Q}$ *be a quasivariety whose equationally defined commutator is additive. Suppose that* $\mathbf{Q}$ *possesses a finite generating set* $\Delta(x, y, z, w, \underline{u})$ *of quaternary commutator equations. Then* $\boldsymbol{SP}(\mathbf{Q}_{\mathrm{PRIME}}) \subseteq \boldsymbol{S}(\mathbf{Q}_{\mathrm{SEMIPRIME}})$ *and* $\boldsymbol{S}(\mathbf{Q}_{\mathrm{SEMIPRIME}})$ *is a quasivariety contained in* $\mathbf{Q}$.

Proof. Let $\Delta(x, y, z, w, \underline{u})$ be as above. Corollary 7.3.3 states that the class $\mathbf{Q}_{\mathrm{SEMIPRIME}}$ is axiomatized by the quasi-identities of $\mathbf{Q}$ and the universal-existential sentence

$$(\forall xy)((\forall \underline{u}) \bigwedge \Delta(x, y, x, y, \underline{u}) \rightarrow x \approx y). \qquad (*)$$

$\mathbf{Q}_{\mathrm{SEMIPRIME}}$ is therefore an elementary class and hence closed under the formation of ultraproducts $\boldsymbol{P}_u$. But Corollary 7.3.3.(vi) also implies that $\mathbf{Q}_{\mathrm{SEMIPRIME}}$ is closed under the formation of subdirect products (as one can directly check) and hence direct products. Therefore $\boldsymbol{SPP}_u(\mathbf{Q}_{\mathrm{SEMIPRIME}}) = \boldsymbol{S}(\mathbf{Q}_{\mathrm{SEMIPRIME}})$. It follows that $\boldsymbol{S}(\mathbf{Q}_{\mathrm{SEMIPRIME}})$ is a quasivariety.

As $\mathbf{Q}_{\mathrm{PRIME}} \subseteq \mathbf{Q}_{\mathrm{SEMIPRIME}}$, we get that $\boldsymbol{SP}(\mathbf{Q}_{\mathrm{PRIME}}) \subseteq \boldsymbol{SP}(\mathbf{Q}_{\mathrm{SEMIPRIME}}) = \boldsymbol{S}(\mathbf{Q}_{\mathrm{SEMIPRIME}})$. Hence $\boldsymbol{SP}(\mathbf{Q}_{\mathrm{PRIME}}) \subseteq \boldsymbol{S}(\mathbf{Q}_{\mathrm{SEMIPRIME}})$. □

Corollary 7.3.3 implies that the class $\mathbf{Q}_{\mathrm{SEMIPRIME}}$ validates quasi-identities of $\mathbf{Q}$ and a single quasi-equation σ_{idem} of the form $\bigwedge \Delta(x, y, x, y, \underline{T}) \rightarrow x \approx y$ obtained from $(*)$ by deleting the block of quantifiers "$(\forall \underline{u})$" and substituting for $\underline{u}$ k-tuples of terms from an appropriate finite set $\underline{T}$. (k is the length of $\underline{u}$.) As $\boldsymbol{S}(\mathbf{Q}_{\mathrm{SEMIPRIME}})$ validates σ and, as shown above, $\boldsymbol{SP}(\mathbf{Q}_{\mathrm{PRIME}}) \subseteq \boldsymbol{S}(\mathbf{Q}_{\mathrm{SEMIPRIME}})$, the class $\boldsymbol{SP}(\mathbf{Q}_{\mathrm{PRIME}})$ validates σ_{idem} as well. In fact, we obtain

Corollary 7.3.5. *The quasivariety $\boldsymbol{S}(\mathbf{Q}_{\mathrm{SEMIPRIME}})$ is axiomatized relative to $\mathbf{Q}$ by the single quasi-equation σ_{idem}.* □

In other words, any inferential base of $\mathbf{Q}^{\models}$ enriched with the standard rule $\bigcup\{\Delta(x, y, x, y, \underline{t} \in \underline{T}\}/x \approx y$ (being a finitarization of $(\forall \underline{u})\ \Delta(x, y, x, y, \underline{u})/x \approx y$) forms a base of the equational logic $\boldsymbol{S}(\mathbf{Q}_{\mathrm{SEMIPRIME}})^{\models}$.

Examples

1. **Groups.** According to Section 6.4, Example 3, the set $\Delta_0(x, y, z, w, \underline{u})$ consisting of equations

$$t(x, z, \underline{u}) \cdot t(y, z, \underline{u})^{-1} \approx t(x, w, \underline{u}) \cdot t(y, w, \underline{u})^{-1}$$

with $t(x, y, \underline{u})$ ranging over arbitrary terms form a generating set for the (congruence) commutator of the class $\mathbf{G}$ of groups. But we can also take the single equation, viz.

$$z^{-1}wu^{-1}x^{-1}yu \approx u^{-1}x^{-1}yuz^{-1}w$$

as a generating set.

It follows from Corollary 7.2.3 that a group $\boldsymbol{A}$ is prime (in $\mathbf{G}$) if and only if it validates the universal-existential sentence

$$(\forall xyzw)((\forall \underline{u})\ z^{-1}wu^{-1}x^{-1}yu \approx u^{-1}x^{-1}yuz^{-1}w \rightarrow x \approx y \vee z \approx w).$$

The class $\boldsymbol{SP}(\mathbf{G}_{\mathrm{PRIME}})$ is the largest RCD quasivariety included in $\mathbf{G}$.
In view of Corollary 7.3.3, a group $\boldsymbol{A}$ is semiprime (in $\mathbf{G}$) if and only if it validates the following universal-existential sentence

$$(\forall xy)((\forall \underline{u})\ x^{-1}yu^{-1}x^{-1}yu \approx u^{-1}x^{-1}yux^{-1}y \rightarrow x \approx y).$$

2. **Equivalence algebras.** According to Section 6.4, the set $\Delta_0(x, y, z, w, \underline{u})$ consisting of all equations of the form

$$t(x, z, \underline{u})\ t(y, z, \underline{u})\ t(y, w, \underline{u})\ (t(x, z, \underline{u})\ t(y, w, \underline{u})\ t(y, w, \underline{u})\ t(x, z, \underline{u})) \approx t(x, w, \underline{u})$$

where $t(x, y, \underline{u})$ ranges over arbitrary terms, is a generating set for the (congruence) commutator of $\mathbf{EA}$.

An equivalence algebra $\boldsymbol{A}$ is prime (in **EA**) if and only if it validates the (infinite) sentence $(\forall xyzw)((\forall \underline{u}) \bigwedge \Delta_0(x, y, z, w, \underline{u}) \rightarrow x \approx y \vee z \approx w)$. $\boldsymbol{SP}(\mathbf{EA}_{\mathrm{PRIME}})$ is the largest RCD quasivariety included in the variety **EA**. (In fact $\boldsymbol{SP}(\mathbf{EA}_{\mathrm{PRIME}})$ is a variety.)

An equivalence algebra $\boldsymbol{A}$ is semiprime (in **EA**) if and only if it validates the infinite sentence $(\forall xy)((\forall \underline{u}) \bigwedge \Delta_0(x, y, x, y, \underline{u}) \rightarrow x \approx y)$, where $\Delta_0(x, y, x, y, \underline{u})$ consists of equations of the form

$$t(x, x, \underline{u})\ t(y, x, \underline{u})\ t(y, y, \underline{u})\ (t(x, x, \underline{u})\ t(y, y, \underline{u})\ t(y, y, \underline{u})\ t(x, x, \underline{u})) \approx t(x, y, \underline{u}),$$

with t ranging over arbitrary terms. □

Chapter 8
More on Finitely Generated Quasivarieties

8.1 Generating Sets for the Equationally-Defined Commutator in Finitely Generated Quasivarieties

We begin with the following observation concerning arbitrary finitely generated quasivarieties:

Theorem 8.1.1. *Let* $\mathbf{Q}$ *be a finitely generated quasivariety. Then there exists a finite set* $\Delta_0(x, y, z, w, \underline{u})$ *of quaternary equations (possibly with parameters) such that*

$$\mathbf{Q}^{\models}(x \approx y) \cap \mathbf{Q}^{\models}(z \approx w) = \mathbf{Q}^{\models}(\Delta_0(x, y, z, w, \underline{u})).$$

Consequently, $\mathbf{Q}$ *has a finite generating set of quaternary commutator equations with parameters.*

Proof. $\mathbf{K}$ be a finite class of finite algebras such that $\mathbf{Q} = Qv(\mathbf{K})$. Furthermore, let m be the least positive integer k such that all algebras of $\mathbf{K}$ are of power $\leqslant k$, i.e., $|A| \leqslant m$, for every algebra $A \in \mathbf{K}$. We clearly have that $\mathbf{Q}^{\models} = \mathbf{K}^{\models}$.

Let $\Delta(x, y, z, w, \underline{u})$ be a possibly infinite set of quaternary equations such that $\mathbf{Q}^{\models}(x \approx y) \cap \mathbf{Q}^{\models}(z \approx w) = \mathbf{Q}^{\models}(\Delta(x, y, z, w, \underline{u}))$. In order to define a finite set of quaternary commutator equations for $\mathbf{Q}$ which satisfies the thesis of the theorem we shall apply to $\Delta(x, y, z, w, \underline{u})$ a simple reduction procedure yielding a set of equations $\Delta_0(x, y, z, w, \underline{u})$ with the required property.

We may write $\underline{u} = u_1, u_2, u_3, \ldots$, i.e., the variables of $\underline{u}$ are indexed by consecutive positive integers. The latter form a set $N \subseteq \omega$. (Thus N is either finite or coincides with the set of natural numbers ω.)

For each function $\sigma : N \to \{1, 2, \ldots, m\}$, where m is given as above, we define the strings of variables

$$\underline{u}_\sigma := u_{\sigma(1)}, u_{\sigma(2)}, u_{\sigma(3)}, \ldots$$

J. Czelakowski, *The Equationally-Defined Commutator*,
DOI 10.1007/978-3-319-21200-5_8

Let $\Delta(x, y, z, w, \underline{u}_\sigma)$ be the resulting set of equations obtained from $\Delta(x, y, z, w, \underline{u})$ by uniformly replacing each parameter u_k of $\underline{u}$ by the corresponding parameter $\underline{u}_{\sigma(k)}$, for all $k \in N$.

Let

$$\Delta_1 := \bigcup\{\Delta(x, y, z, w, \underline{u}_\sigma) : \sigma \in \{1, 2, \ldots, m\}^N\}.$$

Δ_1 is a set of equations in the variables x, y, z, w and m parametric variables $u_1, u_2, u_3, \ldots, u_m$. We write

$$\Delta_1 = \Delta_1(x, y, z, w, u_1, u_2, u_3, \ldots, u_m).$$

Lemma 1. $\mathbf{K}^{\models}(x \approx y) \cap \mathbf{K}^{\models}(z \approx w) = \mathbf{K}^{\models}(\Delta_1)$.

Proof. We have:

Claim 1. *Δ_1 is a set of quaternary commutator equations for* $\mathbf{K}$, *i.e.,*

$$\Delta_1 \subseteq \mathbf{K}^{\models}(x \approx y) \cap \mathbf{K}^{\models}(z \approx w).$$

Proof (of the claim). For every mapping $\sigma \in \{1, 2, \ldots, m\}^N$, the set $\Delta(x, y, z, w, \underline{u}_\sigma)$ consists of quaternary commutator equations for $\mathbf{K}$, because the substitution of parameters determined by σ preserves the property of being a commutator equation for $\mathbf{K}$. It follows that Δ_1 is a set of quaternary commutator equations for $\mathbf{Q}$ as well. □

Claim 2. $\Delta \subseteq \mathbf{K}^{\models}(\Delta_1)$.

Proof (of the claim). Let $\boldsymbol{A} \in \mathbf{K}$ and let $h : \boldsymbol{Te}_\tau \to \boldsymbol{A}$ be a homomorphism validating the equations of Δ_1. As $|\boldsymbol{A}| \leqslant m$, it follows that there is a mapping $\sigma \in \{1, 2, \ldots, m\}^N$ such that $h(u_k) = h(u_{\sigma(k)})$, for all $k \in N$. As $\Delta(x, y, z, w, \underline{u}_\sigma) \subseteq \Delta_1$, h validates $\Delta(x, y, z, w, \underline{u}_\sigma)$ as well and, consequently, h validates Δ. □

It follows from Claims 1–2 and the definition of Δ that

$$\mathbf{K}^{\models}(x \approx y) \cap \mathbf{K}^{\models}(z \approx w) = \mathbf{K}^{\models}(\Delta) \subseteq \mathbf{K}^{\models}(\Delta_1) \subseteq \mathbf{K}^{\models}(x \approx y) \cap \mathbf{K}^{\models}(z \approx w).$$

Hence $\mathbf{K}^{\models}(x \approx y) \cap \mathbf{K}^{\models}(z \approx w) = \mathbf{K}^{eq\models}(\Delta_1)$.

The lemma has been proved. □

Having defined the m-parameterized generating set Δ_1 of quaternary commutator equations for $\mathbf{Q}$, we make the next step.

We define the relation $\boldsymbol{\Omega}$ on the algebra of terms $\boldsymbol{Te}_\tau$ as follows:

$$\alpha \equiv \beta \ (\text{mod } \boldsymbol{\Omega}) \text{ iff the equation } \alpha \approx \beta \text{ is } \mathbf{K}\text{-valid (i.e., } \alpha \approx \beta \in \mathbf{K}^{\models}(\emptyset)),$$

for $\alpha, \beta \in Te_\tau$. Trivially, $\boldsymbol{\Omega}$ is a congruence relation and the quotient algebra $\boldsymbol{Te}_\tau/\boldsymbol{\Omega}$ is the ω-generated free algebra $\boldsymbol{F}_{\mathbf{K}}(\omega)$ in the variety $\boldsymbol{Va}(\mathbf{K})$.

Let $\boldsymbol{T}$ be the subalgebra of the term algebra $\boldsymbol{Te}_\tau$ generated by the variables $x, y, z, w, u_1, u_2, u_3, \ldots, u_m$. We consider the subset F of $\boldsymbol{F}_{\mathbf{K}}(\omega)$ consisting of the abstraction classes of $\boldsymbol{\Omega}$ determined by the terms of $\boldsymbol{T}$. Thus

$$F := \{[t] : t \in T\}.$$

F forms a subalgebra $\boldsymbol{F}$ of $\boldsymbol{F}_{\mathbf{K}}(\omega)$ that is the image of the algebra $\boldsymbol{T}$ under the canonical map from $\boldsymbol{Te}_\tau$ to $\boldsymbol{Te}_\tau/\boldsymbol{\Omega}$.

Claim. *The algebra* $\boldsymbol{F}$ *is finite.*

Proof. Let $\boldsymbol{A}_1, \boldsymbol{A}_2, \ldots, \boldsymbol{A}_k$ be a list of pairwise nonisomorphic algebras of $\mathbf{K}$ such that every algebra of $\mathbf{K}$ is isomorphic with some algebra from the list. For each i, $1 \leqslant i \leqslant k$, define the relation $\sim_i$ on T:

$$s \sim_i t \quad \text{iff} \quad \text{every homomorphism } h : \boldsymbol{F} \to \boldsymbol{A}_i \text{ validates the equation } s \approx t.$$

Each $\sim_i$ has a finite index and $\boldsymbol{\Omega} = \bigcap_{1 \leqslant i \leqslant k} \sim_i$. Consequently, $\boldsymbol{\Omega}$ has a finite index too. □

Let S be a selector of the finite set F. S is a mapping which from each equivalence class $[t]$, $t \in T$, picks out a term $t' \in [t]$.

Given an m-parameterized generating set $\Delta_1 = \Delta_1(x, y, z, w, u_1, u_2, \ldots, u_m)$ of quaternary commutator equations for $\mathbf{Q}$, defined as above, we form a new set Δ_0 of equations. We put

$$\Delta_0 := \{\alpha' \approx \beta' : \alpha \approx \beta \in \Delta_1\}.$$

Lemma 2. $\mathbf{Q}^{\models}(x \approx y) \cap \mathbf{Q}^{\models}(z \approx w) = \mathbf{Q}^{\models}(\Delta_0(x, y, z, w, \underline{u}))$.

Proof. It is clear that the set Δ_0 is finite. As each equation $\alpha \approx \beta \in \Delta_1$ is deductively equivalent to $\alpha' \approx \beta'$ on the basis of $\mathbf{K}^{\models}$, i.e., $\mathbf{K}^{\models}(\alpha \approx \beta) = \mathbf{K}^{\models}(\alpha' \approx \beta')$, it follows that Δ_0 is a finite set of quaternary commutator equations for $\mathbf{Q}$. Moreover, Δ_0 is deductively equivalent to Δ_1 with respect to $\mathbf{K}^{\models}$, i.e., $\mathbf{K}^{\models}(\Delta_0) = \mathbf{K}^{\models}(\Delta_1)$. Applying Lemma 1, we get the thesis of Lemma 2. □

The proof of the theorem is concluded (see Definition 4.3.7). □

Note 1. Let $\mathbf{Q}$ be a quasivariety and suppose that a set of equations $\Delta_0(x, y, z, w, \underline{u})$ has the property that

$$\boldsymbol{Va}(\mathbf{Q})^{\models}(x \approx y) \cap \boldsymbol{Va}(\mathbf{Q})^{\models}(z \approx w) = \boldsymbol{Va}(\mathbf{Q})^{\models}(\Delta_0(x, y, z, w, \underline{u})).$$

Then

$$\mathbf{Q}^{\models}(x \approx y) \cap \mathbf{Q}^{\models}(z \approx w) = \mathbf{Q}^{\models}(\Delta_0(x, y, z, w, \underline{u})).$$

Indeed, as $Id(\mathbf{Q}) = Id(\mathbf{Va}(\mathbf{Q}))$, $\Delta_0(x, y, z, w, \underline{u})$ is a set of commutator equations for $\mathbf{Q}$, i.e., $\Delta_0(x, y, z, w, \underline{u}) \subseteq \mathbf{Q}^{\models}(x \approx y) \cap \mathbf{Q}^{\models}(z \approx w)$. Then

$$\mathbf{Va}(\mathbf{Q})^{\models}(x \approx y) \cap \mathbf{Va}(\mathbf{Q})^{\models}(z \approx w) = \mathbf{Q}^{\models}(x \approx y) \cap \mathbf{Q}^{\models}(z \approx w) \supseteq$$
$$\mathbf{Q}^{\models}(\Delta_0(x, y, z, w, \underline{u})) \supseteq \mathbf{Va}(\mathbf{Q})^{\models}(\Delta_0(x, y, z, w, \underline{u})) =$$
$$\mathbf{Va}(\mathbf{Q})^{\models}(x \approx y) \cap \mathbf{Va}(\mathbf{Q})^{\models}(z \approx w).$$

Hence $\mathbf{Q}^{\models}(x \approx y) \cap \mathbf{Q}^{\models}(z \approx w) = \mathbf{Q}^{\models}(\Delta_0(x, y, z, w, \underline{u}))$.

It follows from the above observation that, in terms of congruence generation, if in the free algebra $\boldsymbol{F}_{\mathbf{Q}}(\omega)$ the congruence $\Theta(x, y) \cap \Theta(z, w)$ is finitely generated in the absolute sense (i.e., in the sense of the closure system $Con(\boldsymbol{F}_{\mathbf{Q}}(\omega))$), then it is finitely generated in the relative sense (i.e., in the sense of the closure system $Con_{\mathbf{Q}}(\boldsymbol{F}_{\mathbf{Q}}(\omega))$). Equivalently, if $\Theta(x, y) \cap \Theta(z, w)$ is compact in the lattice $\boldsymbol{Con}(\boldsymbol{F}_{\mathbf{Q}}(\omega))$, then it is compact in $\boldsymbol{Con}_{\mathbf{Q}}(\boldsymbol{F}_{\mathbf{Q}}(\omega))$.

The above implication cannot be reversed. There are varieties of algebras generated by a single, finite algebra $\boldsymbol{A}$ such that the congruence $\Theta(x, y) \cap \Theta(z, w)$ is compact in the lattice $\boldsymbol{Con}_{\mathbf{Q}}(\boldsymbol{F}_{\mathbf{Q}}(\omega))$ but non-compact in the lattice $\boldsymbol{Con}(\boldsymbol{F}_{\mathbf{Q}}(\omega))$, where $\mathbf{Q} = \mathbf{Qv}(\boldsymbol{A})$. An appropriate example is due to Keith Kearnes (personal correspondence). □

Note 2. Applying basically the same argument as in the proof of Theorem 8.1.1, one can prove the following, more general facts:

Corollary 8.1.2. *Let* $\mathbf{Q}$ *be a finitely generated quasivariety. Then for each pair* m, n *of positive natural numbers and any four sequences of different variables* $\underline{x} = x_1, \ldots, x_m$, $\underline{y} = y_1, \ldots, y_m$, $\underline{z} = z_1, \ldots, z_n$, $\underline{w} = w_1, \ldots, w_n$, *there exists a finite set* $\Delta_0(\underline{x}, \underline{y}, \underline{z}, \underline{w}, \underline{u})$ *of commutator equations (with parameters) for* $\mathbf{Q}$ *in the variables* $\underline{x}, \underline{y}$ *and* $\underline{z}, \underline{w}$ *such that*

$$\mathbf{Q}^{\models}(x_1 \approx y_1, \ldots, x_m \approx y_m) \cap \mathbf{Q}^{\models}(z_1 \approx w_1, \ldots, z_n \approx w_n) =$$
$$\mathbf{Q}^{\models}(\Delta_0(\underline{x}, \underline{y}, \underline{z}, \underline{w}, \underline{u})). \quad \square$$

(See also Definition 3.1.1 and Notes following it.)

Corollary 8.1.3. *Let* $\mathbf{Q}$ *be a finitely generated quasivariety. Then for each natural number* $m \geqslant 2$ *and any two sequences of different variables* $\underline{x} = x_1, \ldots, x_m$, $\underline{y} = y_1, \ldots, y_m$ *of length* m, *the theory*

$$\mathbf{Q}^{\models}(x_1 \approx y_1) \cap \ldots \cap \mathbf{Q}^{\models}(x_m \approx y_m)$$

has a finite generating set. □

The following corollary is an intermediate consequence of Theorems 8.1.1 and 4.3.9:

Corollary 8.1.4. *Let* **Q** *be a finitely generated quasivariety with the relative shifting property. Then there exists a finite set* $\Delta_0(x, y, z, w, \underline{u})$ *of quaternary equations (possibly with parameters) such that*

$$[\Phi, \Psi]^A = \Theta^A_{\mathbf{Q}}(\bigcup\{(\forall \underline{e})\ \Delta^A_0(a, b, c, d, \underline{e}) : \langle a, b\rangle \in \Phi, \langle c, d\rangle \in \Psi\})$$

for all algebras $\mathbf{A} \in \mathbf{Q}$ *and all* $\Phi, \Psi \in Con_{\mathbf{Q}}(\mathbf{A})$. □

The following corollary is crucial:

Corollary 8.1.5. *Let* **Q** *be a finitely generated quasivariety such that the lattice* $\boldsymbol{Th}(\mathbf{Q}^{\models})$ *validates the law of restricted distributivity. Then there exists a finite set* $\Delta_0(x, y, z, w, \underline{u})$ *of quaternary equations (with parameters) such that*

$$[\Theta^A_{\mathbf{Q}}(X), \Theta^A_{\mathbf{Q}}(Y)]^A = \Theta^A_{\mathbf{Q}}(\bigcup\{(\forall \underline{e})\ \Delta^A_0(a, b, c, d, \underline{e}) : \langle a, b\rangle \in X, \langle c, d\rangle \in Y\}) \qquad (*)$$

for all algebras **A** *and all sets* $X, Y \subseteq A^2$.

In particular,

$$[\mathbf{Q}^{\models}(X), \mathbf{Q}^{\models}(Y)] = \mathbf{Q}^{\models}(\bigcup\{(\forall \underline{u})\ \Delta^A_0(\alpha, \beta, \gamma, \delta, \underline{u}) : \alpha \approx \beta \in X, \gamma \approx \delta \in Y\}) \qquad (**)$$

for any sets of equations X *and* Y.

Proof. In view of Theorem 5.3.7, the equationally defined commutator for **Q** is additive. As $\mathbf{Q}^{\models}(x \approx y) \cap \mathbf{Q}^{\models}(z \approx w) = \mathbf{Q}^{\models}(\Delta_0(x, y, z, w, \underline{u}))$, $(*)$ follows from the proof of Theorem 5.1.2.(1) ⇒ (2).

$(**)$ follows from $(*)$ and Lemma 5.2.2. □

We also note:

Corollary 8.1.6. *Let* **Q** *be a finitely generated quasivariety whose equationally defined commutator is additive. Then* $\boldsymbol{SP}(\mathbf{Q}_{\mathrm{PRIME}})$ *is the largest RCD quasivariety included in* **Q**.

Proof. Apply Theorems 7.2.4 and 8.1.1. □

8.2 Triangular Irreducibility of Congruences

Let **Q** be a quasivariety, $\boldsymbol{A} \in \mathbf{Q}$ and let $a_1, \ldots, a_m$ be a finite sequence of elements of A (possibly with repetitions) of length $m \geqslant 3$. In what follows we shall make use of the following triangle table of **Q**-congruences on $\boldsymbol{A}$:

$$\begin{array}{cccccc}
\Theta_{\mathbf{Q}}(a_1,a_2), & \Theta_{\mathbf{Q}}(a_1,a_3), & \Theta_{\mathbf{Q}}(a_1,a_4), & \cdots\cdots & \cdots\cdots & \Theta_{\mathbf{Q}}(a_1,a_m), \\
 & \Theta_{\mathbf{Q}}(a_2,a_3), & \Theta_{\mathbf{Q}}(a_2,a_4), & \cdots\cdots & \cdots\cdots & \Theta_{\mathbf{Q}}(a_2,a_m), \\
 & & \ddots & & & \vdots \\
 & & & \Theta_{\mathbf{Q}}(a_i,a_{i+1}), & \cdots\cdots & \Theta_{\mathbf{Q}}(a_i,a_m), \\
 & & & & \ddots & \vdots \\
 & & & & & \Theta_{\mathbf{Q}}(a_{m-1},a_m),
\end{array}$$

This table contains $r := m(m-1)/2$ elements. It is called the *triangular table* of relatively principal congruences corresponding to the sequence $a_1, \ldots, a_m$.

The $\mathbf{Q}$-congruence $\bigcap_{1\leqslant i<j\leqslant m} \Theta^{A}_{\mathbf{Q}}(a_i, a_j)$, being the intersection of the above congruences, is called the *triangular intersection*.

Definition 8.2.1. Let $m \geqslant 3$ be a natural number. Let $\mathbf{Q}$ be a quasivariety, $\boldsymbol{A} \in \mathbf{Q}$ and $\Phi \in Con_{\mathbf{Q}}(A)$. The congruence Φ is said to be *m-triangularily irreducible* in the lattice $\boldsymbol{Con}_{\mathbf{Q}}(\boldsymbol{A})$ if for every sequence $a_1, \ldots, a_m$ of elements of A (possibly with repetitions) of length m, if $\bigcap_{1\leqslant i<j\leqslant m}(\Theta^{A}_{\mathbf{Q}}(a_i, a_j) +_{\mathbf{Q}} \Phi) = \Phi$ then $a_i \equiv a_j$ (Φ) for some i and j, $1 \leqslant i < j \leqslant m$. □

In particular, the congruence $\mathbf{0}_{A}$ is m-triangularily irreducible in the lattice $\boldsymbol{Con}_{\mathbf{Q}}(\boldsymbol{A})$ iff for every sequence $a_1, \ldots, a_m$ of elements of A of length m, if $\bigcap_{1\leqslant i<j\leqslant m} \Theta^{A}_{\mathbf{Q}}(a_i, a_j) = \mathbf{0}_{A}$, then $a_i = a_j$ for some i and j, $1 \leqslant i < j \leqslant m$.

In what follows we shall make use of the following simple observation, being an instance of Proposition 2.9:

Fact. *Let* $\mathbf{Q}$ *be a quasivariety, let* $\boldsymbol{A} \in \mathbf{Q}$*, and* $\Phi \in Con_{\mathbf{Q}}(A)$*. Let* $\boldsymbol{B} := \boldsymbol{A}/\Phi$ *be the quotient algebra and let* $h : \boldsymbol{A} \to \boldsymbol{B}$ *be the canonical homomorphism. Then*

$$h^{-1}(\Theta^{B}_{\mathbf{Q}}(a/\Phi, b/\Phi)) = \Phi +_{\mathbf{Q}} \Theta^{A}_{\mathbf{Q}}(a, b). \qquad \square$$

Lemma 8.2.2. *Let* $\mathbf{Q}$ *be a quasivariety,* $\mathbf{A} \in \mathbf{Q}$ *and* $\Phi \in Con_{\mathbf{Q}}(A)$*.* Φ *is m-triangularily irreducible in the lattice* $\mathbf{Con_Q(A)}$ *if and only if the congruence* $\mathbf{0}_{A/\Phi}$ *is m-triangularily irreducible in the lattice* $\mathbf{Con_Q}(\mathbf{A}/\Phi)$*.*

Proof. Let $h : \boldsymbol{A} \to \boldsymbol{A}/\Phi$ be the canonical homomorphism. Then, by the above fact,

$$h^{-1}(\Theta^{A/\Phi}_{\mathbf{Q}}(a/\Phi, b/\Phi)) = \Phi +_{\mathbf{Q}} \Theta^{A}_{\mathbf{Q}}(a, b) \qquad (*)$$

for all $a, b \in A$.

(*) and the surjectivity of h imply that for any $a_1, \ldots, a_m \in A$ the following conditions are equivalent:

$$\bigcap_{1\leqslant i<j\leqslant m} (\Theta^{A}_{\mathbf{Q}}(a_i, a_j) +_{\mathbf{Q}} \Phi) = \Phi,$$

$$\bigcap_{1\leqslant i<j\leqslant m} (h^{-1}(\Theta^{A/\Phi}_{\mathbf{Q}}(a_i/\Phi, a_j/\Phi)) = h^{-1}(\mathbf{0}_{A/\Phi}),$$

$$h^{-1}\Big(\bigcap_{1\leqslant i<j\leqslant m}\Theta_{\mathbf{Q}}^{A/\Phi}(a_i/\Phi, a_j/\Phi)\Big) = h^{-1}(\mathbf{0}_{A/\Phi}),$$

$$\bigcap_{1\leqslant i<j\leqslant m}\Theta_{\mathbf{Q}}^{A/\Phi}(a_i/\Phi, a_j/\Phi) = \mathbf{0}_{A/\Phi}.$$

From these conditions we get the equivalence of the following statements:

Φ is m-triangularily irreducible in the lattice $\mathbf{Con_Q}(\boldsymbol{A})$,

$$(\forall a_1,\dots,a_m \in A)\Big(\bigcap_{1\leqslant i<j\leqslant m}(\Theta_{\mathbf{Q}}^{A}(a_i, a_j) +_{\mathbf{Q}} \Phi) = \Phi \Rightarrow$$
$$a_i \equiv a_j\ (\Phi)\ \textit{for some } i \textit{ and } j\Big),$$

$$(\forall a_1,\dots,a_m \in A)\Big(\bigcap_{1\leqslant i<j\leqslant m}\Theta_{\mathbf{Q}}^{A/\Phi}(a_i/\Phi, aj/\Phi) = \mathbf{0}_{A/\Phi} \Rightarrow$$
$$a_i \equiv a_j\ (\Phi)\ \textit{for some } i \textit{ and } j\Big),$$

$$(\forall a_1,\dots,a_m \in A)\Big(\bigcap_{1\leqslant i<j\leqslant m}\Theta_{\mathbf{Q}}^{A/\Phi}(a_i/\Phi, a_j/\Phi) = \mathbf{0}_{A/\Phi} \Rightarrow$$
$$a_i/\Phi = a_j/\Phi\ \textit{for some } i \textit{ and } j\Big),$$

$\mathbf{0}_{A/\Phi}$ *is m-triangularily irreducible in the lattice* $\mathbf{Con_Q}(\boldsymbol{A}/\Phi)$. □

We recall that $\mathbf{Q}_{\mathrm{RFSI}}$ is the class of non-trivial relatively finitely subdirectly irreducible algebras of $\mathbf{Q}$.

Lemma 8.2.3. *Let* $\mathbf{Q}$ *be a quasivariety and m a natural number,* $m \geqslant 3$. *If* $\boldsymbol{A} \in \mathbf{Q}_{\mathrm{RFSI}}$, *then* $\mathbf{0}_A$ *is m-triangularily irreducible in* $\mathbf{Con_Q}(\boldsymbol{A})$.

Proof. Immediate. □

We define

$$\mathbf{Q}_{m-\mathrm{TRI}}$$

to be the class of all members $\boldsymbol{A}$ of a quasivariety $\mathbf{Q}$ for which the congruence $\mathbf{0}_A$ is m-triangularily irreducible in $\boldsymbol{Con}_{\mathbf{Q}}(\boldsymbol{A})$.

It follows from the above lemma that every quasivariety $\mathbf{Q}$ has enough algebras $\boldsymbol{A}$ with m-triangularily irreducible zero congruences $\mathbf{0}_A$ in the sense that every algebra of $\mathbf{Q}$ is isomorphic with a subdirect product of a family of algebras from the class $\mathbf{Q}_{m-\mathrm{TRI}}$, for each $m \geqslant 3$.

Let $\mathbf{K}$ be a class of algebras. $\boldsymbol{P_S}(\mathbf{K})$ denotes the class of isomorphic copies of subdirect products of families of algebras from $\mathbf{K}$.

Corollary 8.2.4. *Let* $\mathbf{Q}$ *be a quasivariety. Then for every positive integer* m, $m \geqslant 3$,

$$\mathbf{Q} = \boldsymbol{SP}(\mathbf{Q}_{m-\mathrm{TRI}}) = \boldsymbol{P_S}(\mathbf{Q}_{m-\mathrm{TRI}}). \qquad \square$$

The theorem below provides a characterization of finitely generated quasivarieties in terms of triangular intersections of relatively principal congruences:

Theorem 8.2.5. *Let* $\mathbf{Q}$ *be an arbitrary quasivariety and* $m \geqslant 3$ *a positive integer. The following conditions are equivalent:*

(i) $\mathbf{Q}$ *is generated by a finite class of algebras each of which has at most* $m-1$ *elements.*

(ii) *For every algebra* $\mathbf{A} \in \mathbf{Q}$ *and for any sequence* $a_1, \ldots, a_m$ *of elements of* A *of length* m *(possibly with repetitions) it is the case that*

$$\bigcap_{1 \leqslant i < j \leqslant m} \Theta^{\mathbf{A}}_{\mathbf{Q}}(a_i, a_j) = \mathbf{0}_{\mathbf{A}}.$$

(iii) *For every algebra* $\mathbf{A} \in \mathbf{Q}$*, for any congruence* $\Phi \in Con_{\mathbf{Q}}(\mathbf{A})$*, and any sequence* $a_1, \ldots, a_m$ *of elements of* A *of length* m *(possibly with repetitions) it is the case that*

$$\bigcap_{1 \leqslant i < j \leqslant m} (\Phi +_{\mathbf{Q}} \Theta^{\mathbf{A}}_{\mathbf{Q}}(a_i, a_j)) = \Phi.$$

(iv) *For any sequence* $x_1, \ldots, x_m$ *of* m *different free generators of the free algebra* $\boldsymbol{F} := \boldsymbol{F}_{\mathbf{Q}}(\omega)$ *and any congruence* $\Phi \in Con_{\mathbf{Q}}(\boldsymbol{F})$,

$$\bigcap_{1 \leqslant i < j \leqslant m} (\Phi +_{\mathbf{Q}} \Theta^{\boldsymbol{F}}_{\mathbf{Q}}(x_i, x_j)) = \Phi.$$

Note. In view of Proposition 2.5, condition (iv) is equivalent to

(iv)* *For any sequence* $x_1, x_2, \ldots, x_m$ *of* m *different individual variables and any set of equations* X,

$$\bigcap_{1 \leqslant i < j \leqslant m} \mathbf{Q}^{\models}(X \cup \{x_i \approx x_j\}) = \mathbf{Q}^{\models}(X).$$

Equivalently, (iv)* holds for any *finite* set of equations X. But in view of the above theorem, (iv)* is equivalent to

(iv)** *For any sequence* $\alpha_1, \alpha_2, \ldots, \alpha_m$ *of terms and any set of equations* X,

$$\bigcap_{1 \leqslant i < j \leqslant m} \mathbf{Q}^{\models}(X \cup \{\alpha_i \approx \alpha_j\}) = \mathbf{Q}^{\models}(X).$$

Indeed, (iv)** trivially implies (iv)*. Conversely, assume (iv)*. As (iv)* is equivalent to (iv), the theorem implies that (iii) holds. In particular, we get that (iii) holds for the free algebra $\boldsymbol{F}_{\mathbf{Q}}(\omega)$. But the last condition is equivalent to (iv)**. So (iv)** holds. $\square$

Proof. (i) ⇒ (iv). Assuming (i), we prove (iv)*. There is a finite class of algebras **K** such that each algebra in **K** is of power $< m$ and $\mathbf{Q} = \mathbf{SP}(\mathbf{K})$. Consequently, $\mathbf{Q}^{\models} = \mathbf{K}^{\models}$. Let X be a theory of $\mathbf{Q}^{\models}$. It suffices to show that $\bigcap_{1 \leqslant i < j \leqslant m} \mathbf{Q}^{\models}(X \cup \{x_i \approx x_j\}) \subseteq X$.

Let m be as above and suppose that $\alpha \approx \beta \notin X$. There exists an algebra $\mathbf{A} \in \mathbf{K}$ and a homomorphism $v : \mathbf{Te}_\tau \to \mathbf{A}$ such that v validates X and $v(\alpha) \neq v(\beta)$. As $|A| < m$, we have that $v(x_i) = v(x_j)$ for some $1 \leqslant i, j \leqslant m$, $i < j$. It follows that $\alpha \approx \beta \notin \mathbf{Q}^{\models}(X \cup \{x_i \approx x_j\})$, by the definition of $\mathbf{Q}^{\models}$, and consequently, $\alpha \approx \beta \notin \bigcap_{1 \leqslant i < j \leqslant m} \mathbf{Q}^{\models}(X \cup \{x_i \approx x_j\})$. So (iv)* holds.

(iv) ⇒ (ii). The proof of this implication is based on two claims. Assume (iv).

Claim 1. *For every countably generated algebra $\mathbf{A} \in \mathbf{Q}$ and for any sequence $a_1, \ldots, a_m$ of elements of A of length m (possibly with repetitions),*

$$\bigcap_{1 \leqslant i < j \leqslant m} \Theta^{\mathbf{A}}_{\mathbf{Q}}(a_i, a_j) = \mathbf{0}_{\mathbf{A}}.$$

Proof (of the claim). Assume $\mathbf{A} \in \mathbf{Q}$ is countably generated. Let $a_1, \ldots, a_m$ be a sequence of elements of A.

Let $h : \mathbf{F} \to \mathbf{A}$ be a surjective homomorphism such that $h(x_i) = a_i$ for $i = 1, 2, \ldots, m$. Let Φ be the kernel of h. Φ is a **Q**-congruence of $\mathbf{F}$ because the quotient algebra $\mathbf{F}/\Phi$ is isomorphic with $\mathbf{A}$.

The fact following Definition 8.2.1 and the surjectivity of h imply that the following conditions are equivalent:

$$\bigcap_{1 \leqslant i < j \leqslant m} (\Phi +_{\mathbf{Q}} \Theta^{\mathbf{F}}_{\mathbf{Q}}(x_i, x_j)) = \Phi,$$

$$\bigcap_{1 \leqslant i < j \leqslant m} h^{-1}(\Theta^{\mathbf{F}/\Phi}_{\mathbf{Q}}(x_i/\Phi, x_j/\Phi)) = \Phi,$$

$$\bigcap_{1 \leqslant i < j \leqslant m} h^{-1}(\Theta^{\mathbf{F}/\Phi}_{\mathbf{Q}}(x_i/\Phi, x_j/\Phi)) = h^{-1}(\mathbf{0}_{\mathbf{F}/\Phi}),$$

$$h^{-1}(\bigcap_{1 \leqslant i < j \leqslant m} \Theta^{\mathbf{F}/\Phi}_{\mathbf{Q}}(x_i/\Phi, x_j/\Phi) = h^{-1}(\mathbf{0}_{\mathbf{F}/\Phi}),$$

$$h^{-1}(\bigcap_{1 \leqslant i < j \leqslant m} \Theta^{\mathbf{A}}_{\mathbf{Q}}(a_i, a_j) = h^{-1}(\mathbf{0}_{\mathbf{A}}),$$

$$\bigcap_{1 \leqslant i < j \leqslant m} \Theta^{\mathbf{A}}_{\mathbf{Q}}(a_i, a_j) = \mathbf{0}_{\mathbf{A}},$$

As the first equality holds by (iv), the last one holds as well. This proves the claim. □

Claim 1 continues to hold for arbitrary algebras of **Q**.

Claim 2. *For every algebra* $\mathbf{A} \in \mathbf{Q}$ *and for any sequence* $a_1, \ldots, a_m$ *of elements of* A *of length* m *(possibly with repetitions),*

$$\bigcap_{1 \leqslant i < j \leqslant m} \Theta^{\mathbf{A}}_{\mathbf{Q}}(a_i, a_j) = \mathbf{0}_{\mathbf{A}}.$$

Proof (of the claim). Let $\mathbf{A} \in \mathbf{Q}$ and let $a_1, \ldots, a_m$ be a sequence of elements of A. Suppose $\langle a, b\rangle \in \bigcap_{1 \leqslant i < j \leqslant m} \Theta^{\mathbf{A}}_{\mathbf{Q}}(a_i, a_j)$. By Theorem 2.1 for each pair $i < j$ there exists a finite generating sequence

$$\langle c_{ij,k}, d_{ij,k}\rangle, \qquad k = 1, \ldots, n_{ij}, \qquad (*)_{ij}$$

of the pair $\langle a, b\rangle$ from the pair $\langle a_i, a_j\rangle$ in the algebra $\mathbf{A}$.

Let $\mathbf{B}$ be the subalgebra of $\mathbf{A}$ generated by the elements of A that are involved in the definition of the sequence $(*)_{ij}$, for all pairs $i < j$. (In particular, $\mathbf{B}$ contains the elements that occur in the pairs $(*)_{ij}$. But $\mathbf{B}$ also contains elements of A that were employed by the quasi-identities applied in the definition of $(*)_{ij}$.) $\mathbf{B}$ is countably generated. It follows from the definition of $\mathbf{B}$ that $(*)_{ij}$ is also a generating sequence of $\langle a, b\rangle$ from the pair $\langle a_i, a_j\rangle$ in the algebra $\mathbf{B}$, for all pairs $i < j$. Consequently, $\langle a, b\rangle \in \bigcap_{1 \leqslant i < j \leqslant m} \Theta^{\mathbf{B}}_{\mathbf{Q}}(a_i, a_j)$. But Claim 1 implies that $a = b$. □

This proves (ii).

(ii) ⇒ (iii). Suppose $\mathbf{A} \in \mathbf{Q}$, $\Phi \in Con_{\mathbf{Q}}(\mathbf{A})$, $a_1, \ldots, a_m \in A$. Let $\mathbf{B} := \mathbf{A}/\Phi$. Let $h : \mathbf{A} \to \mathbf{B}$ be the canonical homomorphism. Hence $\ker(h) = \Phi$ and $\mathbf{B} \in \mathbf{Q}$. Then by the above Fact,

$$\bigcap_{1 \leqslant i < j \leqslant m} (\Phi +_{\mathbf{Q}} \Theta^{\mathbf{A}}_{\mathbf{Q}}(a_i, a_j)) = \bigcap_{1 \leqslant i < j \leqslant m} (h^{-1}(\Theta^{\mathbf{B}}_{\mathbf{Q}}(a_i/\Phi, a_j/\Phi)) =$$

$$h^{-1}(\bigcap_{1 \leqslant i < j \leqslant m} \Theta^{\mathbf{B}}_{\mathbf{Q}}(a_i/\Phi, a_j/\Phi)) = h^{-1}(\mathbf{0}_{\mathbf{B}}) = \Phi.$$

It follows that $\bigcap_{1 \leqslant i < j \leqslant m}(\Phi +_{\mathbf{Q}} \Theta^{\mathbf{A}}_{\mathbf{Q}}(a_i, a_j)) = \Phi$.

(iii) ⇒ (ii). This is obvious.

(ii) ⇒ (i). Assume (ii). To prove (i) it suffices to show the following

Claim 3. *Every algebra in* $\mathbf{Q}_{m-\mathrm{TRI}}$ *is of power* $< m$.

Proof (of the claim). Let $\mathbf{A} \in \mathbf{Q}_{m-\mathrm{TRI}}$ is and suppose *a contrario* that there are m distinct elements in A, say $a_1, \ldots, a_m$. (ii) gives that $\bigcap_{1 \leqslant i < j \leqslant m} \Theta^{\mathbf{A}}_{\mathbf{Q}}(a_i, a_j) = \mathbf{0}_{\mathbf{A}}$. As $\mathbf{A} \in \mathbf{Q}_{m-\mathrm{TRI}}$, it follows that $a_i = a_j$ for some i and j, $1 \leqslant i < j \leqslant m$. So $\mathbf{A}$ has fewer than m elements. A contradiction. □

This concludes the proof of the theorem. □

Corollary 8.2.6. *Suppose that* $\mathbf{Q}$ *is a quasivariety generated by a finite class of algebras each of which has at most* $m-1$ *elements,* $m \geqslant 3$. *For every algebra* $\mathbf{A} \in \mathbf{Q}$, *the following conditions are equivalent:*

(1) $\mathbf{A} \in \mathbf{Q}_{m-\mathrm{TRI}}$,
(2) $\mathbf{A}$ *has at most* $m-1$ *elements.*

Proof. Let $\mathbf{A} \in \mathbf{Q}$.

(1) $\Rightarrow$ (2). Assume (1). Suppose *a contrario* that there are m distinct elements in A, say $a_1, \ldots, a_m$. Theorem 7.1.5 implies that $\bigcap_{1 \leqslant i<j \leqslant m} \Theta^{\mathbf{A}}_{\mathbf{Q}}(a_i, a_j) = \mathbf{0}_A$. It follows by (1) that $a_i = a_j$ for some i and j, $1 \leqslant i < j \leqslant m$. So $\mathbf{A}$ has fewer than m elements.

(2) $\Rightarrow$ (1). Assume (2). Let $a_1, \ldots, a_m$ be a sequence of elements A of length m (possibly with repetitions). (2) trivially implies $a_i = a_j$ for some i and j, $1 \leqslant i < j \leqslant m$. Hence the implication

$$\bigcap_{1 \leqslant i<j \leqslant m} \Theta^{\mathbf{A}}_{\mathbf{Q}}(a_i, a_j) = \mathbf{0}_A \quad \Rightarrow \quad a_i = a_j \text{ for some } i \text{ and } j,\ 1 \leqslant i < j \leqslant m$$

is true, i.e., (1) holds. □

Theorem 8.2.7. *Suppose that* $\mathbf{Q}$ *is a quasivariety in a finite signature generated by a finite class* $\mathbf{K}$ *of algebras each of which has at most* $m-1$ *elements,* $m \leqslant 3$. *Then* $\mathbf{Q}_{m-\mathrm{TRI}}$ *is a finitely axiomatizable class.*

Proof. In view of Corollary 8.2.6, $\mathbf{Q}_{m-\mathrm{TRI}}$ is the class of all at most $(m-1)$-element algebras of $\mathbf{Q}$. It follows that $\mathbf{Q}_{m-\mathrm{TRI}}$ is axiomatized by any set of quasi-identities which axiomatizes $\mathbf{Q}$ together with a single universal sentence

$$(\forall x_1)(\forall x_2)\ldots(\forall x_m) \bigvee \{x_i \approx x_j : 1 \leqslant i < j \leqslant m\},$$

To show that $\mathbf{Q}_{m-\mathrm{TRI}}$ is indeed finitely axiomatizable, we notice that, up to isomorphism, $\mathbf{Q}_{m-\mathrm{TRI}}$ consists of finitely many finite algebras. Evidently, $\mathbf{Q}_{m-\mathrm{TRI}}$ and its complement are algebraic classes, i.e., they are closed under isomorphisms. The trivial ultraproduct argument shows that both $\mathbf{Q}_{m-\mathrm{TRI}}$ and the complement of $\mathbf{Q}_{m-\mathrm{TRI}}$ are closed under the formation of ultraproducts. It follows from Corollary 6.1.16 in Chang and Keisler (1973) that $\mathbf{Q}_{m-\mathrm{TRI}}$ is a finitely axiomatizable class. □

8.3 Equationally-Definable m-Triangular Meets of Relatively Principal Congruences (m-EDTPM)

The following observation is modelled after Lemma 2.1 in Pigozzi (1988), see also Proposition 7.1.3.

Proposition 8.3.1. *Let $m \geqslant 3$ be a natural number and $\Lambda = \Lambda(x_1, x_2, \ldots, x_m, \underline{u})$ a set of equations in variables $x_1, x_2, \ldots, x_m$ (and possibly some parameters $\underline{u}$). Let* **Q** *be a quasivariety. The following conditions are equivalent:*

(1) *For all* $\mathbf{A} \in \mathbf{Q}$ *and for any sequence $a_1, \ldots, a_m$ of elements A of length m,*

$$\bigcap_{1 \leqslant i < j \leqslant m} \Theta^{A}_{\mathbf{Q}}(a_i, a_j) = \Theta^{A}_{\mathbf{Q}}((\forall \underline{e})\ \Lambda^{A}(a_1, a_2, \ldots, a_m, \underline{e}));$$

(2) *For all* $\mathbf{A} \in \mathbf{Q}$ *and for any sequence $a_1, \ldots, a_m$ of elements A of length m,*

$$\bigcap_{1 \leqslant i < j \leqslant m} \Theta^{A}_{\mathbf{Q}}(a_i, a_j) = \mathbf{0}_A \quad \Leftrightarrow$$

$$\mathbf{A} \models (\forall \underline{u}) \bigwedge \Lambda(x_1, x_2, \ldots, x_m, \underline{u})[a_1, a_2, \ldots, a_m];$$

(3) $\mathbf{Q}_{m-\mathrm{TRI}}$ *validates the first-order sentence*

$$(\forall x_1)(\forall x_2)\ldots(\forall x_m)((\forall \underline{u}) \bigwedge \Lambda(x_1, x_2, \ldots, x_m, \underline{u})$$
$$\leftrightarrow \bigvee \{x_i \approx x_j : 1 \leqslant i < j \leqslant m\};$$

(4) *For every pair i, j, $(1 \leqslant i < j \leqslant m)$,* **Q** *validates the equations*

$$\Lambda(x_i/x, x_j/x)$$

and for any algebra $\mathbf{A} \in \mathbf{Q}$ *and any sequence $a_1, \ldots, a_m$ of elements A,*

$$\bigcap_{1 \leqslant i < j \leqslant m} \Theta^{A}_{\mathbf{Q}}(a_i, a_j) \subseteq \Theta^{A}_{\mathbf{Q}}((\forall \underline{e})\ \Lambda^{A}(a_1, a_2, \ldots, a_m, \underline{e})).$$

$\Lambda(x_i/x, x_j/x)$ results from $\Lambda(x_1, x_2, \ldots, x_m, \underline{u})$ by the uniform substitution of the variable x for the variables x_i and x_j, where x is different from $x_1, x_2, \ldots, x_m$ and $\underline{u}$. (If the string $\underline{u}$ is infinite, we can always rename the parametric variables $\underline{u}$ so that there are still infinitely many variables different from those of $\underline{u}$.)

In (3), the symbol "$\leftrightarrow$" is the equivalence connective from the first-order language associated with the signature of **Q**.

Proof. We recall that

$$\Theta^{A}_{\mathbf{Q}}((\forall \underline{e})\ \Lambda^{A}(a_1, a_2, \ldots, a_m, \underline{e})) :=$$
$$\Theta^{A}_{\mathbf{Q}}(\{\langle \alpha(a_1, a_2, \ldots, a_m, \underline{e})), \beta(a_1, a_2, \ldots, a_m, \underline{e})\rangle : \alpha \approx \beta \in \Lambda, \underline{e} \in A^{k}\}).$$

The implications (1) $\Rightarrow$ (2) $\Rightarrow$ (3) are immediate.

(3) ⇒ (4). (3) implies that for each pair i,j ($1 \leqslant i < j \leqslant m$), $\mathbf{Q}_{m-\mathrm{TRI}}$ validates the equations $\Lambda(x_i/x, x_j/x)$ and hence so does $\mathbf{Q}$.

Now let $\mathbf{A} \in \mathbf{Q}$ and let $a_1, \ldots, a_m$ be a sequence of elements of A. To prove the inclusion

$$\bigcap_{1 \leqslant i < j \leqslant m} \Theta^{\mathbf{A}}_{\mathbf{Q}}(a_i, a_j) \subseteq \Theta^{\mathbf{A}}_{\mathbf{Q}}((\forall \underline{e})\ \Lambda^{\mathbf{A}}(a_1, a_2, \ldots, a_m, \underline{e})),$$

it suffices to show that for every m-triangularily irreducible congruence Φ in the lattice $\mathbf{Con}_{\mathbf{Q}}(\mathbf{A})$,

$$\Phi \supseteq \Theta^{\mathbf{A}}_{\mathbf{Q}}((\forall \underline{e})\ \Lambda^{\mathbf{A}}(a_1, a_2, \ldots, a_m, \underline{e})) \ \text{ implies that}$$

$$\Phi \supseteq \bigcap_{1 \leqslant i < j \leqslant m} \Theta^{\mathbf{A}}_{\mathbf{Q}}(a_i, a_j). \qquad (*)$$

Suppose the first inclusion holds. Then

$$\alpha(b_1, c_1, b_2, c_2, \ldots, b_r, c_r, \underline{e})/\Phi = \beta(b_1, c_1, b_2, c_2, \ldots, b_r, c_r, \underline{e})/\Phi$$

for all $\alpha \approx \beta \in \Lambda$ and all sequences $\underline{e} \in A_k$. Thus

$$\mathbf{A}/\Phi \models (\forall \underline{u}) \bigwedge \Lambda(x_1, x_2, \ldots, x_m, \underline{u})[a_1, a_2, \ldots, a_m].$$

As $\mathbf{A}/\Phi \in \mathbf{Q}_{m-\mathrm{TRI}}$, it follows by (3) that $a_i/\Phi = a_j/\Phi$ for some $1 \leqslant i < j \leqslant m$. So $\Theta^{\mathbf{A}}_{\mathbf{Q}}(a_i, a_j) \subseteq \Phi$ for some $i < j$. Consequently, $\Phi \supseteq \bigcap_{1 \leqslant i < j \leqslant m} \Theta^{\mathbf{A}}_{\mathbf{Q}}(a_i, a_j)$. So $(*)$ holds.

(4) ⇒ (1). Let $\mathbf{A} \in \mathbf{Q}$ and let $a_1, \ldots, a_m$ be a sequence of elements A. Fix a pair i,j ($1 \leqslant i < j \leqslant m$) and let $\Phi_0 := \Theta^{\mathbf{A}}_{\mathbf{Q}}(a_i, a_j)$. The quotient algebra $\mathbf{A}/\Phi_0$ belongs to $\mathbf{Q}$. By the first conjunct of (4) we get that $\alpha(a_1, a_2, \ldots, a_m, \underline{e})/\Phi_0 = \beta(a_1, a_2, \ldots, a_m, \underline{e})/\Phi_0$ for all $\alpha \approx \beta \in \Lambda$ and all sequences $\underline{e} \in A^k$, It follows that $\Theta^{\mathbf{A}}_{\mathbf{Q}}((\forall \underline{e})\ \Lambda^{\mathbf{A}}(a_1, a_2, \ldots, a_m, \underline{e})) \subseteq \Theta^{\mathbf{A}}_{\mathbf{Q}}(a_i, a_j)$. This proves the "$\supseteq$"-inclusion of the two inclusions of (1).

As the other inclusion is assumed by (4), condition (1) follows. □

Definition 8.3.2. Let $m \geqslant 3$ be a natural number and $\Lambda(x_1, x_2, \ldots, x_m, \underline{u})$ a set of equations. A quasivariety $\mathbf{Q}$ is said to have *equationally definable m-triangular meets* of (relatively) *principal congruences by* $\Lambda(x_1, x_2, \ldots, x_m, \underline{u})$ if $\mathbf{Q}$ satisfies any of the equivalent conditions of the above proposition.

$\mathbf{Q}$ is said to have *equationally definable m-triangular meets* of (relatively) *principal congruences* (m-EDTPM, for short) if there is a set of equations $\Lambda(x_1, x_2, \ldots, x_m, \underline{u})$ which defines m-triangular meets of (relatively) principal congruences in the algebras of $\mathbf{Q}$. □

Theorem 8.3.3. *Let $m \geqslant 3$ be a natural number. For any quasivariety* $\mathbf{Q}$ *the following conditions are equivalent:*

(i) $\mathbf{Q}$ *has m-EDTPM.*

(ii) *For every algebra $\boldsymbol{A} \in \mathbf{Q}$, for any sequence $a_1, \dots, a_m$ of elements of A of length m (possibly with repetitions) and any congruence $\Phi \in Con_{\mathbf{Q}}(\boldsymbol{A})$,*

$$\Phi +_{\mathbf{Q}} \bigcap_{1 \leqslant i < j \leqslant m} \Theta^{\boldsymbol{A}}_{\mathbf{Q}}(a_i, a_j) = \bigcap_{1 \leqslant i < j \leqslant m} (\Phi +_{\mathbf{Q}} \Theta^{\boldsymbol{A}}_{\mathbf{Q}}(a_i, a_j)).$$

(iii) *For any sequence $x_1, \dots, x_m$ of m different free generators of the free algebra $\boldsymbol{F} := \boldsymbol{F}_{\mathbf{Q}}(\omega)$ and any congruence $\Phi \in Con_{\mathbf{Q}}(\boldsymbol{F})$,*

$$\Phi +_{\mathbf{Q}} \bigcap_{1 \leqslant i < j \leqslant m} \Theta^{\boldsymbol{F}}_{\mathbf{Q}}(x_i, x_j) = \bigcap_{1 \leqslant i < j \leqslant m} (\Phi +_{\mathbf{Q}} \Theta^{\boldsymbol{F}}_{\mathbf{Q}}(x_i, x_j)).$$

Note. Condition (iii) of the above theorem is equivalently expressed in terms of the consequence $\mathbf{Q}^{\models}$ as:

(iii)* $\mathbf{Q}^{\models}(X \cup \bigcap_{1 \leqslant i < j \leqslant m} \mathbf{Q}^{\models}(x_i \approx x_j)) = \bigcap_{1 \leqslant i < j \leqslant m} \mathbf{Q}^{\models}(X \cup \{x_i \approx x_j\})$,
for any set of equations X.

(Here $x_1, x_2, \dots, x_m$ is an arbitrary but fixed sequence of m different individual variables.) But it is easy to see that (iii)* is equivalent to

(iii)** $\mathbf{Q}^{\models}(X \cup \bigcap_{1 \leqslant i < j \leqslant m} \mathbf{Q}^{\models}(\alpha_i \approx \alpha_j)) = \bigcap_{1 \leqslant i < j \leqslant m} \mathbf{Q}^{\models}(X \cup \{\alpha_i \approx \alpha_j\})$,
for any set of equations X and any sequence of terms $\alpha_1, \alpha_2, \dots, \alpha_m$.

Trivially (iii)** implies (iii)*. To prove the converse, assume (iii)*. But (iii)* is equivalent to condition (iii) of the above theorem, and hence to condition (ii). Thus (iii)* implies (ii). But (ii) holds in particular for the free algebra $\boldsymbol{F}_{\mathbf{Q}}(\omega)$. The last fact implies (iii)**. □

Proof. (i) ⇒ (ii). Assume $\mathbf{Q}$ has m-EDTPM with respect to a set of equations $\Lambda = \Lambda(x_1, x_2, \dots, x_m, \underline{u})$. Let $\boldsymbol{A}$ be an algebra in $\mathbf{Q}$, $\Phi \in Con_{\mathbf{Q}}(\boldsymbol{A})$, and $a_1, \dots, a_m$ sequence of elements of A of length m. Let $\boldsymbol{B} := \boldsymbol{A}/\Phi$. As $\boldsymbol{B} \in \mathbf{Q}$, (i) implies that

$$\bigcap_{1 \leqslant i < j \leqslant m} \Theta^{\boldsymbol{B}}_{\mathbf{Q}}(a_i/\Phi, a_j/\Phi) = \Theta^{\boldsymbol{B}}_{\mathbf{Q}}((\forall \underline{e})\ \Lambda^{\boldsymbol{B}}(a_1/\Phi, a_2/\Phi, \dots, a_m/\Phi, \underline{e}/\Phi)). \tag{a}$$

Let $h : \boldsymbol{A} \to \boldsymbol{B}$ be the canonical homomorphism. Then

$$h^{-1}(\Theta^{\boldsymbol{B}}_{\mathbf{Q}}(a/\Phi, b/\Phi)) = \Phi +_{\mathbf{Q}} \Theta^{\boldsymbol{A}}_{\mathbf{Q}}(a, b) \tag{b}$$

for all $a, b \in A$.

(b) and the surjectivity of h imply that

$$h^{-1}\Big(\bigcap_{1\leqslant i<j\leqslant m} \Theta^{B}_{\mathbf{Q}}(a_i/\Phi, a_j/\Phi)\Big) = \tag{c}$$

$$\bigcap_{1\leqslant i<j\leqslant m} h^{-1}(\Theta^{B}_{\mathbf{Q}}(a_i/\Phi, a_j/\Phi)) =$$

$$\bigcap_{1\leqslant i<j\leqslant m} (\Phi +_{\mathbf{Q}} \Theta^{a}_{\mathbf{Q}}(a_i, a_j)).$$

On the other hand, we also get

$$h^{-1}(\Theta^{B}_{\mathbf{Q}}((\forall \underline{e})\ \Lambda^{B}(a_1/\Phi, a_2/\Phi, \ldots, a_m/\Phi, \underline{e}/\Phi))) = \tag{d}$$

$$h^{-1}(\sup{}_{\mathbf{Q}}\{\Theta^{B}_{\mathbf{Q}}(\Lambda^{B}(a_1/\Phi, a_2/\Phi, \ldots, a_m/\Phi, \underline{e}/\Phi)) : \underline{e} \in A^k\}) =$$

$$\sup{}_{\mathbf{Q}}\{h^{-1}(\Theta^{B}_{\mathbf{Q}}(\Lambda^{B}(a_1/\Phi, a_2/\Phi, \ldots, a_m/\Phi, \underline{e}/\Phi))) : \underline{e} \in A^k\} =$$

$$\sup{}_{\mathbf{Q}}\{\Phi +_{\mathbf{Q}} \Theta^{A}_{\mathbf{Q}}(\Lambda^{A}(a_1, a_2, \ldots, a_m, \underline{e})) : \underline{e} \in A^k\} =$$

$$\Phi +_{\mathbf{Q}} \sup{}_{\mathbf{Q}}\{\Theta^{A}_{\mathbf{Q}}(\Lambda^{A}(a_1, a_2, \ldots, a_r, \underline{e})) : \underline{e} \in A^k\} =$$

$$\Phi +_{\mathbf{Q}} \Theta^{A}_{\mathbf{Q}}((\forall \underline{e})\ \Lambda^{A}(a_1, a_2, \ldots, a_m, \underline{e})) =$$

$$\Phi +_{\mathbf{Q}} \bigcap_{1\leqslant i<j\leqslant m} \Theta^{A}_{\mathbf{Q}}(a_i, a_j).$$

(Here $\sup_{\mathbf{Q}}$ is the supremum in the lattice $\boldsymbol{Con}_{\mathbf{Q}}(\boldsymbol{A})$.) But in view of (a), the first congruences of (c) and (d) are identical. Consequently, the last congruences of (c) and (d) are the same. Thus

$$\bigcap_{1\leqslant i<j\leqslant m} (\Phi +_{\mathbf{Q}} \Theta^{A}_{\mathbf{Q}}(a_i, a_j)) = \Phi +_{\mathbf{Q}} \bigcap_{1\leqslant i<j\leqslant m} \Phi^{A}_{\mathbf{Q}}(a_i, a_j).$$

So (ii) holds.

The implication (ii) $\Rightarrow$ (iii) is immediate.

(iii) $\Rightarrow$ (i). Let $x_1, x_2, \ldots, x_m$ be a finite sequence of pairwise different individual variables. Let $\Lambda(x_1, x_2, \ldots, x_m, \underline{u})$ be a set of equations such that

$$\bigcap_{1\leqslant i<j\leqslant m} \mathbf{Q}^{\models}(x_i \approx x_j) = \mathbf{Q}^{\models}(\Lambda(x_1, x_2, \ldots, x_m, \underline{u})).$$

Passing to the algebra $\boldsymbol{F}$ we therefore have that

$$\bigcap_{1\leqslant i<j\leqslant m} \Theta^{F}_{\mathbf{Q}}(x_i, x_j) = \Theta^{A}_{\mathbf{Q}}(\Lambda^{F}(x_1, x_2, \ldots, x_m, \underline{u})).$$

(We identify here terms of the language of $\mathbf{Q}$ with elements of the free algebra $\boldsymbol{F}$.)

Let $\boldsymbol{A}$ be a countably generated algebra in $\mathbf{Q}$, let $a_1, a_2, \ldots, a_m$ be a fixed sequence of elements of A. Moreover, let $\underline{e} \in A^k$ be an arbitrary sequence of length k.

Let $h : \boldsymbol{F} \to \boldsymbol{A}$ be a surjective homomorphism such that $h(x_1) = a_1, \ldots,$ $h(x_m) = a_m$ and $h(\underline{u}) = \underline{e}$. ($h$ is arbitrarily defined for the remaining free generators of $\boldsymbol{F}$.) Let $\Phi \in Con_{\mathbf{Q}}(\boldsymbol{F})$ be the relation-kernel of h. Then, assuming (iii), we have:

$$h^{-1}\Big(\bigcap_{1 \leqslant i < j \leqslant m} \Theta^{\boldsymbol{A}}_{\mathbf{Q}}(a_i, a_j)\Big) = \tag{e}$$

$$\bigcap_{1 \leqslant i < j \leqslant m} h^{-1}(\Theta^{\boldsymbol{A}}_{\mathbf{Q}}(a_i, a_j)) =$$

$$\bigcap_{1 \leqslant i < j \leqslant m} (\Phi +_{\mathbf{Q}} \Theta^{\boldsymbol{F}}_{\mathbf{Q}}(x_i, x_j)) = \quad \text{(by (iii))}$$

$$\Theta^{\boldsymbol{F}}_{\mathbf{Q}}(\Phi \cup \bigcap_{1 \leqslant i < j \leqslant m} \Theta^{\boldsymbol{F}}_{\mathbf{Q}}(x_i, x_j)) =$$

$$\Theta^{\boldsymbol{F}}_{\mathbf{Q}}(\Lambda(x_1, x_2, \ldots, x_m, \underline{u})) +_{\mathbf{Q}} \Phi =$$

$$h^{-1}(\Theta^{\boldsymbol{A}}_{\mathbf{Q}}(\Lambda(a_1, a_2, \ldots, a_m, \underline{e}))).$$

(e) gives that

$$\bigcap_{1 \leqslant i < j \leqslant m} \Theta^{\boldsymbol{A}}_{\mathbf{Q}}(a_i, a_j) = \Theta^{\boldsymbol{A}}_{\mathbf{Q}}(\Lambda(a_1, a_2, \ldots, a_m, \underline{e})) \text{ for any sequence } \underline{e} \in A^k.$$

It follows that

For every countably generated $\boldsymbol{A} \in \mathbf{Q}$ and all $a_1, a_2, \ldots, a_m \in A$:

$$\bigcap_{1 \leqslant i < j \leqslant m} \Theta^{\boldsymbol{A}}_{\mathbf{Q}}(a_i, a_j) = \mathbf{0}_{\boldsymbol{A}} \quad \Leftrightarrow \quad \Theta^{\boldsymbol{A}}_{\mathbf{Q}}(\Lambda(a_1, a_2, \ldots, a_m, \underline{e})) = \mathbf{0}_{\boldsymbol{A}}$$

$$\text{for all sequences } \underline{e} \in A^k. \tag{f}$$

We then observe that (f) continues to hold for algebras $\boldsymbol{A} \in \mathbf{Q}$ of arbitrary cardinality.

For any algebra $\boldsymbol{A} \in \mathbf{Q}$ and any $a_1, a_2, \ldots, a_m \in A$: (g)

$$\bigcap_{1 \leqslant i < j \leqslant m} \Theta^{\boldsymbol{A}}_{\mathbf{Q}}(a_i, a_j) = \mathbf{0}_{\boldsymbol{A}} \quad \Leftrightarrow \quad \Theta^{\boldsymbol{A}}_{\mathbf{Q}}(\Lambda(a_1, a_2, \ldots, a_m, \underline{e})) = \mathbf{0}_{\boldsymbol{A}} \text{ for all } \underline{e} \in A^k.$$

For let $a_1, a_2, \ldots, a_m \in A$. Let us first assume that $\bigcap_{1 \leqslant i < j \leqslant m} \Theta^{\boldsymbol{A}}_{\mathbf{Q}}(a_i, a_j) = \mathbf{0}_{\boldsymbol{A}}$. Let $\underline{e} \in A^k$ be an arbitrary sequence. $\bigcap_{1 \leqslant i < j \leqslant m} \Theta^{\boldsymbol{A}}_{\mathbf{Q}}(a_i, a_j) = \mathbf{0}_{\boldsymbol{A}}$ implies that $\bigcap_{1\leqslant i<j\leqslant m} \Theta^{\boldsymbol{B}}_{\mathbf{Q}}(a_i, a_j) = \mathbf{0}_{\boldsymbol{B}}$ for some countably generated subalgebra $\boldsymbol{B}$ of $\boldsymbol{A}$ that contains $a_1, a_2, \ldots, a_m$ and $\underline{e}$. Hence, by (f), we have that $\Theta^{\boldsymbol{B}}_{\mathbf{Q}}(\Lambda(a_1, a_2, \ldots, a_m, \underline{e})) = \mathbf{0}_{\boldsymbol{B}}$. It follows that $\Theta^{\boldsymbol{A}}_{\mathbf{Q}}(\Lambda(a_1, a_2, \ldots, a_m, \underline{e})) = \mathbf{0}_{\boldsymbol{A}}$ because $\boldsymbol{B}$ is a subalgebra of $\boldsymbol{A}$.

Conversely, suppose that $\Theta^{\boldsymbol{A}}_{\mathbf{Q}}(\Lambda(a_1, a_2, \ldots, a_m, \underline{e})) = \mathbf{0}_{\boldsymbol{A}}$, for all $\underline{e} \in A^k$. Let $\boldsymbol{B}$ be an arbitrary countably generated subalgebra of $\boldsymbol{A}$ which contains $a_1, a_2, \ldots, a_m$. Then evidently, $\Theta^{\boldsymbol{B}}_{\mathbf{Q}}(\Lambda(a_1, a_2, \ldots, a_m, \underline{e})) = \mathbf{0}_{\boldsymbol{B}}$ for all $\underline{e} \in B^k$. (f) thus gives that

$$\bigcap_{1\leqslant i<j\leqslant m} \Theta^{\boldsymbol{B}}_{\mathbf{Q}}(a_i, a_j) = \mathbf{0}_{\boldsymbol{B}} \text{ in every countably generated subalgebra} \qquad (*)$$

$$\boldsymbol{B} \text{ of } \boldsymbol{A} \text{ which contains } a_1, a_2, \ldots, a_m.$$

We then get that

$$\bigcap_{1\leqslant i<j\leqslant m} \Theta^{\boldsymbol{A}}_{\mathbf{Q}}(a_i, a_j) = \mathbf{0}_{\boldsymbol{A}}. \qquad (**)$$

Indeed, if $\langle c, d\rangle \in \bigcap_{1\leqslant i<j\leqslant m} \Theta^{\boldsymbol{A}}_{\mathbf{Q}}(a_i, a_j)$, then Theorem 2.1 implies that there is a countably generated subalgebra $\boldsymbol{B}$ of $\boldsymbol{A}$ which includes $a_1, a_2, \ldots, a_m$ and c, d such that $\langle c, d\rangle \in \bigcap_{1\leqslant i<j\leqslant m} \Theta^{\boldsymbol{B}}_{\mathbf{Q}}(a_i, a_j)$. It follows by $(*)$ that $c = d$. Hence $(**)$ holds. This proves (g).

But (g) is equivalent to condition (2) of Proposition 8.3.1. It follows that $\mathbf{Q}$ has m-EDTPM. □

The following corollary is immediate (cf. Corollary 8.2.6):

Corollary 8.3.4. *Let $m \geqslant 3$ be a natural number. Let $\mathbf{Q}$ be a quasivariety with m-EDTPM with respect to a set of equations $\Lambda = \Lambda(x_1, x_2, \ldots, x_m, \underline{u})$. Then for any algebra $\boldsymbol{A} \in \mathbf{Q}$, the following conditions are equivalent:*

(1) $\boldsymbol{A} \in \mathbf{Q}_{m-\mathrm{TRI}}$.
(2) *For every sequence $a_1, \ldots, a_m$ of elements of A, if $\Theta^{\boldsymbol{A}}_{\mathbf{Q}}(\Lambda(a_1, a_2, \ldots, a_m, \underline{e}) = \mathbf{0}_{\boldsymbol{A}}$, for all $\underline{e} \in A^k$, then $a_i = a_j$ for some $1 \leqslant i < j \leqslant m$.*

Proof. (1) $\Rightarrow$ (2). Assume (1). Let $a_1, \ldots, a_m$ be a sequence of elements of A such that $\Theta^{\boldsymbol{A}}_{\mathbf{Q}}(\Lambda(a_1, a_2, \ldots, a_m, \underline{e}) = \mathbf{0}_{\boldsymbol{A}}$, for all $\underline{e} \in A^k$. But m-EDTPM gives that $\Theta^{\boldsymbol{A}}_{\mathbf{Q}}((\forall \underline{e})\ \Lambda(a_1, a_2, \ldots, a_m, \underline{e})) = \bigcap_{1\leqslant i<j\leqslant m} \Theta^{\boldsymbol{A}}_{\mathbf{Q}}(a_i, a_j)$. It follows that $\bigcap_{1\leqslant i<j\leqslant m} \Theta^{\boldsymbol{A}}_{\mathbf{Q}}(a_i, a_j) = \mathbf{0}_{\boldsymbol{A}}$. As $\boldsymbol{A} \in \mathbf{Q}_{m-\mathrm{TRI}}$, we get that $a_i = a_j$, for some i, j, $(1 \leqslant i < j \leqslant m)$. So (2) holds.

(2) $\Rightarrow$ (1). The proof of this implication is similar. Assume (2). We must show that $\mathbf{0}_{\boldsymbol{A}}$ is m-triangularily irreducible in $\boldsymbol{Con}_{\mathbf{Q}}(\boldsymbol{A})$. Let $a_1, \ldots, a_m$ be a sequence of elements of A such that $\bigcap_{1\leqslant i<j\leqslant m} \Theta^{\boldsymbol{A}}_{\mathbf{Q}}(a_i, a_j) = \mathbf{0}_{\boldsymbol{A}}$. But by m-EDTPM, $\bigcap_{1\leqslant i<j\leqslant m} \Theta^{\boldsymbol{A}}_{\mathbf{Q}}(a_i, a_j) = \Theta^{\boldsymbol{A}}_{\mathbf{Q}}((\forall \underline{e})\ \Lambda(a_1, a_2, \ldots, a_m, \underline{e}))$. Hence $\Theta^{\boldsymbol{A}}_{\mathbf{Q}}((\forall \underline{e})\ \Lambda(a_1, a_2, \ldots, a_m, \underline{e})) = \mathbf{0}_{\boldsymbol{A}}$. (2) implies that $a_i = a_j$, for some i, j, $(1 \leqslant i < j \leqslant m)$. Thus (1) holds. □

It follows from Theorems 8.2.5 and 8.3.3 that:

Every finitely generated quasivariety has m-EDTPM for some $m \geqslant 3$.

Indeed, assume $\mathbf{Q}$ is generated by a finite set of finite algebras, each of power less than m. Suppose $\boldsymbol{A} \in \mathbf{Q}$, $\Phi \in Con_{\mathbf{Q}}(\boldsymbol{A})$, and $a_1, \ldots, a_m$ is a sequence of elements of $\boldsymbol{A}$. By Theorem 8.2.5 we have that $\bigcap_{1 \leqslant i<j \leqslant m} \Theta^{\boldsymbol{A}}_{\mathbf{Q}}(a_i, a_j) = \mathbf{0}_{\boldsymbol{A}}$ and $\bigcap_{1 \leqslant i<j \leqslant m}(\Phi +_{\mathbf{Q}} \Theta^{\boldsymbol{A}}_{\mathbf{Q}}(a_i, a_j)) = \Phi$. Consequently, $\Phi +_{\mathbf{Q}} \bigcap_{1 \leqslant i<j \leqslant m} \Theta^{\boldsymbol{A}}_{\mathbf{Q}}(a_i, a_j) = \Phi = \bigcap_{1 \leqslant i<j \leqslant m}(\Phi +_{\mathbf{Q}} \Theta^{\boldsymbol{A}}_{\mathbf{Q}}(a_i, a_j))$. So $\mathbf{Q}$ has m-EDTPM, by Theorem 8.3.3.

The following observations supplement Theorem 8.2.5:

Theorem 8.3.5. *Let* $\mathbf{Q}$ *be an arbitrary quasivariety and let* m, $m \geqslant 3$, *be a fixed natural number. The following conditions are equivalent:*

(1) $\mathbf{Q}$ *is generated by a finite class consisting of algebras each of which has at most* $m-1$ *elements.*
(2) $\mathbf{Q}$ *has m-EDTPM and* $\bigcap_{1 \leqslant i<j \leqslant m} \Theta^{\boldsymbol{F}}_{\mathbf{Q}}(x_i, x_j) = \mathbf{0}_{\boldsymbol{F}}$ *for any sequence* $x_1, \ldots, x_m$ *of m different free generators of the free algebra* $\boldsymbol{F} := \boldsymbol{F}_{\mathbf{Q}}(\omega)$.

Proof. (1) $\Rightarrow$ (2). Assume (1). The first conjunct of (2) follows from Theorem 8.2.5 and the second conjunct is a particular case of condition (iii) of Theorem 8.2.5.

(2) $\Rightarrow$ (1). Assume (2). Let $x_1, \ldots, x_m$ be a sequence of m different free generators of the free algebra $\boldsymbol{F} = \boldsymbol{F}_{\mathbf{Q}}(\omega)$. The fact that $\mathbf{Q}$ has m-EDTPM and Theorem 8.3.3 imply that

$$\Phi +_{\mathbf{Q}} \bigcap_{1 \leqslant i<j \leqslant m} \Theta^{\boldsymbol{F}}_{\mathbf{Q}}(x_i, x_j) = \bigcap_{1 \leqslant i<j \leqslant m} (\Phi +_{\mathbf{Q}} \Theta^{\boldsymbol{F}}_{\mathbf{Q}}(x_i, x_j)),$$

for any $\mathbf{Q}$-congruence Φ of $\boldsymbol{F}$. This and the second conjunct of (2) give that

$$\Phi = \bigcap_{1 \leqslant i<j \leqslant m} (\Phi +_{\mathbf{Q}} \Theta^{\boldsymbol{F}}_{\mathbf{Q}}(x_i, x_j)),$$

for any $\mathbf{Q}$-congruence Φ of $\boldsymbol{F}$. So (1) holds by Theorem 8.2.5. □

Notes. 1. The second conjunct of (2) is equivalently formulated in terms of the consequence operation $\mathbf{Q}^{eq\models}$ as

$$\bigcap_{1 \leqslant i<j \leqslant m} \mathbf{Q}^{\models}(x_i \approx x_j) = \mathbf{Q}^{\models}(\emptyset). \tag{a}$$

In the presence of the first conjunct, it is equivalent to the condition:

$$\bigcap_{1 \leqslant i<j \leqslant m} \mathbf{Q}^{\models}(\alpha_i \approx \alpha_j) = \mathbf{Q}^{\models}(\emptyset), \quad \textit{for any sequence of terms } \alpha_1, \alpha_2, \ldots, \alpha_m. \tag{b}$$

In view of the above theorems, the property of having m-EDTPM for some m is essentially weaker than the property of being a finitely generated quasivariety.

2. It follows from Theorem 8.3.3.(ii) that *every* RCD quasivariety $\mathbf{Q}$ has m-EDTPM for all $m \geqslant 3$. ($\mathbf{Q}$ need not be finitely generated.) This observation does not extend onto RCM quasivarieties.

More facts concerning m-EDTPM one may found in Czelakowski (2014). □

Theorem 8.3.6. *Let $m \geqslant 3$ be a natural number. Let $\mathbf{Q}$ be a quasivariety with m-EDTPM with respect to a set of equations $\Lambda = \Lambda(x_1, x_2, \ldots, x_m, \underline{u})$. The following conditions are equivalent:*

(1) *$\mathbf{Q}$ is generated by a finite class of algebras each of which has at most $m - 1$ elements.*

(2) *$\mathbf{Q}$ validates the equations $\Lambda(x_1, x_2, \ldots, x_m, \underline{u})$.*

Proof. (1) $\Rightarrow$ (2). Assume (1). It follows from Theorem 8.3.5.(ii) that $\bigcap_{1 \leqslant i < j \leqslant m} \Theta^{F}_{\mathbf{Q}}(x_i, x_j) = \mathbf{0}_F$, where $x_1, \ldots, x_m$ is an arbitrary but fixed sequence of m different free generators of the free algebra $\boldsymbol{F} := \boldsymbol{F}_{\mathbf{Q}}(\omega)$. As $\mathbf{Q}$ has m-EDTPM with respect to Λ we therefore get that $\Theta^{F}_{\mathbf{Q}}((\forall \underline{e})\ \Lambda^{F}(x_1, x_2, \ldots, x_m, \underline{e})) = \mathbf{0}_F$. This is equivalent to $\mathbf{Q}^{\models}((\forall \underline{u})\ \Lambda(x_1, x_2, \ldots, x_m, \underline{u})) = \mathbf{Q}^{\models}(\emptyset)$. It follows that $\mathbf{Q}^{\models}(\Lambda(x_1, x_2, \ldots, x_m, \underline{u})) = \mathbf{Q}^{\models}(\emptyset)$. So (2) holds.

(2) $\Rightarrow$ (1). (2) implies that $\Theta^{A}_{\mathbf{Q}}((\forall \underline{e})\ \Lambda(a_1, a_2, \ldots, a_m, \underline{e})) = \mathbf{0}_A$, for any $\boldsymbol{A} \in \mathbf{Q}$ and any sequence $a_1, \ldots, a_m$ of elements of A. But $\bigcap_{1 \leqslant i < j \leqslant m} \Theta^{A}_{\mathbf{Q}}(a_i, a_j) = \Theta^{A}_{\mathbf{Q}}((\forall \underline{e})\ \Lambda(a_1, a_2, \ldots, a_m, \underline{e}))$ because $\mathbf{Q}$ has m-EDTPM with respect to Λ. It follows that $\bigcap_{1 \leqslant i < j \leqslant m} \Theta^{A}_{\mathbf{Q}}(a_i, a_j) = \mathbf{0}_A$, for any $\boldsymbol{A} \in \mathbf{Q}$ and any sequence $a_1, \ldots, a_m$ of elements of A. This gives (1), by Theorem 8.2.5. □

We already know that *every* finitely generated quasivariety $\mathbf{Q}$ has the m-EDTPM property for sufficiently large m. But, more interestingly, $\mathbf{Q}$ has m-EDTPM with respect to a finite set of *trivial* equations:

Theorem 8.3.7. *Let $m \geqslant 3$ be a natural number. Suppose that $\mathbf{Q}$ is a quasivariety generated by a finite class consisting of algebras each of which has at most $m - 1$ elements. Then $\mathbf{Q}$ has m-EDTPM with respect to the following finite set of trivial equations*

$$\Lambda(x_1, x_2, \ldots, x_m) := \{x_1 \approx x_1, x_2 \approx x_2, \ldots, x_m \approx x_m\}.$$

Proof. Let $\mathbf{Q}$ be as above. It is clear that for any $\boldsymbol{A} \in \mathbf{Q}$ and any $a_1, \ldots, a_m \in A$, $\Theta^{A}_{\mathbf{Q}}(\Lambda(a_1, a_2, \ldots, a_m)) = \mathbf{0}_A$. This means that $\boldsymbol{A} \models \bigwedge \Lambda(x_1, x_2, \ldots, x_m)[a_1, a_2, \ldots, a_m]$. On the other hand, $\bigcap_{1 \leqslant i < j \leqslant m} \Theta^{A}_{\mathbf{Q}}(a_i, a_j) = \mathbf{0}_A$, by Theorem 8.2.5.(ii). It follows that any $\boldsymbol{A} \in \mathbf{Q}$ and any $a_1, \ldots, a_m \in A$, the equivalence

$$\bigcap_{1 \leqslant i < j \leqslant m} \Theta^{A}_{\mathbf{Q}}(a_i, a_j) = \mathbf{0}_A \quad \Leftrightarrow \quad \boldsymbol{A} \models \bigwedge \Lambda(x_1, x_2, \ldots, x_m)[a_1, a_2, \ldots, a_m].$$

is true because its both sides are true.

Thus any algebra $\mathbf{A} \in \mathbf{Q}$ validates condition (2) of Proposition 8.3.1. It follows that $\mathbf{Q}$ has m-EDTPM with respect to $\Lambda(x_1, x_2, \ldots, x_m)$. □

The above proof shows that one cannot expect much from the m-EDTPM in general while studying specific properties of *finitely generated* quasivarieties $\mathbf{Q}$—the equations of $\Lambda(x_1, x_2, \ldots, x_m)$ from Theorem 8.3.7 are not conjoined with the intrinsic structure of the algebras of $\mathbf{Q}$. Consequently, the m-EDTPM property trivializes for $\mathbf{Q}$. But if one imposes a further constraint on m-EDTPM viz. by requiring that $\mathbf{Q}$ has m-EDTPM *with respect to a certain specific set of equations* $\Lambda(x_1, x_2, \ldots, x_m)$, the problem becomes less trivial.

On the other hand, the fact that a quasivariety $\mathbf{Q}$ has m-EDTPM with respect to a set of equations $\Lambda(x_1, x_2, \ldots, x_m, \underline{u})$ need not imply that $\mathbf{Q}$ has m-EDTPM with respect to trivial equations $\{x_1 \approx x_1, x_2 \approx x_2, \ldots, x_m \approx x_m\}$. For instance, as $\mathbf{Q}$ one may take any non-finitely generated RCD quasivariety.

The following characterization of the m-EDTPM property in terms of equational consequence operations has some interesting consequences:

Theorem 8.3.8. *Let $m \geqslant 3$ be a natural number and $\Lambda = \Lambda(x_1, x_2, \ldots, x_m, \underline{u})$ a set of equations. Then for any quasivariety $\mathbf{Q}$ the following conditions are equivalent:*

(1) *$\mathbf{Q}$ has m-EDTPM with respect to $\Lambda(x_1, x_2, \ldots, x_m, \underline{u})$,*

(2) $\mathbf{Q}^{\models}(X \cup (\forall \underline{u})\, \Lambda(x_1, x_2, \ldots, x_m, \underline{u})) = \bigcap_{1 \leqslant i<j \leqslant m} \mathbf{Q}^{\models}(X \cup \{x_i \approx x_j\})$

for any set of equations X (equivalently, for any finite set of equations X),

(3) $\mathbf{Q}^{\models}(X \cup (\forall \underline{u})\, \Lambda(\alpha_1, \alpha_2, \ldots, \alpha_m, \underline{u})) = \bigcap_{1 \leqslant i<j \leqslant m} \mathbf{Q}^{\models}(X \cup \{\alpha_i \approx \alpha_j\})$

for any sequence of terms $\alpha_1, \alpha_2, \ldots, \alpha_m$ and any set of equations X (equivalently, any finite set of equations X).

Proof. (1) implies (3) (see Note following Theorem 8.3.3) and (3) trivially implies (2). In turn, (2) implies (1) because (2) is equivalent to condition (iii) of Theorem 8.3.3. □

8.4 Properties Related to m-EDTPM

In view of Theorem 8.3.3 and the remarks following it, the fact that a quasivariety has m-EDTPM is equivalent to the following equality holding for the consequence operation $\mathbf{Q}^{\models}$:

$$\mathbf{Q}^{\models}(X \cup \bigcap_{1 \leqslant i<j \leqslant m} \mathbf{Q}^{\models}(x_i \approx x_j)) = \bigcap_{1 \leqslant i<j \leqslant m} \mathbf{Q}^{\models}(X \cup \{x_i \approx x_j\}), \qquad (1)_m$$

for any set of equations X and any string of m different individual variables $x_1, x_2, \ldots, x_m$. One may relax m-EDTPM by assuming additionally that the variables $x_1, x_2, \ldots, x_m$ are separated from the variables occurring in the equations of X. This leads to the following definition:

Definition 8.4.1. Let m be a positive integer, $m \geqslant 3$. A quasivariety $\mathbf{Q}$ is said to have *equationally definable separated m-triangular meets* of (relatively) *principal congruences* (in short: $\mathbf{Q}$ has *the separated* m-EDTPM) if $(1)_m$ holds for any set of equations X and any string of m different individual variables $x_1, x_2, \ldots, x_m$ not occurring in the equations of X. □

Theorem 8.4.2. *Let* $\mathbf{Q}$ *be any quasivariety. Then* $\mathbf{Q}$ *has separated m-EDTPM for all* $m \geqslant 3$.

Proof. Apply Corollary 5.2.11. □

We thus see that the above property trivializes for all quasivarieties. But one may also consider the dual property to the separated m-EDTPM.

Definition 8.4.3. Let m be a positive integer, $m \geqslant 3$. A quasivariety $\mathbf{Q}$ is said to satisfy *the dual separated* m-EDTPM if the consequence operation $\mathbf{Q}^{\models}$ obeys the equality dual to $(1)_m$, viz.,

$$\mathbf{Q}^{\models}(X) \cap \mathbf{Q}^{\models}\Big(\bigcup_{1 \leqslant i < j \leqslant m} \mathbf{Q}^{\models}(x_i \approx x_j)\Big) = \mathbf{Q}^{\models}\Big(\bigcup_{1 \leqslant i < j \leqslant m} \mathbf{Q}^{\models}(X) \cap \mathbf{Q}^{\models}(x_i \approx x_j)\Big), \quad (2)_m$$

for any set of equations X and any string of m different individual variables $x_1, x_2, \ldots, x_m$ not occurring in the equations of X. □

Theorem 8.4.4. *Let* $\mathbf{Q}$ *be an arbitrary quasivariety such that the lattice* $\boldsymbol{Th}(\mathbf{Q}^{\models})$ *validates the restricted distributivity property. Then* $\mathbf{Q}$ *has the dual separated m-EDTPM, for all* $m \geqslant 3$.

Proof. This directly follows from Corollary 5.3.2. □

Corollary 8.4.5. *Let* $\mathbf{Q}$ *be an arbitrary RCM quasivariety. Then* $\mathbf{Q}$ *has the dual separated m-EDTPM, for all* $m \geqslant 3$.

Proof. Use Theorems 6.3.2 and 8.4.4. □

Chapter 9
Commutator Laws in Finitely Generated Quasivarieties

9.1 Iterations of Generating Sets

Let **K** be a class of algebras. Suppose $\Delta = \Delta(x, y, z, w, \underline{u})$ is a set of quaternary commutator equations for **K** such that

$$\mathbf{K}^{\models}(x \approx y) \cap \mathbf{K}^{\models}(z \approx w) = \mathbf{K}^{\models}(\Delta(x, y, z, w, \underline{u})). \tag{1}$$

If Δ satisfies (1), it is a generating set for the equationally defined commutator for **Q** (see Definition 4.3.7).

We partition the set of individual variables *Var* into infinite (disjoint) subsets $V, U_1, U_2, \ldots, U_n, \ldots$. We assume that x, y, z, and w belong to V. Then, for each positive integer i we select a string $\underline{u}_i = u_{i,1}, u_{i,2}, \ldots$ of distinct variables in U_i of length k (= the length of $\underline{u}$). k may be equal to ω. But if Δ is finite, then $k \in \omega$. For each positive integer i we also select two different variables x_i and y_i in V different from the fixed variables x, y, z and w. We thus have two infinite sequences of variables

$$x_1, x_2, x_3, \ldots, x_n, x_{n+1}, \ldots$$

$$y_1, y_2, y_3, \ldots, y_n, y_{n+1}, \ldots.$$

We then inductively define an infinite sequence of sets of equations

$$\Delta_1, \Delta_2, \ldots$$

such that

(a) Δ_n is built from the variables $x_1, \ldots, x_{n+1}, y_1, \ldots, y_{n+1}$ and the variables $\underline{v}_n := \underline{u}_1 \cup \ldots \cup \underline{u}_n$ for $n = 1, 2, \ldots$. We therefore write

$$\Delta_n = \Delta_n(x_1, y_1, \ldots, x_{n+1}, y_{n+1}, \underline{v}_n).$$

J. Czelakowski, *The Equationally-Defined Commutator*,
DOI 10.1007/978-3-319-21200-5_9

The variables occurring in $\underline{v}_n$ are called *parametric* variables of Δ_n. The length of $\underline{v}_n$ is $n \cdot k$.

We put:

(b) $\Delta_1(x_1, y_1, x_2, y_2, \underline{v}_1)$ is equal to $\Delta(x/x_1, y/y_1, z/x_2, w/y_2, \underline{u}/\underline{v}_1)$, i.e., $\Delta_1(x_1, y_1, x_2, y_2, \underline{v}_1)$ is the result of the uniform replacing of x by x_1, y by y_1, z by x_2, w by y_2, and the parametric variables $\underline{u}$ in Δ by the consecutive variables of $\underline{v}_1$. (Note that $\underline{v}_1 = \underline{u}_1$ and $\underline{v}_n := \underline{v}_{n-1} \cup \underline{u}_n$ for $n \geqslant 2$.)

(c) $\Delta_{n+1} := \bigcup\{\Delta(x/\alpha, y/\beta, z/x_{n+2}, w/y_{n+2}, \underline{u}/\underline{u}_{n+1}) : \alpha \approx \beta \in \Delta_n\}$, for all $n \geqslant 1$.

Lemma 9.1.1. *If the set Δ is finite, then so are the sets $\Delta_n(x_1, y_1, \ldots, x_{n+1}, y_{n+1}, \underline{v}_n)$, for all n.*

Proof. Immediate. □

Let E be a collection of substitutions $e : \boldsymbol{Te}_\tau \to \boldsymbol{Te}_\tau$ such that

$$e(x_2) = x_3, \quad e(x_3) = x_4, \quad \ldots, \quad e(x_n) = x_{n+1}, \ldots$$

$$e(y_2) = y_3, \quad e(y_3) = y_4, \quad \ldots, \quad e(y_n) = y_{n+1}, \ldots$$

$$e(\underline{u}_1) = \underline{u}_2, \quad e(\underline{u}_2) = \underline{u}_3, \quad \ldots, \quad e(\underline{u}_n) = \underline{u}_{n+1}, \ldots$$

and whose values for x_1 any y_1 range over the equations of Δ_1, i..e.,

(*) for each $e \in E$, the equation $e(x_1) \approx e(y_1)$ belongs to Δ_1,

and

(**) for every equation $\alpha \approx \beta \in \Delta_1$ there is a substitution $e \in E$ such that $e(x_1) = \alpha$, $e(y_1) = \beta$.

It is clear that such a set E exists.

For each positive integer n we define recursively the following sequence of equations:

$$\Sigma_1 := \Delta_1(x_1, y_1, x_2, y_2, \underline{v}_1)$$

$$\Sigma_{n+1} := \bigcup\{e[\Sigma_n] : e \in E\}.$$

Lemma 9.1.2. $\Delta_n = \Sigma_n$*, for all* $n \geqslant 1$.

Proof. The lemma is proved by induction on n.

Induction base.

$\Delta_1 = \Sigma_1$, by definition. We show $\Delta_2 = \Sigma_2$. We have:

$$\Delta_2 := \bigcup\{\Delta(x/\alpha, y/\beta, z/x_3, w/y_3, \underline{u}/\underline{u}_2) : \alpha \approx \beta \in \Delta_1(x_1, y_1, x_2, y_2, \underline{v}_1)\} =$$

$$\bigcup\{\Delta_1(x_1/\alpha, y_1/\beta, x_2/x_3, y_2/y_3, \underline{v}_1/\underline{u}_2) : \alpha \approx \beta \in \Delta_1(x_1, y_1, x_2, y_2, \underline{v}_1)\} =$$

$$\bigcup\{e[\Delta_1] : e \in E\} = \bigcup\{e[\Sigma_1] : e \in E\} = \Sigma_2.$$

Inductive step. Let n be a positive integer, $n \geqslant 2$. We assume that $\Delta_k = \Sigma_k$, for all positive $k \leqslant n$. We claim that $\Delta_{n+1} = \Sigma_{n+1}$. We compute:

$$\Delta_{n+1} =$$

$$\bigcup\{\Delta(x/\alpha, y/\beta, z/x_{n+2}, w/y_{n+2}, \underline{u}/\underline{u}_{n+1}) : \alpha \approx \beta \in \Delta_n\} = \quad \text{(by (IH))}$$

$$\bigcup\{\Delta(x/\alpha, y/\beta, z/x_{n+2}, w/y_{n+2}, \underline{u}/\underline{u}_{n+1}) : \alpha \approx \beta \in \Sigma_n\} =$$

$$\bigcup\{\Delta(\alpha, \beta, x_{n+2}, y_{n+2}, \underline{u}_{n+1}) : \alpha \approx \beta \in \bigcup\{e[\Sigma_{n-1}] : e \in E\}\} =$$

$$\bigcup\{\bigcup\{\Delta(\alpha, \beta, x_{n+2}, y_{n+2}, \underline{u}_{n+1}) : \alpha \approx \beta \in e[\Sigma_{n-1}]\} : e \in E\} =$$

$$\bigcup\{\bigcup\{\Delta(\alpha, \beta, ex_{n+1}, ey_{n+1}, e\underline{u}_n) : \alpha \approx \beta \in e[\Sigma_{n-1}]\} : e \in E\} =$$

$$\bigcup\{\bigcup\{\Delta(e\gamma, e\delta, ex_{n+1}, ey_{n+1}, e\underline{u}_n) : \gamma \approx \delta \in \Sigma_{n-1}\} : e \in E\} =$$

$$\bigcup\{\bigcup\{e[\Delta(\gamma, \delta, x_{n+1}, y_{n+1}, \underline{u}_n)] : \gamma \approx \delta \in \Sigma_{n-1}\} : e \in E\} =$$

$$\bigcup\{e[\bigcup\{\Delta(\gamma, \delta, x_{n+1}, y_{n+1}, \underline{u}_n) : \gamma \approx \delta \in \Sigma_{n-1}\}] : e \in E\} = \quad \text{(by (IH))}$$

$$\bigcup\{e[\bigcup\{\Delta(\gamma, \delta, x_{n+1}, y_{n+1}, \underline{u}_n) : \gamma \approx \delta \in \Delta_{n-1}\}] : e \in E\} =$$

$$\bigcup\{e[\Delta_n] : e \in E\} = \quad \text{(by (IH))}$$

$$\bigcup\{e[\Sigma_n] : e \in E\} = \Sigma_{n+1}.$$

The lemma is proved. □

Note. The above lemma is the equational counterpart of a simple algebraic fact. Suppose $\vee$ is a binary operation symbol. Let x be an individual variable and let $x_1, x_2, x_3, \ldots, x_n, x_{n+1}, \ldots$ be an infinite sequence of pairwise distinct variables, all different from x. Define recursively the following sequence of terms:

$$t_1 := (x_1 \vee x_2),$$

$$t_{n+1} := (t_n \vee x_{n+2}), \qquad \text{for all } n \geqslant 1.$$

Thus t_n equals $((\ldots((x_1 \vee x_2) \vee x_3) \vee \ldots) \vee x_{n+1})$, for all $n \geqslant 1$.

The same sequence can be defined in a different way. Let $e : \boldsymbol{Te}_\tau \to \boldsymbol{Te}_\tau$ be a substitution such that

$$e(x_1) := (x_1 \vee x_2),$$

$$e(x_{n+1}) = x_{n+2}, \qquad \text{for all } n \geqslant 1.$$

We define the following sequence of terms:

$$s_1 := e(x_1),$$

$$s_{n+1} := e(s_n), \qquad \text{for all } n \geqslant 1.$$

The terms t_n and s_n coincide for all $n \geqslant 1$. (We recall that if $t(z_1, \ldots, z_n)$ is a term in variables $z_1, \ldots, z_n$ and e is a substitution, then $e(t)$ is the term $t(z_1/e(z_1), \ldots, z_n/e(z_n))$.) □

Lemma 9.1.3. *Let* **K** *be a class of algebras. Let* $\Delta(x, y, z, w, \underline{u})$ *be a set of quaternary commutator equations for* **K** *satisfying* (1). *Define the sets* $\Delta_1, \Delta_2, \ldots$ *as above. Then*

(a) $\mathbf{K}^{\models}(\Delta_1) = \mathbf{K}^{\models}(x_1 \approx y_1) \cap \mathbf{K}^{\models}(x_2 \approx y_2)$,
(b) $\mathbf{K}^{\models}(\Delta_n) \subseteq \mathbf{K}^{\models}(x_1 \approx y_1) \cap \ldots \cap \mathbf{K}^{\models}(x_{n+1} \approx y_{n+1})$
and
(c) $\Delta_{n+1} \subseteq \mathbf{K}^{\models}(\Delta_n)$ *for* $n = 1, 2, \ldots$

Proof. As $\mathbf{K}^{\models}(x \approx y) \cap \mathbf{K}^{\models}(z \approx w) = \mathbf{K}^{\models}(\Delta(x, y, z, w, \underline{u}))$, we get that $\Delta(x, y, z, w, \underline{u}) \subseteq \mathbf{K}^{\models}(x \approx y)$ and $\Delta(x, y, z, w, \underline{u}) \subseteq \mathbf{K}^{\models}(z \approx w)$. Then the structurality of $\mathbf{K}^{\models}$ and the definition of $\underline{u}_1$ give that $\Delta(x_1, y_1, x_2, y_2, \underline{u}_1) \subseteq \mathbf{K}^{\models}(x_1 \approx y_1)$ and $s\Delta(x_1, y_1, x_2, y_2, \underline{u}_1) \subseteq \mathbf{K}^{\models}(x_2 \approx y_2)$. Hence $\mathbf{K}^{\models}(\Delta_1) \subseteq \mathbf{K}^{\models}(x_1 \approx y_1) \cap \mathbf{K}^{\models}(x_2 \approx y_2)$.

Now assume that $s \approx t \in \mathbf{K}^{\models}(x_1 \approx y_1) \cap \mathbf{K}^{\models}(x_2 \approx y_2)$, i.e., $s \approx t \in \mathbf{K}^{\models}(x_1 \approx y_1)$ and $s \approx t \in \mathbf{K}^{\models}(x_2 \approx y_2)$. Let e_1 be a substitution which is a permutation of individual variables such that $e_1(x_1) = x$, $e_1(y_1) = y$, $e_1(x_2) = z$, $e_1(y_2) = w$, and $e_1(\underline{u}_1) = \underline{u}$. Then, by structurality, $e_1(s \approx t) \in \mathbf{K}^{\models}(x \approx y)$ and $e_1(s \approx t) \in \mathbf{K}^{\models}(z \approx w)$. It follows that

$$e_1(s) \approx e_1(t) \in \mathbf{K}^{\models}(x \approx y) \cap \mathbf{K}^{\models}(z \approx w) = \mathbf{K}^{\models}(\Delta(x, y, z, w, \underline{u})).$$

Hence

$$e_1(s) \approx e_1(t) \in \mathbf{K}^{\models}(\Delta(x, y, z, w, \underline{u})). \qquad (*)$$

Let e_2 be the inverse of e_1. $(*)$ and $e_1(\underline{u}_1) = \underline{u}$ imply that $e_2(\underline{u}) = \underline{u}_1$ and

$$s \approx t = e_2(e_1(s)) \approx e_2(e_1(t)) \in \mathbf{K}^{\models}(e_2(\Delta(x, y, z, w, \underline{u}))) =$$

$$\mathbf{K}^{\models}(\Delta(x/e_2(x), y/e_2(y), z/e_2(z), w/e_2(w), \underline{u}/e_2(\underline{u}))) =$$

$$\mathbf{K}^{\models}(\Delta(x_1, y_1, x_2, y_2, \underline{u}_1)).$$

This proves that $\mathbf{K}^{\models}(x_1 \approx y_1) \cap \mathbf{K}^{\models}(x_2 \approx y_2) \subseteq \mathbf{K}^{\models}(\Delta(x_1, y_1, x_2, y_2, \underline{u}_1))$. So $\mathbf{K}^{\models}(\Delta_1) = \mathbf{K}^{\models}(x_1 \approx y_1) \cap \mathbf{K}^{\models}(x_2 \approx y_2)$, i.e., (a) holds.

(b) follows from the definition of Δ_n.

To prove (c), fix $n \geqslant 1$. Let $\boldsymbol{A}$ be an algebra in $\mathbf{K}$ and let $h : \boldsymbol{Te}_\tau \to \boldsymbol{A}$ be a homomorphism such that h validates every equation in $\Delta_n = \Delta_n(x_1, y_1, \ldots, x_{n+1}, y_{n+1}, \underline{v}_n)$. We claim h validates every equation in $\Delta_{n+1} = \Delta_{n+1}(x_1, y_1, \ldots, x_{n+1}, y_{n+1}, x_{n+2}, y_{n+2}, \underline{v}_{n+1})$. As $h\alpha = h\beta$ for all $\alpha \approx \beta \in \Delta_n$, the fact that $\Delta(x, y, z, w, \underline{u})$ is a set of commutator equations for $\mathbf{K}$ implies that every equation in $\Delta(x/\alpha, y/\beta, z/z_{n+1}, w/w_{n+1}, \underline{u}/\underline{u}_{n+1})$ is validated by h, for all $\alpha \approx \beta \in \Delta_n$. Consequently, h validates the set Δ_{n+1}. □

In view of the above lemma,

$$\mathbf{K}^{\models}(\Delta_1) \supseteq \mathbf{K}^{\models}(\Delta_2) \supseteq \ldots \supseteq \mathbf{K}^{\models}(\Delta_n) \supseteq \ldots \tag{2}$$

i.e., $\mathbf{K}^{\models}(\Delta_1), \mathbf{K}^{\models}(\Delta_2), \ldots$ is a descending chain of theories of the equational consequence $\mathbf{K}^{\models}$ and Δ_1 is a generating set of quaternary commutator equations.

Lemma 9.1.4. *Let* $\mathbf{K}$ *be a class of algebras. Let* $\Delta(x, y, z, w, \underline{u})$ *be a set of quaternary commutator equations for* $\mathbf{K}$ *satisfying* (1). *Define the sets* $\Delta_1, \Delta_2, \ldots$ *as above. Suppose that there exists a positive integer* m *such that* $\mathbf{K}^{\models}(\Delta_{m+1}) = \mathbf{K}^{\models}(\Delta_m)$. *Then* $\mathbf{K}^{\models}(\Delta_n) = \mathbf{K}^{\models}(\Delta_m)$ *for all* $n \geqslant m$.

Proof. Define the set E of substitutions as in the proof of Lemma 9.1.2.

We let $\underline{v}_n - \underline{u}$ denote, for each positive $n \geqslant 2$, the truncated string of variables obtained from $\underline{v}_n$ by deletion of the variables occurring in $\underline{u}$. Thus $\underline{v}_n - \underline{u} = \underline{v}_n - \underline{v}_1$, for all $n \geqslant 2$. Note that $e(\underline{v}_n) = \underline{v}_{n+1} - \underline{u}$, for all $n \geqslant 1$, and all $e \in E$.

We prove by induction that $\mathbf{K}^{\models}(\Delta_{n+1}) = \mathbf{K}^{\models}(\Delta_n)$ for all positive integers $n \geqslant m$. The induction base (for $n = m$) holds in virtue of the assumption. To make the inductive step, fix $n \geqslant m$ and assume that $\mathbf{K}^{\models}(\Delta_{n+1}) = \mathbf{K}^{\models}(\Delta_n)$ holds. We claim that $\mathbf{K}^{\models}(\Delta_{n+2}) = \mathbf{K}^{\models}(\Delta_{n+1})$ holds as well. It suffices to show that $\Delta_{n+1} \subseteq \mathbf{K}^{\models}(\Delta_{n+2})$. We have that

$$\Delta_n(x_1, y_1, x_2, y_2, \ldots, x_{n+1}, y_{n+1}, \underline{v}_n) \subseteq \mathbf{K}^{\models}(\Delta_{n+1}(x_1, y_1, x_2, y_2, \ldots, x_{n+1}, y_{n+1}, x_{n+2}, y_{n+2}, \underline{v}_{n+1})), \tag{a}$$

by the induction hypothesis.

(a) and the structurality of $\mathbf{K}^{\models}$ imply that

$$\Delta_n(ex_1, ey_1, ex_2, ey_2, \ldots, ex_{n+1}, ey_{n+1}, e\underline{v}_n) \subseteq \mathbf{K}^{\models}(\Delta_{n+1}(ex_1, ey_1, ex_2, ey_2, \ldots, ex_{n+1}, ey_{n+1}, ex_{n+2}, ey_{n+2}, e\underline{v}_{n+1})).$$

for all $e \in E$. (Parentheses are suppressed as much as possible here.) This means that

$$\Delta_n(\alpha, \beta, x_3, y_3, \ldots, x_{n+2}, y_{n+2}, \underline{v}_{n+1} - \underline{u}) \subseteq \mathbf{K}^{\models}(\Delta_{n+1}(\alpha, \beta, x_3, y_3, \ldots, x_{n+2}, y_{n+2}, x_{n+3}, y_{n+3}, \underline{v}_{n+2} - \underline{u})),$$

for all $\alpha \approx \beta \in \Delta_1(x_1, y_1, x_2, y_2, \underline{v}_1)$. Consequently,

$$\bigcup\{\Delta_n(\alpha, \beta, x_3, y_3, \ldots, x_{n+2}, y_{n+2}, \underline{v}_{n+1} - \underline{u}) : \tag{b}$$
$$\alpha \approx \beta \in \Delta_1(x_1, y_1, x_2, y_2, \underline{v}_1)\} \subseteq$$
$$\mathbf{K}^{\models}(\bigcup\{\Delta_{n+1}(\alpha, \beta, x_3, y_3, \ldots, x_{n+2}, y_{n+2}, x_{n+3}, y_{n+3}, \underline{v}_{n+2} - \underline{u}) :$$
$$\alpha \approx \beta \in \Delta_1(x_1, y_1, x_2, y_2, \underline{v}_1)\}).$$

But according to Lemma 9.1.2,

$$\bigcup\{\Delta_n(\alpha, \beta, x_3, y_3, \ldots, x_{n+2}, y_{n+2}, \underline{v}_{n+1} - \underline{u}) : \tag{c}$$
$$\alpha \approx \beta \in \Delta_1(x_1, y_1, x_2, y_2, \underline{v}_1)\} = \Delta_{n+1}, \qquad \text{for all positive } n.$$

(b) and (c) imply that $\Delta_{n+1} \subseteq \mathbf{K}^{\models}(\Delta_{n+2})$. This completes the proof of the lemma. □

Theorem 9.1.5. *Let* **K** *be a class of algebras. Let* $\Delta(x, y, z, w, \underline{u})$ *be a set of quaternary commutator equations for* **K** *satisfying* (1). *Define the sets* $\Delta_1, \Delta_2, \ldots$ *as above. Then either*

$$\mathbf{K}^{\models}(\Delta_1), \mathbf{K}^{\models}(\Delta_2), \ldots$$

is a strictly descending infinite chain of theories or there exists a positive integer m *such that*

$$\mathbf{K}^{\models}(\Delta_n) = \mathbf{K}^{\models}(\Delta_m) \qquad \textit{for all } n \geqslant m.$$

Proof. In view of Lemma 9.1.3, $\mathbf{K}^{\models}(\Delta_1), \mathbf{K}^{\models}(\Delta_2), \ldots$ is a descending chain of theories. Suppose that the first disjunct of the statement of the theorem does not hold. Thus the above chain is not strictly descending. Hence there exists a positive integer m such that $\mathbf{K}^{\models}(\Delta_{m+1}) = \mathbf{K}^{\models}(\Delta_m)$. This, in view of Lemma 9.1.4, implies that $\mathbf{K}^{\models}(\Delta_n) = \mathbf{K}^{\models}(\Delta_m)$ for all $n \geqslant m$. So the second disjunct of the statement holds. □

Theorem 9.1.5 and the lemmas preceding it neither assume the additivity of the equationally defined commutator nor the finiteness of the set $\Delta(x, y, z, w, \underline{u})$. (The above proof encompasses the case $\Delta(x, y, z, w, \underline{u})$ is infinite. But in this general case, the sets Δ_n, $n \in \omega$, are infinite as well.)

However, the assumption that the set Δ is finite together with additivity of the equationally defined commutator will be to a large extent explored in the next section.

Example 9.1.6. The symbol **2** stands for a two-element Boolean algebra. **BA** denotes the variety of Boolean algebras, **BA** $=$ ***ISP*(2)**. As **BA** is congruence-distributive, the commutator of any two congruences on a Boolean algebra coincides with the meet of the two congruences. Let $\alpha \approx \beta$ be the equation

$$(x \leftrightarrow y) \vee (z \leftrightarrow w) \approx 1.$$

(Here $\leftrightarrow, \vee$ and 1 stand for the operation symbols of Boolean operations of equivalence and join, respectively. 1 stands for the unit element.) It is clear that $\alpha \approx \beta$ is a quaternary commutator equation for **BA**. Moreover, the singleton set $\Delta(x, y, z, w, \underline{u}) := \{\alpha \approx \beta\}$ generates the (equational) commutator for **BA**, i.e.,

$$\mathbf{BA}^{\models}(\alpha \approx \beta) = \mathbf{BA}^{\models}(x \approx y) \cap \mathbf{BA}^{\models}(z \approx w).$$

($\Delta(x, y, z, w)$ does not contain parametric variables.)

We define the sets $\Delta_n(x_1, y_1, x_2, y_2, \ldots, x_{n+1}, y_{n+1})$ for $n = 1, 2, \ldots$ as above. Each set Δ_n is a singleton, $\Delta_n = \{\alpha_n \approx \beta_n\}$, where β_n is the constant 1 and the term $\alpha_n(x_1, y_1, x_2, y_2, \ldots, x_{n+1}, y_{n+1})$ is inductively defined as follows:

$$\alpha_1(x_1, y_1, x_2, y_2) := (x_1 \leftrightarrow y_1) \vee (x_2 \leftrightarrow y_2),$$

$$\alpha_{n+1}(x_1, y_1, \ldots, x_{n+2}, y_{n+2}) := (\alpha_n \leftrightarrow 1) \vee (x_{n+2} \leftrightarrow y_{n+2})$$

for $n = 1, 2, \ldots$

We claim that

$$\mathbf{BA}^{\models}(\Delta_1), \mathbf{BA}^{\models}(\Delta_2), \ldots$$

is a strictly descending infinite chain of equational theories.

In view of the above remarks, we have that

$$\mathbf{BA}^{\models}(\Delta_1) \supseteq \mathbf{BA}^{\models}(\Delta_2) \supseteq \ldots \supseteq \mathbf{BA}^{\models}(\Delta_n) \supseteq \ldots$$

Fix $n \geqslant 1$. To prove that $\mathbf{BA}^{\models}(\Delta_{n+1})$ is a proper subset of $\mathbf{BA}^{\models}(\Delta_n)$, it suffices to show that $\alpha_n \approx \beta_n \notin \mathbf{BA}^{\models}(\alpha_{n+1} \approx \beta_{n+1})$. Select a homomorphism $h : \boldsymbol{Te}_\tau \to \mathbf{2}$ so that $h(x_1) = h(x_2) = \ldots = h(x_{n+1}) = 0$, $h(y_1) = h(y_2) = \ldots = h(y_{n+1}) = 1$, and $h(x_{n+2}) = h(y_{n+2})$. Then $h(\alpha_i) = 0$ for $i = 1, 2, \ldots, n$ but $h(\alpha_{n+1}) = 1$. It follows that $h(\alpha_{n+1}) = h(\beta_{n+1}) = 1$ but $0 = h(\alpha_n) \neq h(\beta_n) = 1$.

We also note that

$$\mathbf{BA}^{\models}(\Delta_n) = \mathbf{BA}^{\models}(x_1 \approx y_1) \cap \ldots \cap \mathbf{BA}^{\models}(x_{n+1} \approx y_{n+1})$$

for all $n \geqslant 1$. □

9.2 Pure Commutator Terms and Their Interpretations

In this section $\boldsymbol{x}_1, \ldots, \boldsymbol{x}_n, \ldots$ is a countably infinite set of congruence variables. They range over the set of congruences of algebras.

The set of pure commutator terms is defined as follows:

(i) every congruence variable is a pure commutator term,
(ii) if $\boldsymbol{t}_1$ and $\boldsymbol{t}_2$ are pure commutator terms, then so is the expression $[\boldsymbol{t}_1, \boldsymbol{t}_2]$,
(iii) nothing else is a pure commutator term.

We shall often use the well-known convention of suppressing parentheses if it does not lead to misunderstanding.

By a *pure commutator law* for the commutator we shall mean any formula of the form $t_1 \approx t_2$, where t_1 and t_2 are commutator terms. E.g., $[[\boldsymbol{x}_1, \boldsymbol{x}_2], \boldsymbol{x}_3] \approx [\boldsymbol{x}_1, [\boldsymbol{x}_2, \boldsymbol{x}_3]]$ and $[\boldsymbol{x}_1, \boldsymbol{x}_2] \approx [\boldsymbol{x}_2, \boldsymbol{x}_1]$ are pure commutator laws.

Let $\mathbf{Q}$ be a quasivariety and $\boldsymbol{A} \in \mathbf{Q}$. $\langle \boldsymbol{Con}_{\mathbf{Q}}(\boldsymbol{A}), [\,\cdot\,]^{A}\rangle$ stands for the algebraic lattice $\boldsymbol{Con}_{\mathbf{Q}}(\boldsymbol{A})$ enriched with the equationally defined commutator operation $[\,\cdot\,]^{A}$ defined on the set of $\mathbf{Q}$-congruences of $\boldsymbol{A}$.

If $\boldsymbol{t}(\boldsymbol{x}_1, \ldots, \boldsymbol{x}_n)$ is a commutator term containing at most the variables among $\boldsymbol{x}_1, \ldots, \boldsymbol{x}_n$, and $\Phi_1, \ldots, \Phi_n \in Con_{\mathbf{Q}}(\boldsymbol{A})$, then $\boldsymbol{t}^{A}(\Phi_1, \ldots, \Phi_n)$ is the value of $\boldsymbol{t}$ computed for $\Phi_1, \ldots, \Phi_n$ in the commutator lattice $\langle \boldsymbol{Con}_{\mathbf{Q}}(\boldsymbol{A}), [\cdot]^{A}\rangle$. $\boldsymbol{t}_{A}(\Phi_1, \ldots, \Phi_n)$ is defined by induction on the length of $\boldsymbol{t}$. If $\boldsymbol{t}$ is $[\boldsymbol{x}_1, \boldsymbol{x}_2]$, then $\boldsymbol{t}^{A}(\Phi_1, \Phi_2) = [\Phi_1, \Phi_2]^{A}$.

The fact that the equationally defined commutator for $\mathbf{Q}$ is additive implies that for any pure commutator term $\boldsymbol{t}(\boldsymbol{x}_1, \ldots, \boldsymbol{x}_n)$, the operation $\boldsymbol{t}^{A}$ is additive in each argument for any algebra $\boldsymbol{A} \in \mathbf{Q}$.

The following theorem provides syntactical tools enabling one to express various properties of the equationally defined commutator in terms of sets of equations of $Eq(\tau)$.

Theorem 9.2.1. *Let* $\mathbf{Q}$ *be a quasivariety with the additive equationally defined commutator. For every positive integer n and for any pure n-ary commutator term* $\boldsymbol{t}(\boldsymbol{x}_1, \ldots, \boldsymbol{x}_n)$ *there exists a set of equations* $\Sigma_t(x_1, y_1, \ldots, x_n, y_n, \underline{u})$ *of* $Eq(\tau)$ *in* $2n$ *variables* $x_1, y_1, \ldots, x_n, y_n$ *(and possibly with some parameters* $\underline{u}$*) such that for any algebra* $\boldsymbol{A} \in \mathbf{Q}$ *and any sets* $X_1, \ldots, X_n \subseteq A^2$,

$$\boldsymbol{t}^{A}(\Theta_{\mathbf{Q}}(X_1), \ldots, \Theta_{\mathbf{Q}}(X_n)) = \qquad (\boldsymbol{t})^*$$

$$\Theta_{\mathbf{Q}}(\bigcup\{(\forall \underline{e})\ \Sigma_t(a_1, b_1, \ldots, a_n, b_n, \underline{e}) : \langle a_1, b_1\rangle \in X_1, \ldots, \langle a_n, b_n\rangle \in X_n\}).$$

Moreover, if $\mathbf{Q}$ *is finitely generated, then the sets* Σ_t *can be assumed to be finite.*

Proof. Let $\Delta = \Delta(x, y, z, w, \underline{u})$ be a set of quaternary commutator equations for $\mathbf{Q}$ so that condition (1) of Section 9.1 is satisfied. We select two infinite sequences of pairwise different individual variables

$$x_1, x_2, x_3, \ldots, x_n, x_{n+1}, \ldots$$

$$y_1, y_2, y_3, \ldots, y_n, y_{n+1}, \ldots.$$

It is assumed that there are also infinitely many individual variables lying *outside* the set of those which belong to the two sequences. We therefore assume that V is an infinite set of variables occurring neither in $x_1, x_2, x_3, \ldots, x_n, x_{n+1}, \ldots$ nor in $y_1, y_2, y_3, \ldots, y_n, y_{n+1}, \ldots$. Moreover, the variables $x, y, z, w, \underline{u}$ all belong to V. (By renaming the parameters $\underline{u}$, we assume that there are still infinitely many variables in V that are different from those of $\underline{u}$.)

The proof of $(t)^*$ is by induction on the complexity of commutator terms.

If $(t)^*$ is the i-th term variable $\boldsymbol{x}_i$, then we put $\Sigma_t := \{x_i \approx y_i\}$. Then $(t)^*$ trivially holds.

Let t_1 and t_2 be pure commutator terms and Σ_1, Σ_2 the corresponding sets of equations. We assume that $(t_1)^*$ and $(t_2)^*$ hold. Let $t := [t_1, t_2]$. We define:

$$\Sigma_t := \bigcup\{\Delta(\alpha, \beta, \gamma, \delta, \underline{u}) : \alpha \approx \beta \in \Sigma_1, \gamma \approx \delta \in \Sigma_2\}.$$

We claim that $(t)^*$ holds.

Suppose that the term $t = [t_1, t_2]$ is n-ary, $t = t(x_1, \ldots, x_n)$. Then t_1 and t_2 are at most in the variables $x_1, \ldots, x_n$. We first prove:

Claim. *Let $(t)^*$ be as above. For any $\boldsymbol{A} \in \mathbf{Q}$ and $a_1, b_1, \ldots, a_n, b_n \in A$,*

$$t^{\boldsymbol{A}}(\Theta_{\mathbf{Q}}(a_1, b_1), \ldots, \Theta_{\mathbf{Q}}(a_n, b_n)) = \Theta_{\mathbf{Q}}((\forall \underline{e})\ \Sigma_t(a_1, b_1, \ldots, a_n, b_n, \underline{e})). \qquad (t)^{**}$$

Proof (of the claim). Fix $\boldsymbol{A} \in \mathbf{Q}$ and assume that $a_1, b_1, \ldots, a_n, b_n \in A$. By the induction hypothesis we have that

$$t_1^{\boldsymbol{A}}(\Theta_{\mathbf{Q}}(a_1, b_1), \ldots, \Theta_{\mathbf{Q}}(a_n, b_n)) = \Theta_{\mathbf{Q}}((\forall \underline{e}')\ \Sigma_1(a_1, b_1, \ldots, a_n, b_n, \underline{e}'))$$

and

$$t_2^{\boldsymbol{A}}(\Theta_{\mathbf{Q}}(a_1, b_1), \ldots, \Theta_{\mathbf{Q}}(a_n, b_n)) = \Theta_{\mathbf{Q}}((\forall \underline{e}'')\ \Sigma_2(a_1, b_1, \ldots, a_n, b_n, \underline{e}'')).$$

To simplify notation, we mark the sequence $a_1, b_1, \ldots, a_n, b_n$ as $\underline{a}, \underline{b}$. We work in the lattice $\boldsymbol{Con}_{\mathbf{Q}}(\boldsymbol{A})$. Then

$$\begin{gathered} t^{\boldsymbol{A}}(\Theta_{\mathbf{Q}}(a_1, b_1), \ldots, \Theta_{\mathbf{Q}}(a_n, b_n)) = \\ [t_1^{\boldsymbol{A}}(\Theta_{\mathbf{Q}}(a_1, b_1), \ldots, \Theta_{\mathbf{Q}}(a_n, b_n)), t_2^{\boldsymbol{A}}(\Theta_{\mathbf{Q}}(a_1, b_1), \ldots, \Theta_{\mathbf{Q}}(a_n, b_n))]^{\boldsymbol{A}} = \\ [\Theta_{\mathbf{Q}}((\forall \underline{e}')\ \Sigma_1(a_1, b_1, \ldots, a_n, b_n, \underline{e}')), \Theta_{\mathbf{Q}}((\forall \underline{e}'')\ \Sigma_2(a_1, b_1, \ldots, a_n, b_n, \underline{e}''))]^{\boldsymbol{A}} = \\ \text{(by additivity)} \end{gathered}$$

$$\begin{aligned} \sup{}_{\mathbf{Q}}\{[\Theta_{\mathbf{Q}}(\alpha(\underline{a}, \underline{b}, \underline{e}'), \beta(\underline{a}, \underline{b}, \underline{e}')), \Theta_{\mathbf{Q}}(\gamma(\underline{a}, \underline{b}, \underline{e}''), \delta(\underline{a}, \underline{b}, \underline{e}''))]^{\boldsymbol{A}} : \qquad \text{(a)} \\ \alpha \approx \beta \in \Sigma_1, \gamma \approx \delta \in \Sigma_2, \underline{e}' \in A^{<\omega}, \underline{e}'' \in A^{<\omega}\}. \end{aligned}$$

(The supremum is taken in $\boldsymbol{Con}_{\mathbf{Q}}(\boldsymbol{A})$.) Since $[\Theta_{\mathbf{Q}}(a', b'), \Theta_{\mathbf{Q}}(c', d')]^{\boldsymbol{A}} = \Theta_{\mathbf{Q}}((\forall \underline{f})\ \Delta(a', b', c', d', \underline{f}))$ for any $a', b', c', d' \in A$, the element (a) is equal to

$$\begin{aligned} \sup{}_{\mathbf{Q}}\{\Theta_{\mathbf{Q}}((\forall \underline{f})\ \Delta(\alpha(\underline{a}, \underline{b}, \underline{e}'), \beta(\underline{a}, \underline{b}, \underline{e}'), \gamma(\underline{a}, \underline{b}, \underline{e}''), \delta(\underline{a}, \underline{b}, \underline{e}''), \underline{f})) : \qquad \text{(b)} \\ \alpha \approx \beta \in \Sigma_1, \gamma \approx \delta \in \Sigma_2, \underline{e}' \in A^{<\omega}, \underline{e}'' \in A^{<\omega}\}. \end{aligned}$$

But the element (b) is equal to $\Theta_{\mathbf{Q}}((\forall \underline{e})\ \Sigma_t(\underline{a}, \underline{b}, \underline{e}))$, by the definition of Σ_t. Consequently,

$$t^{\mathbf{A}}(\Theta_{\mathbf{Q}}(a_1, b_1), \ldots, \Theta_{\mathbf{Q}}(a_n, b_n)) = \Theta_{\mathbf{Q}}((\forall \underline{e})\ \Sigma_t(\underline{a}, \underline{b}, \underline{e})). \quad \square$$

Since the operation $\boldsymbol{t}^{\mathbf{A}}$ is additive in each argument, $(\boldsymbol{t})^*$ follows from the above claim (see also the proof of Proposition 9.2.6).

This concludes the proof of the theorem. □

We shall now extend the set of pure commutator terms by adding to the vocabulary two constants $\mathbf{0}, \mathbf{1}$, representing the zero and the full congruences respectively and the symbol "$\cdot$" representing the intersection of congruences. The recursive definition of the new, extended set of terms is obvious.

The properties of a set of quaternary commutator equations $\Delta_0(x, y, z, w, \underline{u})$ have an impact on the behaviour of the equationally defined commutator itself. In the simplest cases one may additionally assume that Δ_0 is finite or it has no parameters at all. The first case is investigated in Chapter 8 in the context of finitely generated quasivarieties. Freese and McKenzie (1987) made the following list of commutator identities they investigate for congruence modular varieties of algebras:

$$\boldsymbol{x} \cdot [\boldsymbol{y}, \boldsymbol{y}] \approx [\boldsymbol{x} \cdot \boldsymbol{y}, \boldsymbol{y}]. \tag{CI1}$$

$$[\boldsymbol{x}, \boldsymbol{y}] \approx \boldsymbol{x} \cdot \boldsymbol{y} \cdot [\mathbf{1}, \mathbf{1}]. \tag{CI2}$$

$$[\boldsymbol{x}, \boldsymbol{y}] \approx \boldsymbol{x} \cdot \boldsymbol{y}. \tag{CI3}$$

$$[\boldsymbol{x}, \boldsymbol{y}] \approx \mathbf{0}. \tag{CI4}$$

$$[\mathbf{1}, \mathbf{1}] \approx \mathbf{0}. \tag{CI5}$$

$$[\mathbf{1}, [\mathbf{1}, \mathbf{1}]] \approx \mathbf{0}. \tag{CI6}$$

$$[[\mathbf{1}, \mathbf{1}], [\mathbf{1}, \mathbf{1}]] \approx \mathbf{0}. \tag{CI7}$$

$$[\boldsymbol{x}, \mathbf{1}] \approx \boldsymbol{x}. \tag{CI8}$$

1 represents the largest congruence on the algebra, **0** is interpreted as the identity congruence, and the variables $\boldsymbol{x}, \boldsymbol{y}, \boldsymbol{z}$ represent arbitrary congruences. The identity (CI3), which is equivalent to relative congruence distributivity, is also investigated in Chapter 6 of this paper. If (CI5) holds in an algebra $\mathbf{A}$, then $\mathbf{A}$ is called *Abelian*.

The crucial observation made by Kearnes and McKenzie (1992, Thm. 3.1) is that the commutator in every finitely generated RCM quasivariety obeys (CI1).

Commutator equations as well as other derived equations are a convenient tool enabling one to characterize syntactically various properties of the commutator. Below we shall investigate some of such commutator properties.

Given a sequence of variables $\boldsymbol{x}_1, \boldsymbol{x}_2, \ldots$ ranging over congruences, we define the following sequence of pure commutator terms $\boldsymbol{c}_n(\boldsymbol{x}_1, \boldsymbol{x}_2, \ldots, \boldsymbol{x}_{n+1})$, $n = 1, 2, \ldots$:

$$\boldsymbol{c}_1 \text{ is } [\boldsymbol{x}_1, \boldsymbol{x}_2],$$
$$\boldsymbol{c}_{n+1} := [\boldsymbol{c}_n, \boldsymbol{x}_{n+2}], \quad \text{for } n = 1, 2, \ldots$$

$\boldsymbol{c}_n$ is thus the term $[[\ldots[[\boldsymbol{x}_1,\boldsymbol{x}_2],\boldsymbol{x}_3],\ldots],\boldsymbol{x}_{n+1}]$, for $n = 1, 2, \ldots$.

The following observation establishes the relationship between equations $\Delta_n(x_1, y_1, x_2, y_2, \ldots, x_{n+1}, y_{n+1}, \underline{v}_n)$ defined earlier and the commutator term $\boldsymbol{c}_n$ for all n.

Theorem 9.2.2. *Let* $\mathbf{Q}$ *be a quasivariety whose equationally defined commutator is additive. Define the set* $\Delta_n = \Delta_n(x_1, y_1, x_2, y_2, \ldots, x_{n+1}, y_{n+1}, \underline{v}_n)$ *for each positive n as in Section* 9.1 *above. Then for every algebra* $\boldsymbol{A}$ *in* $\mathbf{Q}$*, for every* $n \geqslant 1$*, and any* $c_1, d_1, \ldots, c_{n+1}, d_{n+1} \in A$*,*

$$c_n^{\boldsymbol{A}}(\Theta_{\mathbf{Q}}(c_1, d_1), \Theta_{\mathbf{Q}}(c_2, d_2), \ldots, \Theta_{\mathbf{Q}}(c_{n+1}, d_{n+1})) =$$
$$\sup{}_{\mathbf{Q}}\{\Theta_{\mathbf{Q}}(\Delta_n(c_1, d_1, c_2, d_2, \ldots, c_{n+1}, d_{n+1}, \underline{f}_n)) : \underline{f}_n \in A^{nk}\}.$$

(The supremum is taken in the lattice $\boldsymbol{Con}_{\mathbf{Q}}(\boldsymbol{A})$.)

The proof is based on two lemmas.

Given an algebra $\boldsymbol{A}$ in $\mathbf{Q}$ and two sequences $a_1, \ldots, a_k, \ldots$ and $b_1, \ldots, b_k, \ldots$ of elements of A (finite or infinite but of the same length) we inductively define the following sequence of $\mathbf{Q}$-congruences of $\boldsymbol{A}$:

$$\Psi_1 := [\Theta_{\mathbf{Q}}(a_1, b_1), \Theta_{\mathbf{Q}}(a_2, b_2)]^{\boldsymbol{A}},$$
$$\Psi_{n+1} := [\Psi_n, \Theta_{\mathbf{Q}}(a_{n+2}, b_{n+2})]^{\boldsymbol{A}}, \quad \text{for any } n \geqslant 1.$$

The first lemma follows immediately from the definition of the sequence Ψ_n, $n \geqslant 1$:

Lemma 9.2.3. *Let* $\boldsymbol{A}$ *be an algebra in* $\mathbf{Q}$*, and let* $a_1, \ldots, a_k, \ldots$ *and* $b_1, \ldots, b_k, \ldots$ *be sequences of elements of A of the same length. Define the congruences* Ψ_n*,* $n \geqslant 1$*, as above. Then, for all* $n \geqslant 1$*,*

$$\Psi_{n+1} \subseteq \Psi_n. \tag{3_n}$$

Proof. Fix n. (3_n) follows from the properties of the commutator. (Additivity is not needed here.)

We have: $\Psi_{n+1} = [\Psi_n, \Theta_{\mathbf{Q}}(a_{n+2}, b_{n+2})]^{\boldsymbol{A}} \subseteq \Psi_n \cap \Theta_{\mathbf{Q}}(a_{n+2}, b_{n+2}) \subseteq \Psi_n$. □

Lemma 9.2.4. *Let* $\boldsymbol{A}$ *be an algebra in* $\mathbf{Q}$ *and let* $a_1, \ldots, a_k, \ldots$ *and* $b_1, \ldots, b_k, \ldots$ *be sequences of elements of A of the same length. Then for every n,*

$$\Psi_n = \sup{}_{\mathbf{Q}}\{\Theta_{\mathbf{Q}}(\Delta_n(a_1, b_1, \ldots, a_{n+1}, b_{n+1}, \underline{f}_n)) : \underline{f}_n \in A^{nk}\}. \tag{4_n}$$

Proof. Induction on n.

Induction base. For $n = 1$ we have

$$\Psi_1 = [\Theta_{\mathbf{Q}}(a_1, b_1), \Theta_{\mathbf{Q}}(a_2, b_2)]^{\mathbf{A}} = \sup{}_{\mathbf{Q}}\{\Theta_{\mathbf{Q}}(\Delta(a_1, b_1, a_2, b_2, \underline{e})) : \underline{e} \in A^k\},$$

by the fact that Δ is a generating set of quaternary commutator equations. But $\Theta_{\mathbf{Q}}(\Delta(a_1, b_1, a_2, b_2, \underline{e})) = \Theta_{\mathbf{Q}}(\Delta_1(a_1, b_1, a_2, b_2, \underline{e}))$, for all sequences $\underline{e} \in A^k$. It follows that $\sup{}_{\mathbf{Q}}\{\Theta_{\mathbf{Q}}(\Delta(a_1, b_1, a_2, b_2, \underline{e})) : \underline{e} \in A^k\} = \sup{}_{\mathbf{Q}}\{\Theta_{\mathbf{Q}}(\Delta_1(a_1, b_1, a_2, b_2, \underline{e})) : \underline{e} \in A^k\}$. So (4_1) holds.

Inductive step. Assume (4_n) holds. We prove (4_{n+1}). Letting $\underline{a}_{n+1}, \underline{b}_{n+1}$ denote the sequence $a_1, b_1, \ldots, a_{n+1}, b_{n+1}$, respectively, we have:

$$\Psi_{n+1} = [\Psi_n, \Theta_{\mathbf{Q}}(a_{n+2}, b_{n+2})]^{\mathbf{A}} = \quad \text{(by IH)}$$

$$[\sup{}_{\mathbf{Q}}\{\Theta_{\mathbf{Q}}(\Delta_n(\underline{a}_{n+1}, \underline{b}_{n+1}, \underline{f}_n)) : \underline{f}_n \in A^{nk}\}, \Theta_{\mathbf{Q}}(a_{n+2}, b_{n+2})]^{\mathbf{A}} = \quad \text{(additivity)}$$

$$\sup{}_{\mathbf{Q}}\{[\Theta_{\mathbf{Q}}(\Delta_n(\underline{a}_{n+1}, \underline{b}_{n+1}, \underline{f}_n)), \Theta_{\mathbf{Q}}(a_{n+2}, b_{n+2})]^{\mathbf{A}} : \underline{f}_n \in A^{nk}\} =$$

$$\sup{}_{\mathbf{Q}}\{[\sup{}_{\mathbf{Q}}\{\Theta_{\mathbf{Q}}(\alpha(\underline{a}_{n+1}, \underline{b}_{n+1}, \underline{f}_n), \beta(\underline{a}_{n+1}, \underline{b}_{n+1}, \underline{f}_n)) :$$
$$\alpha \approx \beta \in \Delta_n\}, \Theta_{\mathbf{Q}}(a_{n+2}, b_{n+2})]^{\mathbf{A}} : \underline{f}_n \in A^{nk}\}) = \quad \text{(additivity)}$$

$$\sup{}_{\mathbf{Q}}\{[\Theta_{\mathbf{Q}}(\alpha(\underline{a}_{n+1}, \underline{b}_{n+1}, \underline{f}_n), \beta(\underline{a}_{n+1}, \underline{b}_{n+1}, \underline{f}_n)), \Theta_{\mathbf{Q}}(a_{n+2}, b_{n+2})]^{\mathbf{A}} :$$
$$\alpha \approx \beta \in \Delta_n, \underline{f}_n \in A^{nk}\} =$$

$$\sup{}_{\mathbf{Q}}\{\sup{}_{\mathbf{Q}}\{\Theta_{\mathbf{Q}}(\Delta(\alpha(\underline{a}_{n+1}, \underline{b}_{n+1}, \underline{f}_n), \beta(\underline{a}_{n+1}, \underline{b}_{n+1}, \underline{f}_n), a_{n+2}, b_{n+2}, \underline{e}) :$$
$$\underline{e} \in A^k\} : \alpha \approx \beta \in \Delta_n, \underline{f}_n \in A^{nk}\} =$$

$$\sup{}_{\mathbf{Q}}\{\Theta_{\mathbf{Q}}(\Delta(\alpha(\underline{a}_{n+1}, \underline{b}_{n+1}, \underline{f}_n), \beta(\underline{a}_{n+1}, \underline{b}_{n+1}, \underline{f}_n), a_{n+2}, b_{n+2}, \underline{e}) :$$
$$\alpha \approx \beta \in \Delta_n, \underline{f}_n \in A^{nk}, \underline{e} \in A^k\} =$$

$$\sup{}_{\mathbf{Q}}\{\Theta_{\mathbf{Q}}(\Delta_{n+1}(\underline{a}_{n+1}, \underline{b}_{n+1}, a_{n+2}, b_{n+2}, \underline{f}_n, \underline{e})) : \underline{f}_n \in A^{nk}, \underline{e} \in A^k\} =$$

$$\sup{}_{\mathbf{Q}}\{\Theta_{\mathbf{Q}}(\Delta_{n+1}(a_1, b_1, \ldots, a_{n+1}, b_{n+1}, a_{n+2}, b_{n+2}, \underline{f}_{n+1})) : \underline{f}_{n+1} \in A^{(n+1)k}\}.$$

So (4_{n+1}) holds.

This proves the lemma. □

The theorem follows from Lemma 9.2.4 and the definition of the commutator terms $\boldsymbol{c}_n$, $n = 1, 2, \ldots$. □

The suitably modified versions of the above lemmas also hold for equational theories of the consequence operation $\mathbf{Q}^{\models}$. To simplify notation, we let C denote the consequence $\mathbf{Q}^{\models}$.

Corollary 9.2.5. *Let* **Q** *be a quasivariety whose equationally defined commutator is additive. Then for all n and any two sequences of terms* $s_1, \ldots, s_{n+1}, t_1, \ldots, t_{n+1}$ *of* Te_τ

$$c_n(C(s_1 \approx t_1), C(s_2 \approx t_2), \ldots, C(s_{n+1} \approx t_{n+1})) = \qquad (4_n)$$

$$C((\forall \underline{v}_n)\ \Delta_n(s_1, t_1, \ldots, s_{n+1}, t_{n+1}, \underline{v}_n)). \quad \square$$

Note. If **Q** is relatively congruence-distributive or Abelian, then for any $n \geqslant 1$ and any different variables $x_1, \ldots, x_{n+1}, y_1, \ldots, y_{n+1}$,

$$c_n(C(x_1 \approx y_1), C(x_2 \approx y_2), \ldots, C(x_{n+1} \approx y_{n+1})) = C(x_1 \approx y_1) \cap C(x_2 \approx y_2) \cap \ldots \cap C(x_{n+1} \approx y_{n+1}). \quad (5_n)$$

(5_1) holds for any quasivariety. For $n = 2$, we get

$$[[C(x_1 \approx y_1), C(x_2 \approx y_2)], C(x_3 \approx y_3)] = C(x_1 \approx y_1) \cap C(x_2 \approx y_2) \cap C(x_3 \approx y_3),$$

i.e.,

$$[C(x_1 \approx y_1) \cap C(x_2 \approx y_2), C(x_3 \approx y_3)] = C(x_1 \approx y_1) \cap C(x_2 \approx y_2) \cap C(x_3 \approx y_3).$$

The last identity implies the following form of associativity:

$$[[C(x_1 \approx y_1), C(x_2 \approx y_2)], C(x_3 \approx y_3)] = [C(x_1 \approx y_1), [C(x_2 \approx y_2), C(x_3 \approx y_3)]].$$

It is an open problem if (5_n) $(n \geqslant 2)$ holds for other algebraically interesting quasivarieties. $\square$

We recall that $(\forall \underline{v}_n)\ \Delta_n(s_1, t_1, \ldots, s_{n+1}, t_{n+1}, \underline{v}_n)$ equals

$$\bigcup\{\Delta_n(s_1, t_1, \ldots, s_{n+1}, t_{n+1}, \underline{e}_n) : \underline{e}_n \text{ is a sequence of terms of the length of } \underline{v}_n\}.$$

Proposition 9.2.6. *Let* **Q** *be a quasivariety whose equationally defined commutator is additive. Let* $\mathbf{A} \in \mathbf{Q}$. *Let n be a positive integer n,* $\Phi_1, \ldots, \Phi_{n+1} \in Con_{\mathbf{Q}}(\mathbf{A})$ *and let* $X_1, \ldots, X_{n+1}$ *be sets such that* $\Phi_1 = \Theta_{\mathbf{Q}}(X_1), \ldots, \Phi_{n+1} = \Theta_{\mathbf{Q}}(X_{n+1})$. *Then*

$$c_n^A(\Phi_1, \ldots, \Phi_{n+1}) = \sup{}_{\mathbf{Q}}\{c_n^A(\Theta_{\mathbf{Q}}(a_1, b_1), \ldots, \Theta_{\mathbf{Q}}(a_{n+1}, b_{n+1})) : \langle a_1, b_1\rangle \in X_1, \ldots, \langle a_{n+1}, b_{n+1}\rangle \in X_{n+1}\}. \qquad (*)_n$$

A formula analogous to $(*)_n$ *holds for the term algebra* $\boldsymbol{Te}_\tau$ *and for the equational theories of* $\mathbf{Q}^{\models}$.

(The supremum is taken in the lattice $\boldsymbol{Con}_{\mathbf{Q}}(\boldsymbol{A})$.)

Proof. Induction on n.

Base step. $n = 1$.

$$\begin{aligned}
\boldsymbol{c}_1(\Phi_1, \Phi_2) = [\Phi_1, \Phi_2]^{\boldsymbol{A}} = [\Theta_{\mathbf{Q}}(X_1), \Theta_{\mathbf{Q}}(X_2)]^{\boldsymbol{A}} = \quad &\text{(by additivity)} \\
\sup{}_{\mathbf{Q}}\{[\Theta_{\mathbf{Q}}(a_1, b_1), \Theta_{\mathbf{Q}}(a_2, b_2)]^{\boldsymbol{A}} : \langle a_1, b_1\rangle \in X_1, \langle a_2, b_2\rangle \in X_2\} = \\
\sup{}_{\mathbf{Q}}\{\boldsymbol{c}_1(\Theta_{\mathbf{Q}}(a_1, b_1), \Theta_{\mathbf{Q}}(a_2, b_2)) : \langle a_1, b_1\rangle \in X_1, \langle a_2, b_2\rangle \in X_2\}.
\end{aligned}$$

Inductive step. Suppose $(*)_n$ holds. We show $(*)_{n+1}$. We have:

$$\begin{aligned}
\boldsymbol{c}_{n+1}(\Phi_1, \ldots, \Phi_{n+1}, \Phi_{n+2}) = [\boldsymbol{c}_n(\Phi_1, \ldots, \Phi_{n+1}), \Phi_{n+2}]^{\boldsymbol{A}} = \quad &\text{(IH)} \\
[\sup{}_{\mathbf{Q}}\{\boldsymbol{c}_n(\Theta_{\mathbf{Q}}(a_1, b_1), \ldots, \Theta_{\mathbf{Q}}(a_{n+1}, b_{n+1})) : \\
\langle a_1, b_1\rangle \in X_1, \ldots, \langle a_{n+1}, b_{n+1}\rangle \in X_{n+1}\}, \Phi_{n+2}]^{\boldsymbol{A}} = \\
\sup{}_{\mathbf{Q}}\{[\boldsymbol{c}_n(\Theta_{\mathbf{Q}}(a_1, b_1), \ldots, \Theta_{\mathbf{Q}}(a_{n+1}, b_{n+1})), \Phi_{n+2}]^{\boldsymbol{A}} : \\
\langle a_1, b_1\rangle \in X_1, \ldots, \langle a_{n+1}, b_{n+1}\rangle \in X_{n+1}\} = \\
\sup{}_{\mathbf{Q}}\{[\boldsymbol{c}_n(\Theta_{\mathbf{Q}}(a_1, b_1), \ldots, \Theta_{\mathbf{Q}}(a_{n+1}, b_{n+1})), \Theta_{\mathbf{Q}}(a_{n+2}, b_{n+2})]^{\boldsymbol{A}} : \\
\langle a_1, b_1\rangle \in X_1, \ldots, \langle a_{n+1}, b_{n+1}\rangle \in X_{n+1}, \langle a_{n+2}, b_{n+2}\rangle \in X_{n+2}\} = \\
\sup{}_{\mathbf{Q}}\{\boldsymbol{c}_{n+1}(\Theta_{\mathbf{Q}}(a_1, b_1), \ldots, \Theta_{\mathbf{Q}}(a_{n+1}, b_{n+1}), \Theta_{\mathbf{Q}}(a_{n+2}, b_{n+2}) : \\
\langle a_1, b_1\rangle \in X_1, \ldots, \langle a_{n+1}, b_{n+1}\rangle \in X_{n+1}, \langle a_{n+2}, b_{n+2}\rangle \in X_{n+2}\}.
\end{aligned}$$

So $(*)_{n+1}$ holds. □

Theorem 9.2.7. *Let* $\mathbf{Q}$ *be a quasivariety whose equationally defined commutator is additive. Let* m *be a positive integer. The following conditions are equivalent:*

(1) $\mathbf{Q}^{\models}(\Delta_m) = \mathbf{Q}^{\models}(\Delta_{m+1})$,
(2) Δ_{m+1}/Δ_m *is a set of rules of* $\mathbf{Q}^{\models}$,
(3) *The lattices of* $\mathbf{Q}$*-congruences on the algebras of* $\mathbf{Q}$ *obey the commutator law:*

$$\boldsymbol{c}_{m+1} \approx \boldsymbol{c}_m.$$

Consequently, if at least one of (1)–(3) *holds, the above lattices also obey the laws:* $\boldsymbol{c}_n \approx \boldsymbol{c}_m$*, for all* $n \geqslant m$.

Proof. (1) ⇒ (2). This is trivial.

(2) ⇒ (1). Use Lemma 9.1.3.

(1) ⇒ (3). Assume (1). Let $\boldsymbol{A}$ be an algebra in $\mathbf{Q}$ and let $\Psi_1, \ldots, \Psi_{m+1}, \Psi_{m+2}$ be $\mathbf{Q}$-congruences on $\boldsymbol{A}$. We must show that

$$\boldsymbol{c}_{m+1}(\Psi_1, \ldots, \Psi_{m+1}, \Psi_{m+2}) = \boldsymbol{c}_m(\Psi_1, \ldots, \Psi_{m+1}).$$

Select sets of pairs $Y_1, \ldots, Y_{m+1}, Y_{m+2} \subseteq A^2$ such that $\Psi_k = \Theta_{\mathbf{Q}}(Y_k)$ for $k = 1, 2, \ldots, m+2$. We work in the lattice $\boldsymbol{Con}_{\mathbf{Q}}(\boldsymbol{A})$.

(1) implies that

$$\sup{}_{\mathbf{Q}}\{\Theta_{\mathbf{Q}}(\Delta_{m+1}(a_1, b_1, \ldots, a_{m+2}, b_{m+2}, \underline{f}_{m+1})) : \underline{f}_{m+1} \in A^{(m+1)k}\} =$$
$$\sup{}_{\mathbf{Q}}\{\Theta_{\mathbf{Q}}(\Delta_m(a_1, b_1, \ldots, a_{m+1}, b_{m+1}, \underline{f}_m)) : \underline{f}_m \in A^{mk}\},$$

for all $a_1, \ldots, a_{m+1}, a_{m+2}, b_1, \ldots, b_{m+1}, b_{m+2} \in A$. Hence, applying Theorem 9.2.2, we get that

$$\boldsymbol{c}_{m+1}(\Theta_{\mathbf{Q}}(a_1, b_1), \ldots, \Theta_{\mathbf{Q}}(a_{m+1}, b_{m+1}), \Theta_{\mathbf{Q}}(a_{m+2}, b_{m+2})) =$$
$$\sup{}_{\mathbf{Q}}\{\Theta_{\mathbf{Q}}(\Delta_{m+1}(a_1, b_1, \ldots, a_{m+2}, b_{m+2}, \underline{f}_{m+1})) : \underline{f}_{m+1} \in A^{(m+1)k}\} =$$
$$\sup{}_{\mathbf{Q}}\{\Theta_{\mathbf{Q}}(\Delta_m(a_1, b_1, \ldots, a_{m+1}, b_{m+1}, \underline{f}_m)) : \underline{f}_m \in A^{mk}\} =$$
$$\boldsymbol{c}_m(\Theta_{\mathbf{Q}}(a_1, b_1), \ldots, \Theta_{\mathbf{Q}}(a_{m+1}, b_{m+1})),$$

for all $\langle a_1, b_1 \rangle \in Y_1, \ldots, \langle c_{m+1}, d_{m+1} \rangle \in Y_{m+1}$, and $\langle c_{m+2}, d_{m+2} \rangle \in Y_{m+2}$. Then, by Proposition 9.2.6, (3) follows.

(3) $\Rightarrow$ (1). Assume (3). (3) implies that in any algebra $\boldsymbol{A}$ in $\mathbf{Q}$ and for all $a_1, \ldots, a_{m+1}, a_{m+2}, b_1, \ldots, b_{m+1}, b_{m+2} \in A$ it is the case that

$$\boldsymbol{c}_{m+1}(\Theta_{\mathbf{Q}}(a_1, b_1), \ldots, \Theta_{\mathbf{Q}}(a_{m+1}, b_{m+1}), \Theta_{\mathbf{Q}}(a_{m+2}, b_{m+2})) =$$
$$\boldsymbol{c}_m(\Theta_{\mathbf{Q}}(a_1, b_1), \ldots, \Theta_{\mathbf{Q}}(a_{m+1}, b_{m+1})).$$

Taking into account Theorem 9.2.2, we see that the above equality is equivalent to

$$\sup{}_{\mathbf{Q}}\{\Theta_{\mathbf{Q}}(\Delta_{m+1}(a_1, b_1, \ldots, a_{m+2}, b_{m+2}, \underline{f}_{m+1})) : \underline{f}_{m+1} \in A^{(m+1)k}\} =$$
$$\sup{}_{\mathbf{Q}}\{\Theta_{\mathbf{Q}}(\Delta_m(a_1, b_1, \ldots, a_{m+1}, b_{m+1}, \underline{f}_m)) : \underline{f}_m \in A^{mk}\}.$$

But the last equality implies (1). So (1) holds.

As to the last statement, it is proved by induction on i that the (universal) validity of $\boldsymbol{c}_{m+1} \approx \boldsymbol{c}_m$ for the $\mathbf{Q}$-congruences on the algebras of $\mathbf{Q}$ implies the universal validity of $\boldsymbol{c}_{m+i} \approx \boldsymbol{c}_m$, for all positive integers i. Use Lemma 9.1.4 and Theorem 9.1.5.

We shall give a short proof of this statement without resort to Lemma 9.1.4. Assume

$$\boldsymbol{c}_{n-1}(\boldsymbol{x}_1, \boldsymbol{x}_2, \ldots, \boldsymbol{x}_n) \approx \boldsymbol{c}_n(\boldsymbol{x}_1, \boldsymbol{x}_2, \ldots, \boldsymbol{x}_n, \boldsymbol{x}_{n+1}) \tag{a}$$

is $\mathbf{Q}$-valid, i.e., it is validated by the lattices of $\mathbf{Q}$-congruences on the algebras of $\mathbf{Q}$. We claim that

$$\boldsymbol{c}_n(\boldsymbol{x}_1, \boldsymbol{x}_2, \ldots, \boldsymbol{x}_n, \boldsymbol{x}_{n+1}) \approx \boldsymbol{c}_{n+1}(\boldsymbol{x}_1, \boldsymbol{x}_2, \ldots, \boldsymbol{x}_n, \boldsymbol{x}_{n+1}, \boldsymbol{x}_{n+2}) \tag{b}$$

is $\mathbf{Q}$-valid as well.

As $\boldsymbol{c}_{n-1}(\boldsymbol{x}_1, \boldsymbol{x}_2, \ldots, \boldsymbol{x}_n) \approx [\boldsymbol{c}_{n-1}(\boldsymbol{x}_1, \boldsymbol{x}_2, \ldots, \boldsymbol{x}_n), \boldsymbol{x}_{n+1}]$ is **Q**-valid, substituting a variable $\boldsymbol{u}$ for $\boldsymbol{x}_{n+1}$, we see that

$$\boldsymbol{c}_{n-1}(\boldsymbol{x}_1, \boldsymbol{x}_2, \ldots, \boldsymbol{x}_n) \approx [\boldsymbol{c}_{n-1}(\boldsymbol{x}_1, \boldsymbol{x}_2, \ldots, \boldsymbol{x}_n), \boldsymbol{u}] \tag{c}$$

is **Q**-valid too. Hence, multiplying both sides of (c) by $\boldsymbol{x}_{n+2}$, we get that

$$[\boldsymbol{c}_{n-1}(\boldsymbol{x}_1, \boldsymbol{x}_2, \ldots, \boldsymbol{x}_n), \boldsymbol{x}_{n+2}] \approx [[\boldsymbol{c}_{n-1}(\boldsymbol{x}_1, \boldsymbol{x}_2, \ldots, \boldsymbol{x}_n), \boldsymbol{u}], \boldsymbol{x}_{n+2}] \tag{d}$$

is **Q**-valid. ($\boldsymbol{u}$ is a congruence variables different from $\boldsymbol{x}_1, \boldsymbol{x}_2, \ldots, \boldsymbol{x}_{n+1}, \boldsymbol{x}_{n+2}$.)

On the other hand, substituting $\boldsymbol{x}_{n+2}$ for $\boldsymbol{u}$ in (c) and applying (a), we get that

$$[\boldsymbol{c}_{n-1}(\boldsymbol{x}_1, \boldsymbol{x}_2, \ldots, \boldsymbol{x}_n), \boldsymbol{x}_{n+2}] \approx \boldsymbol{c}_{n-1}(\boldsymbol{x}_1, \boldsymbol{x}_2, \ldots, \boldsymbol{x}_n) \approx \boldsymbol{c}_n(\boldsymbol{x}_1, \boldsymbol{x}_2, \ldots, \boldsymbol{x}_n, \boldsymbol{x}_{n+1}) \tag{e}$$

are **Q**-valid.

In turn, substituting $\boldsymbol{x}_{n+1}$ for $\boldsymbol{u}$ in (c) gives that

$$[\boldsymbol{c}_{n-1}(\boldsymbol{x}_1, \boldsymbol{x}_2, \ldots, \boldsymbol{x}_n), \boldsymbol{x}_{n+2}] \approx [[\boldsymbol{c}_{n-1}(\boldsymbol{x}_1, \boldsymbol{x}_2, \ldots, \boldsymbol{x}_n), \boldsymbol{x}_{n+1}], \boldsymbol{x}_{n+2}] \tag{f}$$

is **Q**-valid.

(e) and (f) give that $\boldsymbol{c}_n(\boldsymbol{x}_1, \boldsymbol{x}_2, \ldots, \boldsymbol{x}_n, \boldsymbol{x}_{n+1}) \approx [[\boldsymbol{c}_{n-1}(\boldsymbol{x}_1, \boldsymbol{x}_2, \ldots, \boldsymbol{x}_n), \boldsymbol{x}_{n+1}], \boldsymbol{x}_{n+2}]$ is **Q**-valid. So (b) holds. □

9.3 Nilpotency

We recall that the constants **1** and **0** denote the universal and zero congruences on any algebra, respectively.

Let m be a positive integer. A quasivariety **Q** is $(m-1)$-*nilpotent*, if there exists a positive integer m such that the equational commutator for **Q** obeys the law

$$\boldsymbol{c}_m(\mathbf{1}, \ldots, \mathbf{1}) \approx \mathbf{0}. \tag{Nil$_m$}$$

(**1** occurs here $(m+1)$-times.) Thus if **Q** is $(m-1)$-nilpotent, it is also n-nilpotent for all $n \geqslant m-1$.

$(\mathrm{Nil})_m$ implies the identity:

$$\boldsymbol{c}_m \approx \mathbf{0}$$

(i.e., $\boldsymbol{c}_m(\boldsymbol{x}_1, \boldsymbol{x}_2, \ldots, \boldsymbol{x}_{m+1}) \approx \mathbf{0}$ for all (**Q**-congruences) $\boldsymbol{x}_1, \boldsymbol{x}_2, \ldots, \boldsymbol{x}_{m+1}$). Moreover, if the equationally defined commutator for **Q** is additive, then the converse also holds, i.e., $\boldsymbol{c}_m \approx \mathbf{0}$ implies $(\mathrm{Nil})_m$.

Theorem 9.3.1. *Let* **Q** *be a quasivariety whose equationally defined commutator is additive. Suppose that* $\Delta(x, y, z, w, \underline{u})$ *satisfies condition* (1) *of Section* 9.1, *i.e.,*

$\mathbf{Q}^{\models}(x \approx y) \cap \mathbf{Q}^{\models}(z \approx w) = \mathbf{Q}^{\models}(\Delta(x, y, z, w, \underline{u}))$. *Then* $\mathbf{Q}$ *is* $(m-1)$*-nilpotent if and only if the equations* $\Delta_m(x_1, y_1, x_2, y_2, \ldots, x_{m+1}, y_{m+1}, \underline{v}_m)$ *are valid in* $\mathbf{Q}$.

Proof. Use Theorem 9.2.1. □

The following observation is immediate:

Proposition 9.3.2. *If* $\mathbf{Q}$ *is an* $(m-1)$*-nilpotent quasivariety, then its equationally defined commutator satisfies the law* $\boldsymbol{c}_{m+1} \approx \boldsymbol{c}_m$ *and, consequently, it also satisfies the laws:* $\boldsymbol{c}_n \approx \boldsymbol{c}_m$*, for all* $n \geqslant m$. □

0-nilpotent quasivarieties are also called *Abelian* quasivarieties. Thus a quasivariety $\mathbf{Q}$ is Abelian if it satisfies the identity $\boldsymbol{c}_1(\mathbf{1}, \mathbf{1}) \approx \mathbf{0}$, i.e., $[\mathbf{1}, \mathbf{1}] \approx \mathbf{0}$. Equivalently, a quasivariety with the additive equationally defined commutator is Abelian if and only if $[\Theta_{\mathbf{Q}}(a, b), \Theta_{\mathbf{Q}}(c, d)]$ is the identity congruence in any algebra $\boldsymbol{A} \in \mathbf{Q}$, for all $a, b, c, d \in A$.

Theorem 9.3.3. *Let* $\mathbf{Q}$ *be a quasivariety whose equationally defined commutator is additive. Suppose that a set* $\Delta(x, y, z, w, \underline{u})$ *satisfies condition* (1) *of Section* 9.1*, i.e.,* $\mathbf{Q}^{\models}(x \approx y) \cap \mathbf{Q}^{\models}(z \approx w) = \mathbf{Q}^{\models}(\Delta(x, y, z, w, \underline{u}))$. *Then* $\mathbf{Q}$ *is Abelian if and only if* $\mathbf{Q}$ *validates the equations* $\Delta(x, y, z, w, \underline{u})$. □

The above corollary states that Abelian quasivarieties whose equationally defined commutator is additive are characterized by the condition that $\mathbf{Q}^{\models}(x \approx y) \cap \mathbf{Q}^{\models}(z \approx w)$is the trivial equational theory.

Example 9.1.6 shows that the variety **BA** of Boolean algebras obeys the law $\boldsymbol{c}_{m+1} \approx \boldsymbol{c}_m$ for *no* positive integer m. Hence **BA** is $(m-1)$-nilpotent for no m.

9.4 Identities Weaker than Nilpotency

The identity $\boldsymbol{c}_{m+1} \approx \boldsymbol{c}_m$ is of rather a special character. By modifying the above reasoning we shall formulate weaker commutator identities which hold in relative congruence lattices of finitely generated quasivarieties with the additive equationally defined commutator.

Let $\Delta(x, y, z, w, \underline{u})$ be a set of parameterized quaternary equations. We define for any $n \geqslant 1$ the following sets of equations:

$$\Gamma_1(x, y, z, w, \underline{u}) := \Delta(x, y, z, w, \underline{u}),$$

$$\Gamma_{n+1}(x, y, z, w, \underline{u}) := \bigcup \{\Delta(\alpha, \beta, z, w, \underline{u}) : \alpha \approx \beta \in \Gamma_n(x, y, z, w, \underline{u})\}.$$

Proposition 9.4.1. *Let* $\mathbf{Q}$ *be a quasivariety. Let* $\Delta(x, y, z, w, \underline{u})$ *be a set of quaternary commutator equations for* $\mathbf{Q}$*. Then:*

(1) $\Gamma_n(x, y, z, w, \underline{u})$ *is a set of quaternary commutator equations for* $\mathbf{Q}$*, for all* $n \geqslant 1$,
(2) *If* Δ *is finite, then so are the sets* Γ_n*, for all* n,

(3) $\mathbf{Q}^{\models}(\Gamma_{n+1}) \subseteq \mathbf{Q}^{\models}(\Gamma_n)$, *for all* $n \geqslant 1$,
(4) *Suppose that* $\mathbf{Q}^{\models}(\Gamma_{m+1}) = \mathbf{Q}^{\models}(\Gamma_m)$ *for some* $m \geqslant 1$. *Then* $\mathbf{Q}^{\models}(\Gamma_n) = \mathbf{Q}^{\models}(\Gamma_m)$ *for all* $n \geqslant m$.

Proof. Straightforward. □

Theorem 9.4.2. *Let* $\mathbf{Q}$ *be a finitely generated quasivariety whose equationally defined commutator is additive. Let* $\Delta(x, y, z, w, \underline{u})$ *be a finite set of commutator equations which satisfies condition* (1) *of Section* 9.1. *Then there exists a positive integer* m *such that* $\mathbf{Q}^{\models}(\Gamma_n) = \mathbf{Q}^{\models}(\Gamma_m)$ *for all* $n \geqslant m$.

Proof. We have: $\mathbf{Q} = \boldsymbol{SP}(\mathbf{K})$ for some finite set $\mathbf{K}$ of finite algebras. In view of Theorem 8.1.1, there exists a finite set $\Delta(x, y, z, w, \underline{u})$ which satisfies condition (1) of Section 9.1, where $\underline{u} = u_1, u_2, \ldots, u_k$ for some $k \geqslant 0$. Proposition 9.4.1 states that

$$\mathbf{Q}^{\models}(\Gamma_1) \supseteq \mathbf{Q}^{\models}(\Gamma_2) \supseteq \ldots \supseteq \mathbf{Q}^{\models}(\Gamma_n) \supseteq \ldots$$

is a descending chain of theories Suppose *a contrario* that $\mathbf{Q}^{\models}(\Gamma_{n+1})$ is a proper subset of $\mathbf{Q}^{\models}(\Gamma_n)$ for all n. Hence for each $n \geqslant 1$ there exists an equation $\alpha_n \approx \beta_n \in \Gamma_n$ such that $\alpha_n \approx \beta_n \notin \mathbf{Q}^{\models}(\Gamma_{n+1})$. Let $\boldsymbol{T}$ be the subalgebra of $\boldsymbol{Te}_\tau$ generated by the variables x, y, z, w and $u_1, u_2, \ldots, u_k$. It follows that for each $n \geqslant 1$ there exists a homomorphism h_n of $\boldsymbol{T}$ into an algebra $\boldsymbol{A} \in \mathbf{K}$ such that h_n validates Γ_{n+1} and falsifies $\alpha_n \approx \beta_n$. But the above assumptions made about $\mathbf{K}$ and $\Delta(x, y, z, w, \underline{u})$ imply that the set $\{h_n : n \geqslant 1\}$ is actually finite. We may therefore write $\{h_n : n \geqslant 1\} = \{h_n : 1 \leqslant n \leqslant m_0\}$ for some sufficiently large m_0. It follows that there exists i, $1 \leqslant i \leqslant m_0$, such that the set $\{n \geqslant 1 : h_n = h_i\}$ is infinite. Select two numbers m and n from this set, where $m < n$. We have $h_m = h_n$. Let us consider the equation $\alpha_n \approx \beta_n$. According to the definition of h_n we have that $h_n(\alpha_n) \neq h_n(\beta_n)$. Hence $h_m(\alpha_n) \neq h_m(\beta_n)$, because $h_m = h_n$. On the other hand, $\alpha_n \approx \beta_n \in \Gamma_n \subset \mathbf{Q}^{\models}(\Gamma_m)$, because $m < n$. This gives that $\alpha_n \approx \beta_n \in \mathbf{Q}^{\models}(\Gamma_m)$. But as h_m validates Γ_m, we get $h_m(\alpha_n) = h_m(\beta_n)$. A contradiction.

It follows from the above reasoning that there exists $m \geqslant 1$ such that $\mathbf{Q}^{\models}(\Gamma_m) = \mathbf{Q}^{\models}(\Gamma_{m+1})$. Consequently, $\mathbf{Q}^{\models}(\Gamma_n) = \mathbf{Q}^{\models}(\Gamma_m)$ for all $n \geqslant m$. □

We define the following sequence of binary commutator terms $\boldsymbol{d}_n(\boldsymbol{x}, \boldsymbol{y})$, $n = 1, 2, \ldots$:

$$\boldsymbol{d}_1 := [\boldsymbol{x}, \boldsymbol{y}],$$
$$\boldsymbol{d}_{n+1} := [\boldsymbol{d}_n, \boldsymbol{y}], \quad \text{for } n = 1, 2, \ldots$$

$\boldsymbol{d}_n$ is thus the term $[[\ldots[[\boldsymbol{x}, \boldsymbol{y}], \boldsymbol{y}], \ldots], \boldsymbol{y}]$, where $\boldsymbol{y}$ occurs n times, for $n = 1, 2, \ldots$.

Lemma 9.4.3. *Let* $\mathbf{Q}$ *be a quasivariety whose equationally defined commutator is additive. Define the set* $\Gamma_n = \Gamma_n(x, y, z, w, \underline{u})$ *for each positive* n *as above. Then, for every algebra* $\mathbf{A}$ *in* $\mathbf{Q}$, *for every positive* n, *and for all* $a, b, c, d \in A$,

$$d_n(\Theta_{\mathbf{Q}}(a,b), \Theta_{\mathbf{Q}}(c,d), \ldots, \Theta_{\mathbf{Q}}(c,d))$$
$$= \sup{}_{\mathbf{Q}}\{\Theta_{\mathbf{Q}}(\Gamma_n(a,b,c,d,\underline{e})) : \underline{e} \in A^k\}.$$

Proof. Immediate. □

Theorem 9.4.4. *Let* $\mathbf{Q}$ *be a quasivariety whose equationally defined commutator is additive. Let* m *be a positive integer. Then the following conditions are equivalent:*

(a) $\mathbf{Q}^{\models}(\Gamma_m) = \mathbf{Q}^{\models}(\Gamma_{m+1})$.
(b) *The lattices of* $\mathbf{Q}$*-congruences on the algebras of* $\mathbf{Q}$ *obey the commutator law*

$$\boldsymbol{d}_{m+1} \approx \boldsymbol{d}_m.$$

Consequently, they also obey the laws: $\boldsymbol{d}_n \approx \boldsymbol{d}_m$*, for all* $n \geqslant m$.

(a) can be restated by saying that Γ_{m+1}/Γ_m is a set of rules of $\mathbf{Q}$.

Proof. By Lemma 9.4.3 and Proposition 9.4.1. □

Corollary 9.4.5. *Let* $\mathbf{Q}$ *be a finitely generated quasivariety whose equationally defined commutator is additive. Then there exists a positive integer* m *such that* $\mathbf{Q}^{\models}(\Gamma_m) = \mathbf{Q}^{\models}(\Gamma_{m+1})$*. Consequently,* $\mathbf{Q}$ *obeys the commutator laws* $\boldsymbol{d}_n \approx \boldsymbol{d}_m$*, for all* $n \geqslant m$.

Proof. By Theorems 9.4.2 and 9.4.4. □

We define the following sequence of unary commutator terms $\boldsymbol{e}_n(\boldsymbol{x})$, $n = 1, 2, \ldots$:

$$\boldsymbol{e}_1 := [\boldsymbol{x}, \boldsymbol{x}],$$
$$\boldsymbol{e}_{n+1} := [\boldsymbol{e}_n, \boldsymbol{x}], \qquad \text{for } n = 1, 2, \ldots$$

$\boldsymbol{e}_n$ is thus the term $[[\ldots[[\boldsymbol{x},\boldsymbol{x}],\boldsymbol{x}],\ldots],\boldsymbol{x}]$, where $\boldsymbol{x}$ occurs $n+1$ times, for $n = 1, 2, \ldots$. $\boldsymbol{e}_n$ is called the *nth-potent term.*

It is clear that every quasivariety obeys the laws: $\boldsymbol{e}_{n+1}(\boldsymbol{x}) \leqslant \boldsymbol{e}_n(\boldsymbol{x})$ for all positive n and $\boldsymbol{e}_1(\boldsymbol{x}) \leqslant \boldsymbol{x}$.

Lemma 9.4.6. *Let* $\mathbf{Q}$ *be a quasivariety whose equationally defined commutator is additive. Put* $\Gamma'_n(x,y,\underline{u}) := \Gamma_n(x,y,x,y,\underline{u})$ *for each positive* n*, where* $\Gamma_n(x,y,z,w,\underline{u})$ *is defined as above. Then, for every algebra* $\mathbf{A}$ *in* $\mathbf{Q}$*, for every positive* n*, and for all* $a, b \in A$,

$$\boldsymbol{e}_n(\Theta_{\mathbf{Q}}(a,b), \Theta_{\mathbf{Q}}(a,b), \ldots, \Theta_{\mathbf{Q}}(a,b)) = \sup{}_{\mathbf{Q}}\{\Theta_{\mathbf{Q}}(\Gamma'_n(a,b,\underline{e})) : \underline{e} \in A^k\}.$$

(The congruence $\Theta_{\mathbf{Q}}(a,b)$ on the left-hand side occurs $n+1$ times. $\Gamma'_1(x,y,\underline{u})$ coincides with $\Delta(x,y,x,y,\underline{u})$. The supremum is taken in the lattice $\boldsymbol{Con}_{\mathbf{Q}}(\mathbf{A})$.)

Proof. Immediate. □

Theorem 9.4.7. *Let* $\mathbf{Q}$ *be a quasivariety whose equationally defined commutator is additive. Let m be a positive integer. Then the following conditions are equivalent:*

(a) $\mathbf{Q}^{\models}(\Gamma'_m) = \mathbf{Q}^{\models}(\Gamma'_{m+1})$.
(b) *The lattices of* $\mathbf{Q}$*-congruences on the algebras of* $\mathbf{Q}$ *obey the commutator law* $\boldsymbol{e}_{m+1} \approx \boldsymbol{e}_m$. *Consequently, they also obey the laws:* $\boldsymbol{e}_n \approx \boldsymbol{e}_m$, *for all* $n \geqslant m$.

(a) can be restated by saying that Γ'_{m+1}/Γ'_m is a set of rules of $\mathbf{Q}^{\models}$.

Proof. By Lemma 9.4.6 and Theorem 9.4.4. □

Corollary 9.4.8. *Let* $\mathbf{Q}$ *be a finitely generated quasivariety whose equationally defined commutator is additive. Then there exists a positive integer m such that* $\mathbf{Q}^{\models}(\Gamma'_m) = \mathbf{Q}^{\models}(\Gamma'_{m+1})$. *Consequently,* $\mathbf{Q}$ *obeys the commutator laws* $\boldsymbol{e}_n \approx \boldsymbol{e}_m$, *for all* $n \geqslant m$. *It follows that* $\mathbf{Q}$ *validates the commutator quasi-equation:*

$$\boldsymbol{e}_{m+1}(\boldsymbol{x}) \approx \mathbf{0} \rightarrow \boldsymbol{e}_m(\boldsymbol{x}) \approx \mathbf{0}. \quad \square$$

For every RCD quasivariety $\mathbf{Q}$ it is the case that $\boldsymbol{e}_1(\Phi, \Phi) = \Phi$, for any $\boldsymbol{A} \in \mathbf{Q}$ and $\Phi \in Con_{\mathbf{Q}}(\boldsymbol{A})$. Hence every RCD quasivariety (not necessarily finitely generated) satisfies the commutator quasi-equation $\boldsymbol{e}_1(\boldsymbol{x}) \approx \mathbf{0} \rightarrow \boldsymbol{x} \approx \mathbf{0}$. But this quasi-equation itself does not entail congruence-distributivity—see Kearnes (1990).

9.5 Commutator Identities in Finitely Generated Quasivarieties

We shall use the commutator terms $\boldsymbol{c}_n(\boldsymbol{x}_1, \boldsymbol{x}_2, \ldots, \boldsymbol{x}_{n+1})$, $n = 1, 2, \ldots$ defined as above. $\boldsymbol{c}_n$ is the term $[[\ldots[[\boldsymbol{x}_1, \boldsymbol{x}_2], \boldsymbol{x}_3], \ldots], \boldsymbol{x}_{n+1}]$, for $n = 1, 2, \ldots$.

The following observation is crucial:

Theorem 9.5.1. *Let* $\mathbf{Q}$ *be a finitely generated quasivariety, generated by a class consisting of algebras each of which has at most* $m - 1$ *elements, where* $m \geqslant 3$. *Suppose that the equationally defined commutator of* $\mathbf{Q}$ *is additive. Let* $\boldsymbol{A} \in \mathbf{Q}$ *and let* $a_1, \ldots, a_m$ *be a sequence of elements of A of length m (possibly with repetitions). Let* $\langle b_i, c_i\rangle$ $(1 \leqslant i \leqslant r,\ r = m(m-1)/2)$ *be an enumeration of the pairs* $\langle a_i, a_j\rangle$, *where* $1 \leqslant i < j \leqslant m$. *Then*

$$\boldsymbol{c}_{r-1}(\Theta^{\boldsymbol{A}}_{\mathbf{Q}}(b_1, c_1), \ldots, \Theta^{\boldsymbol{A}}_{\mathbf{Q}}(b_r, c_r)) = \mathbf{0}_{\boldsymbol{A}}.$$

In particular, if $x_1, \ldots, x_m$ *are free generators in the free algebra* $\boldsymbol{F} := \boldsymbol{F}_{\mathbf{Q}}(\omega)$ *and* $\langle y_i, z_i\rangle$ $(1 \leqslant i \leqslant r)$ *is an enumeration of the pairs* $\langle x_i, x_j\rangle$, *where* $1 \leqslant i < j \leqslant m$, *then*

$$\boldsymbol{c}_{r-1}(\Theta^{\boldsymbol{F}}(y_1, z_1), \ldots, \Theta^{\boldsymbol{F}}(y_r, z_r)) = \mathbf{0}^{\boldsymbol{F}}.$$

Proof. In view of Theorem 8.2.5 we have that $\bigcap_{1 \leqslant i < j \leqslant m} \Theta^{\boldsymbol{A}}_{\mathbf{Q}}(a_i, a_j) = \mathbf{0}_{\boldsymbol{A}}$. As

$$c_{r-1}(\Theta^{A}_{\mathbf{Q}}(b_1, c_1), \ldots, \Theta^{A}_{\mathbf{Q}}(b_n, c_n)) \subseteq \bigcap_{1 \leqslant i < j \leqslant m} \Theta^{A}_{\mathbf{Q}}(a_i, a_j) = \mathbf{0}_{\mathbf{A}},$$

the theorem follows. □

It is an open problem how to express the above equality as a commutator identity (cf. Theorem 9.2.1).

Example. Let $\mathbf{Q}$ be as above with $m = 4$, $\mathbf{A} \in \mathbf{Q}$ and a_1, a_2, a_3, a_4 be a sequence of elements of A. Then $r = 6$ and let us assume that the sequence $\Theta_{\mathbf{Q}}(b_i, c_i)$ $(1 \leqslant i \leqslant 6)$ equals $\Theta_{\mathbf{Q}}(a_1, a_2)$, $\Theta_{\mathbf{Q}}(a_1, a_3)$, $\Theta_{\mathbf{Q}}(a_1, a_4)$, $\Theta_{\mathbf{Q}}(a_2, a_3)$, $\Theta_{\mathbf{Q}}(a_2, a_4)$, $\Theta_{\mathbf{Q}}(a_3, a_4)$. Since $\mathbf{Q}$ is generated by a class of algebras each of which has at most 3 elements, it follows that

$$c_5(\Theta_{\mathbf{Q}}(b_1, c_1), \ldots, \Theta_{\mathbf{Q}}(b_6, c_6)) =$$
$$c_5(\Theta_{\mathbf{Q}}(a_1, a_2), \Theta_{\mathbf{Q}}(a_1, a_3), \Theta_{\mathbf{Q}}(a_1, a_4), \Theta_{\mathbf{Q}}(a_2, a_3), \Theta_{\mathbf{Q}}(a_2, a_4), \Theta_{\mathbf{Q}}(a_3, a_4)) = \mathbf{0}_{\mathbf{A}}.$$

The above equality trivially holds because $a_i = a_j$ for some $i < j$. Hence $\Theta^{A}_{\mathbf{Q}}(a_i, a_j) = \mathbf{0}_A$ and, consequently, $c_5(\Theta^{A}_{\mathbf{Q}}(b_1, c_1), \ldots, \Theta^{A}_{\mathbf{Q}}(b_6, c_6)) = \mathbf{0}_{\mathbf{A}}$. □

In the next step we shall use the sets of equations

$$\Delta_n = \Delta_n(x_1, y_1, x_2, y_2, \ldots, x_{n+1}, y_{n+1}, \underline{v}_n),$$

for $n = 1, 2, \ldots$ from Section 9.1 to encode the above corollary in terms of commutator equations.

Let $\mathbf{Q}$ be a quasivariety whose equationally defined commutator is additive and let $\Delta(x, y, z, w, \underline{u})$ be a set of quaternary commutator equations for $\mathbf{Q}$ satisfying condition (1) in Section 9.1. Let $m \geqslant 3$ be a positive integer. Fix a sequence $x_1, \ldots, x_m$ of m different individual variables. (We may use the variables $x_1, x_2, \ldots, x_{n+1}$ occurring in the equations of Δ_n for a sufficiently large n). Let $\langle z_i, w_i \rangle$ $(1 \leqslant i \leqslant r,\ r = m(m-1)/2)$ be an enumeration of the pairs $\langle x_i, x_j \rangle$, where $1 \leqslant i < j \leqslant m$. We put $\underline{v} := \underline{v}_{r-1}$ and

$$\Lambda_m(x_1, x_2, \ldots, x_m, \underline{v}) := \Delta_{r-1}(z_1, w_1, z_2, w_2, \ldots, z_r, w_r, \underline{v}_{r-1}).$$

In other words $\Delta_{r-1}(z_1, w_1, z_2, w_2, \ldots, z_r, w_r, \underline{v}_{r-1})$ results from $\Delta_{r-1} = \Delta_{r-1}(x_1,\ y_1, x_2, y_2, \ldots, x_r, y_r, \underline{v}_{r-1})$ by the uniform substitution of the variables from the enumeration $\langle z_i, w_i \rangle$ $(1 \leqslant i \leqslant r,\ r = m(m-1)/2)$ for the consecutive variables x_i and y_i occurring in Δ_{r-1}.

Different enumerations $\langle z_i, w_i \rangle$ $(1 \leqslant i \leqslant r)$ of the pairs $\langle x_i, x_j \rangle$, where $1 \leqslant i < j \leqslant m$, yield different sets of equations $\Delta_{r1}(z_1, w_1, z_2, w_2, \ldots, z_r, w_r, \underline{v}_{r-1})$ because each time different substitutions are applied to the variables occurring in $\Delta_{r-1}(x_1, y_1, x_2, y_2, \ldots, x_r, y_r, \underline{v}_n)$ to obtain the set of equations $\Lambda_m(x_1, \ldots, x_m, \underline{v}) = \Delta_{r-1}(z_1, w_1, z_2, w_2, \ldots,\ z_r, w_r, \underline{v}_{r-1})$. (The parametric variables $\underline{v}_{r-1}$ are not touched.)

The set $\Lambda_m(x_1, x_2, \ldots, x_m, \underline{v})$ is finite whenever the set $\Delta(x, y, z, w, \underline{u})$ is finite. If $\mathbf{Q}$ is finitely generated, we may assume that $\Delta(x, y, z, w, \underline{u})$ is finite, by Theorem 8.1.1.

Theorem 9.5.2. *Let $\mathbf{Q}$ be a finitely generated quasivariety whose equationally defined commutator is additive. Let $\Delta(x, y, z, w, \underline{u})$ be a finite set of quaternary commutator equations satisfying condition* (1) *in Section* 9.1. *Suppose that $\mathbf{Q}$ is generated by a class consisting of algebras each of which has at most $m-1$ elements, where $m \geqslant 3$ is a positive integer. Define $\Lambda_m(x_1, x_2, \ldots, x_m, \underline{v})$ as above. Let $\mathbf{A} \in \mathbf{Q}$ and let $a_1, \ldots, a_m$ be an arbitrary sequence of elements of A. Then*

$$\Theta^{A}_{\mathbf{Q}}(\Lambda_m(a_1, \ldots, a_m, \underline{e})) = \mathbf{0}_A,$$

for all sequences $\underline{e} \in A^{(r-1)k}$.

In particular, if $x_1, \ldots, x_m$ are free generators of the free algebra $\mathbf{F} := \mathbf{F}_{\mathbf{Q}}(\omega)$, then

$$\Theta^{F}_{\mathbf{Q}}(\Lambda_m(x_1, \ldots, x_m, \underline{e})) = \mathbf{0}_F,$$

for all $\underline{e} \in F^{(r-1)k}$.

Proof. Let $\langle b_i, c_i \rangle$ $(1 \leqslant i \leqslant r)$ be an enumeration of the pairs $\langle a_i, a_j \rangle$, where $1 \leqslant i < j \leqslant m$, compatible with the enumeration $\langle z_i, w_i \rangle$ $(1 \leqslant i \leqslant r)$ of the pairs $\langle x_i, x_j \rangle$, where $1 \leqslant i < j \leqslant m$.) In view of Theorem 9.2.1,

$$\mathbf{c}_{r-1}(\Theta^{A}_{\mathbf{Q}}(b_1, c_1), \ldots, \Theta^{A}_{\mathbf{Q}}(b_r, c_r)) =$$

$$\sup{}_{\mathbf{Q}}\{\Theta^{A}_{\mathbf{Q}}(\Delta_{r-1}(b_1, c_1, b_2, c_2, \ldots, b_r, c_r, \underline{e}_{r-1})) : \underline{e}_{r-1} \in A^{(r-1)k}\} =$$

$$\sup{}_{\mathbf{Q}}\{\Theta^{A}_{\mathbf{Q}}(\Lambda_m(a_1, \ldots, a_m, \underline{e}_{r-1})) : \underline{e} \in A^{(r-1)k}\}.$$

(The suprema are taken in the lattice $\boldsymbol{Con}_{\mathbf{Q}}(\mathbf{A})$.) Then apply Theorem 9.5.1. □

Corollary 9.5.3. *Let $\mathbf{Q}$, Δ, and $\Lambda_m(x_1, \ldots, x_m, \underline{v})$ be defined as above. Then $\boldsymbol{Va}(\mathbf{Q})$ validates the equations $\Lambda_m(x_1, \ldots, x_m, \underline{v})$.* □

Example 9.5.4. For instance, suppose that in the above corollary $m = 3$ and x_1, x_2, x_3 are different variables. Then $r = m(m-1)/2 = 3$. Let $\langle x_1, x_2 \rangle$, $\langle x_1, x_3 \rangle$, $\langle x_2, x_3 \rangle$ be an enumeration of the pairs $\langle x_i, x_j \rangle$, $1 \leqslant i < j \leqslant 3$.

We return to Example 9.1.6. In this case we have that $\Delta(x, y, z, w) = \{(x \leftrightarrow y) \vee (z \leftrightarrow w) \approx 1\}$ (no parameters occur) and, consequently,

$$\Delta_1(x_1, y_1, x_2, y_2) = \{(x_1 \leftrightarrow y_1) \vee (x_2 \leftrightarrow y_2) \approx 1\},$$

$$\Delta_2(x_1, y_1, x_2, y_2, x_3, y_3) = \bigcup\{\Delta(x/\alpha, y/\beta, z/x_3, w/y_3) : \alpha \approx \beta \in \Delta_1\} =$$

$$\{(\alpha \leftrightarrow \beta) \vee (x_3 \leftrightarrow y_3) \approx 1 : \alpha \approx \beta \in \Delta_1\} =$$

$$\{(((x_1 \leftrightarrow y_1) \vee (x_2 \leftrightarrow y_2)) \leftrightarrow 1) \vee (x_3 \leftrightarrow y_3) \approx 1\}.$$

As **BA** is generated (as a quasivariety) by a two-element Boolean algebra, it follows from the above corollary that **BA** validates the equations $\Lambda_3(x_1, x_2, x_3) := \Delta_2(x_1, x_2, x_1, x_3, x_2, x_3)$. (In fact, this set consists of a single equation.) We have:

$$\Lambda_3(x_1, x_2, x_3) = \Delta_2(x_1, x_2, x_1, x_3, x_2, x_3) =$$
$$\{(((x_1 \leftrightarrow x_2) \vee (x_1 \leftrightarrow x_3)) \leftrightarrow 1) \vee (x_2 \leftrightarrow x_3) \approx 1\}.$$

One can directly verify that indeed the equation

$$(((x_1 \leftrightarrow x_2) \vee (x_1 \leftrightarrow x_3)) \leftrightarrow 1) \vee (x_2 \leftrightarrow x_3) \approx 1 \qquad (*)$$

holds in **BA**.

Since **BA** validates the equation $x \leftrightarrow 1 \approx x$, equation $(*)$ is equivalent (in $\mathbf{BA}^{\models}$) to

$$(x_1 \leftrightarrow x_2) \vee (x_1 \leftrightarrow x_3) \vee (x_2 \leftrightarrow x_3) \approx 1$$

(some parentheses are omitted). □

Theorem 9.5.5. *Let $m \geqslant 3$ be a fixed natural number. Suppose that* **Q** *is a (finitely generated) quasivariety generated by a finite class* **K** *of algebras, where each algebra in K has at most $m-1$ elements. If the equationally defined commutator for* **Q** *is additive, then* **Q** *has m-EDTPM with respect to the finite set of equations $\Lambda_m(x_1, x_2, \ldots, x_m, \underline{v})$ defined as above, i.e.,*

$$\bigcap\{\Theta^{A}_{\mathbf{Q}}(a_i, a_j) : 1 \leqslant i < j \leqslant m\} = \Theta^{A}_{\mathbf{Q}}((\forall \underline{e})(\Lambda_m(a_1, a_2, \ldots, a_m, \underline{e})),$$

for any algebra $\boldsymbol{A} \in \mathbf{Q}$ *and any* $a_1, \ldots, a_m \in A$.

Proof. The assumptions and Corollary 9.5.3 imply that **Q** validates the equations $\Lambda_m(x_1, x_2, \ldots, x_m, \underline{v})$. Thus, if $\boldsymbol{A} \in \mathbf{Q}$ and $a_1, \ldots, a_m \in A$, then $\boldsymbol{A} \models (\forall \underline{v}) \bigwedge(\Lambda_m(x_1, x_2, \ldots, x_m, \underline{v})[a_1, a_2, \ldots, a_m]$.

According to Theorem 8.2.5 it is also the case that $\bigcap\{\Theta^{A}_{\mathbf{Q}}(a_i, a_j) : 1 \leqslant i < j \leqslant m\} = \mathbf{0}_A$. It follows that the equivalence

$$\bigcap\{\Theta^{A}_{\mathbf{Q}}(a_i, a_j) : 1 \leqslant i < j \leqslant m\} = \mathbf{0}_A \quad \Leftrightarrow$$
$$\boldsymbol{A} \models (\forall \underline{v}) \bigwedge(\Lambda_m(x_1, x_2, \ldots, x_m, \underline{v})[a_1, a_2, \ldots, a_m].$$

is true because its both components are true.

Thus any algebra $\boldsymbol{A} \in \mathbf{Q}$ validates condition (2) of Proposition 8.3.1. This shows that **Q** has m-EDTPM with respect to $\Lambda_m(x_1, x_2, \ldots, x_m, \underline{v})$, by Definition 8.3.2. □

Theorem 9.5.5 concludes the presentation of the equationally defined commutator for quasivarieties.

Final Remarks

This book outlines the theory of the equationally defined commutator. The approach presented here appears to have not been considered in the literature. The basic notions and results are new. From the logical viewpoint the focus of the book is on central theorems. This focus determines the narrative structure of the book. Secondary threads have not been investigated in more detail; many of them have been left open and identified as problems for further scrutiny. Some of these problems seem to be difficult and they are a challenge for algebraists and logicians. The following problem, referred to as *Pigozzi's conjecture*, is central:

Problem. Let τ be a finite signature. Let $\mathbf{Q}$ be a finitely generated and relatively congruence-modular quasivariety of τ-algebras. Prove $\mathbf{Q}$ is finitely based. □

Appendix A
Algebraic Lattices

A.1 Partially Ordered Sets

Let P be a set. A binary relation $\leqslant$ on P is an *order* (or *partial order*) on P if and only if $\leqslant$ satisfies the following conditions:

(i) $\leqslant$ is *reflexive*, i.e., $a \leqslant a$, for all $a \in A$;
(ii) $\leqslant$ is *transitive*, i.e., $a \leqslant b$ and $b \leqslant c$ implies $a \leqslant c$, for all $a, b, c \in A$;
(iii) $\leqslant$ is *antisymmetric*, i.e., $a \leqslant b$ and $b \leqslant a$ implies $a \leqslant b$, for all $a, b \in A$.

A *partially ordered set*, a *poset*, for short, is a set with an order defined on it.

Each order relation $\leqslant$ on P gives rise to a relation $<$ of *strict order*: $a < b$ in P if and only if $a \leqslant b$ and $a \neq b$.

Let $\boldsymbol{P} = (P, \leqslant)$ be a poset and X a subset of P. Then X inherits an order relation from P: given $x, y \in X$, $x \leqslant y$ in X if and only if $x \leqslant y$ in P. We then also say that the order on X is *induced* by the order from P.

(1) An element $u \in P$ is called *an upper bound* of the set X if $x \leqslant u$, for all $x \in X$.
(2) An element $l \in P$ is called a *lower bound* of the set X if $l \leqslant x$, for all $x \in X$.
(3) An element $a \in P$ is called the *least upper bound* of the set X if a is an upper bound of X and $a \leqslant u$ for every upper bound u of X.
    If X has a least upper bound, this is called the *supremum* of X and is written "sup(X)".
(4) An element $b \in P$ is called the *greatest lower bound* of the set X if b is a lower bound of X and $l \leqslant b$ for every lower bound l of X.
    If X has a greatest lower bound, this is called the *infimum* of X and is written "inf(X)".

Instead of "sup(X)" and "inf(X)" we shall often write "$\bigvee X$" and "$\bigwedge X$"; in particular, we write "$a \vee b$" and "$a \wedge b$" instead of "sup($\{a, b\}$)" and "inf($\{a, b\}$)".

If the poset $\boldsymbol{P}$ itself has an upper bound u, then it is the only upper bound. u is then called the *greatest element* of $\boldsymbol{P}$ or the *top element* of $\boldsymbol{P}$. In an analogous way the

J. Czelakowski, *The Equationally-Defined Commutator*,
DOI 10.1007/978-3-319-21200-5

notion of the *least element* of $\boldsymbol{P}$ is defined. The least element of $\boldsymbol{P}$ is often denoted by $\mathbf{0}$ and is also called the *zero* or the *bottom element* of the poset $\boldsymbol{P}$.

A set $X \subseteq P$ is:

(a) an *upper directed* subset of $\boldsymbol{P}$ if for every pair $a, b \in X$ there exists an element $c \in X$ such that $a \leqslant c$ and $b \leqslant c$ (or, equivalently, if every finite non-empty subset of X has an upper bound which is an element of X);
(b) a *chain* in $\boldsymbol{P}$ if, for every pair $a, b \in X$, either $a \leqslant b$ or $b \leqslant a$ (that is, if any two elements of X are comparable);
(c) a *well-ordered subset* of $\boldsymbol{P}$ (or: a *well-ordered chain* in $\boldsymbol{P}$) if X is a chain in which every non-empty subset $Y \subseteq X$ has the least element (in Y).

Equivalently, in the presence of the Axiom of Dependent Choices from set theory, X is well-ordered if and only if it is a chain and there is no infinite strictly decreasing sequence $c_0 > c_1 > \ldots > c_n > \ldots$ of elements of X.

Every well-ordered set X is isomorphic with a unique ordinal, called the *type* of X.

A subset directed downwards is defined similarly; when nothing to the contrary is said, "directed" will always mean "directed upwards".

If the poset $(P, \leqslant)$ itself is a chain or directed, then it is simply called a chain or a directed poset.

Definition A.1.1. Let $\boldsymbol{P} = (P, \leqslant)$ be a poset.

(1) $\boldsymbol{P}$ is *directed-complete* if for every non-empty directed subset $D \subseteq P$, the supremum $\sup(D)$ exists in $\boldsymbol{P}$.
(2) $\boldsymbol{P}$ is *chain-complete* (or *inductive*) if for every non-empty chain $C \subseteq P$, the supremum $\sup(C)$ exists in $\boldsymbol{P}$.
(3) $\boldsymbol{P}$ is *well-orderably-complete* if for every well-ordered non-empty chain $C \subseteq P$, the supremum $\sup(C)$ exists in $\boldsymbol{P}$. □

Every directed-complete poset is chain-complete and every inductive poset is well-orderably complete. It is known from set theory that in the presence of the Axiom of Choice they are mutually equivalent.

Note. The empty subset of a poset is usually assumed to be well-ordered. However, if $\boldsymbol{P} = (P, \leqslant)$ is well orderably complete in the above sense, then the supremum of the empty subset *may not* exist. But if it exists, it is the least element in $\boldsymbol{P}$.

Analogously, in inductive posets and directed-complete posets the zero may not exist.

In the literature, directed-complete posets are also called *continuous* posets and the often used abbreviation "$(P, \leqslant)$ is a cpo" marks that the poset $(P, \leqslant)$ is continuous.

A.2 Semilattices and Lattices

A non-empty poset $\boldsymbol{P} = (P, \leqslant)$ is a *join-semilattice*, if for any two elements $a, b \in P$, $\sup(\{a, b\})$ exists in $\boldsymbol{P}$.

Join-semilattices are equivalently defined as algebras $\boldsymbol{P} = (P, \vee)$ of signature (2) satisfying the following identities:

$$x \vee x \approx x, \qquad x \vee y \approx y \vee x, \qquad x \vee (y \vee z) \approx (x \vee y) \vee z.$$

A non-empty poset $\boldsymbol{P} = (P, \leqslant)$ is a *lattice*, if for any two elements $a, b \in P$, $\sup(\{a, b\})$ and $\inf(\{a, b\})$ exist in $\boldsymbol{P}$.

Lattices are equivalently defined as algebras $\boldsymbol{P} = (P, \wedge, \vee)$ of signature $(2, 2)$ satisfying the identities:

$$\begin{aligned}
&x \wedge x \approx x, \qquad x \wedge y \approx y \wedge x, \qquad x \wedge (y \wedge z) \approx (x \wedge y) \wedge z,\\
&x \vee x \approx x, \qquad x \vee y \approx y \vee x, \qquad x \vee (y \vee z) \approx (x \vee y) \vee z,\\
&x \wedge (x \vee y) \approx x, \qquad\qquad x \vee (x \wedge y) \approx x.
\end{aligned}$$

Thus for every lattice $\boldsymbol{P} = (P, \wedge, \vee)$, its reduct $(P, \vee)$ is a join-semilattice, and the reduct $(P, \wedge)$ is a *meet-semilattice*, the notion dual to the first one. Moreover the two semilattices are conjoined by the last two axioms.

A non-empty poset $\boldsymbol{P} = (P, \leqslant)$ is a *complete lattice* if every subset $X \subseteq P$ has both a least upper greatest lower bound $\sup(X)$ (the supremum, also called the join) and a greatest lower bound $\inf(X)$ (the infimum, also called the meet) in $\boldsymbol{P}$. It is assumed that $\inf(\emptyset)$ is the top element of $\boldsymbol{P}$ and $\sup(\emptyset)$ is the zero element in $\boldsymbol{P}$. The join and the meet of X are often denoted by

$$\bigvee X \qquad \text{and} \qquad \bigwedge X,$$

respectively.

Every complete lattice is a lattice, and every finite lattice is a complete lattice.

Theorem A.2.1. *Every directed-complete join-semilattice* $\boldsymbol{L} = (L, \leqslant, \boldsymbol{0})$ *with zero is a complete lattice.*

Proof. If E is a non-empty set $E \subseteq L$, then define $D := \{\sup(E_f) : E_f$ is a finite subset of $E\}$. D is well-defined and directed. As $\sup(D)$ exists in $\boldsymbol{L}$ and $\sup(E) = \sup(D)$, it follows that for every non-empty set $E \subseteq L$, $\sup(E)$ exists in $\boldsymbol{L}$. It is then easy to verify that $\inf(E)$ also exists in $\boldsymbol{L}$. For let $B := \{b \in L : b$ is a lower bound of $E\}$. Then $\boldsymbol{0} \in B$ and $\sup(B) = \inf(E)$. □

A sublattice $\boldsymbol{M}$ of a complete lattice $\boldsymbol{L}$ is called a *closed sublattice* of $\boldsymbol{L}$ if for *every non-empty subset* X of M, the elements $\sup(X)$ and $\inf(X)$, as computed in $\boldsymbol{L}$, are actually in M. If moreover, the zero and the unit elements of $\boldsymbol{L}$ belong to M, the sublattice $\boldsymbol{M}$ is called a *complete sublattice* of $\boldsymbol{L}$. Thus a sublattice $\boldsymbol{M}$ of $\boldsymbol{L}$ is a complete sublattice of $\boldsymbol{L}$ if for *every subset* X of M, the elements $\sup(X)$ and $\inf(X)$, as computed in L, belong to M. It follows that every complete sublattice $\boldsymbol{M}$ of $\boldsymbol{L}$ has the same zero and unit elements as $\boldsymbol{L}$ and $\boldsymbol{M}$ itself is a complete lattice. It should be noted that a closed sublattice $\boldsymbol{M}$ of a complete lattice $\boldsymbol{L}$ need not possess the zero and the unit elements, but even if $\boldsymbol{M}$ has them, they need not coincide with their counterparts in $\boldsymbol{L}$.

If $\boldsymbol{P} = (P, \leqslant)$ is a poset, then $(P, \geqslant)$ is also a poset, where $\geqslant$ is the dual of $\leqslant$:

$$a \geqslant b \quad \Leftrightarrow_{df} \quad b \leqslant a,$$

for all $a, b \in P$. The poset $(P, \geqslant)$ is called the *dual* of $\boldsymbol{P}$ and denoted by $\boldsymbol{P}^d$.

If $\boldsymbol{P} = (P, \leqslant)$ is a lattice, then the dual poset $\boldsymbol{P}^d = (P, \geqslant)$ is a lattice as well. If $\boldsymbol{P}$ is viewed as an algebra $(P, \wedge, \vee)$ satisfying the above axioms, then $\boldsymbol{P}^d$ becomes the algebra $(P, \vee, \wedge)$. Thus $\boldsymbol{P}^d$ results from $\boldsymbol{P}$ by interchanging $\wedge$ and $\vee$.

If a lattice $\boldsymbol{P}$ is represented by a Hasse diagram, then the Hasse diagram of $\boldsymbol{P}^d$ is obtained by turning upside down the diagram of $\boldsymbol{P}$.

The above facts yield the *Principle of Duality for lattices*, which states that if σ is a statement that is true in every lattice and σ^d is the statement obtained from σ by interchanging $\leqslant$ and $\geqslant$ and interchanging $\wedge$ and $\vee$, then σ^d is true in every lattice.

In some cases one may obtain a specialized versions of the Duality Principle obtained by restricting the above general quantifier 'in every lattice' to some proper subclasses of lattices. Examples are the classes of distributive and modular lattices. Thus, in particular, the Principle of Duality for modular lattices states that σ is true in *every* modular lattice if and only if the dual statement σ^d is true in all modular lattices.

The equational class of *distributive* lattices is defined by the identity:

$$x \wedge (y \vee z) \approx (x \wedge y) \vee (x \wedge z).$$

Equivalently, the class of distributive lattices is defined by the dual equation:

$$x \vee (y \wedge z) \approx (x \vee y) \wedge (x \vee z).$$

The class of *modular* lattices is defined by the identity:

$$((x \wedge z) \vee y) \wedge z \approx (x \wedge z) \vee (y \wedge z).$$

Equivalently, this class is defined by the dual condition:

$$((x \vee z) \wedge y) \vee z \approx (x \vee z) \wedge (y \vee z).$$

Modular lattices form an equational class. The following theorem is attributed to Dedekind (see McKenzie et al. (1987), Theorem 2.25):

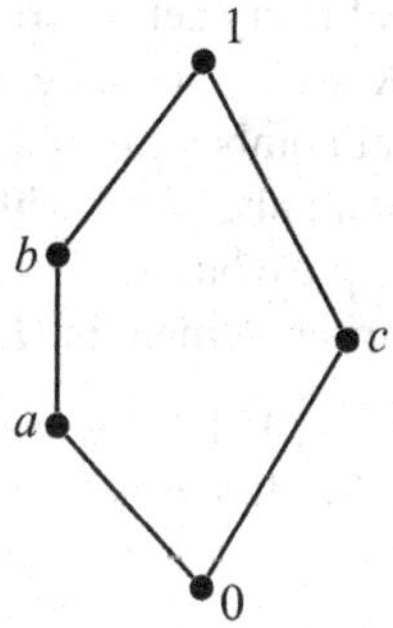

Fig. A.1 The lattice N_5.

Theorem A.2.2. *For any lattice* $\mathbf{L}$ *the following conditions are equivalent:*

(1) $\mathbf{L}$ *is modular.*
(2) $\mathbf{L}$ *validates the quasi-identity* $z \leqslant x \to x \wedge (y \vee z) \approx (x \wedge y) \vee z$.
(3) $\mathbf{L}$ *validates the identity* $((x \wedge z) \vee y) \wedge z \approx (x \wedge z) \vee (y \wedge c)$.
(4) *For any* $a, b, c \in L$, *if* $a \leqslant b$, $a \wedge c = b \wedge c$, *and* $a \vee c = b \vee c$, *then* $a = b$.
(5) $\mathbf{L}$ *has no sublattice isomorphic to* N_5 *(Fig. A.1).*

Each of conditions (2)–(5) can be replaced by their duals.

A.3 Extensions by Ideals. Algebraic Lattices

Let $\mathbf{L} = (L, \leqslant)$ be a complete lattice. An element $x \in L$ is *compact* if, whenever D is a directed and *non-empty* subset of L and $x \leqslant \sup(D)$, there is an element $d \in D$ such that $x \leqslant d$. Equivalently, an element $x \in L$ is compact if, for every non-empty set $E \subseteq L$, if $x \leqslant \sup(E)$, then $x \leqslant \sup(E_f)$ for some finite set $E_f \subseteq E$. The equivalence of these two conditions follows from the (already noted) fact that for any non-empty set $E \subseteq L$, the set $D := \{\sup(E_f) : E_f \text{ is a finite subset of } E\}$ is directed and $\sup(E) = \sup(D)$ in $\mathbf{L}$.

We let $K(\mathbf{L})$ denote the set of compact elements of $\mathbf{L}$.

As $\mathbf{L} = (L, \leqslant)$ contains zero $\mathbf{0}$, we note that $\mathbf{0}$ is compact. Moreover, if x and y are compact, then $x \vee y$ is compact as well.

For every $a \in L$, define the set $K(a) := \{x \in K(\mathbf{L}) : x \leqslant a\}$. The set $K(a)$ is non-empty (because it contains $\mathbf{0}$) and directed. In fact, $K(a)$ is a join-semilattice.

Definition A.3.1. A complete lattice $\mathbf{L} = (L, \leqslant)$ is said to be *algebraic* if for every $a \in L$, $\sup(K(a)) = a$. □

Thus, in any algebraic lattice each element is a directed limit of its compact approximations.

Examples. The power set $\wp(X)$ of every set X forms an algebraic lattice with zero under inclusion relation. The set $K(\wp(X))$ coincides the family of finite subsets of X.

Let I be the unit interval of real numbers, $I := \{x \in \mathbb{R} : 0 \leqslant x \leqslant 1\}$. The lattice $(\mathbb{Q} \cap I, \leqslant)$ formed by the rational numbers of I with the standard ordering $\leqslant$ is not complete. The lattice $(I, \leqslant)$ is complete but it is not algebraic. One can easily check that the number 0 is the only compact element in $(I, \leqslant)$. □

Definition A.3.2. Let $\boldsymbol{P}_0 = (P_0, \leqslant, \boldsymbol{0})$ be a join-semilattice with zero. A subset $\boldsymbol{I} \subseteq P_0$ is called an *ideal* of $\boldsymbol{P}_0$ if, for all $a, b \in P_0$, the following conditions hold:

(i1) If $a, b \in \boldsymbol{I}$, then $a \vee b \in \boldsymbol{I}$;
(i2) If $a \in I$ and $b \leqslant a$, then $b \in \boldsymbol{I}$.
(i3) $\boldsymbol{0} \in \boldsymbol{I}$. □

Ideal$(\boldsymbol{P}_0)$ denotes the set of all ideals of $\boldsymbol{P}_0$.

P_0 is the largest ideal in $\boldsymbol{P}_0$ and $\{\boldsymbol{0}\}$—the smallest one. The intersection of any non-empty family of ideals is an ideal too. This means the family *Ideal*$(\boldsymbol{P}_0)$, ordered by inclusion, is a closure system and hence it forms a complete lattice. It is assumed that $\{\boldsymbol{0}\}$ is the intersection of the empty family of ideals.

It is easy to see that in the lattice $\langle Ideal(\boldsymbol{P}_0), \wedge, \vee\rangle$ of ideals of $\boldsymbol{P}_0$ it is the case that

$$\boldsymbol{I} \wedge \boldsymbol{J} := \boldsymbol{I} \cap \boldsymbol{J},$$
$$\boldsymbol{I} \vee \boldsymbol{J} := \{x \in P_0 : x \leqslant i \vee j, \text{ for some } i \in \boldsymbol{I},\ j \in \boldsymbol{J}\}.$$

for all ideals $\boldsymbol{I}, \boldsymbol{J}$.

For every $a \in P_0$, the set

$$\downarrow a := \{x \in P_0 : x \leqslant a\}$$

is an ideal. $\downarrow a$ is called the *principal ideal* generated by a.

Here are simple observations about the lattices of ideals.

Theorem A.3.3. *Let $\boldsymbol{P}_0 = (P_0, \leqslant, \boldsymbol{0})$ be an arbitrary join-semilattice with zero. The poset $(Ideal(\boldsymbol{P}_0), \subseteq)$ of ideals of $\boldsymbol{P}_0$ is an algebraic lattice. Moreover, the subposet $(K(Ideal(\boldsymbol{P}_0)), \subseteq)$ consisting of compact elements of the lattice $(Ideal(\boldsymbol{P}_0), \subseteq)$ forms a join-semilattice which is isomorphic with $\boldsymbol{P}_0$.*

Proof. As the family *Ideal*$(\boldsymbol{P}_0)$ is a closure system, the poset $(Ideal(\boldsymbol{P}_0), \subseteq)$ is a complete lattice. If $(\boldsymbol{X}, \subseteq)$ is a non-empty directed family of ideals, then the union $\bigcup \boldsymbol{X}$ is also an ideal of $\boldsymbol{P}_0$. Clearly, $\bigcup \boldsymbol{X}$ is the supremum of the family $\boldsymbol{X}$ in the sense of $(Ideal(\boldsymbol{P}_0), \subseteq)$.

We prove that the family $\{\downarrow a : a \in L_0\}$ of principal ideals of $\boldsymbol{P}_0$ coincides with the family $K(Ideal(\boldsymbol{P}_0))$. Let $a \in P_0$ and let $\boldsymbol{X}$ be a directed non-empty family of ideals such that $\downarrow a \subseteq \bigcup \boldsymbol{X}$. It follows that there exists an ideal $\boldsymbol{I} \in \boldsymbol{X}$ such that $a \in \boldsymbol{I}$. Hence $\downarrow a \subseteq \boldsymbol{I}$. This shows that $\downarrow a$ is compact in the poset $(Ideal(\boldsymbol{P}_0), \subseteq)$.

To prove the converse, we notice that $\boldsymbol{I} = \bigcup\{\downarrow a : a \in \boldsymbol{I}\}$ for every ideal $\boldsymbol{I}$. Furthermore, the family $\{\downarrow a : a \in \boldsymbol{I}\}$ is directed and non-empty. Thus every ideal is a supremum of a non-empty family of compact ideals.

Now, if an ideal $\boldsymbol{I}$ is a compact element in $(Ideal(\boldsymbol{P}_0), \subseteq)$, it follows from the above equality that $\boldsymbol{I} \subseteq \downarrow a$ for some $a \in \boldsymbol{I}$. Hence $\boldsymbol{I} = \downarrow a$.

The above facts prove that the lattice $(Ideal(\boldsymbol{P}_0), \subseteq)$ is algebraic.

It is clear that the mapping $\phi : P_0 \to Ideal(\boldsymbol{P}_0)$ given by

$$\phi(a) := \downarrow a, \qquad a \in P_0,$$

establishes an isomorphism between the join-semilattices $\boldsymbol{P}_0$ and $(K(Ideal(\boldsymbol{P}_0)), \subseteq)$. □

The order structure of any algebraic lattice is fully determined by the order structure of the join-semilattice of its compact elements. This fact is established by the following theorem:

Theorem A.3.4. *Let $\boldsymbol{P} = (P, \leqslant)$ and $\boldsymbol{Q} = (Q, \leqslant)$ be algebraic lattices. $\boldsymbol{P}$ and $\boldsymbol{Q}$ are isomorphic if and only if the join-semilattices $(K(\boldsymbol{P}), \leqslant, \boldsymbol{0})$ and $(K(\boldsymbol{Q}), \leqslant, \boldsymbol{0})$ of their compact elements are isomorphic.*

The proof is based on several simple lemmas.

Lemma A.3.5. *Let $\boldsymbol{P}_0 = (P_0, \leqslant, \boldsymbol{0})$, $\boldsymbol{Q}_0 = (Q_0, \leqslant, \boldsymbol{0})$ be isomorphic join-semilattices with zero and let $f : \boldsymbol{P}_0 \cong \boldsymbol{Q}_0$ be an isomorphism between them. The mapping f^* given by the formula*

$$f^*(\boldsymbol{J}) := f^{-1}(\boldsymbol{J}), \quad \textit{for all } \boldsymbol{J} \in Ideal(\boldsymbol{Q}_0),$$

is an isomorphism between the complete lattices $(Ideal(\boldsymbol{Q}_0),\subseteq)$ and $(Ideal(\boldsymbol{P}_0),\subseteq)$, symbolically,

$$f^* : (Ideal(\boldsymbol{Q}_0), \subseteq) \cong (Ideal(\boldsymbol{P}_0), \subseteq).$$

The proof is easy and omitted. □

Lemma A.3.6. *Let $\boldsymbol{L} = (L, \leqslant)$ be an algebraic lattice. $\boldsymbol{L}$ is isomorphic with the lattice $(Ideal(K(\boldsymbol{L}), \subseteq)$ of ideals of the join-semilattice $(K(\boldsymbol{L}), \leqslant, \boldsymbol{0})$ of compact elements of $\boldsymbol{L}$.*

Proof (of the lemma). For every $a \in L$ define $K(a)$ as above, i.e.,

$$K(a) := \{x \in K(\boldsymbol{L}) : x \leqslant a\}.$$

Since $\boldsymbol{L}$ is algebraic, $K(a)$ is an ideal of the join-semilattice $(K(\boldsymbol{L}), \leqslant)$.

Claim 1. *Let $\boldsymbol{I}$ be an ideal of the join-semilattice $(K(\boldsymbol{L}), \leqslant)$. Then $\boldsymbol{I} = K(a)$ for some $a \in L$.*

Proof (of the claim). As $\boldsymbol{I}$ is a (non-empty) subset of $(K(\boldsymbol{L}), \leqslant)$, and hence of $\boldsymbol{L}$, $\sup(\boldsymbol{I})$ exists in $\boldsymbol{L}$. (If $\boldsymbol{I} = \{\boldsymbol{0}\}$, then $\sup(\boldsymbol{I}) = \boldsymbol{0}$.) We define: $a := \sup(\boldsymbol{I})$. Evidently, $\boldsymbol{I} \subseteq K(a)$ because $\boldsymbol{I}$ contains only compact elements and every element of $\boldsymbol{I}$ is equal or smaller than a. Now, let $x \in K(a)$. So x is compact and $x \leqslant a = \sup(\boldsymbol{I})$. Hence $x \leqslant i$ for some $i \in I$, by compactness. Since $\boldsymbol{I}$ is an ideal, we therefore have that $x \in I$. So $K(a) \subseteq \boldsymbol{I}$. Consequently, $\boldsymbol{I} = K(a)$. This proves the claim. □

Claim 2. *Let $a, b \in L$. Then $K(a) \subseteq K(b)$ if and only if $a \leqslant b$.*

Proof (of the claim). Use the fact that $a = \sup(K(a))$ and $b = \sup(K(b))$.

To prove the lemma, we define the mapping $h : L \to Ideal(K(\boldsymbol{L})$ by:

$$h(a) := K(a), \qquad \text{for all } a \in L.$$

Clearly, h is well-defined. By Claim 1, h is surjective. By Claim 2, h is an isomorphism between the lattices $\boldsymbol{L}$ and $(Ideal(K(\boldsymbol{L}), \subseteq)$ and hence it is also one-to-one. □

Lemma A.3.7. *Let $\boldsymbol{P} = (P, \leqslant)$ and $\boldsymbol{Q} = (Q, \leqslant)$ be algebraic lattices. If $\boldsymbol{P}$ and $\boldsymbol{Q}$ are isomorphic, then so are the join-semilattices of compact elements $(K(\boldsymbol{P}), \leqslant, \boldsymbol{0})$ and $(K(\boldsymbol{Q}), \leqslant, \boldsymbol{0})$.*

Proof (of the lemma). Let $f : \boldsymbol{P} \cong \boldsymbol{Q}$ be an isomorphism. It suffices to prove that f maps $K(\boldsymbol{P})$ onto $K(\boldsymbol{Q})$. Thus we need to show that, for every $a \in P$, $a \in K(\boldsymbol{P})$ if and only if $f(a) \in K(\boldsymbol{Q})$. A tedious verification of this condition is left to the reader. □

We pass to the proof of Theorem A.3.4.

($\Leftarrow$). We assume that

$$(K(\boldsymbol{P}), \leqslant, \boldsymbol{0}) \cong (K(\boldsymbol{Q}), \leqslant, \boldsymbol{0}).$$

The second statement of Lemma A.3.5 implies that

$$(Ideal(K(\boldsymbol{P})), \subseteq) \cong (Ideal(K(\boldsymbol{Q})), \subseteq).$$

In turn, Lemma A.3.6 gives that

$$\boldsymbol{P} \cong (Ideal(K(\boldsymbol{P})), \subseteq) \quad \text{and} \quad \boldsymbol{Q} \cong (Ideal(K(\boldsymbol{Q})), \subseteq).$$

Consequently, $\boldsymbol{P} \cong \boldsymbol{Q}$.

($\Rightarrow$). This is the content of Lemma A.3.7.

The theorem has been proved. □

The following result follows from Theorems A.3.3–A.3.4:

Theorem A.3.8. *Let $\boldsymbol{P}_0 = (P_0, \leqslant, \boldsymbol{0})$ be an arbitrary join-semilattice with zero. There exists a unique (up to isomorphism) complete lattice $\boldsymbol{P} = (P, \leqslant)$ with the following properties:*

(1) *$\boldsymbol{P}$ is algebraic.*
(2) *The poset $K(\boldsymbol{P})$ of compact elements of $\boldsymbol{P}$ is isomorphic with $\boldsymbol{P}_0$.*

Proof. In view of Theorem A.3.3, the poset $(Ideal(\boldsymbol{P}_0), \subseteq)$ of ideals of $\boldsymbol{P}_0$ satisfies (1) and (2). In virtue of Theorem A.3.4, there is only one (up to isomorphism) poset $\boldsymbol{P}$ which satisfies (1)–(2). □

Corollary A.3.9. *Let $\boldsymbol{L}_0 = (L_0, \leqslant, \boldsymbol{0})$ be an arbitrary lattice with zero. There exists a unique (up to isomorphism) complete lattice $\boldsymbol{L} = (L, \leqslant)$ with the following properties :*

(1) *$\boldsymbol{L}$ is algebraic.*
(2) *The poset $K(\boldsymbol{L})$ of compact elements of $\boldsymbol{L}$ is a lattice isomorphic with $\boldsymbol{L}_0$.*

The unique algebraic lattice $\boldsymbol{L} = (L, \leqslant)$ satisfying conditions (1)–(2) of the above corollary is called the *algebraic completion* of the lattice $\boldsymbol{L}_0 = (L_0, \leqslant, \boldsymbol{0})$. In view of Theorems A.3.3–A.3.4, the lattice $(Ideal(\boldsymbol{L}_0), \subseteq)$ of ideals of $\boldsymbol{L}_0$ is *the* algebraic extension of $\boldsymbol{L}_0$ since it is unique. $(Ideal(\boldsymbol{L}_0), \subseteq)$ is also called the *algebraic extension of $\boldsymbol{L}_0$ by ideals.*

According to the above corollary, every lattice with zero $\boldsymbol{L}_0 = (L_0, \leqslant, \boldsymbol{0})$ is embeddable in the complete algebraic lattice $(Ideal(\boldsymbol{L}_0), \subseteq)$ via the map $\phi : L_0 \to Ideal(\boldsymbol{L}_0)$ given by

$$\phi(a) := \downarrow a, \qquad a \in L_0.$$

It should be noted ϕ is *not* a complete embedding $\boldsymbol{L}_0$ into $(Ideal(\boldsymbol{L}_0), \subseteq)$ because it does not preserve supremums of infinite subsets of $\boldsymbol{L}_0$, that is, if $A \subseteq L_0$ and $\sup(A)$ exists in $\boldsymbol{L}_0$, then $\phi(\sup(A))$ may be strictly greater than $\sup(\{\phi(a) : a \in A\})$. Indeed, if $\boldsymbol{L}_0 = (L_0, \leqslant, \boldsymbol{0})$ itself is an algebraic lattice, then $(Ideal(\boldsymbol{L}_0), \subseteq)$ is a proper extension of $\boldsymbol{L}_0$, because *every* element of L_0 becomes a compact element in $(Ideal(\boldsymbol{L}_0), \subseteq)$, by the above corollary. If ϕ preserved supremums, then *every non-compact* element of $\boldsymbol{L}_0$ would be non-compact in $(Ideal(\boldsymbol{L}_0), \subseteq)$, which is excluded by Corollary A.3.9.(2).

Lemma A.3.10. *For any lattice with zero $\boldsymbol{L}_0$ the embedding $\phi : \boldsymbol{L}_0 \to Ideal(\boldsymbol{L}_0)$ preserves arbitrary meets in $\boldsymbol{L}_0$.*

Proof. Let B be a non-empty subset of a lattice $\boldsymbol{L}_0$ and assume $\inf(B)$ exists in $\boldsymbol{L}_0$. As ϕ is monotone, we have that $\phi(\inf(B)) \subseteq \inf(\{\phi(b) : b \in B\})$. We shall show that $\phi(\inf(B))$ is the g.l.b. of the set $\{\phi(b) : b \in B\}$ in $(Ideal(\boldsymbol{L}_0), \subseteq)$. Suppose $\boldsymbol{I} \in Ideal(\boldsymbol{L}_0)$ so that $\boldsymbol{I}$ is a lower bound of $\{\phi(b) : b \in B\}$. This means that $\boldsymbol{I} \subseteq \downarrow b$, for all $b \in B$. Hence $\boldsymbol{I} \subseteq \downarrow \inf(B)$, i.e., $\boldsymbol{I} \subseteq \phi(\inf(B))$. □

In this context it is natural to speak of $\boldsymbol{L}_0$-complete ideals (in $\boldsymbol{L}_0$). Let $\boldsymbol{L}_0 = (L_0, \leqslant, \boldsymbol{0})$ be a lattice with zero. An ideal $\boldsymbol{I} \in Ideal(\boldsymbol{L}_0)$ is *$\boldsymbol{L}_0$-complete* if, for any non-empty set $A \subseteq L_0$ such that $\sup(A)$ exists in $\boldsymbol{L}_0$, $\sup(A) \in \boldsymbol{I}$ whenever $A \subseteq I$.

Let $Ideal_c(\boldsymbol{L}_0)$ be the set of all $\boldsymbol{L}_0$-complete ideals. This set is non-empty, because the ideals $\{\boldsymbol{0}\}$, L_0 and $\downarrow a$, for all $a \in L_0$ are $\boldsymbol{L}_0$-complete. As the intersection of any non-empty family of $\boldsymbol{L}_0$-complete ideals is an $\boldsymbol{L}_0$-complete ideal, $Ideal_c(\boldsymbol{L}_0)$ is a closure system on L_0. (The ideal $\{\mathbf{0}\}$ is the intersection of the void family.) It follows that $(Ideal_c, \subseteq)$ is a complete lattice with $\{\mathbf{0}\}$ and L_0 as the bottom and top elements, respectively. If $\boldsymbol{L}_0$ is a complete lattice, $\boldsymbol{L}_0$-complete ideals are exactly the ideals of the form $\downarrow a$, where $a \in L_0$.

Lemma A.3.11. *For any lattice with zero* $\boldsymbol{L}_0$ *the mapping* $\phi : \boldsymbol{L}_0 \to Ideal_c(\boldsymbol{L}_0)$ *is a complete isomorphic embedding, i.e., it is one-to-one and preserves arbitrary joins and meets in* $\boldsymbol{L}_0$.

Proof. It is clear that ϕ is well-defined, one-to-one and monotone. From the proof of Lemma A.3.10 it follows that ϕ preserves meets in $\boldsymbol{L}_0$. We show ϕ preserves arbitrary joins in $\boldsymbol{L}_0$.

Let A be a non-empty subset of L_0 and assume $\sup(A)$ exist in $\boldsymbol{L}_0$. As ϕ is monotone, we have that $\sup(\{\phi(a) : a \in A\}) \subseteq \phi(\sup(A))$. We shall show that $\phi(\sup(A))$ is the l.u.b. of the set $\{\phi(a) : a \in A\}$ in $(Ideal_c(\boldsymbol{L}_0), \subseteq)$. For let $\boldsymbol{I} \in Ideal_c(L_0)$ be an upper bound of $\{\phi(a) : a \in A\}$. This means that $\downarrow a \subseteq \boldsymbol{I}$, for all $a \in A$. It follows that $A \subseteq \boldsymbol{I}$. As $\boldsymbol{I}$ is $\boldsymbol{L}_0$-complete, we get that $\sup(A) \in \boldsymbol{I}$. Hence $\downarrow \sup(A) \subseteq \boldsymbol{I}$. This shows that $\phi(\sup(A))$ is the l.u.b. of $\{\phi(a) : a \in A\}$. □

$(Ideal_c(\boldsymbol{L}_0), \subseteq)$ though being a complete lattice need not be a sublattice of $(Ideal(\boldsymbol{L}_0), \subseteq)$. For instance, given two ideals $\boldsymbol{I}, \boldsymbol{J} \in Ideal_c(\boldsymbol{L}_0)$, the ideal $K := \{x \in L_0 : x \leqslant i \vee j$, for some $i \in \boldsymbol{I}, j \in \boldsymbol{J}\}$ is the join of $\boldsymbol{I}$ and $\boldsymbol{J}$ in the lattice $(Ideal(\boldsymbol{L}_0), \subseteq)$.

If $\boldsymbol{L}_0$ is a complete lattice, $\boldsymbol{L}_0$-complete ideals are exactly the ideals of the form $\downarrow a$, where $a \in L_0$. In this case the embedding ϕ trivializes, because the lattice $(Ideal_c(\boldsymbol{L}_0), \subseteq)$ is identified with $\boldsymbol{L}_0$. This case also shows that $(Ideal_c(\boldsymbol{L}_0), \subseteq)$ need not be an algebraic lattice. On the other hand, $(Ideal(\boldsymbol{L}_0), \subseteq)$ is an algebraic lattice being a proper extension of $\boldsymbol{L}_0$ and hence of $(Ideal_c(\boldsymbol{L}_0), \subseteq)$. From the above remarks it follows that $(Ideal_c(\boldsymbol{L}_0), \subseteq)$, though being a sublattice of $(Ideal(\boldsymbol{L}_0), \subseteq)$ for any complete lattice $\boldsymbol{L}_0$, is not a complete sublattice, that is, meets and joins of infinite subsets of $Ideal_c(\boldsymbol{L}_0)$ may differ from the corresponding meets and joins in $(Ideal(\boldsymbol{L}_0), \subseteq)$.

A.4 Distributive and Modular Algebraic Lattices

Let $\boldsymbol{P}$ is a join-semilattice with $\mathbf{0}$, and $a, b \in P$. We define:

$$M(a,b) \quad \Leftrightarrow_{df} \quad \text{for all } c, d \in P, b \leqslant a \vee c \text{ and } b \leqslant a \vee d \text{ imply that } (\exists e \in P)(b \leqslant a \vee e \ \&\ e \leqslant c \ \&\ e \leqslant a \vee d).$$

Theorem A.4.1. *Let* $\boldsymbol{P}$ *be a join-semilattice with* $\mathbf{0}$. *The lattice* $(Ideal(\boldsymbol{P}), \subseteq)$ *is modular if and only if* $M(a,b)$ *holds for all* $a, b \in P$.

Proof. $(\Leftarrow)$. Let $\boldsymbol{I}, \boldsymbol{J}, \boldsymbol{K}$ be ideals of $\boldsymbol{P}$, and suppose $M(a,b)$ holds, for all $a, b \in P$. We want to show

$$(\boldsymbol{I} \vee \boldsymbol{J}) \wedge (\boldsymbol{I} \vee \boldsymbol{K}) = \boldsymbol{I} \vee ((\boldsymbol{I} \vee \boldsymbol{J}) \wedge \boldsymbol{K}).$$

The inclusion '$\supseteq$' holds in any lattice, so we only need to show the reverse inclusion '$\subseteq$'. Let $x \in (\boldsymbol{I} \vee \boldsymbol{J}) \wedge (\boldsymbol{I} \vee \boldsymbol{K})$. Then $x \leqslant i \vee j$, $x \leqslant i' \vee k$, with $i, i' \in \boldsymbol{I}$, $j \in \boldsymbol{J}$ and $k \in \boldsymbol{K}$. Therefore $i'' := i \vee i' \in \boldsymbol{I}$ and $x \leqslant i'' \vee j$, $x \leqslant i'' \vee k$. By $M(i'', x)$, there exists $l \in P$ such that $l \leqslant k$, $l \leqslant i'' \vee j$, and $x \leqslant i'' \vee l$. But $l \in (\boldsymbol{I} \vee \boldsymbol{J}) \wedge \boldsymbol{K}$ and $i'' \in \boldsymbol{I}$, and hence $x \in \boldsymbol{I} \vee ((\boldsymbol{I} \vee \boldsymbol{J}) \wedge \boldsymbol{K})$.

($\Rightarrow$). Assume $(Ideal(\boldsymbol{P}), \subseteq)$ is modular, and let $a, b \in P$. To show $M(a, b)$, let $c, d \in P$ and suppose $b \leqslant a \vee c$ and $b \leqslant a \vee d$. Then $b \in (\downarrow a \vee \downarrow c) \wedge (\downarrow a \vee \downarrow d) =$ (by modularity) $\downarrow a \vee ((\downarrow d \vee \downarrow a) \wedge \downarrow c)$, that is, $b \leqslant f \vee e$ with $f \leqslant a$ and $e \leqslant (\downarrow d \vee \downarrow a) \wedge \downarrow c$. Thus $b \leqslant a \vee e$, $e \leqslant c$ and $e \leqslant a \vee d$. □

In virtue of Lemma A.3.6, every algebraic lattice $\boldsymbol{L}$ is isomorphic with the lattice of ideals of the join-semilattice of compact elements of $\boldsymbol{L}$. Therefore Theorem A.4.1 implies:

Corollary A.4.2. *Let* $\boldsymbol{L}$ *be an algebraic lattice.* $\boldsymbol{L}$ *is modular if and only if* $M(a, b)$ *holds for all compact elements* a, b *of* $\boldsymbol{L}$. □

The next lemma is useful in commutator theory:

Lemma A.4.3. *Let* $\boldsymbol{P}$ *be a join-semilattice with* $\boldsymbol{0}$ *generated by a set* $X \subseteq P$. *If* $M(a, b)$ *holds for all* $a, b \in X$, *then* $M(a, b)$ *holds for all* $a, b \in P$.

Proof. First we show $M(a, b)$ holds for all $a \in X$ and all $b \in P$. Let $a \in X$, and let

$$b = \bigvee_{0<i<n} x_i,$$

where $x_i \in X$, for $0 < i < n$. We show by induction on n that $M(a, b)$ holds. If $n = 1$, then $b \in X$ and $M(a, b)$ holds by assumption. Now let $n \geqslant 2$, $b' := \bigvee_{0<i<n-1} x_i$, and assume $M(a, b')$ holds. To prove that $M(a, b)$ holds, let c, d be elements of P so that $b \leqslant a \vee c$, $b \leqslant a \vee d$. Then trivially $b' \leqslant a \vee c$, $b' \leqslant a \vee d$, so by $M(a, b')$ there exists $e' \in P$ such that

$$b' \leqslant a \vee e', \qquad e' \leqslant c, \qquad e' \leqslant a \vee d. \tag{1}$$

Also $x_{n-1} \leqslant a \vee c$, $x_{n-1} \leqslant a \vee d$, and since $M(a, x_{n-1})$ holds, there exists $e'' \in P$ such that

$$e'' \leqslant c, \qquad e'' \leqslant a \vee d, \qquad x_{n-1} \leqslant a \vee e''. \tag{2}$$

Let $e := e' \vee e''$; then by (1) and (2), $e \leqslant c$, $e \leqslant a \vee d$ and $b = b' \vee x_{n-1} \leqslant a \vee e' \vee a \vee e'' = a \vee e$.

Next we prove that $M(a, b)$ holds for all $a, b \in P$. Let

$$a = \bigvee_{0<i<n} x_i,$$

where $x_i \in X$ for $i < n$, and $b \in P$. If $n = 1$, then $a \in X$ and $M(a, b)$ holds by the first part of the proof. If $n \geqslant 2$, then write $a' := \bigvee_{0<i<n-1} x_i$, and assume $M(a', b)$ holds. Let $c, d \in P$ such that $b \leqslant a \vee c$, $b \leqslant a \vee d$. Then $b \leqslant a' \vee x_{n-1} \vee c$, $b \leqslant a' \vee x_{n-1} \vee d$, so by $M(a', b)$ there exists $e' \in P$ such that

$$b \leqslant a' \vee e', \qquad e' \leqslant x_{n-1} \vee c, \qquad e' \leqslant a' \vee x_{n-1} \vee d. \tag{3}$$

Since, by the first part of the proof, $M(x_{n-1}, e')$ holds, there is an $e \in P$ such that

$$e' \leqslant x_{n-1} \vee e, \qquad e \leqslant c, \qquad e \leqslant x_{n-1} \vee (a' \vee d). \tag{4}$$

(3) and (4) imply that $b \leqslant a' \vee e' \leqslant a' \vee x_{n-1} \vee e = a \vee e$. Since $e \leqslant c$ and $e \leqslant x_{n-1} \vee (a' \vee d) = a \vee d$, we thus see that $M(a, b)$ holds. □

Let $\boldsymbol{P}$ be a join-semilattice with $\mathbf{0}$ and $a, b \in P$. We also define:

$D(a, b) \quad \Leftrightarrow_{df} \quad$ for all $c, d \in L$, $b \leqslant a \vee c$ and $b \leqslant a \vee d$ imply $(\exists e \in L)(e \leqslant c \;\&\; e \leqslant d \;\&\; b \leqslant a \vee e)$.

Theorem A.4.4. *Let $\boldsymbol{P}$ be a join-semilattice with $\mathbf{0}$. The lattice $(Ideal(\boldsymbol{P}), \subseteq)$ is distributive if and only if $D(a, b)$ holds for all $a, b \in P$.*

Proof. Suitably modify the above proof of Theorem A.4.1. □

Theorem A.4.5. *Let $\boldsymbol{L}_0 = (L_0, \leqslant, \mathbf{0})$ be an arbitrary distributive (modular) lattice with zero. The unique algebraic completion $\boldsymbol{L} = (L, \leqslant)$ of $\boldsymbol{L}_0$ is distributive (modular) as well.*

Proof. We assume $\boldsymbol{L}_0$ is modular. It is easy to see that $M(a, b)$ holds in $\boldsymbol{L}_0$ (treated as a join-semilattice with zero), for all $a, b \in L_0$, by modularity. Hence, by Theorem A.4.1, the algebraic lattice $(Ideal(\boldsymbol{L}_0), \subseteq)$ is modular. As $(Ideal(\boldsymbol{L}_0), \subseteq)$ is the unique algebraic completion of $\boldsymbol{L}_0$, the theorem follows.

A similar argument applies when $\boldsymbol{L}_0$ is distributive. □

Theorem A.4.1 and the results following it were proved in Czelakowski (1998).

Appendix B
A Proof of Theorem 3.3.4 for Relatively Congruence-Modular Quasivarieties

This appendix contains a proof of a special case of Theorem 5.3.4 in which conditions $(\mathrm{EqDistr})_{m,n}$ are directly computed for any RCM quasivariety. The proof involves some purely syntactical techniques which are useful in other contexts and it does not make use of von Neumann's Theorem (Theorem 6.3.1).

Theorem 5.3.4*. *Let* $\mathbf{Q}$ *be an RCM quasivariety. The consequence* $\mathbf{Q}^{\models}$ *validates* $(\mathrm{EqDistr})_{m,n}$ *for any positive integers m and n.*

Proof. To simplify notation, we mark the consequence $\mathbf{Q}^{\models}$ by C. We are to show that

$$(\mathrm{EqDistr})_{m,n} \qquad C(x_1 \approx y_1, \dots, x_m \approx y_m) \cap C(z_1 \approx w_1, \dots, z_n \approx w_n) = C\Big(\bigcup_{1 \leqslant i \leqslant m,\ 1 \leqslant j \leqslant n} C(x_i \approx y_i) \cap C(z_j \approx w_j)\Big),$$

where m and n are arbitrary positive integers and $\underline{x} = x_1, \dots, x_m$, $\underline{y} = y_1, \dots, y_m$, $\underline{z} = z_1, \dots, z_n$, $\underline{w} = w_1, \dots, w_n$ are disjoint sequences of pairwise different individual variables. The proof of $(\mathrm{EqDistr})_{m,n}$ is on double induction on m and n.

To give an insight into the proof, we shall first consider a simple special case with $m = 2$ and $n = 1$:

$$(\mathrm{EqDistr})_{2,1} \qquad C(x_1 \approx y_1, x_2 \approx y_2) \cap C(z \approx w) = C(C(x_2 \approx y_2) \cap C(z \approx w) \cup C(x_1 \approx y_1) \cap C(z \approx w)).$$

Let us put: $B_1 := C(x_1 \approx y_1)$, $B_2 := C(x_2 \approx y_2)$, $A := C(z \approx w)$. Then $(\mathrm{EqDistr})_{2,1}$ takes the form

$$(B_1 \vee B_2) \cap A = (B_2 \cap A) \vee (B_1 \cap A).$$

J. Czelakowski, *The Equationally-Defined Commutator*,
DOI 10.1007/978-3-319-21200-5

We have:

$$(B_1 \cap A) \vee (B_2 \cap A) = A \cap (B_2 \vee (A \cap B_1)),$$

by modularity. Therefore it suffices to show that

$$(B_1 \vee B_2) \cap A \subseteq A \cap (B_2 \vee (A \cap B_1)). \qquad (1)_{2,1}$$

Suppose $\alpha(x_1, y_1, x_2, y_2, z, w, \underline{u}) \approx \beta(x_1, y_1, x_2, y_2, z, w, \underline{u})$ belongs to the set on the left-hand side of $(1)_{2,1}$. So $\alpha \approx \beta \in C(z \approx w)$ and $\alpha \approx \beta \in C(x_1 \approx y_1, x_2 \approx y_2)$. Let $e : \boldsymbol{Te}_\tau \to \boldsymbol{Te}_\tau$ be an endomorphism (a substitution) such that $ey_2 = x_2$ and e is the identity map on the remaining variables. As $\alpha \approx \beta \in C(x_1 \approx y_1, x_2 \approx y_2)$, we get that $e\alpha \approx e\beta \in C(ex_1 \approx ey_1, ex_2 \approx ey_2) = C(x_1 \approx y_1)$ and $e\alpha \approx e\beta \in C(ez \approx ew) = C(z \approx w)$. Hence $e\alpha \approx e\beta \in C(z \approx w) \cap C(x_1 \approx y_1)$, i.e.,

$$e\alpha \approx e\beta \in A \cap B_1. \qquad (*)$$

Claim 1. $\alpha \approx \beta \in B_2 \vee (A \cap B_1)$.

Proof (of the claim). We have: $B_2 \vee (A \cap B_1) = C(\{x_2 \approx y_2\} \cup A \cap B_1)$.

Let $\mathbf{A} \in \mathbf{Q}$ and $h : \boldsymbol{Te}_\tau \to \mathbf{A}$ be a homomorphism which validates the equations of $\{x_2 \approx y_2\} \cup A \cap B_1$. Trivially $hx_2 = hy_2$ and h validates $A \cap B_1$. But then $(*)$ yields that h validates $e\alpha \approx e\beta$, i.e., $he\alpha = he\beta$. We also that $he\alpha = h\alpha$ and $he\beta = h\beta$, because $hx_2 = hy_2$. It follows that $h\alpha = h\beta$. This proves the claim. □

As $\alpha \approx \beta \in A$, $(1)_{2,1}$ follows from the claim.

The general case is a modification of the above proof. We first prove:

Lemma B.1. $C = \mathbf{Q}^{\models}$ *validates*

$(\mathrm{EqDistr})_{m,1}$

$$C(x_1 \approx y_1, \ldots, x_m \approx y_m) \cap C(z \approx w) = C\Big(\bigcup_{1 \leqslant i \leqslant m} C(x_i \approx y_i) \cap C(z \approx w)\Big),$$

for all $m \geqslant 2$.

Proof (of the lemma). For a given $m \geqslant 2$ we adopt the following notation:

$$\underline{x} \approx \underline{y} := \{x_1 \approx y_1, \ldots, x_m \approx y_m\},$$
$$A := C(z \approx w),$$
$$B_1 := C(x_1 \approx y_1),$$
$$\vdots$$
$$B_m := C(x_m \approx y_m),$$
$$B_{m+1} := C(x_{m+1} \approx y_{m+1}).$$

$(\text{EqDistr})_{m,1}$ then takes the form:

$$(B_1 \vee \ldots \vee B_m) \cap A = C(\bigcup_{1 \leqslant i \leqslant m} C(x_i \approx y_i) \cap C(z \approx w)),$$

But in view of Theorem 4.1.4,

$$C(\bigcup_{1 \leqslant i \leqslant m} C(x_i \approx y_i) \cap C(z \approx w)) = A * (B_1, \ldots, B_m).$$

It therefore suffices to show that

$$(B_1 \vee \ldots \vee B_m) \cap A \subseteq A * (B_1, \ldots, B_m), \qquad (1)_{m,1}$$

for all $m \geqslant 2$, because the reverse inclusion trivially holds.

The proof of $(1)_{m,1}$ for $m = 2$ has just been given. To make an inductive step, assume m is a positive integer, $m \geqslant 2$, and $(1)_{m,1}$ holds. We show

$$(B_1 \vee \ldots \vee B_m \vee B_{m+1}) \cap A \subseteq A * (B_1, \ldots, B_m, B_{m+1}), \qquad (1)_{m+1,1}$$

But, by the definition of $*$, $A*(B_1, \ldots, B_m, B_{m+1}) := A \cap (B_{m+1} \vee (A*(B_1, \ldots, B_m)))$. We therefore must show that

$$(B_1 \vee \ldots \vee B_m \vee B_{m+1}) \cap A \subseteq A \cap (B_{m+1} \vee (A * (B_1, \ldots, B_m))). \qquad (2)$$

Suppose $\alpha \approx \beta$ belongs to the set on the left-hand side of 2. So $\alpha \approx \beta \in C(z \approx w)$ and $\alpha \approx \beta \in C(\underline{x} \approx \underline{y}, x_{m+1} \approx y_{m+1})$. Let $e : \boldsymbol{Te}_\tau \to \boldsymbol{Te}_\tau$ be the endomorphism such that $ey_{m+1} = x_{m+1}$ and e is the identity map on the remaining variables. As $\alpha \approx \beta \in C(\underline{x} \approx \underline{y}, x_{m+1} \approx y_{m+1})$, we get that $e\alpha \approx e\beta \in C(e\underline{x} \approx e\underline{y}, ex_{m+1} \approx ey_{m+1}) = C(\underline{x} \approx \underline{y})$ and $e\alpha \approx e\beta \in C(ez \approx ew) = C(z \approx w)$. Hence $e\alpha \approx e\beta \in C(z \approx w) \cap C(\underline{x} \approx \underline{y}) = (B_1 \vee \ldots \vee B_m) \cap A$. This implies, by IH, that

$$e\alpha \approx e\beta \in A * (B_1, \ldots, B_m), \qquad (**)$$

Claim 2. $\alpha \approx \beta \in B_{m+1} \vee (A * (B_1, \ldots, B_m))$.

Proof (of the claim). We must show that

$$\alpha \approx \beta \in C(\{x_{m+1} \vee y_{m+1}\} \cup (A * (B_1, \ldots, B_m))).$$

Let $\boldsymbol{A} \in \mathbf{Q}$ and $h : \boldsymbol{Te}_\tau \to \boldsymbol{A}$ be a homomorphism which validates the equations of $\{x_{m+1} \approx y_{m+1}\} \cup A * (B_1, \ldots, B_m)$. Trivially $hx_{m+1} = hy_{m+1}$ and h validates $A * (B_1, \ldots, B_m)$. But $(**)$ yields that h validates $e\alpha \approx e\beta$, i.e., $he\alpha = he\beta$. We therefore have that $he\alpha = h\alpha$ and $he\beta = h\beta$, because $hx_{m+1} = hy_{m+1}$. It follows that $h\alpha = h\beta$. This proves the claim. □

(2) follows from the claim and the assumption that $\alpha \approx \beta \in C(z \approx w) = A$.

So $(2)_{m+1,1}$ holds. We have thus proved $(2)_{m,1}$ for all m, which means that $(\mathrm{EqDistr})_{m,1}$ holds for all m. This proves Lemma B.1. □

Lemma B.2. *Fix m. Then $C = \mathbf{Q}^{\models}$ validates* $(\mathrm{EqDistr})_{m,n}$ *for all n.*

Proof (of the lemma). The base of induction, viz. $(\mathrm{EqDistr})_{m,1}$ has been proved. To make an the inductive step, assume

$$(\mathrm{EqDistr})_{m,n} \qquad C(x_1 \approx y_1, \ldots, x_m \approx y_m) \cap C(z_1 \approx w_1, \ldots, z_n \approx w_n) = C\Big(\bigcup_{1 \leqslant i \leqslant m,\ 1 \leqslant j \leqslant n} C(x_i \approx y_i) \approx C(z_j \approx w_j)\Big),$$

We are to show that

$$(\mathrm{EqDistr})_{m,n+1} \qquad C(x_1 \approx y_1, \ldots, x_m \approx y_m) \cap C(z_1 \approx w_1, \ldots, z_n \approx w_n, z_{n+1} \approx w_{n+1}) = C\Big(\bigcup_{1 \leqslant i \leqslant m,\ 1 \leqslant j \leqslant n+1} C(x_i \approx y_i) \cap C(z_j \approx w_j)\Big).$$

To simplify notation, we introduce the following abbreviations:

$$\begin{aligned}
\underline{x} \approx \underline{y} &:= \{x_1 \approx y_1, \ldots, x_m \approx y_m\},\\
\underline{z} \approx \underline{w} &:= \{z_1 \approx w_1, \ldots, z_n \approx w_n\},\\
A &:= C(x_1 \approx y_1, \ldots, x_m \approx y_m) \quad (= C(\underline{x} \approx \underline{y})),\\
B_1 &:= C(z_1 \approx w_1, \ldots, z_n \approx w_n) \quad (= C(\underline{z} \approx \underline{w})),\\
B_2 &:= C(z_{n+1} \approx w_{n+1}).
\end{aligned}$$

$(\mathrm{EqDistr})_{m,n+1}$ then takes the form

$$A \cap (B_1 \vee B_2) = C\Big(\bigcup_{1 \leqslant i \leqslant m,\ 1 \leqslant j \leqslant n+1} C(x_i \approx y_i) \cap C(z_j \approx w_j)\Big). \qquad (2)_{m,n+1}$$

By modularity, $(\mathrm{EqDistr})_{m,n}$ and $(\mathrm{EqDistr})_{m,1}$ we get:

$$\begin{aligned}
&A * (B_1, B_2) = (A \cap B_1) \vee (A \cap B_2) =\\
&C(\underline{x} \approx \underline{y}) \cap C(\underline{z} \approx \underline{w}) \vee C(\underline{x} \approx \underline{y}) \cap C(z_{n+1} \approx w_{n+1}) =\\
&C\Big(\bigcup_{1 \leqslant i \leqslant m,\ 1 \leqslant j \leqslant n} C(x_i \approx y_i) \cap C(z_j \approx w_j)\Big) \vee C\Big(\bigcup_{1 \leqslant i \leqslant m} C(x_i \approx y_i) \cap C(z_{n+1} \approx w_{n+1})\Big) =\\
&C\Big(\bigcup_{1 \leqslant i \leqslant m,\ 1 \leqslant j \leqslant n+1} C(x_i \approx y_i) \cap C(z_j \approx w_j)\Big).
\end{aligned}$$

The last theory is the same as the set on the right side of $(2)_{m,n+1}$. Thus in order to prove the lemma it suffices to show:

Claim. $A \cap (B_1 \vee B_2) \subseteq A * (B_1, B_2)$.

Proof (of the claim). Suppose $\alpha \approx \beta \in A \cap (B_1 \vee B_2)$. This means that

$$\alpha \approx \beta \in C(\underline{x} \approx \underline{y}) \tag{a}$$

and

$$\alpha \approx \beta \in C(\underline{z} \approx \underline{w}, z_{n+1} \approx w_{n+1}). \tag{b}$$

Let $e : \boldsymbol{Te}_\tau \to \boldsymbol{Te}_\tau$ be the endomorphism given by $ew_{n+1} = z_{n+1}$ and $ex = x$ for all variables $x \neq w_{n+1}$. (a) and (b) imply that

$$e\alpha \approx e\beta \in C(e\underline{x} \approx e\underline{y}) = C(\underline{x} \approx \underline{y}) = A$$

and

$$e\alpha \approx e\beta \in C(e\underline{z} \approx e\underline{w}, ez_{n+1} \approx ew_{n+1}) = C(\underline{z} \approx \underline{w}) = B_1,$$

because $ez_{n+1} \approx ew_{n+1}$ is the trivial equation $z_{n+1} \approx z_{n+1}$. Consequently,

$$e\alpha \approx e\beta \in A \cap B_1. \tag{c}$$

We want to prove that $\alpha \approx \beta \in A * (B_1, B_2) := A \cap (B_2 \vee (A \cap B_1))$. As $\alpha \approx \beta \in A$, by (a), it remains to show that $\alpha \approx \beta \in B_2 \vee (A \cap B_1)$, i.e.,

$$\alpha \approx \beta \in C(\{z_{n+1} \approx w_{n+1}\} \cup C(\underline{x} \approx \underline{y}) \cap C(\underline{z} \approx \underline{w})). \tag{d}$$

Let $\boldsymbol{A} \in \mathbf{Q}$ and $h : \boldsymbol{Te}_\tau \to \boldsymbol{A}$ be a homomorphism which validates the equations of $\{z_{n+1} \approx w_{n+1}\} \cup C(\underline{x} \approx \underline{y}) \cap C(\underline{z} \approx \underline{w})$. We wish to show that $h\alpha = h\beta$. We have that $hz_{n+1} = hw_{n+1}$ and h validates $C(\underline{x} \approx \underline{y}) \cap C(\underline{z} \approx \underline{w}) = A \cap B_1$. This fact and (c) imply that h validates $e\alpha \approx e\beta$, i.e., $he\alpha = he\beta$. As $hz_{n+1} = hw_{n+1}$, we obviously have that $he\alpha = h\alpha$ and $he\beta = h\beta$. The identities $he\alpha = he\beta$, $he\alpha = h\alpha$ and $he\beta = h\beta$ imply $h\alpha = h\beta$. So (d) holds.

This proves the claim and at the same time concludes the proof of $(2)_{m,n+1}$. □

Thus Lemma B.2 has been proved. □

As the inductive proof of $(\text{EqDistr})_{m,n}$ for all positive m, n is completed, Theorem 5.3.4* has been proved. □

Appendix C
Inferential Bases for Relatively Congruence-Modular Quasivarieties

The focus of this appendix is on inferential bases of the *equational system* $\mathbf{Q}^{\models}$ associated with any RCM quasivariety $\mathbf{Q}$.

Let $\mathbf{Q}$ be a quasivariety of algebras of signature τ. A finite set $P = P(x, y, z, w, \underline{u})$ of equations (where $\underline{u} = u_1, \ldots, u_n$) is said to have the *detachment property* with respect to $\mathbf{Q}^{\models}$ if

$$z \approx w \in \mathbf{Q}^{\models}(P \cup \{x \approx y\}). \qquad (*)_P$$

Any set P with the detachment property is also denoted by $(x \approx y) \Rightarrow_{\underline{u}} (z \approx w)$. $(*)_P$ states that

$$P \cup \{x \approx y\}/z \approx w \qquad (\mathrm{MP})_P$$

is a rule of inference of the equational logic $\mathbf{Q}^{\models}$. The rule $(\mathrm{MP})_P$ is called the *detachment rule* corresponding to P. This rule is also marked as

$$x \approx y, (x \approx y) \Rightarrow_{\underline{u}} (z \approx w)/z \approx w.$$

Any set of equations with the detachment property is also called an *implication system* for $\mathbf{Q}^{\models}$. $\mathbf{Q}$ usually possesses many implication sets. This is a consequence of Theorem C.1 below.

Let $\mathcal{P}$ be a family of finite sets of equations with the detachment property. We shall say that the consequence $\mathbf{Q}^{\models}$ *admits the Parameterized Local Deduction Theorem* (PLDT, for short) with respect to $\mathcal{P}$ if for every set $\Sigma \cup \{p \approx q, r \approx s\}$ of equations the following equivalence holds:

> $r \approx s \in \mathbf{Q}^{\models}(\Sigma \cup \{p \approx q\}) \quad \Leftrightarrow \quad$ there exists a set $P(x, y, z, w, u_1, \ldots, u_n) \in \mathcal{P}$ and a sequence of terms $(t_1, \ldots, t_n)$ of Te_τ such that $\alpha(p, q, r, s, t_1, \ldots, t_n) \approx \beta(p, q, r, s, t_1, \ldots, t_n) \in \mathbf{Q}^{\models}(\Sigma)$ for all equations $\alpha \approx \beta$ of P.

J. Czelakowski, *The Equationally-Defined Commutator*,
DOI 10.1007/978-3-319-21200-5

$(\alpha(p, q, r, s, t_1, \ldots, t_n)$ is the result of simultaneous substituting the terms $p, q, r, s, t_1, \ldots, t_n$ for the variables $x, y, z, w, u_1, \ldots, u_n$ in the term $\alpha(x, y, z, w, u_1, \ldots, u_n)$, respectively.)

The following theorems were proved in Czelakowski and Dziobiak (1996):

Theorem C.1. *For every quasivariety* $\mathbf{Q}$ *there exists a family* $\mathcal{P}$ *of finite sets of equations with the detachment property such that* $\mathbf{Q}^{\models}$ *admits the Parameterized Local Deduction Theorem with respect to* $\mathcal{P}$. □

Theorem C.2. *Let* $\mathbf{Q}$ *be any quasivariety. Suppose* $\mathcal{P} = \{P_i(x, y, z, w, \underline{u}_i) : i \in I\}$ *is a family of finite sets of equations with the detachment property for* $\mathbf{Q}^{\models}$ *such that* $\mathbf{Q}^{\models}$ *admits the Parameterized Local Deduction Theorem with respect to* $\mathcal{P}$. *Then the set* $Id(\mathbf{Q})$ *together with the set of rules*

$$r_i : \qquad \{x \approx y\} \cup P_i(x, y, z, w, \underline{u}_i)/z \approx w$$

$(i \in I)$ *forms an inferential base for* $\mathbf{Q}^{\models}$. □

Theorem C.3. *Let* $\mathbf{Q}$ *be any quasivariety.* $\mathbf{Q}^{\models}$ *has an inferential base consisting of the set of equations* $Id(\mathbf{Q})$, *the rules of Birkhoff's logic, and a countable set of (parameterized) rules of the form* $P_i(x, y, \underline{u}_i)/x \approx y$ $(i \in I)$, *where each* $P_i(x, y, \underline{u}_i)$ *is a finite set of equations in two variables* x *and* y *(and possibly with parameters* $\underline{u}_i$*).*

The theorem states that in order to axiomatize $\mathbf{Q}^{\models}$ it suffices to strengthen any base for the variety $\boldsymbol{Va}(\mathbf{Q})$ by the above rules $P_i(x, y, \underline{u}_i)/x \approx y$ $(i \in I)$.

Proof. We first prove the following lemma:

Lemma C.4. *Let* $\mathbf{Q}$ *be a quasivariety and let* $\mathcal{P} = \{Q_i(x, y, z, w, \underline{v}_i) : i \in I\}$ *be a family of finite sets of equations which determines PLDT for* $\mathbf{Q}^{\models}$. *For each* $i \in I$ *define the set of equations:* $R_i(x, z, w, \underline{v}_i) := Q_i(x, y/x, z, w, \underline{v}_i)$. *Then the set of rules*

$$r_i : \qquad R_i(x, z, w, \underline{v}_i)/z \approx w$$

$(i \in I)$ *together with Birkhoff's rules and the equations of* $Id(\mathbf{Q})$ *forms an inferential base for* $\mathbf{Q}^{\models}$.

Proof (of the lemma). To simplify notation, we put: $C := \mathbf{Q}^{\models}$. As $\mathcal{P}$ determines PLDT for C, Theorem C.2 gives that C is axiomatized by $Id(\mathbf{Q})$ and the set of the following rules:

$$r_i' : \qquad x \approx y, Q_i(x, y, z, w, \underline{v}_i)/z \approx w$$

$(i \in I)$. (Brikhoff's rules are then derivable from $Id(\mathbf{Q})$ and the rules r_i' $(i \in I)$.)

Let C_0 be the equational consequence determined by the rules r_i $(i \in I)$, the rules of Birkhoff and the set of axioms $Id(\mathbf{Q})$. As C_0 validates Birkhoff's rules, C_0 is a finitary equational logic. This implies that C_0 is semantically determined by some quasivariety $\mathbf{Q}_0$, that is $C_0 = \mathbf{Q}_0^{\models}$. (In fact, $\mathbf{Q}_0$ is the largest class of models for C_0.) It is clear that $C_0 \leqslant C$, because each rule r_i is an instantiation of the rule r'_i. Consequently, $\mathbf{Q} \subseteq \mathbf{Q}_0$. To show that $C_0 = C$, and hence $\mathbf{Q} \subseteq \mathbf{Q}_0$, it suffices to prove that C_0 validates each rule r'_i $(i \in I)$. Suppose otherwise that $z \approx w \notin C_0(x \approx y, Q_i(x, y, z, w, \underline{v}_i))$ for some $i \in I$. There exists an algebra $\boldsymbol{A} \in \mathbf{Q}_0$ and a homomorphism $h : \boldsymbol{Te}_\tau \to \boldsymbol{A}$ such that $hx = hy$, h validates the equations of $Q_i(x, y, z, w, \underline{v}_i)$ and $hz \neq hw$. But as $hx = hy$, we get that h validates the equations of $R_i(x, z, w, \underline{v}_i)$ as well. Since r_i is a rule of C_0 and $\boldsymbol{A}$ is a model for C_0, we obtain that $hz = hw$. A contradiction.

This proves the lemma. □

In the final step, for each $i \in I$ we make the substitution x/u_i, z/x, w/y in the set $R_i(x, z, w, \underline{v}_i)$ and define:

$$P_i(x, y, \underline{u}_i) := R_i(x/u_i, z/x, w/y, \underline{v}_i) \qquad (n \in N_0).$$

Here u_i is a variable different from $x, z, w, \underline{v}_i$. u_i is thus a parametric variable added to $\underline{v}_i$. (So $\underline{u}_i = \underline{v}_i \cup \{u_i\}$ for each i.) Moreover, we can safely assume that the new parametric variables u_i and u_j are pairwise different whenever $i \neq j$, for all $i, j \in I$).

The theorem follows from the above lemma and the definition of $P_i(x, y, \underline{u}_i)$, $i \in I$. □

The following theorem, being the main result of this appendix, provides a structural description of an inferential base for the equational logic $\mathbf{Q}^{\models}$ associated with an arbitrary RCM quasivariety $\mathbf{Q}$. The proof makes use of some theorems on PLDT for any RCM quasivariety $\mathbf{Q}$ given in Czelakowski (1998, 2001).

Theorem C.5. *Let* $\mathbf{Q}$ *be any RCM quasivariety.* $\mathbf{Q}^{\models}$ *has an inferential base consisting of the set of equations* $Id(\mathbf{Q})$*, the rules of Birkhoff's logic and a countable set of (parameterized) rules of the form*

$$r_n : \qquad P_n(x, y, \underline{u}_n)/x \approx y$$

($n \in N_0$*, where* N_0 *is an initial segment of* ω*) such that*

(1) *for each* $n \in N_0$, $P_n(x, y, \underline{u}_n)$ *is a finite set of equations in variables* x, y *(and possibly with some other parametric variables* $\underline{u}_n$*);*
(2) *for any different* $m, n \in N_0$ *the parameters* $\underline{u}_m$ *and* $\underline{u}_n$ *of* P_m *and* P_n *are separated;*
(3) *for each* $n \in N_0$*, there exists a sequence of terms* $\underline{t}_n$ *such that*

$$P_n(x, x, \underline{t}_n) \subseteq Id(\mathbf{Q});$$

(4) *for each non-terminal number* n *in* N_0*, there exists a sequence of terms* $\underline{t}_{n+1}$ *such that*

$$P_{n+1}(x, y, \underline{t}_{n+1}) \subseteq \mathbf{Q}^{\models}(P_n(x, y, \underline{u}_n)).$$

The theorem states that a base for $\mathbf{Q}^{\models}$ is obtained from any base for $\mathbf{Va}(\mathbf{Q})$ by adjoining a certain set of rules $P_n(x, y, \underline{u}_n)/x \approx y$ $(n \in N_0)$ that satisfy conditions (1)–(4).

Proof. Let $\mathbf{Q}$ be an arbitrary but fixed RCM quasivariety. To simplify notation we put $C := \mathbf{Q}^{\models}$. We shall need the following observation which is crucial for inferential bases of C:

Lemma C.6. *Every family $\mathcal{P}$ of finite sets of equations which determines PLDT for C satisfies the following condition: for any $P_1(x, y, z, w, \underline{u}_1)$, $P_2(x, y, z, w, \underline{u}_2) \in \mathcal{P}$ there exists a set $P(x, y, z, w, \underline{u}) \in \mathcal{P}$ and a sequence $\underline{t}$ of terms such that*

$$P(x, y, z, w, \underline{t}) \subseteq C(P_1(x, y, z, w, \underline{u}_1) \quad \textit{and} \qquad \text{(mod)}$$
$$P(x, y, z, w, \underline{t}) \subseteq C(P_2(x, y, z, w, \underline{u}_2) \cup \{x \approx y\}). \qquad \square$$

Lemma C.6 is a consequence of Corollary A.4.2 of Appendix A (see also Czelakowski 1998, 2001).

We shall first prove the following fact:

Lemma C.7. *Let $\mathbf{Q}$ be any RCM quasivariety. $\mathbf{Q}^{\models}$ has an inferential base consisting of the set of equations $Id(\mathbf{Q})$, the rules of Birkhoff's logic and a countable set of (parameterized) rules of the form*

$$r_n: \qquad P''_n(x, z, w, \underline{v}_n)/z \approx w$$

($n \in N_0$, where N_0 is an initial segment of ω) such that

(1)″ *for each $n \in N_0$, $P''_n(x, z, w, \underline{v}_n)$ is a finite set of equations in variables x, z, w and possibly some other parametric variables $\underline{v}_n$;*
(2)″ *for any different $m, n \in N_0$ the parameters $\underline{v}_m$ and $\underline{v}_n$ of P''_m and P''_n are separated;*
(3)″ *for each $n \in N_0$, there exists a sequence of terms $\underline{t}_n$ such that*

$$P''_n(x, z, z, \underline{t}_n) \subseteq Id(\mathbf{Q});$$

(4)″ *for each non-terminal number n in N_0, there exists a sequence of terms t_{n+1} such that*

$$P''_{n+1}(x, z, w, \underline{t}_{n+1}) \subseteq \mathbf{Q}^{\models}(P''_n(x, z, w, \underline{v}_n)).$$

The lemma states that a base for $\mathbf{Q}^{\models}$ is obtained from any base for $\mathbf{Va}(\mathbf{Q})$ by adjoining a certain set of rules $r_n : P''_n(x, z, w, \underline{v}_n)/z \approx w$ $(n \in N_0)$ that satisfy conditions (1)″–(4)″.

Proof (of the lemma). Let $\mathcal{P}_0$ be a family of finite sets of equations which determines PLDT for C. We assume that a Day implication system $(x \approx y) \Rightarrow_D (z \approx w)$

for **Q** is in $\mathcal{P}_0$ since otherwise it may always be adjoined to $\mathcal{P}_0$. We may also assume without loss of generality that different sets in $\mathcal{P}_0$ have separated parametric variables. (This can be achieved by making appropriate substitutions for parametric variables.) We then define:

$$\mathcal{Q} := \{R(x, y, z, w, \underline{u}_R) \in \mathcal{P}_0 : (\exists \underline{t}_R)\ R(x, y, z, w, \underline{u}_R/\underline{t}_R) \\ \subseteq C(x \approx y, (x \approx y) \Rightarrow_D (z \approx w))\}.$$

The family $\mathcal{Q}$ is non-empty, because $(x \approx y) \Rightarrow_D (z \approx w) \in \mathcal{Q}$. (The set $(x \approx y) \Rightarrow_D (z \approx w)$ does not involve parameters.)

Claim 1. *$\mathcal{Q}$ determines PLDT for C.*

Proof (of the claim). Let p, q, r, s be arbitrary terms and X a set of equations. Assume $r \approx s \in C(X, p \approx q)$. We claim that $R(x, y, z, w, \underline{t}) \subseteq C(X)$ for some $R(x, y, z, w, \underline{u}) \in \mathcal{Q}$ and a sequence of terms $\underline{t}$. As $\mathcal{P}_0$ determines PLDT for C, there exists $R'(x, y, z, w, \underline{u}') \in \mathcal{P}_0$ and a sequence of terms $\underline{t}'$ such that $R'(p, q, r, s, \underline{t}') \subseteq C(X)$. But according to Lemma C.6, there also exists $R(x, y, z, w, \underline{u}) \in \mathcal{P}_0$ and a sequence of terms t'' such that

$$R(x, y, z, w, \underline{t}'') \subseteq C(R'(x, y, z, w, \underline{u}')) \tag{a}$$

and

$$R(x, y, z, w, \underline{t}'') \subseteq C(x \approx y, (x \approx y) \Rightarrow_D (z \approx w)). \tag{b}$$

For the above sets R and R', we select a substitution $e : \boldsymbol{Te}_\tau \to \boldsymbol{Te}_\tau$ which sends the parametric variables $\underline{u}'$ to $\underline{t}'$ and e is the identity map on the remaining variables. (a) implies that

$$R(x, y, z, w, e\underline{t}'') \subseteq C(R'(x, y, z, w, e\underline{u}')) = C(R'(x, y, z, w, \underline{t}')) \subseteq C(X).$$

Putting $\underline{t} := e\underline{t}''$, we have that $R(x, y, z, w, \underline{t}) \subseteq C(X)$. Moreover, (b) yields that $R(x, y, z, w, \underline{t}) \subseteq C(x \approx y, (x \approx y) \Rightarrow_D (z \approx w))$, which means that $R(x, y, z, w, \underline{u})$ is in $\mathcal{Q}$. This proves the claim. □

As $\mathcal{Q}$ is countable, we may enumerate it by natural numbers from an initial segment N_0 of ω,

$$\mathcal{Q} = \{Q_n(x, y, z, w, \underline{v}_n) : n \in N_0\}.$$

(Thus either $N_0 = \omega$ if $\mathcal{Q}$ is infinite or $N_0 = \{0, \dots, k\}$ for some k if $\mathcal{Q}$ is finite.) Moreover we can assume that $Q_0(x, y, z, w, \underline{v}_0)$ coincides with the Day implication system $(x \approx y) \Rightarrow_D (z \approx w)$.

According to Lemma C.6 and the above claim, for any two numbers $m, n \in N_0$ there exists $k \in N_0$, $m \leqslant k$, $n \leqslant k$, and a sequence of terms $\underline{t}_k$ such that $Q_k(x, y, z, w, \underline{t}_k) \subseteq C(Q_m(x, y, z, w, \underline{v}_m))$ and $Q_k(x, y, z, w, \underline{t}_k) \subseteq C(x \approx y, Q_n(x, y, z, w, \underline{v}_n))$. We then recursively define a subfamily

$$\mathcal{P} := \{P'_n(x, y, z, w, \underline{v}_n) : n \in N_0\}$$

of $\mathcal{Q}$ as follows. We put: $P'_0(x, y, z, w, \underline{v}_0) := Q_0(x, y, z, w, \underline{v}_0)$ $(= (x \approx y) \Rightarrow_D (z \approx w))$. (The set of parameters $\underline{v}_0$ is empty.)

Suppose $n \in N_0$ and the sets $P'_0(x, y, z, w, \underline{v}_0), \ldots, P'_n(x, y, z, w, \underline{v}_n)$ have been defined. If n is not the greatest number in N_0, we consider the sets $P'_n(x, y, z, w, \underline{v}_n)$ and $Q_{n+1}(x, y, z, w, \underline{v}_{n+1})$. As P'_n and Q_{n+1} are in $\mathcal{Q}$, there is a set $Q(x, y, z, w, \underline{u})$ in $\mathcal{Q}$ and a sequence of terms $\underline{t}$ such that $Q(x, y, z, w, \underline{t}) \subseteq C(x \approx y, P'_n(x, y, z, w, \underline{v}_n))$ and $Q(x, y, z, w, \underline{t}) \subseteq C(Q_{n+1}(x, y, z, w, \underline{v}_{n+1}))$. We then put: $P'_{n+1}(x, y, z, w, \underline{v}_{n+1}) := Q(x, y, z, w, \underline{u})$. If n is the greatest number in N_0, the recursive procedure terminates.

We therefore have that $P'_0(x, y, z, w, \underline{v}_0) := Q_0(x, y, z, w, \underline{v}_0)$ and for any positive $n \in N_0$ there is a sequence of terms $\underline{t}_n$ such that

$$P'_n(x, y, z, w, \underline{t}_n) \subseteq C(Q_n(x, y, z, w, \underline{v}_n)). \tag{c}$$

We therefore get that for any non-terminal $n \in N_0$ there is a sequence of terms $\underline{s}_{n+1}$ such that

$$P'_{n+1}(x, y, z, w, \underline{s}_{n+1}) \subseteq C(x \approx y, P'_n(x, y, z, w, \underline{v}_n)). \tag{d}$$

Indeed, (d) holds for $n = 0$ directly from the definition of P'_1. Assuming that (d) holds for n, we show it holds for $n + 1$. We have: $P'_{n+2}(x, y, z, w, \underline{v}_{n+1}) = Q(x, y, z, w, \underline{u})$, where $Q(x, y, z, w, \underline{u})$ is a set in $\mathcal{Q}$ such that $Q(x, y, z, w, \underline{t}) \subseteq C(x \approx y, P'_{n+1}(x, y, z, w, \underline{v}_{n+1}))$ for some sequence of terms $\underline{t}$. It follows, by structurality, that there is a sequence of terms $\underline{s}_{n+2}$ such that $P'_{n+2}(x, y, z, w, \underline{s}_{n+2}) = Q(x, y, z, w, \underline{t}) \subseteq C(x \approx y, P'_{n+1}(x, y, z, w, \underline{v}_{n+1}))$. So (d) holds.

Claim 2. *$\mathcal{P}$ determines PLDT for C.*

Proof (of the claim). Indeed, suppose that $r \approx s \in C(X, p \approx q)$ for some terms p, q, r, s and a set of equations X. We claim that there is $n \in N_0$ and a sequence of terms $\underline{t}'_n$ such that $P'_n(p, q, r, s, \underline{t}'_n) \subseteq C(X)$.

According to Claim 1, there exists $n \in N_0$ and a sequence of terms $\underline{s}_n$ such that

$$Q_n(p, q, r, s, \underline{s}_n) \subseteq C(X). \tag{e}$$

If $n = 0$, $\underline{s}_n$ is empty, then $P'_0(x, y, z, w) = Q_0(x, y, z, w) \subseteq C(X)$ and we are done. If n is positive, we apply to (c) a substitution such that x/p, y/q, z/r, w/s and $\underline{v}_n/\underline{s}_n$. Then, by structurality, there is a sequence of terms t'_n such that $P'_n(p, q, r, s, \underline{t}'_n) \subseteq C(Q_n(p, q, r, s, \underline{s}_n))$. Consequently, $P'_n(p, q, r, s, \underline{t}'_n) \subseteq C(X)$, by (e). This proves the claim. □

In the next step we define the family of sets

$$\{P'_n(x, y/x, z, w, \underline{v}_n) : P'_n(x, y, z, w, \underline{v}_n) \in \mathcal{P}\}$$

and mark this family as

$$\{P''_n(x, z, w, \underline{v}_n) : n \in N_0\}.$$

As $\mathcal{P}$ determines PLDT for C, Lemma C.4 yields that C is axiomatized by $Id(\mathbf{Q})$, the rules of Birkhoff's logic, and the rules

$$r_n : \qquad P''_n(x, z, w, \underline{v}_n)/z \approx w,$$

$n \in N_0$.

We shall check conditions (1)″–(4)″. (1)″ and (2)″ are immediate.

As to (3)″, the definitions of $\mathcal{P}$ and of the family $\{P''_n(x, z, w, \underline{v}_n) : n \in N_0\}$ yield that for each $n \in N_0$ there is a sequence of terms $\underline{t}_n$ such that

$$P''_n(x, z, w, \underline{t}_n) \subseteq C((x \approx x) \Rightarrow_D (z \approx w)). \tag{f}$$

((f) is proved by induction on n with the help of (d).) But $(x \approx x) \Rightarrow_D (z \approx w) \subseteq C(z \approx w)$. Indeed, according to the definition of $\Rightarrow_D$, we have that $(x \approx y) \Rightarrow_D (z \approx w) \subseteq C(x \approx y, z \approx w) \cap C(x \approx z, y \approx w)$. After identifying y with x, we see that structurality gives that $(x \approx x) \Rightarrow_D (z \approx w) \subseteq C(x \approx x, z \approx w) \cap C(x \approx z, x \approx w) \subseteq C(z \approx w)$. We therefore have that

$$P''_n(x, z, w, \underline{t}_n) \subseteq C(z \approx w).$$

(3) then follows.

As to (4)″, the definition of $\mathcal{P}$ gives that for each non-terminal $n \in N_0$ there exists a sequence of terms $\underline{s}_{n+1}$ such that $P'_{n+1}(x, y, z, w, \underline{s}_{n+1}) \subseteq C(x \approx y, P'_n(x, y, z, w, \underline{v}_n))$. Hence, by structurality, there exists a sequence of terms $\underline{t}'_{n+1}$ such that $P'_{n+1}(x, y/x, z, w, \underline{t}'_{n+1}) \subseteq C(P'_n(x, y/x, z, w, \underline{v}_n))$. From this inclusion condition (4)″ follows.

The lemma has been proved. □

In the last step, for each $n \in N_0$ we define the substitution $x/u_n, z/x, w/y$ for the set $P''_n(x, z, w, \underline{v}_n)$ and put:

$$P_n(x, y, \underline{u}_n) := P'_n(x/u_n, z/x, w/y, \underline{v}_n) \qquad (n \in N_0).$$

Here u_n is a variable different from $x, z, w, \underline{v}_n$. u_n is thus a parametric variable added to $\underline{v}_n$. (So $\underline{u}_n = \underline{v}_n \cup \{u_n\}$.) Moreover, we can safely assume that the new parametric variables u_m and u_n are pairwise different whenever $m \neq n$, for all $m, n \in N_0$).

The theorem follows from the above lemma and the definition of $P_n(x, y, \underline{u}_n)$, $n \in N_0$. □

Corollary C.8. *Let* **Q** *be any RCM quasivariety. Let R be the set of rules defined as in the above theorem. Then for each pair of natural numbers* $n, m \in N_0$ *such that* $n < m$ *there exists a sequence of terms* $\underline{s}_{m,n}$ *for which* $P_m(x, y, \underline{s}_{m,n}) \subseteq \mathbf{Q}^{\models}(P_n(x, y, \underline{u}_n))$.

Proof. Fix $n \in N_0$. Then the induction on k with $n + k \in N_0$ gives that there is a sequence of terms $\underline{r}_{n+k}$ such that

$$P_{n+k}(x, y, \underline{r}_{n+k}) \subseteq C(P_n(x, y, \underline{u}_n)). \qquad (*)_{n+k}$$

The case $k = 1$ follows from (4). Assume $(*)_{n+k}$ holds. We show $(*)_{n+(k+1)}$. (4) gives that there is a sequence of terms $\underline{t}_{n+k+1}$ such that

$$P_{n+k+1}(x, y, \underline{t}_{n+k+1}) \subseteq C(P_{n+k}(x, y, \underline{u}_{n+k})).$$

Taking a suitable substitution e we then get that

$$P_{n+k+1}(x, y, e(\underline{t}_{n+k+1})) \subseteq C(P_{n+k}(x, y, \underline{r}_{n+k})). \qquad (1)$$

Hence putting $\underline{r}_{n+k+1} := e(\underline{t}_{n+k+1})$, we obtain that

$$P_{n+k+1}(x, y, \underline{r}_{n+k+1}) \subseteq C(P_n(x, y, \underline{u}_n)),$$

by (1) and $(*)_{n+k}$. So $(*)_{n+(k+1)}$ holds.

The corollary thus follows. □

A quasivariety **Q** has the *relative congruence extension property* (RCEP), if for any algebras $\mathbf{A}_0, \mathbf{A}$ with $\mathbf{A} \in \mathbf{Q}$ and $\mathbf{A}_0$ being a subalgebra of $\mathbf{A}$, every congruence $\Phi_0 \in Con_{\mathbf{Q}}(\mathbf{A}_0)$ can be extended to a **Q**-congruence on $\mathbf{A}$, that is, there is a congruence $\Phi \in Con_{\mathbf{Q}}(\mathbf{A})$ such that the restriction of Φ to $\mathbf{A}_0$ is equal to Φ_0.

Corollary C.9. *Let* **Q** *be an RCM quasivariety. If* **Q** *has RCEP, then* $\mathbf{Q}^{\models}$ *is axiomatized by the set of equations* $Id(\mathbf{Q})$, *the rules of Birkhoff's logic and a countable set of rules of the form*

$$r_n : \qquad P_n(x, y)/x \approx y$$

($n \in N_0$), *where* N_0 *is an initial segment of* ω, *such that*

(1)* *for each* $n \in N_0$, $P_n(x, y)$ *is a finite set of equations in variables x and y only;*
(1)* *for each* $n \in N_0$,

$$P_n(x, x) \subseteq Id(\mathbf{Q});$$

(2)* *for each non-terminal number* $n \in N_0$

$$P_{n+1}(x, y) \subseteq \mathbf{Q}^{\models}(P_n(x, y)).$$

Proof. As $\mathbf{Q}$ has RCEP, there exists a family $\mathcal{P}_0$ of finite sets of equations in the four variables x, y, z, w only which determines LDT for C. We assume that a Day implication system $(x \approx y) \Rightarrow_D (z \approx w)$ for $\mathbf{Q}$ is in $\mathcal{P}_0$. Then proceed as in the proof of the above theorem. □

Bibliography

Agliano, P., Ursini, A.: On subtractive varieties (II): general properties. Algebra Univers. **36**, 222–259 (1996)

Agliano, P., Ursini, A.: On subtractive varieties (III): from ideals to congruences. Algebra Univers. **37**, 296–333 (1997a)

Agliano, P., Ursini, A.: On subtractive varieties (IV): definability of principal ideals. Algebra Univers. **38**, 355–389 (1997b)

Baker, K.: Primitive satisfaction and equational problems for lattices and other algebras. Trans. Am. Math. Soc. **190**, 125–150 (1974)

Baker, K.: Finite equational bases for finite algebras in a congruence distributive equational class. Adv. Math. **24**, 207–243 (1977)

Balbes, R.: A note on distributive sublattices of a modular lattice. Fundam. Math. **65**(2), 219–222 (1969)

Bergman, C.: Universal Algebra: Fundamentals and Selected Topics. CRC, Boca Raton/London/New York (2011)

Birkhoff, G.: Lattice Theory. American Mathematical Society Colloquium Publications, 3rd edn., vol. 25. American Mathematical Society, Providence (1967)

Blok, W.J., Pigozzi, D.: A finite basis theorem for quasivarieties. Algebra Univers. **22**, 1–13 (1986)

Blok, W.J., Pigozzi, D.: Algebraizable logics. Memoirs of the American Mathematical Society, No. 396. Amer. Math. Soc., Providence (1989)

Blok, W.J., Pigozzi, D.: Algebraic semantics for universal Horn logic without equality. In: Romanowska, A., Smith, J.D.H. (eds.) Universal Algebra and Quasigroups, pp. 11–56. Heldermann, Berlin (1992)

Burris, S., Sankappanavar, H.P.: A Course in Universal Algebra. Graduate Texts in Mathematics, vol. 78. Springer, Berlin/New York (1981)

Chang, C.C., Keisler, H.J.: Model Theory. North-Holland/American Elsevier, Amsterdam/London/New York (1973)

Czédli, G., Horváth, E.K.: Congruence distributivity and modularity permit tolerances. Acta Universitatis Palackianae Olomucensis. Mathematica **41**, 39–42 (2002)

Czédli, G., Horváth, E.K., Lipparini, P.: Optimal Mal'tsev conditions for cogruence modular varieties. Algebra Univers. **53**, 267–279 (2005)

Czelakowski, J.: Filter-distributive logics. Stud. Logica **43**, 353–377 (1985)

Czelakowski, J.: Relative principal congruences in congruence-modular quasivarieties. Algebra Univers. **39**, 81–101 (1998)

Czelakowski, J.: Protoalgebraic Logics. Kluwer, Dordrecht (2001)

J. Czelakowski, *The Equationally-Defined Commutator*,
DOI 10.1007/978-3-319-21200-5

Czelakowski, J.: General theory of the commutator for deductive systems. Part I. Basic facts. Stud. Logica **83**, 183–214 (2006)

Czelakowski, J.: Additivity of the commutator and residuation. Rep. Math. Log. **43**, 109–132 (2008)

Czelakowski, J.: Triangular irreducibilty of conguences in quasivarieties. Algebra Univers. **71**, 261–283 (2014)

Czelakowski, J., Dziobiak, W.: Congruence distributive quasivarieties whose subdirectly irreducible members form a universal class. Algebra Univers. **27**, 128–149 (1990)

Czelakowski, J., Dziobiak, W.: A single quasi-identity for a quasivariety with the Fraser-Horn property. Algebra Univers. **29**, 10–15 (1992)

Czelakowski, J., Dziobiak, W.: The parametrized local deduction theorem for quasivarieties of algebras and its applications. Algebra Univers. **35**, 373–419 (1996)

Day, A.: A characterization of modularity for congruence lattices of algebras. Can. Math. Bull. **12**, 167–173 (1969)

Day, A., Freese, R.: A characterization of identities implying congruence modularity. Can. J. Math. **32**, 1140–1167 (1980)

Day, A., Gumm, H.P.: Some characterizations of the commutator. Algebra Univers. **29**, 61–78 (1992)

Dent, T., Kearnes, K.A., Szendrei, Á.: An easy test for congruence modularity. Algebra Univers. **67**, 375–392 (2012)

Dziobiak, W.: Relative congruence distributivity within quasivarieties of nearly associative Φ-algebras. Fundam. Math. **135**(2), 77–95 (1990)

Dziobiak, W., Maróti, M., McKenzie, R.N., Nurakunov, A.: The weak extension property and finite axiomatizability of quasivarieties. Fundam. Math. **202**(3), 199–222 (2009)

Erné, M.: Weak distributive laws and their role in lattices of congruences and equational theories. Algebra Univers. **25**, 290–321 (1988)

Font, J.M.: Abstract Algebraic Logic: An Introductory Textbook, (2016, forthcoming)

Font, J.M.: Abstract algebraic logic. an introductory chapter. In: Galatos, N., Terui, K. (eds.) Hiroakira Ono on Residuated Lattices and Substructural Logics. The Series: Outstanding Contributions to Logic. Springer, New York (2014)

Font, J.M., Jansana, R.: A General Algebraic Semantics for Sentential Logics. Lecture Notes in Logic, 2nd edn., vol. 7. Springer, New York (2009)

Font, J.M., Jansana, R., Pigozzi, D.: A survey of abstract algebraic logic. Stud. Logica **74**, 13–97 (2003)

Font, J.M., Jansana, R., Pigozzi, D.: Update to "a survey of abstract algebraic logic". Stud. Logica **91**, 125–130 (2009)

Freese, R., McKenzie, R.N.: Commutator Theory for Congruence Modular Varieties. London Mathematical Society Lecture Note Series, vol. 125. Cambridge University Press, Cambridge/New York (1987). The second edition is available online

Gorbunov, V.: A characterization of residually small quasivarieties (in Russian). Doklady Akademii Nauk SSSR **275**, 204–207 (1984)

Grätzer, G.: General Lattice Theory. Akademie, Berlin (1978)

Grätzer, G.: Universal Algebra, 2nd edn. Springer, New York (2008)

Gumm, H.P.: Geometrical Methods in Congruence Modular Algebras. Mem. Am. Math. Soc. **45**, 286 (1983)

Gumm, H.P., Ursini, A.: Ideals in universal algebras. Algebra Univers. **19**, 45–54 (1984)

Hagemann, J., Herrmann, C.: A concrete ideal multiplication for algebraic systems and its relation to congruence distributivity. Archiv der Mathematik (Basel) **32**, 234–245 (1979)

Herrmann, C.: Affine algebras in congruence modular varieties. Acta Sci. Math. (Seged) **41**, 119–125 (1979)

Hobby, D., McKenzie, R.N.: The Structure of Finite Algebras. Contemporary Mathematics, vol. 76. American Mathematical Society, Providence (1988)

Idziak, P.: Elementary theory of finite equivalential algebras. Rep. Math. Log. **25**, 81–89 (1991)

Idziak, P.M., Słomczyńska, K., Wroński, A.: Equivalential algebras: a study of Fregean varieties. In: Font, J., Jansana, R., Pigozzi, D. (eds.) Proc. Workshop on Abstract Algebraic Logic, Spain, July 1–5, 1997, pp. 95–100. CRM Quaderns 10, Barcelona (1998, preprint)

Idziak, P.M., Słomczyńska, K., Wroński, A.: Commutator in equivalential algebra and Fregean varieties. Algebra Univers. **65**, 331–340 (2011)

Idziak, P.M., Wroński, A.: Definability of principal congruences in equivalential algebras. Colloq. Math. **74**(2), 225–238 (1997)

Ježek, J.: Universal Algebra. First edition. Available on the internet: http:/www.karlin.mff.cuni.cz/~jezek/ua.pdf (2008)

Jónsson, B.: Distributive sublattices of a modular lattice. Proc. Am. Math. Soc. **6**, 682–688 (1955)

Kearnes, K.A.: Relatively congruence distributive subquasivarieties of a congruence modular variety. Bull. Aust. Math. Soc. **41**, 87–96 (1990)

Kearnes, K.A.: An order-theoretic property of the commutator. Int. J. Algebra Comput. **3**, 491–533 (1993)

Kearnes, K.A.: Varieties with a difference term. J. Algebra **177**, 926–960 (1995)

Kearnes, K.A.: Almost all minimal idempotent varieties are congruence modular. Algebra Univers. **44**, 39–45 (2000)

Kearnes, K.A.: Congruence join-semidistributivity is equivalent to a congruence identity. Algebra Univers. **46**, 373–387 (2001)

Kearnes, K.A., Kiss, E.W.: The shape of congruence lattices. Mem. Am. Math. Soc. **222**, 1046 (2013)

Kearnes, K.A., McKenzie, R.N.: Commutator theory for relatively modular quasivarieties. Trans. Am. Math. Soc. **331**(2), 465–502 (1992)

Kearnes, K.A., Szendrei, Á.: The relationship between two commutators. Int. J. Algebra Comput. **8**, 497–531 (1998)

Kiss, E.W.: Three remarks on the modular commutator. Algebra Univers. **29**, 455–476 (1992)

Lyndon, R.: Identities in two-valued calculi. Trans. Am. Math. Soc. **71**, 457–465 (1951)

Lyndon, R.: Identities in finite algebras. Proc. Am. Math. Soc. **5**, 8–9 (1954)

Maróti, M., McKenzie, R.N.: Finite basis problems and results for quasivarieties. Stud. Logica **78**, 293–320 (2004)

McKenzie, R.N.: Finite equational bases for congruence modular varieties. Algebra Univers. **24**, 224–250 (1987)

McKenzie, R.N., McNulty, G.F., Taylor, W.F.: Algebras, Lattices, Varieties, vol I. Wadsworth & Brooks/Cole, Monterey (1987)

Nurakunov, A.: Quasi-identities of congruence-distributive quasivarieties of algebras. Sib. Math. J. **42**, 108–118 (2001)

Pigozzi, D.: Finite basis theorems for relatively congruence-distributive quasivarieties. Trans. Am. Math. Soc. **310**(2), 499–533 (1988)

Quackenbush, R.: Quasi-affine algebras. Algebra Univers. **20**, 318–327 (1985)

Scott, D.S.: Continuous lattices. In: Lawvere, F.W. (ed.) Toposes, Algebraic Geometry and Logic. Lecture Notes in Mathematics, vol. 274, pp. 97–136. Springer, Berlin/New York (1972)

Słomczyńska, K.: Equivalential algebras. Part I: Representation Algebra Univers. 35, 524–547 (1996)

Tamura, S.: A note on distributive sublattices of a lattice. Proc. Jpn. Acad. **47**, 603–605 (1971)

Taylor, W.: Some applications of the term condition. Algebra Univers. **14**, 11–24 (1982)

Ursini, A.: Prime ideals in universal algebra. Acta Univ. Carol. – Math. et Phys. **25**(1), 75–87 (1984)

Ursini, A.: On subtractive varieties (I). Algebra Univers. **31**, 204–222 (1994)

Ursini, A.: On subtractive varieties (V): congruence modularity and the commutator. Algebra Univers. **43**, 51–78 (2000)

von Neumann, J.: Lectures on Continuous Geometries. Institute of Advanced Studies, Princeton (1936–1937)
Willard, R.: A finite basis theorem for residually small, congruence meet-semidistributive varieties. J. Symb. Log. **65**, 187–200 (2000)
Willard, R.: Extending Baker's theorem. Algebra Univers. **45**, 335–344 (2001)
Wójcicki, R.: Theory of Logical Calculi. Basic Theory of Consequence Operations. Kluwer, Dordrecht/Boston/London (1988)

Symbol Index

J. Czelakowski, *The Equationally-Defined Commutator*,
DOI 10.1007/978-3-319-21200-5

Index of Definitions and Terms

J. Czelakowski, *The Equationally-Defined Commutator*,
DOI 10.1007/978-3-319-21200-5

Index of Names

J. Czelakowski, *The Equationally-Defined Commutator*,
DOI 10.1007/978-3-319-21200-5

Printed in the United States
By Bookmasters